Advancements in Developing Abiotic Stress-Resilient Plants

Advancements in Developing Abiotic Stress-Resilient Plants

Basic Mechanisms to Trait Improvements

Edited by
M. Iqbal R. Khan, Palakolanu Sudhakar Reddy, and Ravi Gupta

CRC Press
Taylor & Francis Group
Boca Raton London New York

CRC Press is an imprint of the
Taylor & Francis Group, an informa business

First edition published 2022
by CRC Press
6000 Broken Sound Parkway NW, Suite 300, Boca Raton, FL 33487-2742

and by CRC Press
4 Park Square, Milton Park, Abingdon, Oxon, OX14 4RN

CRC Press is an imprint of Taylor & Francis Group, LLC

ISBN: 9780367747725 (hbk)
ISBN: 9780367748043 (pbk)
ISBN: 9781003159636 (ebk)

DOI: 10.1201/9781003159636

Typeset in Times
by Newgen Publishing UK

Contents

Editors

Dr. M. Iqbal R. Khan is an assistant professor of Botany at Jamia Hamdard, New Delhi, India. His current research interests are elucidation physiological and molecular mechanisms associated with abiotic stress tolerance. Dr. Khan has found a significant role of phytohormones and suggested that phytohormones play an important in controlling stress responses, and interact in coordination with each other for defense signal networking to fine-tune tolerance mechanisms. He is also exploring the role of plant signaling molecules and their impact on plant homeostasis under abiotic stresses. He has been recognized as a Young Scientist Award by the Indian Society of Plant Physiology, Young Scientist Platinum Jubilee Award by The National Academy of Sciences, India (NASI), and Junior Scientist of the Year from National Environmental Science Academy New Delhi, India.

Dr. Palakolanu Sudhakar Reddy is a Scientist (Cell, Molecular biology and Genetic engineering) at International Crops Research Institute for the Semi-Arid Tropics (ICRISAT), Telengala, India. He has a PhD from International Centre for Genetic Engineering and Biotechnology (ICGEB), New Delhi, where he focused on "Identification and characterization of the Heat Shock Proteins from Pennisetum glaucum". He was a primary team member for developing high-throughput, low-cost methodologies for the isolation of promoters and genes, sequencing for phage DNA and identification of stable reference genes for qRT-PCR studies. He has been awarded with several National and International awards, including Leibniz-DAAD Postdoctoral Fellowship, Young Scientist and INSPIRE Faculty Awards from the Department of Science and Technology (DST), Government of India. He is an associate fellow of Andhra Pradesh Akademi of Sciences from 2015. He worked in Germany and Philippines on functional genomics and molecular aspects of abiotic stress tolerance and genome editing. He has published more than 39 research articles in international peer reviewed journals, edited one book, written eight book chapters, and has a patent. He is a reviewer of several reputed international journals, including Frontiers in plant science, Plant Physiology and Biochemistry, BMC Genomics, Environmental and Experimental Botany, Gene and PLoS ONE.

Dr. Ravi Gupta earned his Ph.D. in Plant Physiology and Biochemistry from the University of Delhi, India. After gaining his Ph.D., he spent more than five years in South Korea working as a postdoctoral fellow and Research Professor at Pusan National University, Busan, Korea. Dr. Gupta is the author of more than 60 scientific journal publications and several book publications. He is a recipient of several prestigious national and international awards such as Ramalingaswami re-entry fellowship, Ramanujan fellowship, and ITS fellowship from Government of India and Korea Research Fellowship from Government of South Korea. Dr. Gupta has been invited to give many lectures at national and international conferences and workshops and has received independent funding from the Governments of India and South Korea. His current objectives include unravelling the biotic and abiotic stress signaling in plants, especially rice and soybean, using gel-based and gel-free proteomics approaches. Moreover, He is also working towards the development of novel techniques for the depletion of high-abundance proteins from different plant tissues to increase their proteome coverage, which would help analyse the "Hidden Proteome" of the plants.

1 Physiological, Molecular, and Biochemical Responses of Rice to Drought Stress

Ashish B. Rajurkar[1]*, *Dhananjay Gotarkar*[2], *and Seema Rana*[3]
[1]Plant Science Division, University of Missouri, Columbia, MO, USA
[2]IRRI South Asia Regional Centre (ISARC), Varanasi, UP, India
[3]Tamil Nadu Agricultural University, Coimbatore, TN, India
*Corresponding Author, ashu.raj123@gmail.com

CONTENTS

1.1 INTRODUCTION

Drought or water deficit is one of the major environmental constraints on rice productivity, particularly in rainfed ecosystems and largely in major rice -growing ecosystems, as groundwater, a valuable resource for irrigation during drought, is continuously declining (Ray *et al.*, 2015). Erratic rainfall across the globe may exacerbate droughts, with increasing frequency of water stress during the cropping season. Increased drought will increase water stress. Exposure of plants to drought adversely affects them at every stage from germination to reproduction and finally limits yield (Pandey and Shukla, 2015; Khan *et al.*, 2015). Rice is highly susceptible to water stress. Drought stress reduces accumulation of biomass, in general, and causes yield reduction. The magnitude of yield loss depends on timing, plant growth stage and duration, and the severity of drought stress. Drought stress affects rice plants throughout their lifecycle, but drought stress from the intermittent (during maximum tillering, flowering-reproductive growth) to the late (after panicle initiation) stage may greatly affect the yield. The drought-mediated yield losses in the reproductive stage in rice are given in Table 1.1.

DOI: 10.1201/9781003159636-1

TABLE 1.1
Yield Losses in Rice as a Result of Reproductive-Stage Drought

Severity of Reproductive Drought Stress	Yield Reduction (%)	Reference
Lowland moderate stress	45–60	Vikram *et al.*, 2011 Dixit *et al.*, 2012, Rajurkar *et al.*, 2019, 2021
Lowland severe stress	65–91	Vikram *et al.*, 2011 Ghimire *et al.*, 2012 Rajurkar *et al.*, 2019, 2021
Upland mild stress	18–39	Vikram *et al.*, 2011 Sandhu *et al.*, 2014
Upland moderate stress	70–75	Vikram *et al.*, 2011 Sandhu *et al.*, 2014
Upland severe stress	80–97	Bernier *et al.*, 2007 Lafitte *et al.*, 2007 Dixit *et al.*, 2012

Extreme climate change causing lower precipitation and drought has negative effects in many growing areas of the world (Lobell *et al.*, 2011). Drought is frequent in many parts of South and Southeast Asia, affecting 46 Mha rainfed lowland and 10 Mha upland rice area in Asia (Pandey *et al.*, 2007). In India alone, 14.4 and 6.0 Mha of the rainfed lowland and upland rice area, respectively, are affected by drought (Mahajan *et al.*, 2017). Recent predictions suggest further increased frequency and intensity of drought and increase concern over water deficit problems in the coming decades (Wassmann *et al.*, 2009). Given the increasing severity, it is necessary to develop cultivars with inbuilt mechanisms for drought stress tolerance and to deliver adapted varieties to improve productivity in drought-affected environments.

Plants adapt different strategies, such as drought escape, recovery and resistance, which can be further divided into drought avoidance and tolerance (Fukai and Cooper, 1995). Drought avoidance is usually associated with reduced water loss, extensive root system for water uptake and reduced leaf area to avoid evaporation. During drought tolerance, plants maintain their normal functioning even with low water potential within the tissues, and this is associated with accumulation of compatible solutes and protoplasmic resistance (Price *et al.*, 2002). Improving drought resistance is a complex and difficult task to achieve, because sometimes it comes with limitations such as short lifecycle, leading to reduced grain yield, and lower carbon assimilation with ultimate reduction in grain yield is seen in varieties using drought escape and avoidance strategies. In the case of drought tolerance mechanisms, increased solute concentration for osmotic adjustment may have a negative impact on plant growth. Therefore, adaptation of crops to drought stress must maintain a balance between the drought resistance mechanisms introduced to guarantee sustainable productivity (Mitra, 2001; Yang *et al.*, 2010). Plants sense their environments by adaptive morphological, anatomical, cellular, physiological and biochemical changes. And thus, multidisciplinary approaches are needed to understand responses and mechanisms and develop strategies to bridge the yield gaps under various environmental stresses. A detailed understanding of drought stress responses is desirable for the development of resilient cultivars suitable for drought conditions (Khan *et al.*, 2021; Ratnakumar *et al.*, 2016). In the past decades, our understanding of molecular and cellular mechanisms involved in drought stress resistance has strongly improved, enabling us to integrate drought avoidance and tolerance strategies into the desired crop cultivars for improving their performance under stress conditions. The current chapter reviews different physiological, molecular and biochemical responses of plants to drought stress and also provides molecular insight into the adaptive mechanisms occurring in plants as a defensive plan to combat complex stress like drought.

1.2 PHYSIOLOGICAL RESPONSES AND MECHANISMS UNDER DROUGHT STRESS

Molecular, cellular, physiological, biochemical and developmental responses to abiotic stress involve several traits and the underlying genes controlling drought tolerance. To address the complexity of plant responses to drought, it is vital to understand drought stress and its effects on physiological processes or component traits, and their role in yield improvement. However, efforts to dissect drought resistance, by identifying and characterizing component traits and transferring them into cultivars with high-yielding genetic backgrounds, have had very limited success. Nevertheless, there are a few instances in which trait-based selection for drought resistance has resulted in actual yield improvement (Richards, 2006; Sinclair *et al.*, 2004).

Understanding the physiological basis of crop performance under drought conditions will contribute to the identification and manipulation of traits associated with improved water use efficiency and yield under water deficit conditions and will complement conventional and molecular breeding programs. Different traits may play important roles at different drought severity levels and at different crop developmental stages. The drought resistance traits are mainly divided into constitutive traits (e.g., root traits: root diameter, root thickness) and induced/adaptive traits (e.g., osmotic adjustment). In evaluating traits for improved adaptation to drought, both constitutive and adaptive (inducible) traits are considered important. Constitutive traits are expressed under normal conditions and do not require water stress for their expression, whereas adaptive traits are expressed only in response to water deficit (Kamoshita *et al.*, 2008). All the traits have either positive or negative influence on yield, depending on the existing drought situation (timing, severity and duration) and depending on whether a survival or production mechanism is necessary. The physiological traits whose importance in drought resistance in rice has been demonstrated are summarized in the following subsections.

1.2.1 Leaf Rolling and Leaf Area Index

The turgidity of the cells is lost with severe drought, resulting in wilting of leaves, which is expressed as leaf rolling in rice and most of the cereals. Leaf rolling is the first visual symptom of drought reaction and occurs due to the inability of leaves to sustain the transpiration demand of the plant (Blum, 1988). This trait has been found to be a useful criterion in assessing levels of drought tolerance in large-scale screening (Chang and Loresto, 1986). Leaf rolling is an adaptive response to water deficit, which helps in maintaining favorable water balance within plant tissues and thus maintains cell growth and activity (Sellammal *et al.*, 2014). Bhattarai and Subudhi (2018) observed higher leaf rolling as a major phenotypic change in rice plants under drought stress. Leaf rolling reduces the transpiration rate and canopy temperature, thereby improving water use efficiency (WUE; Townley-Smith and Hurd 1979). Xu *et al.* (2002) reported better WUE in genotypes with partially rolled or folded leaves but reduced leaf area, while a recent study by Cal *et al.*, 2019 suggested that leaf rolling under drought was more affected by leaf morphology than by stomatal conductance, leaf water status, or maintenance of shoot biomass and grain yield. QTLs for leaf rolling have been identified in rice (Nguyen *et al.*, 2013; Subashri *et al.*, 2009). Subashri *et al.* (2009) also found collocation of the leaf rolling QTL region with QTLs for panicle exertion, panicle length, plant height and biological yield.

Evaporative demand is controlled by a decrease in leaf growth that affects leaf area either by reducing individual leaf growth or by reducing the number of leaves (Tardieu, 1996). A plant reduces its leaf area by drying older leaves and thereby reducing water loss, overall reducing WUE. Leaf area is usually given in terms of Leaf Area Index (LAI), which is leaf surface area per unit soil surface area and affects plant transpiration. Plants with fewer leaves transpire less without significantly changing their net primary production. Also, the greater the LAI, the lower is the evapotranspiration from soil, and vice versa. Grain yield has been reported to be positively correlated with green leaf area under terminal drought, and negative relationships were observed between grain yield and leaf senescence (Borell *et al.,* 2000).

1.2.2 Leaf Water Potential (LWP) and Relative Water Content (RWC)

Exposure of plants to drought stress substantially decreases the LWP, RWC and transpiration rate, with a concomitant increase in leaf temperature (Siddique *et al.,* 2001). RWC and LWP have long been associated with rice performance under water deficit (O'Toole and Moya, 1978) and were found to be correlated with yield at the flowering stage under drought stress (Lafitte, 2002; Jongdee *et al.,* 2006). RWC is also useful in better characterization of plant root and shoot physiology under drought stress (Anupama *et al.,* 2019). Drought-resistant varieties show higher LWP and maintain a higher RWC as compared with drought-sensitive varieties under moisture deficit conditions (Swamy *et al.,* 1983; Anupama *et al.,* 2019). Maintenance of higher LWP under drought is empirically related to better stem extension and panicle exertion (Jearakongman, 2005) as well as to reduced delay in flowering (Pantuwan *et al.,* 2002).

1.2.3 Osmotic Adjustment

Osmotic adjustment (OA) is a metabolic process associated with the synthesis of various compatible solutes within plant cells in response to water stress (Serraj and Sinclair, 2002) and is a prime drought stress adaptive engine in support of plant production and crop yield under water-limiting conditions (Blum, 2017). As soil moisture declines, OA causes a reduction in osmotic potential, favoring turgor maintenance and protecting the integrity of metabolic functions. The role of OA in maintaining yield under drought has been reported in rice (Praba *et al.,* 2009). Under severe water stress, higher OA capacity may help plants to withstand a prolonged drought spell and recover more promptly upon rehydration. Significant variations in OA were reported among rice lines (Babu *et al.,* 2001). Indica rice varieties are known to have higher OA capacity compared with japonica varieties (Kole, 2006). OA has a positive effect on leaf rolling, tissue death and retention of leaf area in rice under drought stress (Fukai and Cooper, 1995). OA is recognized as an effective cellular and metabolic mechanism that can be manipulated to produce water deficit-resistant crop varieties (Blum, 2017; Abdelrahman *et al.,* 2018). Swapna and Shylaraj (2017) identified rice varieties under polyethylene glycol (PEG)-mediated osmotic stress and drought stress using hydroponics and found that the identified varieties had less leaf rolling, better drought recovery ability and better RWC, increased membrane stability index with osmolyte accumulation, and antioxidant enzymatic activity. The role of different osmolytes in OA is detailed in Zivcak *et al.* (2016) and is briefly discussed in the later biochemical section. QTLs for OA have been detected by several studies (Lilley *et al.,* 1996; Zhang *et al.,* 2001; Robin *et al.,* 2003).

1.2.4 Stomatal Density, Aperture Size and Stomatal Conductance

Stomata play a central role in the pathways of both carbon uptake and water loss by plants (Chaves *et al.,* 2016). Stomata maintain leaf temperature through water loss in transpiration and regulate the gaseous flow needed in the process of photosynthesis. Stomata plays an adaptive role under drought by stomatal pore opening or closing, aperture size, stomatal density and distribution pattern (Hetherington and Woodward, 2003). Stomatal density and aperture size are the two main factors determining stomatal conductance and thus affect photosynthetic ability. Stomatal conductance, by reducing transpiration and soil water absorption, plays an essential role in regulating the water balance of the plant (Sinclair *et al.*, 2011). Canopy temperature measured by thermal imaging is a proxy for stomatal conductance and has already proven to be a good indicator of drought stress in the field (Leinonen *et al.*, 2006; Munns *et al.*, 2010) and has been used in several studies (Melandri *et al.*, 2020). Reducing stomatal conductance under water-limited conditions may lead to a greater yield by improving intrinsic WUE (iWUE) (Leakey *et al.*, 2019). Lower stomatal density in drought-tolerant rice varieties showed a reduced transpiration rate (Anupama *et al.*, 2019), while Kulya *et al.* (2018) reported higher stomatal density and reduced stomatal length in chromosomal segment substitution lines (CSSLs) of rice cultivar KhaoDawk Mali 105 (KDML105) carrying drought-tolerant DT-QTL segments from drought-tolerant donors, DH103 and DH212. Henry *et al.* (2019) showed an interesting trend in a large effect drought yield QTL (qDTY12.1) near-isogenic lines (NILs) and the donor parent Way Rarem for conserving water through low and high stomatal conductance under high and low evapo-transpirational demand, respectively, compared with the recipient parent Vandana. Pantuwan *et al.* (2002) found an association of the process of plant water conservation through stomatal regulation with reduced spikelet sterility and increased grain yield under reproductive-stage drought.

1.2.5 Root Traits

Roots are the first organs exposed to water stress. A quick strategic response in root morphological traits is the important physiological parameter for drought adaptation (Ingram *et al.*, 1994; Niones *et al.*, 2012). Root growth and distribution are modulated by the stress and assist in providing drought resistance through avoidance mechanisms, either by growing deeper and exploring deep soil moisture or maintaining root growth. The possession of a deep, thick and larger root system is generally considered favorable, allowing the crop to maintain its water status even under water deficit conditions (Nguyen *et al.*, 1997), and is considered important in determining drought resistance in upland rice. Deep-rooting cultivars are more resistant to drought than shallow-rooting ones (Farooq *et al.*, 2009a).

Under different types of drought stress, plasticity in root density, total root length (Tran *et al.*, 2015), and lateral root length and/or branching (Kano-Nakata *et al.*, 2013) has been observed to improve shoot biomass, water uptake and photosynthesis under drought stress in rice (Sandhu *et al.*, 2016). Furthermore, root architecture is also considered to be a key trait for dissecting the genotypic differences in rice responses to water deficit cues (Henry *et al.*, 2011). The spatial distribution of roots largely determines the genotypic potential for extraction of water and influences productivity under drought stress. The deep root–shoot ratio is well correlated and affects the ability of a variety to absorb water from the deep soil layers, determining a variety's resistance to drought (Yoshida and Hasegawa, 1982), while thick roots persist longer and produce more and larger branch roots, thereby increasing root density and water uptake capacity (Ingram *et al.*, 1994). A deep and thick root system has been thought to be advantageous for improved drought tolerance in rainfed ecosystems (Fukai and Cooper, 1995, Comas *et al.*, 2013). Thick roots also exert greater penetration ability in hard soil under drought (Babu *et al.*, 2001). Jeong *et al.* (2013) observed increased yield (9–26%) in transgenic rice over-expressing *OsNAC5* through increased root diameter under drought. In a number of species, root characteristics such as increased root density, maximum root length, root thickness, rooting depth, number of nodal roots, root:shoot ratio and branching of root systems have all been associated with plant adaptation to water stress. A positive correlation between root traits and yield and its related components under drought stress was reported by Babu *et al.* (2003). Kumar *et al.* (2004) reported an association between higher root pulling resistance (RPR) and maintenance of higher LWP under severe drought stress, as well as a positive correlation with grain yield and RWC with RPR. Various QTLs for root length, root thickness and root penetration ability have been detected and incorporated into rice varieties to enhance drought tolerance; a yield advantage of 1 t ha^{-1} was found in introgressed plants with longer root length QTLs compared with controls, with

60% less water consumption than traditional varieties (Steele *et al.*, 2006, 2007). Uga *et al.* (2013) identified, cloned and characterized the QTL *Deeper Rooting 1* (*DRO1*), controlling root growth angle. The NILs developed by introgression of DRO1 into shallow-rooting IR64 exhibited enhanced drought tolerance by growing deeper roots, resulting in high-yield performance under drought (Uga *et al.*, 2013; Arai-Sanoh *et al.*, 2014).

1.2.6 Molecular Breeding for Physiological and Secondary Traits

Over the last decade, the drought breeding program at the International Rice Research Institute (IRRI, Philippines) has made significant progress in developing abiotic stress-tolerant lines in rice, targeting grain yield, and led to the development of more than 20 high-yielding, drought-tolerant lines with release of varieties across South and Southeast Asia and Africa since 2009 (Sandhu and Kumar, 2017). There are reports of several grain yield QTLs, i.e., *qDTY1.1*, *qDTY2.1*, *qDTY2.2*, *qDTY3.1*, *qDTY3.2*, *qDTY9.1* and *qDTY12.1*, showing large effects under drought (Khan *et al.*, 2021). However, studies targeting secondary traits are scarce. Though a number of secondary traits have been determined to be associated with drought resistance, very few have been used in rice breeding programs to improve yield under water limited conditions, which is still in progress. This is because selection for secondary traits in breeding programs needs extensive investment in phenotyping these traits and is prone to problems of repeatability due to high genotype × environment (G × E) interaction. Progress in high-throughput precision phenotyping (phenomics) and genomics technologies is now overcoming the phenotyping and genotyping bottlenecks, enabling a more precise detection of QTLs for "hard-to phenotype" complex secondary/physiological traits, a problem that is particularly relevant for the selection of drought-resistant genotypes. Molecular marker technology or marker assisted breeding has been identified as a powerful tool for selection of traits that are otherwise difficult to screen. Molecular markers allow breeders to track the genetic loci controlling drought resistance without measuring the phenotype, thus reducing the need for extensive field testing over space and time (Nguyen *et al.*, 1997). With this advancement, recent mapping approaches have already provided closely linked markers for complex traits such as stay-green (Borrell *et al.*, 2014), OA (Babu *et al.*, 2001; Robin *et al.*, 2003), root traits (Champoux *et al.*, 1995; Uga *et al.*, 2013; Wade *et al.*, 2015), and QTLs controlling drought avoidance mechanisms (such as leaf rolling, leaf drying, RWC of leaves, LWP, stomatal conductance and relative growth rate under stress) in rice (Courtois *et al.*, 2000; Babu *et al.*, 2003; Khowaja and Price, 2008; Barik *et al.*, 2018, 2019). The QTLs identified for various physiological and secondary traits using different mapping populations are summarized in Table 1.2.

1.3 MOLECULAR RESPONSES AND MECHANISMS UNDER DROUGHT STRESS

Tolerance to water stress is a quantitatively controlled trait in plants. Drought stress induces multiple molecular and biochemical changes within plants, leading to alterations in morpho-physiological characters that are important in order to survive. Drought changes the expression of genes regulating water transport, oxidative damage, osmotic balance and damage repair mechanisms. Recently developed molecular tools, e.g., RNA-Seq and bioinformatics, have accelerated the discovery of stress-responsive genes and TFs in many plant species (Khan *et al.*, 2018; Kaur and Asthir, 2017; Xiong *et al.*, 2005).

Some of the changes that are transduced are in the form of signals such as protein phosphorylation/de-phosphorylation, calcium signaling, reactive oxygen species, abscisic acid (ABA) biosynthesis and cross -talk with other phytohormones (Ali *et al.*, 2020; Tiwari *et al.*, 2017; Khan and Khan, 2017). Phytohormones play a central role in regulating plant growth and development in response to environmental changes (Cutler *et al.*, 2010; Khan *et al.*, 2015). Phytohormonal cross-talk occurs

TABLE 1.2
Recent QTLs Identified for Physiological and Secondary Traits under Drought Stress in Rice

Population	Environment	Physiological/Secondary Traits	References
IR64/Azucena DH (Double Haploid) population	Upland fields	Leaf rolling, leaf drying, relative water content	Courtois *et al.*, 2000
	PVC cylinders	Leaf rolling, root length, root number, root volume, root weight, root thickness, drought score	Hemamalini *et al.*, 2000
CT9993/IR62266	Greenhouse	Cell membrane stability	Tripathy *et al.*, 2000
DH population	Field conditions	Osmotic adjustment, root thickness, root weight, root length, root pulling force	Zhang *et al.*, 2001
	Field conditions	Relative water content, leaf rolling, canopy temperature, leaf drying, root morphology	Babu *et al.*, 2003
Bala/Azucena Recombinant Inbred lines (RILs)	Field conditions	Leaf rolling, leaf drying and leaf relative water content	Price *et al.*, 2002a
	Growth room with controlled environment conditions	Leaf morphological traits, leaf area, leaf dry weight, leaf water relations and rolling	Khowaja and Price, 2008
IAC165/Co39	Greenhouse	Root traits, root thickness, root weight, maximum root length	Courtois *et al.*, 2003
IR1552/Azucena RIL Population	Controlled environment and PVC pots	Seminal root length, adventitious root number, lateral root length and number	Zheng *et al.*, 2003
IR62266/IR60080	Greenhouse	Osmotic adjustment	Robin *et al.*, 2003
IRAT109/ Yuefu DH Lines	Field, PVC pipes, aerobic conditions	Root traits, basal root thickness, total root number, maximum root length, root weight	Li *et al.*, 2005
Zhenshan 97/IRAT109	Field conditions	Leaf rolling time and leaf drying score, canopy temperature, root traits	Yue *et al.*, 2005, Yue *et al.*, 2006
	Rainproof drought screen facility	Canopy temperature, leaf water potential	Liu *et al.*, 2005
Akihikari × IRAT109	Hydroponics drought stress using PEG	Relative growth rate and specific water use	Kato *et al.*, 2008
Teqing/Lemont BC introgression lines (ILs)	Field conditions	Photosynthetic rate, stomatal conductance, transpiration rate, stomata frequency, chlorophyll content	Zhao *et al.*, 2008
IR20/Nootripathu RILs	Rainfed fields	Leaf rolling and leaf drying, canopy temperature, relative water content, drought score	Gomez *et al.*, 2010, Salunkhe *et al.*, 2011, Prince *et al.*, 2015, Rajurkar *et al.*, 2019
Zhenshan 97/IRAT109 NILs	PVC pipes	Root growth rate, root volume, deep root volume	Ding *et al.*, 2011

(continued)

TABLE 1.2 (Continued)
Recent QTLs Identified for Physiological and Secondary Traits under Drought Stress in Rice

Population	Environment	Physiological/Secondary Traits	References
IR64/INRC10192 RILs, Backcross population	Hydroponics drought stress using PEG, field condition	Root length, root dry weight	Srividya *et al.*, 2011, Patil *et al.*, 2017
Haogelao/Shennong265	Field conditions	Photosynthesis parameters, Net photosynthesis rate, Stomatal conductance, Transpiration rate	Gu *et al.*, 2012
Khaunoongmo/Q5 F2 population	Field conditions	Leaf rolling, leaf drying, plant recovery	Nguyen *et al.*, 2013
IR64/Cabacu RILs	Field conditions	Leaf rolling, root pulling force	Trijatmiko *et al.*, 2014
Swarna/WAB450	Field conditions	Canopy temperature, grain yield under drought	Saikumar *et al.*, 2014
Association mapping population, collections of RILs from Zhenshan97/IRAT109 and Chinese landraces	Field conditions	Ratio of deep rooting	Lou *et al.*, 2015
Dular/IR64 RILs	Upland fields and greenhouse	Leaf rolling, root growth angle, seedling-stage root length, root dry weight and crown root number	Catolos *et al.*, 2017
Samgang/Nagdong DH population	Field and greenhouse conditions	Visual drought tolerance and relative water content	Kim *et al.*, 2017
Cocodrie/N-22 RILs	Greenhouse, pots	Leaf rolling score	Bhattarai and Subudhi, 2018
CR 143–2-2/Krishnahamsa RILs	Rain out shelter	Relative water content	Barik *et al.*, 2018
IAPAR-9/Akihikari and IAPAR-9/Liaoyan241	Hydroponics drought stress using PEG, field managed drought	leaf rolling index, rooty number and root length	Han *et al.*, 2018
Dongnong422/Kongyu131, Xiaobaijingzi/Kongyu131 RIL population	Field conditions	Leaf area, chlorophyll content	Yang *et al.*, 2018
Koshihikari/Takanari//Koshihikari, or Koshihikari/Takanari//Takanari, BC1F1 plants	Field condition	Canopy temperature difference	Fukuda *et al.*, 2018
CR 143–2-2/Krishnahamsa	Field condition, rain out shelter	Leaf rolling, leaf drying	Barik *et al.*, 2019
GWAS, Vietnamese landraces panel	Net house, plastic trays	Leaf relative water content, drought sensitivity score, recovery ability	Hoang *et al.*, 2019
IR55419- 04/Super Basmati F2 population	Greenhouse, PVC pipes	Leaf drying score, leaf dry weight, leaf area, root dry weight, deep root length, volume, surface area and deep root diameter	Sabar *et al.*, 2019

TABLE 1.2 (Continued)
Recent QTLs Identified for Physiological and Secondary Traits under Drought Stress in Rice

Population	Environment	Physiological/Secondary Traits	References
KDML105/ DH212 or DH103 CSSL population	Field condition	Traits relevant to drought tolerance and avoidance	Shearman 2020
Rice diversity panel consisting of 293 *indica* accessions	Field condition	Canopy temperature	Melandri *et al.*, 2020

at the biosynthesis and degradation which are triggered after the occurrence of drought and acts as a signaling pathway to trigger adaptive responses. The role of ABA in drought stress response is well known, and several ABA-responsive genes controlling drought stress responses have already been reported elsewhere (Khan *et al.*, 2021). ABA is a major player in cellular growth reduction, stomatal closure and reduced transpiration rate under drought stress (Shinozaki and Yamaguchi-Shinozaki, 2007). However, several studies report that many drought-related genes do not respond to ABA, suggesting the existence of ABA-independent signal transduction pathways and cross-talk or involvement of other hormones like salicylic acid (SA), jasmonic acid (JA), cytokinins (CK) and ethylene (ET) in drought stress responses (Khan *et al.*, 2021; Per et al., 2018). Li *et al.* (2017) isolated the *ABA stress* and *ripening* (*ASR*) gene from the upland rice variety IRAT 109 and demonstrated that over-expression of *OsASR5* enhanced osmotic and drought tolerance in rice by regulating leaf water status and ABA biosynthesis and promoting stomatal closure under drought stress conditions.

In the last decades, hundreds of drought-inducible genes in rice have been identified using microarrays, expressed sequence tags and quantitative reverse transcription polymerase chain reaction (qRT-PCR) (Rabbani *et al.*, 2003; Degenkolbe *et al.*, 2009; Todaka *et al.*, 2012; Borah *et al.*, 2017; Sharma *et al.*, 2019). Among these identified genes, some are involved in the protection and repair mechanism of cells, while other genes are involved in regulation of signal transduction, activation of TFs and biosynthesis of signaling molecules (regulatory proteins) (Shinozaki and Yamaguchi-Shinozaki, 2007). The expression of stress-induced genes is largely regulated by specific TFs, and it has been estimated that the rice genome contains 1,611 TF genes that belong to 37 gene families (Xiong *et al.*, 2005). Stress-related TF families exhibit distinctive binding domains such as dehydration-responsive element binding (DREB)/C- repeat-binding factor (CBF), ethylene-responsive factor (ERF), basic-leucine zipper (bZip), ABA-responsive element binding (AREB)/ ABRE binding factor (ABF), NAM (no apical meristem), and CUC (cup-shaped cotyledon) (NAC) conserved domain and homeodomain-leucine zipper (HD-Zip) (Xiong *et al.*, 2005). Some of the important TFs identified and characterized in rice are listed in Table 1.3.

TF binding to a DNA sequence can result in activation or repression of transcription. TFs bind to cis-regulatory elements, which are mainly found within the promoter region of stress-inducible genes. The most common cis-regulatory element in ABA-regulated genes is the ABRE, which is recognized by the bZip family (Hossain *et al.*, 2010). The DRE/C-repeat (CRT) element, which is recognized by the DREB/ERF family, regulates function in ABA-independent gene expression (Nakashima *et al.*, 2009). One particular large group of TFs involved in stress responses in several plant species is that of the HD-Zip genes (Bhattacharjee *et al.*, 2016), which are a subgroup of the homeobox genes. There are several reports of drought-tolerant genes that have been characterized in rice using genetic engineering technology, and their exact roles and phenotypes have been listed earlier (Yoo *et al.*, 2017; Oladosu *et al.*, 2019); furthermore, recently, a network-based supervised

TABLE 1.3
Transcription Factors Functionally Characterized under Stresses in Rice

Transcription Factor (TF)	Stress Tolerance	Reference
OsRab7	Drought and heat	El-Esawi and Alayafi, 2019
SNAC1	Drought and salinity	Hu *et al.*, 2006
OsNAC5	Drought	Jeong *et al.*, 2013
OsNAC6	Abiotic and biotic	Nakashima *et al.*, 2007
OsNAC10	Drought	Jeong *et al.*, 2010
OsNAC14	Drought	Shim *et al.*, 2018
OsNAC045	Drought and salinity	Zheng *et al.*, 2009
OsERF71	Drought	Lee *et al.*, 2016
SUB1A	Drought and flooding	Fukao *et al.*, 2011
OsbZIP23	Drought and salinity	Xiang *et al.*, 2008
OsbZIP16	Drought	Chen *et al.*, 2012
OsbZIP71	Drought and salinity	Liu *et al.*, 2014
OsbZIP46	Drought	Tang *et al.*, 2012
OsbZIP72	Drought	Lu *et al.*, 2009
OsMYB2	Dehydration, salinity and cold	Yang *et al.*, 2012
OaMYB55	Drought and heat	Casaretto *et al.*, 2016
OsDREB1F	Drought, salinity and low temperature	Wang *et al.*, 2008
OsDREB2A	Drought	Cui *et al.*, 2011
OsABF2	Drought, salinity and cold	Hossain *et al.*, 2010
Oshox4	Drought	Agalou *et al.*, 2008
ZFP245	Drought, cold and oxidative stress	Huang *et al.*, 2009
ZFP252	Drought and salinity	Xu *et al.*, 2008
OsCDPK7	Drought, salinity and cold	Saijo *et al.*, 2000
AP37	Drought	Oh *et al.*, 2009
OsbHLH148	Drought	Seo *et al.*, 2011
ARAG1	Drought	Zhao *et al.*, 2010
OsWR1	Drought	Wang *et al.*, 2012
OsTZF1	Drought and salinity	Jan *et al.*, 2013

machine learning framework that accurately predicts and ranks all rice TFs in the genome according to their potential association with drought tolerance has been reported (Gupta *et al.*, 2020).

1.4 BIOCHEMICAL RESPONSES AND MECHANISMS UNDER DROUGHT STRESS

Diversified groups of compatible solutes and biochemicals have been identified in various crop species under stress. They mainly act as osmoprotectants and go through a biochemical process of osmotic adjustment under stress conditions. Several solutes can be accumulated for osmoprotection, such as free amino acids like proline, nitrogenous compounds like glycine betaine, and sugars and other solutes, to balance osmotic pressure in plant cells. Among all the solutes, proline takes the lead when it comes to drought tolerance in plants, with others including the sugars and antioxidants, which are known as reactive oxygen species (ROS) scavengers (Hasanuzzaman *et al.*, 2018; Hanif *et al.*, 2020).

1.4.1 Proline

The role of proline in stress tolerance was first reported in rye grasses (Kemble and Macpherson, 1954). Changes in the concentration of proline compared with other amino acids have been observed in rice under water stress (Mansour and Salama, 2020), and it is used as a biochemical marker to select resistance in such conditions (Fahramand *et al.*, 2014). Proline is proposed to screen germplasm for drought adaptation in various crop species. It plays a highly beneficial role by acting as an osmolyte (Verbruggen and Hermans, 2008; Mansour and Salama, 2020; Per *et al.*, 2017), a metal chelator, an anti-oxidative defense molecule and a signaling molecule during stress conditions (Dar *et al.*, 2016). Earlier studies have reported the role of proline as a detoxifier or osmoprotectant by scavenging ROS (Liang *et al.*, 2013; Per *et al.*, 2017). The accumulation of proline is predicted to be involved in plant damage repair ability by boosting antioxidant activity during drought stress (Lum *et al.*, 2014; Mansour and Salama, 2020).

1.4.2 Polyamines

Polyamines (PAs) are positively charged molecules involved in drought tolerance response (Takahashi and Kakehi, 2010). The most ubiquitous PAs in plants are putrescine (Put), spermidine (Spd) and spermine (Spm). They can regulate osmotic and ionic homeostasis, stabilize membranes and act as antioxidants along with interaction among other signaling molecules (Pandey and Shukla, 2015; Asgher *et al.*, 2018). Under drought stress conditions, higher PA contents in plants are related to increased photosynthetic capacity, reduced water loss, improved OA and detoxification. Other roles of polyamines include enhancing the DNA-binding activity of TFs, prevention of senescence (Bouchereau *et al.*, 1999; Panagiotidis *et al.*, 1995), protein phosphorylation and conformational transition of DNA (Martin-Tanguy, 2001). PA accumulation is the immediate response observed after exposure to drought conditions in rice leaves (Yang *et al.*, 2007). The exogenous application of PAs alleviates drought stress by improving the WUE, production of proline, anthocyanin, maintenance of LWC and reduction in oxidative damage in rice (Farooq *et al.*, 2009b). Studies indicate that foliar application is more effective than seed priming, and among PAs, Spm is the most effective in improving drought tolerance in rice (Farooq *et al.*, 2009a; Do *et al.*, 2013; Pandey and Shukla, 2015).

1.4.3 Allantoin

Allantoin is a major purine metabolite in plants under drought stress conditions (Silvente *et al.*, 2012). Recent studies on knockout mutants in the allantoinase gene have shown an accumulation of allantoin and enhanced seed survivability and growth in drought and osmotic stress conditions (Watanabe *et al.*, 2014; Takagi *et al.*, 2016). Allantoin plays a role in activating ABA metabolism, which is known as a stress hormone and regulates various drought stress responses in plants (Sah *et al.*, 2016). A metabolomics study on drought-tolerant rice shoots and roots revealed the accumulation of more allantoin in shoots compared with the roots (Casartelli *et al.*, 2018). Recently, scientists at IRRI, Philippines have discovered the first plant guanine deaminase (OsGDA1) gene in rice and shown its importance for plant survival under drought stress by maintaining the xanthine pull, which is also required for accumulation of the allantoin that will induce ABA synthesis during stress conditions (unpublished data).

1.5 CONCLUSION AND FUTURE PERSPECTIVES

Rice improvement for environmental stresses such as drought is complex and challenging. Considering the future global climate change, increased intensity and frequency of drought, and its effects on cereal crops, development of crop cultivars resilient to specific environmental stress is needed for future food security. Genetically, drought tolerance of rice is a complex trait under polygenic control and involves complex morpho-physiological mechanisms. Drought tolerance involves various aspects such as OA, plant signaling to control growth, transpiration, shoot and root system architectures, and several phytohormonal feedbacks and/or cross-talks. The responses of plants to tissue water deficit determine their level of drought tolerance. To address the complexity of plant responses to drought, it is vital to understand the physiological and genetic basis of this response with comprehensive information and a better understanding of mechanisms with a multi-disciplinary approach, i.e., physiology, molecular biology, genomics and breeding, which are believed to address the multigenic nature of abiotic stresses, including drought tolerance. Considerable work has been undertaken to understand the genetic basis of putative drought-adaptive traits in rice. However, breeding for drought tolerance is extremely challenging due to the complexity and responses associated with various stress-adaptive mechanisms, uncertainty in onset of stress, large G × E interactions, and various molecular, biochemical and physiological phenomena affecting plant growth and development. Large effect QTLs for grain yield under drought have been identified in rice. However, the use of these QTLs in molecular breeding is limited, with a lack of repeatability across environments and genetic background. Comprehensive understanding of these QTLs is needed in order to improve their efficacy in marker assisted breeding (MAB). Understanding the physiological and molecular mechanisms associated with these yield QTLs will hasten MAB for drought resistance. Also, the majority of responses are mediated by cellular and biochemical mechanisms, involving cross-talk between phytohormones, which in turn triggers the stress-responsive mechanisms. Understanding phytohormonal cross-talk and hormonal dynamics and identifying gene regulatory networks of complex traits will help in developing a realistic framework to uncover the drought responses that will make progress in understanding how rice productivity can be maximized under limited water conditions. The increased access to rice genome information and transcriptomic datasets has provided a path to identify the gene regulatory networks of complex traits like drought tolerance. The QTLs, genes, TFs and hormones that modulate the physiology and morphology of plants under stress are indicative developers of next-generation "drought-proof" rice varieties.

REFERENCES

Abdelrahman, M., Burritt, D. J., & Tran, L. S. P. (2018). The use of metabolomic quantitative trait locus mapping and osmotic adjustment traits for the improvement of crop yields under environmental stresses. In: *Seminars in Cell & Developmental Biology* (Vol. 83, pp. 86–94). Academic Press. Doi:10.1016/j.semcdb.2017.06.020

Agalou, A., Purwantomo, S., Övernäs, E., Johannesson, H., Zhu, X., Estiati, A., ... & Ouwerkerk, P. B. (2008). A genome-wide survey of HD-Zip genes in rice and analysis of drought-responsive family members. *Plant Molecular Biology*, *66*(1–2), 87–103.

Ali, S., Hayat, K., Iqbal, A., & Xie, L. (2020). Implications of abscisic acid in the drought stress tolerance of plants. *Agronomy*, *10*(9), 1323.

Anupama, A., Bhugra, S., Lall, B., Chaudhury, S., & Chugh, A. (2019). Morphological, transcriptomic and proteomic responses of contrasting rice genotypes towards drought stress. *Environmental and Experimental Botany*, *166*, 103795.

Arai-Sanoh, Y., Takai, T., Yoshinaga, S., Nakano, H., Kojima, M., Sakakibara, H., ... & Uga, Y. (2014). Deep rooting conferred by DEEPER ROOTING 1 enhances rice yield in paddy fields. *Scientific Reports*, *4*(1), 1–6.

Asgher, M., Khan, M. I. R., Anjum, N. A., Verma, S., Vyas, D., Per, T. S., Masood, A., Khan, N. A. (2018). Ethylene and polyamines in counteracting heavy metal phytotoxicity: a crosstalk perspective. *Journal of Plant Growth Regulation*, *37*(4), 1050–1065.

Babu, R. C., Nguyen, B. D., Chamarerk, V., Shanmugasundaram, P., Chezhian, P., Jeyaprakash, P., & Wade, L. J. (2003). Genetic analysis of drought resistance in rice by molecular markers. *Crop Science*, *43*(4), 1457–1469.

Babu, R. C., Shashidhar, H. E., Lilley, J. M., Thanh, N. D., Ray, J. D., Sadasivam, S., ... & Nguyen, H. T. (2001). Variation in root penetration ability, osmotic adjustment and dehydration tolerance among accessions of rice adapted to rainfed lowland and upland ecosystems. *Plant Breeding*, *120*(3), 233–238.

Barik, S. R., Pandit, E., Pradhan, S. K., Mohanty, S. P., & Mohapatra, T. (2019). Genetic mapping of morpho-physiological traits involved during reproductive stage drought tolerance in rice. *PLoS One*, *14*(12), e0214979.

Barik, S. R., Pandit, E., Pradhan, S. K., Singh, S., Swain, P., & Mohapatra, T. (2018). QTL mapping for relative water content trait at reproductive stage drought stress in rice. *Indian Journal of Genetics*, *78*(4), 401–408.

Bernier, J., Kumar, A., Ramaiah, V., Spaner, D., & Atlin, G. (2007). A large-effect QTL for grain yield under reproductive-stage drought stress in upland rice. *Crop Science*, *47*(2), 507–516.

Bhattacharjee, A., Khurana, J. P., & Jain, M. (2016). Characterization of rice homeobox genes, OsHOX22 and OsHOX24, and over-expression of OsHOX24 in transgenic Arabidopsis suggest their role in abiotic stress response. *Frontiers in Plant Science*, *7*, 627.

Bhattarai, U., & Subudhi, P. K. (2018). Genetic analysis of yield and agronomic traits under reproductive-stage drought stress in rice using a high-resolution linkage map. *Gene*, *669*, 69–76.

Blum, A. (1988). Drought resistance. In: *Plant Breeding for Stress Environments* (pp. 43–47). CRC Press, Boca Raton, FL.

Blum, A. (2017). Osmotic adjustment is a prime drought stress adaptive engine in support of plant production. *Plant, Cell & Environment*, *40*(1), 4–10.

Borah, P., Sharma, E., Kaur, A., Chandel, G., Mohapatra, T., Kapoor, S., & Khurana, J. P. (2017). Analysis of drought-responsive signalling network in two contrasting rice cultivars using transcriptome-based approach. *Scientific Reports*, *7*(1), 1–21.

Borrell, A. K., Hammer, G. L., & Henzell, R. G. (2000). Does maintaining green leaf area in sorghum improve yield under drought? II. Dry matter production and yield. *Crop Science*, *40*(4), 1037–1048.

Borrell, A. K., Mullet, J. E., George-Jaeggli, B., van Oosterom, E. J., Hammer, G. L., Klein, P. E., & Jordan, D. R. (2014). Drought adaptation of stay-green sorghum is associated with canopy development, leaf anatomy, root growth, and water uptake. *Journal of Experimental Botany*, *65*(21), 6251–6263.

Bouchereau, A., Aziz, A., Larher, F., & Martin-Tanguy, J. (1999). Polyamines and environmental challenges: recent development. *Plant Science*, *140*(2), 103–125.

Cal, A. J., Sanciangco, M., Rebolledo, M. C., Luquet, D., Torres, R. O., McNally, K. L., & Henry, A. (2019). Leaf morphology, rather than plant water status, underlies genetic variation of rice leaf rolling under drought. *Plant, Cell & Environment*, *42*(5), 1532–1544.

Casaretto, J. A., El-Kereamy, A., Zeng, B., Stiegelmeyer, S. M., Chen, X., Bi, Y. M., & Rothstein, S. J. (2016). Expression of OsMYB55 in maize activates stress-responsive genes and enhances heat and drought tolerance. *BMC Genomics*, *17*(1), 1–15.

Casartelli, A., Riewe, D., Hubberten, H. M., Altmann, T., Hoefgen, R., & Heuer, S. (2018). Exploring traditional aus-type rice for metabolites conferring drought tolerance. *Rice*, *11*(1), 1–16.

Catolos, M., Sandhu, N., Dixit, S., Shamsudin, N. A., Naredo, M. E., McNally, K. L., ... & Kumar, A. (2017). Genetic loci governing grain yield and root development under variable rice cultivation conditions. *Frontiers in Plant Science*, *8*, 1763.

Champoux, M. C., Wang, G., Sarkarung, S., Mackill, D. J., O'Toole, J. C., Huang, N., & McCouch, S. R. (1995). Locating genes associated with root morphology and drought avoidance in rice via linkage to molecular markers. *Theoretical and Applied Genetics*, *90*(7–8), 969–981.

Chang, T. T, & Loresto, G. C. (1986). Screening techniques for drought resistance in rice. In: *Approaches for Incorporating Drought and Salinity Resistance in Crop Plants* (pp. 108–129). Ed. V. L. Chopra and R. S. Paroda. Oxford IBH, New Delhi.

Chaves, M. M., Costa, J. M., Zarrouk, O., Pinheiro, C., Lopes, C. M., & Pereira, J. S. (2016). Controlling stomatal aperture in semi-arid regions—the dilemma of saving water or being cool? *Plant Science*, *251*, 54–64.

Chen, H., Chen, W., Zhou, J., He, H., Chen, L., Chen, H., & Deng, X. W. (2012). Basic leucine zipper transcription factor OsbZIP16 positively regulates drought resistance in rice. *Plant Science*, *193*, 8–17.

Comas, L., Becker, S., Cruz, V. M. V., Byrne, P. F., & Dierig, D. A. (2013). Root traits contributing to plant productivity under drought. *Frontiers in Plant Science*, *4*, 442.

Courtois, B., McLaren, G., Sinha, P. K., Prasad, K., Yadav, R., & Shen, L. (2000). Mapping QTLs associated with drought avoidance in upland rice. *Molecular Breeding*, *6*(1), 55–66.

Courtois, B., Shen, L., Petalcorin, W., Carandang, S., Mauleon, R., & Li, Z. (2003). Locating QTLs controlling constitutive root traits in the rice population IAC 165× Co39. *Euphytica*, *134*(3), 335–345.

Cui, M., Zhang, W., Zhang, Q., Xu, Z., Zhu, Z., Duan, F., & Wu, R. (2011). Induced over-expression of the transcription factor OsDREB2A improves drought tolerance in rice. *Plant Physiology and Biochemistry*, *49*(12), 1384–1391.

Cutler, S. R., Rodriguez, P. L., Finkelstein, R. R., & Abrams, S. R. (2010). Abscisic acid: emergence of a core signaling network. *Annual Review of Plant Biology*, *61*, 651–679.

Dar, M. I., Naikoo, M. I., Rehman, F., Naushin, F., & Khan, F. A. (2016). Proline accumulation in plants: roles in stress tolerance and plant development. In: *Osmolytes and Plants Acclimation to Changing Environment: Emerging Omics Technologies* (pp. 155–166). Springer, New Delhi.

Degenkolbe, T., Do, P. T., Zuther, E., Repsilber, D., Walther, D., Hincha, D. K., & Köhl, K. I. (2009). Expression profiling of rice cultivars differing in their tolerance to long-term drought stress. *Plant Molecular Biology*, *69*(1–2), 133–153.

Ding, X., Li, X., & Xiong, L. (2011). Evaluation of near-isogenic lines for drought resistance QTL and fine mapping of a locus affecting flag leaf width, spikelet number, and root volume in rice. *Theoretical and Applied Genetics*, *123*(5), 815.

Dixit, S., Swamy, B. M., Vikram, P., Ahmed, H. U., Cruz, M. S., Amante, M., ... & Kumar, A. (2012). Fine mapping of QTLs for rice grain yield under drought reveals sub-QTLs conferring a response to variable drought severities. *Theoretical and Applied Genetics*, *125*(1), 155–169.

Do, P. T., Degenkolbe, T., Erban, A., Heyer, A. G., Kopka, J., Köhl, K. I., ... & Zuther, E. (2013). Dissecting rice polyamine metabolism under controlled long-term drought stress. *PLoS One*, *8*(4), e60325.

El-Esawi, M. A., & Alayafi, A. A. (2019). Overexpression of rice Rab7 gene improves drought and heat tolerance and increases grain yield in rice (*Oryza sativa* L.). *Genes*, *10*(1), 56.

Fahramand, M., Mahmoody, M., Keykha, A., Noori, M., & Rigi, K. (2014). Influence of abiotic stress on proline, photosynthetic enzymes and growth. *International Research Journal of Applied and Basic Science*, *8*(3), 257–265.

Farooq, M., Kobayashi, N., Wahid, A., Ito, O., & Basra, S. M. (2009a). 6 Strategies for producing more rice with less water. *Advances in Agronomy*, *101*(4), 351–388.

Farooq, M., Wahid, A., & Lee, D. J. (2009b). Exogenously applied polyamines increase drought tolerance of rice by improving leaf water status, photosynthesis and membrane properties. *Acta Physiologiae Plantarum*, *31*(5), 937–945.

Fukai, S., & Cooper, M. (1995). Development of drought-resistant cultivars using physiomorphological traits in rice. *Field Crops Research*, *40*(2), 67–86.

Fukao, T., Yeung, E., & Bailey-Serres, J. (2011). The submergence tolerance regulator SUB1A mediates crosstalk between submergence and drought tolerance in rice. *The Plant Cell*, *23*(1), 412–427.

Fukuda, A., Kondo, K., Ikka, T., Takai, T., Tanabata, T., & Yamamoto, T. (2018). A novel QTL associated with rice canopy temperature difference affects stomatal conductance and leaf photosynthesis. *Breeding Science*, *68*(3), 305–315.

Ghimire, K. H., Quiatchon, L. A., Vikram, P., Swamy, B. M., Dixit, S., Ahmed, H., ... & Kumar, A. (2012). Identification and mapping of a QTL (qDTY1. 1) with a consistent effect on grain yield under drought. *Field Crops Research*, *131*, 88–96.

Gomez, S. M., Boopathi, N. M., Kumar, S. S., Ramasubramanian, T., Chengsong, Z., Jeyaprakash, P., ... & Babu, R. C. (2010). Molecular mapping and location of QTLs for drought-resistance traits in indica rice (*Oryza sativa* L.) lines adapted to target environments. *Acta Physiologiae Plantarum*, *32*(2), 355–364.

Gu, J., Yin, X., Struik, P. C., Stomph, T. J., & Wang, H. (2012). Using chromosome introgression lines to map quantitative trait loci for photosynthesis parameters in rice (*Oryza sativa* L.) leaves under drought and well-watered field conditions. *Journal of Experimental Botany*, *63*(1), 455–469.

Gupta, C., Ramegowda, V., Basu, S., & Pereira, A. (2020). Prediction and characterization of transcription factors involved in drought stress response. *bioRxiv*.

Han, B., Wang, J., Li, Y., Ma, X., Jo, S., Cui, D., … Han, L. (2018). Identification of quantitative trait loci associated with drought tolerance traits in rice (*Oryza sativa* L.) under PEG and field drought stress. *Euphytica*, *214*(74).

Hanif, S., Saleem, M. F., Sarwar, M., Irshad, M., Shakoor, A., Wahid, M. A., & Khan, H. Z. (2020). Biochemically triggered heat and drought stress tolerance in rice by proline application. *Journal of Plant Growth Regulation*, *40*, 305–312.

Hasanuzzaman, M., Nahar, K., Anee, T. I., Khan, M. I. R., & Fujita, M. (2018). Silicon-mediated regulation of antioxidant defense and glyoxalase systems confers drought stress tolerance in *Brassica napus L. South African Journal of Botany*, *115*, 50–57.

Hemamalini, G. S., Shashidhar, H. E., & Hittalmani, S. (2000). Molecular marker assisted tagging of morphological and physiological traits under two contrasting moisture regimes at peak vegetative stage in rice (*Oryza sativa* L.). *Euphytica*, *112*(1), 69–78.

Henry, A., Gowda, V. R., Torres, R. O., McNally, K. L., & Serraj, R. (2011). Variation in root system architecture and drought response in rice (*Oryza sativa*): phenotyping of the OryzaSNP panel in rainfed lowland fields. *Field Crops Research, 120*(2), 205–214.

Henry, A., Stuart-Williams, H., Dixit, S., Kumar, A., & Farquhar, G. (2019). Stomatal conductance responses to evaporative demand conferred by rice drought-yield quantitative trait locus qDTY12. 1. *Functional Plant Biology, 46*(7), 660–669.

Hetherington, A. M., & Woodward, F. I. (2003). The role of stomata in sensing and driving environmental change. *Nature, 424*(6951), 901–908.

Hoang, G. T., Van Dinh, L., Nguyen, T. T., Ta, N. K., Gathignol, F., Mai, C. D., ... & Lebrun, M. (2019). Genome-wide association study of a panel of Vietnamese rice landraces reveals new QTLs for tolerance to water deficit during the vegetative phase. *Rice, 12*(1), 4.

Hossain, M. A., Cho, J. I., Han, M., Ahn, C. H., Jeon, J. S., An, G., & Park, P. B. (2010). The ABRE-binding bZIP transcription factor OsABF2 is a positive regulator of abiotic stress and ABA signaling in rice. *Journal of Plant Physiology, 167*(17), 1512–1520.

Hu, H., Dai, M., Yao, J., Xiao, B., Li, X., Zhang, Q., & Xiong, L. (2006). Overexpressing a NAM, ATAF, and CUC (NAC) transcription factor enhances drought resistance and salt tolerance in rice. *Proceedings of the National Academy of Sciences, 103*(35), 12987–12992.

Huang, J., Sun, S. J., Xu, D. Q., Yang, X., Bao, Y. M., Wang, Z. F., ... & Zhang, H. (2009). Increased tolerance of rice to cold, drought and oxidative stresses mediated by the overexpression of a gene that encodes the zinc finger protein ZFP245. *Biochemical and Biophysical Research Communications, 389*(3), 556–561.

Ingram, K. T., Bueno, F. D., Namuco, O. S., Yambao, E. B., & Beyrouty, C. A. (1994). Rice root traits for drought resistance and their genetic variation. In *Rice roots: nutrient and water use: selected papers from the International Rice Research Conference* (pp. 67–77). Kirk, G. J. D. (ed.). Los Banos, Laguna (Philippines): IRRI, 1994. ISBN 971-22-0050-7.

Jan, A., Maruyama, K., Todaka, D., Kidokoro, S., Abo, M., Yoshimura, E., ... & Yamaguchi-Shinozaki, K. (2013). OsTZF1, a CCCH-tandem zinc finger protein, confers delayed senescence and stress tolerance in rice by regulating stress-related genes. *Plant Physiology, 161*(3), 1202–1216.

Jearakongman, S. 2005. Validation and discovery of quantitative trait loci for drought tolerance in backcross introgression lines in rice (Oryza sativa L.) cultivar IR64. PhD Thesis. Kasetsart University, pp. 95.

Jeong, J. S., Kim, Y. S., Baek, K. H., Jung, H., Ha, S. H., Do Choi, Y., ... & Kim, J. K. (2010). Root-specific expression of OsNAC10 improves drought tolerance and grain yield in rice under field drought conditions. *Plant Physiology, 153*(1), 185–197.

Jeong, J. S., Kim, Y. S., Redillas, M. C., Jang, G., Jung, H., Bang, S. W., ... & Kim, J. K. (2013). OsNAC5 overexpression enlarges root diameter in rice plants leading to enhanced drought tolerance and increased grain yield in the field. *Plant Biotechnology Journal, 11*(1), 101–114.

Jongdee, B., Pantuwan, G., Fukai, S., & Fischer, K. (2006). Improving drought tolerance in rainfed lowland rice: an example from Thailand. *Agricultural Water Management, 80*(1–3), 225–240.

Kamoshita, A., Babu, R. C., Boopathi, N. M., & Fukai, S. (2008). Phenotypic and genotypic analysis of drought-resistance traits for development of rice cultivars adapted to rainfed environments. *Field Crops Research, 109*, 1–23.

Kano-Nakata, M., Gowda, V. R., Henry, A., Serraj, R., Inukai, Y., Fujita, D., ... & Yamauchi, A. (2013). Functional roles of the plasticity of root system development in biomass production and water uptake under rainfed lowland conditions. *Field Crops Research, 144*, 288–296.

Kato, Y., Hirotsu, S., Nemoto, K., & Yamagishi, J. (2008). Identification of QTLs controlling rice drought tolerance at seedling stage in hydroponic culture. *Euphytica, 160*(3), 423–430.

Kaur, G., & Asthir, B. (2017). Molecular responses to drought stress in plants. *Biologia Plantarum, 61*(2), 201–209.

Kemble, A. R., & Macpherson, H. T. (1954). Liberation of amino acids in perennial rye grass during wilting. *Biochemical Journal, 58*(1), 46.

Khan, A., Pan, X., Najeeb, U., Tan, D. K. Y., Fahad, S., Zahoor, R., & Luo, H. (2018). Coping with drought: stress and adaptive mechanisms, and management through cultural and molecular alternatives in cotton as vital constituents for plant stress resilience and fitness. *Biological Research, 51*.

Khan, M. I. R., Asgher, M., Fatma, M., Per, T. S., & Khan, N. A. (2015). Drought stress vis a vis plant functions in the era of climate change. *Climate Change and Environmental Sustainability, 3*(1), 13–25.

Khan, M. I. R., & Khan, N. A. (Eds.) (2017). *Reactive Oxygen Species and Antioxidant System in Plants: Role and Regulation under Abiotic Stress*. Springer, Singapore.

Khan, M. I. R., Palakolanu, S. R., Chopra, P., Rajurkar, A. B., Gupta, R., Iqbal, N., & Maheshwari, C. (2021). Improving drought tolerance in rice: Ensuring food security through multi-dimensional approaches. Physiologia Plantarum, *172*(2), 645–668.

Khowaja, F. S., & Price, A. H. (2008). QTL mapping rolling, stomatal conductance and dimension traits of excised leaves in the Bala× Azucena recombinant inbred population of rice. *Field Crops Research*, *106*(3), 248–257.

Kim, T. H., Hur, Y. J., Han, S. I., Cho, J. H., Kim, K. M., Lee, J. H., ... & Shin, D. (2017). Drought-tolerant QTL qVDT11 leads to stable tiller formation under drought stress conditions in rice. *Plant Science*, *256*, 131–138.

Kole, C. (Ed.). (2006). *Cereals and Millets* (Vol. 1). Springer Science & Business Media, Springer, Germany.

Kulya, C., Siangliw, J. L., Toojinda, T., Lontom, W., Pattanagul, W., Sriyot, N., ... & Theerakulpisut, P. (2018). Variation in leaf anatomical characteristics in chromosomal segment substitution lines of KDML105 carrying drought tolerant QTL segments. *ScienceAsia*, *44*, 197–211.

Kumar, R., Malaiya, S., & Srivastava, M. N. (2004). Evaluation of morphophysiological traits associated with drought tolerance in rice. *Indian Journal of Plant Physiology*, *9*, 305–307.

Lafitte, H. R., Yongsheng, G., Yan, S., & Li, Z. K. (2007). Whole plant responses, key processes, and adaptation to drought stress: the case of rice. *Journal of Experimental Botany*, *58*(2), 169–175.

Lafitte, R. (2002). Relationship between leaf relative water content during reproductive stage water deficit and grain formation in rice. *Field Crops Research*, *76*(2–3), 165–174.

Leakey, A. D., Ferguson, J. N., Pignon, C. P., Wu, A., Jin, Z., Hammer, G. L., & Lobell, D. B. (2019). Water use efficiency as a constraint and target for improving the resilience and productivity of C3 and C4 crops. *Annual Review of Plant Biology*, *70*, 781–808.

Lee, D. K., Jung, H., Jang, G., Jeong, J. S., Kim, Y. S., Ha, S. H., ... & Kim, J. K. (2016). Overexpression of the OsERF71 transcription factor alters rice root structure and drought resistance. *Plant Physiology*, *172*(1), 575–588.

Leinonen, I., Grant, O. M., Tagliavia, C. P. P., Chaves, M. M., & Jones, H. G. (2006). Estimating stomatal conductance with thermal imagery. *Plant, Cell & Environment*, *29*(8), 1508–1518.

Li, J., Li, Y., Yin, Z., Jiang, J., Zhang, M., Guo, X., ... & Shao, Y. (2017). Os ASR 5 enhances drought tolerance through a stomatal closure pathway associated with ABA and H_2O_2 signalling in rice. *Plant Biotechnology Journal*, *15*(2), 183–196.

Li, Z., Mu, P., Li, C., Zhang, H., Li, Z., Gao, Y., & Wang, X. (2005). QTL mapping of root traits in a doubled haploid population from a cross between upland and lowland japonica rice in three environments. *Theoretical and Applied Genetics*, *110*(7), 1244–1252.

Liang, X., Zhang, L., Natarajan, S. K., & Becker, D. F. (2013). Proline mechanisms of stress survival. *Antioxidants & Redox Signaling*, *19*(9), 998–1011.

Lilley, J. M., Ludlow, M. M., McCouch, S. R., & O'Toole, J. C. (1996). Locating QTL for osmotic adjustment and dehydration tolerance in rice. *Journal of Experimental Botany*, *47*(9), 1427–1436.

Liu, C., Mao, B., Ou, S., Wang, W., Liu, L., Wu, Y., ... & Wang, X. (2014). OsbZIP71, a bZIP transcription factor, confers salinity and drought tolerance in rice. *Plant Molecular Biology*, *84*(1–2), 19–36.

Liu, H., Zou, G., Liu, G., Hu, S., Li, M., Yu, X., ... & Luo, L. (2005). Correlation analysis and QTL identification for canopy temperature, leaf water potential and spikelet fertility in rice under contrasting moisture regimes. *Chinese Science Bulletin*, *50*(4), 317–326.

Lobell, D. B., Schlenker, W., & Costa-Roberts, J. (2011). Climate trends and global crop production since 1980. *Science*, *333*(6042), 616–620.

Lou, Q., Chen, L., Mei, H., Wei, H., Feng, F., Wang, P., ... & Luo, L. (2015). Quantitative trait locus mapping of deep rooting by linkage and association analysis in rice. *Journal of Experimental Botany*, *66*(15), 4749–4757.

Lu, G., Gao, C., Zheng, X., & Han, B. (2009). Identification of OsbZIP72 as a positive regulator of ABA response and drought tolerance in rice. *Planta*, *229*(3), 605–615.

Lum, M. S., Hanafi, M. M., Rafii, Y. M., & Akmar, A. S. N. (2014). Effect of drought stress on growth, proline and antioxidant enzyme activities of upland rice. *Journal of Animal and Plant Sciences*, *24*(5), 1487–1493.

Mahajan, G., Kumar, V., & Chauhan, B. S. (2017). Rice production in India. In: *Rice Production Worldwide* (pp. 53–91). Springer, Cham.

Mansour, M. M. F., & Salama, K. H. A. (2020). Proline and abiotic stresses: responses and adaptation. In: *Plant Ecophysiology and Adaptation under Climate Change: Mechanisms and Perspectives II* (pp. 357–397). Springer, Singapore.

Martin-Tanguy, J. (2001). Metabolism and function of polyamines in plants: recent development (new approaches). *Plant Growth Regulation*, *34*(1), 135–148.

Melandri, G., Prashar, A., McCouch, S. R., Van Der Linden, G., Jones, H. G., Kadam, N., ... & Ruyter-Spira, C. (2020). Association mapping and genetic dissection of drought-induced canopy temperature differences in rice. *Journal of Experimental Botany*, *71*(4), 1614–1627.

Mitra, J. (2001). Genetics and genetic improvement of drought resistance in crop plants. *Current Science*, *8*, 758–763.

Munns, R., James, R. A., Sirault, X. R., Furbank, R. T., & Jones, H. G. (2010). New phenotyping methods for screening wheat and barley for beneficial responses to water deficit. *Journal of Experimental Botany*, *61*(13), 3499–3507.

Nakashima, K., Ito, Y., & Yamaguchi-Shinozaki, K. (2009). Transcriptional regulatory networks in response to abiotic stresses in Arabidopsis and grasses. *Plant Physiology*, *149*(1), 88–95.

Nakashima, K., Tran, L. S. P., Van Nguyen, D., Fujita, M., Maruyama, K., Todaka, D., ... & Yamaguchi-Shinozaki, K. (2007). Functional analysis of a NAC-type transcription factor OsNAC6 involved in abiotic and biotic stress-responsive gene expression in rice. *The Plant Journal*, *51*(4), 617–630.

Nguyen, H. T., Babu, R. C., & Blum, A. (1997). Breeding for drought resistance in rice: physiology and molecular genetics considerations. *Crop Science*, *37*(5), 1426–1434.

Nguyen, N. M., Hoang, L. H., Furuya, N., Tsuchiya, K., & Nguyen, T. T. (2013). Quantitative trait loci (QTLs) associated with drought tolerance in Vietnamese local rice (*Oryza sativa* L.). *Journal of the Faculty of Agriculture, Kyushu University*, *58*(1), 1–6.

Niones, J. M., Suralta, R. R., Inukai, Y., & Yamauchi, A. (2012). Field evaluation on functional roles of root plastic responses on dry matter production and grain yield of rice under cycles of transient soil moisture stresses using chromosome segment substitution lines. *Plant and Soil*, *359*(1–2), 107–120.

Oh, S. J., Kim, Y. S., Kwon, C. W., Park, H. K., Jeong, J. S., & Kim, J. K. (2009). Overexpression of the transcription factor AP37 in rice improves grain yield under drought conditions. *Plant Physiology*, *150*(3), 1368–1379.

Oladosu, Y., Rafii, M. Y., Samuel, C., Fatai, A., Magaji, U., Kareem, I., ... & Kolapo, K. (2019). Drought resistance in rice from conventional to molecular breeding: a review. *International Journal of Molecular Sciences*, *20*(14), 3519.

O'Toole, J. C., & Moya, T. B. (1978). Genotypic variation in maintenance of leaf water potential in rice 1. *Crop Science*, *18*(5), 873–876.

Panagiotidis, C. A., Artandi, S., Calame, K., & Silverstein, S. J. (1995). Polyamines alter sequence-specific DNA-protein interactions. *Nucleic Acids Research*, *23*(10), 1800–1809.

Pandey, S., Bhandari, H. S., & Hardy, B. (2007). Economic costs of drought and rice farmers' coping mechanisms: a cross-country comparative analysis. International Rice Research Institute.

Pandey, V., & Shukla, A. (2015). Acclimation and tolerance strategies of rice under drought stress. *Rice Science*, *22*(4), 147–161.

Pantuwan, G., Fukai, S., Cooper, M., Rajatasereekul, S., & O'toole, J. C. (2002). Yield response of rice (*Oryza sativa* L.) genotypes to drought under rainfed lowlands: 2. Selection of drought resistant genotypes. *Field Crops Research*, *73*(2–3), 169–180.

Patil, S., Srividhya, A., Veeraghattapu, R., Deborah, D. A. K., Kadambari, G. M., Nagireddy, R., ... & Vemireddy, L. R. (2017). Molecular dissection of a genomic region governing root traits associated with drought tolerance employing a combinatorial approach of QTL mapping and RNA-seq in rice. *Plant Molecular Biology Reporter*, *35*(4), 457–468.

Per, T. S., Khan, N. A., Reddy, P. R., Masood, A., Hasanuzzaman, M., Khan, M. I. R., & Anjum, N. A. (2017). Approaches in modulating proline metabolism in plants for salt and drought stress tolerance: phytohormones, mineral nutrients and transgenics. *Plant Physiology and Biochemistry*, *115*, 126–140.

Per, T. S., Khan, M. I. R., Anjum, N. A., Masood, A., Hussain, S. J., & Khan, N. A. (2018). Jasmonates in plants under abiotic stresses: crosstalk with other phytohormones matters. *Environmental and Experimental Botany*, *145*, 104–120.

Praba, M. L., Cairns, J. E., Babu, R. C., & Lafitte, H. R. (2009). Identification of physiological traits underlying cultivar differences in drought tolerance in rice and wheat. *Journal of Agronomy and Crop Science*, *195*, 30–46.

Price, A. H., Steele, K. A., Moore, B. J., & Jones, R. G. W. (2002). Upland rice grown in soil-filled chambers and exposed to contrasting water-deficit regimes: II. Mapping quantitative trait loci for root morphology and distribution. *Field Crops Research*, *76*(1), 25–43.

Price, A. H., Townend, J., Jones, M. P., Audebert, A., & Courtois, B. (2002a). Mapping QTLs associated with drought avoidance in upland rice grown in the Philippines and West Africa. *Plant Molecular Biology*, *48*(5–6), 683–695.

Prince, S. J., Beena, R., Gomez, S. M., Senthivel, S., & Babu, R. C. (2015). Mapping consistent rice (*Oryza sativa* L.) yield QTLs under drought stress in target rainfed environments. *Rice*, *8*(1), 25.

Rabbani, M. A., Maruyama, K., Abe, H., Khan, M. A., Katsura, K., Ito, Y., ... & Yamaguchi-Shinozaki, K. (2003). Monitoring expression profiles of rice genes under cold, drought, and high-salinity stresses and abscisic acid application using cDNA microarray and RNA gel-blot analyses. *Plant Physiology*, *133*(4), 1755–1767.

Rajurkar, A. B., Muthukumar, C., Ayyenar, B., Thomas, H. B., & Chandra Babu, R. (2021). Mapping consistent additive and epistatic QTLs for plant production traits under drought in target populations of environment using locally adapted landrace in rice (*Oryza sativa* L.). *Plant Production Science*, *24*(3), 388–403.

Rajurkar, A. B., Muthukumar, C., Bharathi, A., Thomas, H. B., & Babu, R. C. (2019). Saturation mapping of consistent QTLs for yield and days to flowering under drought using locally adapted landrace in rice (*Oryza sativa* L.). *NJAS-Wageningen Journal of Life Sciences*, *88*, 66–75.

Ratnakumar, P., Khan, M. I. R., Minhas, P. S., Farooq, M. A., Sultana, R., Per, T. S., Deokate, P. P., Nafees Khan, A., Singh, Y., & Rane, J. (2016). Can plant bio-regulators minimize crop productivity losses caused by drought, salinity and heat stress? An integrated review. *Journal of Applied Botany and Food Quality*, *89*, 113–125.

Ray, D. K., Gerber, J. S., MacDonald, G. K., & West, P. C. (2015). Climate variation explains a third of global crop yield variability. *Nature Communications*, *6*(1), 1–9.

Richards, R. A. (2006). Physiological traits used in the breeding of new cultivars for water-scarce environments. *Agricultural Water Management*, *80*(1–3), 197–211.

Robin, S., Pathan, M. S., Courtois, B., Lafitte, R., Carandang, S., Lanceras, S., ... & Li, Z. (2003). Mapping osmotic adjustment in an advanced back-cross inbred population of rice. *Theoretical and Applied Genetics*, *107*(7), 1288–1296.

Sabar, M., Shabir, G., Shah, S. M., Aslam, K., Naveed, S. A., & Arif, M. (2019). Identification and mapping of QTLs associated with drought tolerance traits in rice by a cross between Super Basmati and IR55419–04. *Breeding Science*, *69*(1), 169–178.

Sah, S. K., Reddy, K. R., & Li, J. (2016). Abscisic acid and abiotic stress tolerance in crop plants. *Frontiers in Plant Science*, *7*, 571.

Saijo, Y., Hata, S., Kyozuka, J., Shimamoto, K., & Izui, K. (2000). Over-expression of a single Ca2+-dependent protein kinase confers both cold and salt/drought tolerance on rice plants. *The Plant Journal*, *23*(3), 319–327.

Saikumar, S., Gouda, P. K., Saiharini, A., Varma, C. M. K., Vineesha, O., Padmavathi, G., & Shenoy, V. V. (2014). Major QTL for enhancing rice grain yield under lowland reproductive drought stress identified using an *O. sativa/O. glaberrima* introgression line. *Field Crops Research*, *163*, 119–131.

Salunkhe, A. S., Poornima, R., Prince, K. S. J., Kanagaraj, P., Sheeba, J. A., Amudha, K., ... & Babu, R. C. (2011). Fine mapping QTL for drought resistance traits in rice (*Oryza sativa* L.) using bulk segregant analysis. *Molecular Biotechnology*, *49*(1), 90–95.

Sandhu, N., Singh, A., Dixit, S., Cruz, M. T. S., Maturan, P. C., Jain, R. K., & Kumar, A. (2014). Identification and mapping of stable QTL with main and epistasis effect on rice grain yield under upland drought stress. *BMC Genetics*, *15*(1), 1–15.

Sandhu, N., Raman, K. A., Torres, R. O., Audebert, A., Dardou, A., Kumar, A., & Henry, A. (2016). Rice root architectural plasticity traits and genetic regions for adaptability to variable cultivation and stress conditions. *Plant Physiology*, *171*(4), 2562–2576.

Sandhu, N., & Kumar, A. (2017). Bridging the rice yield gaps under drought: QTLs, genes, and their use in breeding programs. *Agronomy*, *7*(2), 27.

Sellammal, R., Robin, S., & Raveendran, M. (2014). Association and heritability studies for drought resistance under varied moisture stress regimes in backcross inbred population of rice. *Rice Science*, *21*(3), 150–161.

Seo, J. S., Joo, J., Kim, M. J., Kim, Y. K., Nahm, B. H., Song, S. I., ... & Choi, Y. D. (2011). OsbHLH148, a basic helix-loop-helix protein, interacts with OsJAZ proteins in a jasmonate signaling pathway leading to drought tolerance in rice. *The Plant Journal*, *65*(6), 907–921.

Serraj, R. and Sinclair, T. R. (2002). Osmolyte accumulation: can it really help increase crop yield under drought conditions? *Plant, Cell & Environment*, *25*, 335–341.

Sharma, E., Jain, M., & Khurana, J. P. (2019). Differential quantitative regulation of specific gene groups and pathways under drought stress in rice. *Genomics*, *111*(6), 1699–1712.

Shearman, J. R. (2020). A single segment substitution line population for identifying traits relevant to drought tolerance and avoidance. *Genomics*. (In Press).

Shim, J. S., Oh, N., Chung, P. J., Kim, Y. S., Choi, Y. D., & Kim, J. K. (2018). Overexpression of OsNAC14 improves drought tolerance in rice. *Frontiers in Plant Science*, *9*, 310.

Shinozaki, K., & Yamaguchi-Shinozaki, K. (2007). Gene networks involved in drought stress response and tolerance. *Journal of Experimental Botany*, *58*(2), 221–227.

Siddique, M. R. B., Hamid, A. I. M. S., & Islam, M. S. (2000). Drought stress effects on water relations of wheat. *Botanical Bulletin of Academia Sinica*, *41*(1), 35–39.

Silvente, S., Sobolev, A. P., & Lara, M. (2012). Metabolite adjustments in drought tolerant and sensitive soybean genotypes in response to water stress. *PLoS One*, *7*(6), e38554.

Sinclair, T. R., Purcell, L. C., & Sneller, C. H. (2004). Crop transformation and the challenge to increase yield potential. *Trends in Plant Science*, *9*(2), 70–75.

Srividya, A., Ramanarao, P. V., Sridhar, S., Jayaprada, M., Anuradha, G., Srilakshmi, B., ... & Vemireddy, L. R. (2011). Molecular mapping of QTLs for drought related traits at seedling stage under PEG induced stress conditions in rice. *American Journal of Plant Sciences*, *2*(02), 190.

Steele, K. A., Price, A. H., Shashidhar, H. E., & Witcombe, J. R. (2006). Marker-assisted selection to introgress rice QTLs controlling root traits into an Indian upland rice variety. *Theoretical and Applied Genetics*, *112*(2), 208–221.

Steele, K. A., Virk, D. S., Kumar, R., Prasad, S. C., & Witcombe, J. R. (2007). Field evaluation of upland rice lines selected for QTLs controlling root traits. *Field Crops Research*, *101*(2), 180–186.

Subashri, M., Robin, S., Vinod, K. K., Rajeswari, S., Mohanasundaram, K., & Raveendran, T. S. (2009). Trait identification and QTL validation for reproductive stage drought resistance in rice using selective genotyping of near flowering RILs. *Euphytica*, *166*(2), 291–305.

Swamy, N. R., Murthy, M. S., & Reddy, P. R. (1983). Screening of rice cultivars for drought prone areas. Short communication. *Oryza*, 20(4), 249–251.

Swapna, S., & Shylaraj, K. S. (2017). Screening for osmotic stress responses in rice varieties under drought condition. *Rice Science*, *24*(5), 253–263.

Takagi, H., Ishiga, Y., Watanabe, S., Konishi, T., Egusa, M., Akiyoshi, N., ... & Sakamoto, A. (2016). Allantoin, a stress-related purine metabolite, can activate jasmonate signaling in a MYC2-regulated and abscisic acid-dependent manner. *Journal of Experimental Botany*, *67*(8), 2519–2532.

Takahashi, T., & Kakehi, J. I. (2010). Polyamines: ubiquitous polycations with unique roles in growth and stress responses. *Annals of Botany*, *105*(1), 1–6.

Tang, N., Zhang, H., Li, X., Xiao, J., & Xiong, L. (2012). Constitutive activation of transcription factor OsbZIP46 improves drought tolerance in rice. *Plant Physiology*, *158*(4), 1755–1768.

Tardieu, F. (1996). Drought perception by plants. Do cells of droughted plants experience water stress? In: *Drought Tolerance in Higher Plants: Genetical, Physiological and Molecular Biological Analysis* (pp. 93–104). Springer, Dordrecht.

Tiwari, S., Lata, C., Singh Chauhan, P., Prasad, V., & Prasad, M. (2017). A functional genomic perspective on drought signalling and its crosstalk with phytohormone-mediated signalling pathways in plants. *Current Genomics*, *18*(6), 469–482.

Todaka, D., Nakashima, K., Shinozaki, K., & Yamaguchi-Shinozaki, K. (2012). Toward understanding transcriptional regulatory networks in abiotic stress responses and tolerance in rice. *Rice*, *5*(1), 1–9.

Townley-Smith, T. F., & Hurd, E. A. (1979).Testing and selecting for drought resistance in wheat. In: *Stress Physiology in Crop Plants* (pp. 447–464). Ed. H. Mussell, & R.C. Staples. John Wiley and Sons, New York.

Tran, T. T., Kano-Nakata, M., Suralta, R. R., Menge, D., Mitsuya, S., Inukai, Y., & Yamauchi, A. (2015). Root plasticity and its functional roles were triggered by water deficit but not by the resulting changes in the forms of soil N in rice. *Plant and Soil*, 386(1–2), 65–76.

Trijatmiko, K. R., Prasetiyono, J., Thomson, M. J., Cruz, C. M. V., Moeljopawiro, S., & Pereira, A. (2014). Meta-analysis of quantitative trait loci for grain yield and component traits under reproductive-stage drought stress in an upland rice population. *Molecular Breeding*, *34*(2), 283–295.

Tripathy, J. N., Zhang, J., Robin, S., Nguyen, T. T., & Nguyen, H. T. (2000). QTLs for cell-membrane stability mapped in rice (*Oryza sativa* L.) under drought stress. *Theoretical and Applied Genetics*, *100*(8), 1197–1202.

Uga, Y., Sugimoto, K., Ogawa, S., Rane, J., Ishitani, M., Hara, N., ... & Inoue, H. (2013). Control of root system architecture by DEEPER ROOTING 1 increases rice yield under drought conditions. *Nature Genetics*, *45*(9), 1097.

Verbruggen, N., & Hermans, C. (2008). Proline accumulation in plants: a review. *Amino Acids*, *35*(4), 753–759.

Vikram, P., Swamy, B. M., Dixit, S., Ahmed, H. U., Cruz, M. T. S., Singh, A. K., & Kumar, A. (2011). qDTY 1.1, a major QTL for rice grain yield under reproductive-stage drought stress with a consistent effect in multiple elite genetic backgrounds. *BMC Genetics*, *12*(1), 1–15.

Wade, L. J., Bartolome, V., Mauleon, R., Vasant, V. D., Prabakar, S. M., Chelliah, M., ... & Henry, A. (2015). Environmental response and genomic regions correlated with rice root growth and yield under drought in the OryzaSNP panel across multiple study systems. *PloS One*, *10*(4), e0124127.

Wang, Q., Guan, Y., Wu, Y., Chen, H., Chen, F., & Chu, C. (2008). Overexpression of a rice OsDREB1F gene increases salt, drought, and low temperature tolerance in both Arabidopsis and rice. *Plant Molecular Biology*, *67*(6), 589–602.

Wang, Y., Wan, L., Zhang, L., Zhang, Z., Zhang, H., Quan, R., ... & Huang, R. (2012). An ethylene response factor OsWR1 responsive to drought stress transcriptionally activates wax synthesis related genes and increases wax production in rice. *Plant Molecular Biology*, *78*(3), 275–288.

Wassmann, R., Jagadish, S. V. K., Heuer, S., Ismail, A., Redona, E., Serraj, R., ... & Sumfleth, K. (2009). Climate change affecting rice production: the physiological and agronomic basis for possible adaptation strategies. *Advances in Agronomy*, *101*, 59–122.

Watanabe, S., Matsumoto, M., Hakomori, Y., Takagi, H., Shimada, H., & Sakamoto, A. (2014). The purine metabolite allantoin enhances abiotic stress tolerance through synergistic activation of abscisic acid metabolism. *Plant, Cell & Environment*, *37*(4), 1022–1036.

Xiang, Y., Tang, N., Du, H., Ye, H., & Xiong, L. (2008). Characterization of OsbZIP23 as a key player of the basic leucine zipper transcription factor family for conferring abscisic acid sensitivity and salinity and drought tolerance in rice. *Plant Physiology*, *148*(4), 1938–1952.

Xiong, Y., Liu, T., Tian, C., Sun, S., Li, J., & Chen, M. (2005). Transcription factors in rice: a genome-wide comparative analysis between monocots and eudicots. *Plant Molecular Biology*, *59*(1), 191–203.

Xu, D. Q., Huang, J., Guo, S. Q., Yang, X., Bao, Y. M., Tang, H. J., & Zhang, H. S. (2008). Overexpression of a TFIIIA-type zinc finger protein gene ZFP252 enhances drought and salt tolerance in rice (*Oryza sativa* L.). *FEBS Letters*, *582*(7), 1037–1043.

Xu, J. L., Zhong, D. B., Yu, S. B., Luo, L. J., & Li, Z. K. (2002). *QTLS Affecting Leaf Rolling and Folding in Rice*. International Rice Research Institute, Philippines.

Yang, A., Dai, X., & Zhang, W. H. (2012). A R2R3-type MYB gene, OsMYB2, is involved in salt, cold, and dehydration tolerance in rice. *Journal of Experimental Botany*, *63*(7), 2541–2556.

Yang, J., Zhang, J., Liu, K., Wang, Z., & Liu, L. (2007). Involvement of polyamines in the drought resistance of rice. *Journal of Experimental Botany*, *58*(6), 1545–1555.

Yang, L., Wang, J., Lei, L., Wang, J., JunaidSubhani, M., Liu, H., ... & Zou, D. (2018). QTL mapping for heading date, leaf area and chlorophyll content under cold and drought stress in two related recombinant inbred line populations (Japonica rice) and meta-analysis. *Plant Breeding*, *137*(4), 527–545.

Yang, S., Vanderbeld, B., Wan, J., & Huang, Y. (2010). Narrowing down the targets: towards successful genetic engineering of drought-tolerant crops. *Molecular Plant*, *3*(3), 469–490.

Yoo, Y. H., NaliniChandran, A. K., Park, J. C., Gho, Y. S., Lee, S. W., An, G., & Jung, K. H. (2017). OsPhyB-mediating novel regulatory pathway for drought tolerance in rice root identified by a global RNA-Seq transcriptome analysis of rice genes in response to water deficiencies. *Frontiers in Plant Science*, *8*, 580.

Yoshida, S., & Hasegawa, S. (1982). The rice root system: its development and function. *Drought Resistance in Crops with Emphasis on Rice*, *10*, 97–134.

Yue, B., Xiong, L., Xue, W., Xing, Y., Luo, L., & Xu, C. (2005). Genetic analysis for drought resistance of rice at reproductive stage in field with different types of soil. *Theoretical and Applied Genetics*, *111*(6), 1127–1136.

Yue, B., Xue, W., Xiong, L., Yu, X., Luo, L., Cui, K., ... & Zhang, Q. (2006). Genetic basis of drought resistance at reproductive stage in rice: separation of drought tolerance from drought avoidance. *Genetics*, *172*(2), 1213–1228.

Zhang, J., Zheng, H. G., Aarti, A., Pantuwan, G., Nguyen, T. T., Tripathy, J. N., ... & Sarkarung, S. (2001). Locating genomic regions associated with components of drought resistance in rice: comparative mapping within and across species. *Theoretical and Applied Genetics*, *103*(1), 19–29.

Zhao, L., Hu, Y., Chong, K., & Wang, T. (2010). ARAG1, an ABA-responsive DREB gene, plays a role in seed germination and drought tolerance of rice. *Annals of Botany*, *105*(3), 401–409.

Zhao, X. Q., Xu, J. L., Zhao, M., Lafitte, R., Zhu, L. H., Fu, B. Y., ... & Li, Z. K. (2008). QTLs affecting morph-physiological traits related to drought tolerance detected in overlapping introgression lines of rice (*Oryza sativa* L.). *Plant Science*, *174*(6), 618–625.

Zheng, B. S., Yang, L., Zhang, W. P., Mao, C. Z., Wu, Y. R., Yi, K. K., ... & Wu, P. (2003). Mapping QTLs and candidate genes for rice root traits under different water-supply conditions and comparative analysis across three populations. *Theoretical and Applied Genetics*, *107*(8), 1505–1515.

Zheng, X., Chen, B., Lu, G., & Han, B. (2009). Overexpression of a NAC transcription factor enhances rice drought and salt tolerance. *Biochemical and Biophysical Research Communications*, *379*(4), 985–989.

Zivcak, M., Brestic, M., & Sytar, O. (2016). Osmotic adjustment and plant adaptation to drought stress. In: *Drought Stress Tolerance in Plants, Vol 1* (pp. 105–143). Springer, Cham.

2 Coordinated Functions of Reactive Oxygen Species Metabolism and Defense Systems in Abiotic Stress Tolerance

Swati Sachdev[1], *Priyanka Jaiswal*[2], *and*
Mohammad Israil Ansari[3]*
[1]Department of Applied Sciences and Humanities, Faculty of Engineering and Technology, Rama University, Kanpur, India
[2]Dr. D. Y Patil Biotechnology and Bioinformatics Institute, Dr. D. Y Patil Vidyapeeth, Pune, India
[3]Department of Botany, University of Lucknow, Lucknow, India
*Corresponding author: E-mail: ansari_mi@lkouniv.ac.in

CONTENTS

DOI: 10.1201/9781003159636-2

2.1 INTRODUCTION

Maintaining plant growth and productivity to sustain the needs of a growing population under perturbed environmental scenarios is a major challenge for agricultural and plant scientists. Morphological, physiological, and metabolic activities of plants during development are highly influenced by abiotic factors. The aberrant changes in climatic phenomena have altered the abiotic factors and aggravated the duration and strength of stresses (Raza et al. 2019). Incidences of abiotic stresses, including drought, waterlogging, heat, chilling/freezing, salinity, UV radiation, light, and pollutants (heavy metals, organic compounds), diminish plant yield and increase economic losses (Ahanger et al. 2017; Bhuyan et al. 2020; Iqbal et al. 2021). In normal field conditions, a number of stresses often occur in combination, which may be due to interrelated pathways (Bulgari et al. 2019; Sharma et al. 2019; Zandalinas et al. 2020). For instance, drought elevates salt stress and encourages over-production of reactive oxygen species (ROS) in plant cells, which react with biomolecules and modify their structure and function, causing oxidative stress (Sachdev et al. 2021). Multiple stresses acting together as a consortium on plants severely damage their growth and development (Bulgari et al. 2019; Sharma et al. 2019).

Members of the ROS family comprise radical and non-radical forms of molecular oxygen (O_2) (Mitler 2017). Plants under normal circumstances convert 1–2% of consumed O_2 into ROS as a by-product of aerobic processes (Roychowdhury et al. 2019; Shah et al. 2019). However, unfavorable conditions exacerbate ROS production, resulting in an ROS burst in a plant cell. ROS are considered as toxic molecules, and their excess accumulation affects cellular and molecular components of plants (Kerchev et al. 2020). Generally, plants contain enzymatic and non-enzymatic antioxidant-based defense systems that scavenge or detoxify ROS. Nevertheless, if the ROS level remains unchecked or persists higher than the antioxidants' quenching threshold, this triggers oxidative stress and may lead to programmed cell death (PCD) (Kerchev et al. 2020). Due to their severe impacts on plant cells, ROS in the past have always been referred to as "bad" or toxic molecules; however, in the last few decades, studies have established the role of ROS in the regulation of developmental processes and facilitation of defense response against abiotic stresses, designating them as "good" molecules also (Mittler 2017; Kerchev et al. 2020). Therefore, ROS are addressed as a double-edged sword.

ROS, on the one hand, incite cell death, on the other hand, the transient increase upregulates genes and causes a surge in concentration of proteins and metabolites, enhancing the plant's tolerance against stress (Kerchev et al. 2020). As ROS perform a dual role within the plant cell, understanding the pathways responsible for triggering oxidative stress or switching tolerance can be employed as a strategic approach to reduce oxidative damage and reinforce their defense mechanism under projected climatic scenarios. The present chapter attempts to summarize the knowledge about the metabolism of ROS within the plant cell, the activity of different antioxidants in maintaining their level under equilibrium, and the mechanism of stimulating plant defense systems under abiotic stresses.

2.2 DIFFERENT KINDS OF ROS GENERATED UNDER ABIOTIC STRESSES

ROS primarily include superoxide radical ($O_2^{\cdot -}$), hydroxyl radical ($\cdot OH$), hydrogen peroxide (H_2O_2), and singlet oxygen ($^1O_2$) generated through partial reduction or excitation of O_2 (Figure 2.1) in different cell organelles like chloroplasts, mitochondria, peroxisomes, and others (Mittler 2017; Jalil and Ansari 2018; Maurya 2020). Some of the ROS species are very toxic and highly reactive, while others are less toxic. However, less toxic or reactive species undergo conversion leading to the production of reactive ROS, which finally affect cellular components such as cellular and organelle membranes, photosynthetic activity, etc. (Das and Roychowdhury 2014).

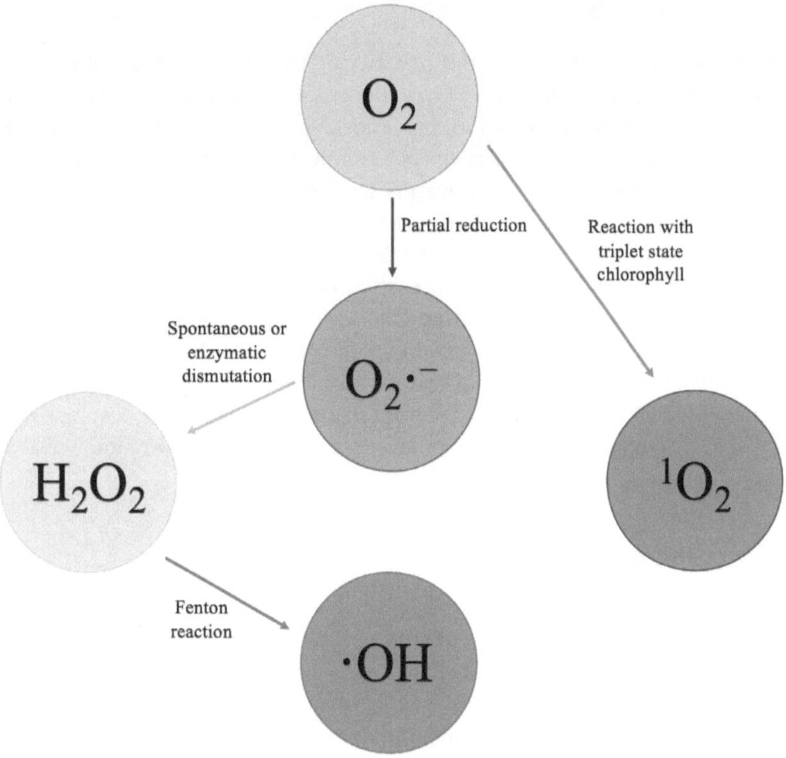

FIGURE 2.1 Different types of ROS generated within plant cell from molecular oxygen (O_2). O_2 on partial reduction forms superoxide radical ($O_2^{\cdot-}$), which undergoes spontaneous or enzymatic dismutation by superoxide dismutase, resulting in the formation of hydrogen peroxide (H_2O_2). H_2O_2 in the presence of redox metals produces hydroxyl radical via Fenton reaction. Singlet oxygen (·OH) is generated on reaction of O_2 with triplet state chlorophyll.

2.2.1 Superoxide Radical ($O_2^{\cdot-}$)

Superoxide radical ($O_2^{\cdot-}$) is the first or primary ROS generated within the plant cell under stress (Das and Roychowdhury 2014). $O_2^{\cdot-}$ is generated at photosystem (PS) I in chloroplasts through partial reduction of O_2 by the photosynthetic electron transport chain (ETC) (Kohli et al. 2019). In addition to the chloroplast, $O_2^{\cdot-}$ is also produced in mitochondria, peroxisomes, apoplast, and cell wall (Kohli et al. 2019). It is a moderately reactive species with a half-life time of 2–4 μs. $O_2^{\cdot-}$ can migrate to a distance of 30 nm and is present in a concentration lower than 0.5% (~ 0.3%) in the cytosol (De Grey 2002; Mittler 2017; Kohli et al. 2019). Due to moderate toxicity, these molecules do not induce extensive damage; however, they undergo chemical or enzymatic dismutation and generate H_2O_2, which further transforms to yield highly toxic ·OH radicals, which cause oxidative damage (Das and Roychowdhury 2014). The dismutation reaction is catalyzed by the enzymatic antioxidant superoxide dismutase (SOD); it occurs at the rate ~$2 \times 10^9\ M^{-1}\ s^{-1}$, or 10^4 times the rate constant for spontaneous dismutation (Kohli et al. 2019). Moreover, the protonated form of $O_2^{\cdot-}$, known as hydroperoxyl radical (HOO· or $HO_2\cdot$), is comparatively more toxic and induces oxidative stress in the plant cell (Kohli et al. 2019).

2.2.2 Singlet Oxygen ($^1O_2$)

These ROS molecules are produced by the reaction of triplet state chlorophyll (^{3}Chl) with O_2. Insufficient availability of carbon dioxide (CO_2) under stress due to the closure of stomata favors the formation of $^1O_2$. The half-time of $^1O_2$ is 3 μs, but it possesses the ability to diffuse to the distance of 100 nm and thus causes damage to a wide range of biomolecules, such as proteins, lipids, etc. $^1O_2$ can easily oxidize the C-C double bond; thus, it mainly targets the double bonds of amino acid residues of proteins, polyunsaturated fatty acids (PUFA) of lipid membranes, guanine bases in nucleic bases, and thiol groups (Dmitrieva et al. 2020). This leads to the formation of hydroperoxides, which are responsible for initiating a free radical chain reaction (Dmitrieva et al. 2020). $^1O_2$ induces severe damage to photosynthetic machinery, affecting PS I and PS II, causing loss of activity and sometimes leading to cell death (Das and Roychowdhury 2014). An *Arabidopsis* mutant with enhanced $^1O_2$ production ability shows increased lipid peroxidation under photo-oxidative stress and induced cell death (Triantaphylides et al. 2008). Apart from damaging effects, $^1O_2$ also has been identified as instrumental in the upregulation of genes that protect against photo-oxidative stress (Das and Roychowdhury 2014).

2.2.3 Hydrogen Peroxide (H_2O_2)

H_2O_2 is a moderately reactive species. Enzymatic or spontaneous dismutation of $O_2^{\cdot-}$ results in the production of H_2O_2. Spontaneous dismutation of $O_2^{\cdot-}$ is favored by low pH, whereas enzymatic dismutation reaction is catalyzed by the enzymatic antioxidant SOD. Under stress conditions, due to limited availability of CO_2, the ribulose 1,5-bisphosphate (RuBP) oxygenation process is favored, leading to photorespiration and production of H_2O_2. Apart from photorespiration, photo-oxidation of nicotinamide adenine dinucleotide phosphate (NADPH) oxidase and xanthine oxidase (XOD) result in H_2O_2 production. The major organelles that participate in H_2O_2 formation include chloroplasts and mitochondria via ETC, cytosol, apoplast, plasma membrane, and peroxisomes (Mittler 2017; Smirnoff and Arnaud 2019). H_2O_2 is moderately reactive and displays both toxicity and defense activity. At low concentration, it regulates signaling pathways for processes like photorespiration, stomatal movement, cell cycle, growth, and development (Das and Roychowdhury 2014). H_2O_2 as compared with other ROS has a significantly long half-life time of 1 ms; hence, it can travel to a longer distance (more than 1 μm) and can cross plant cell membranes through aquaporins (Das and Roychowdhury 2014). On the contrary, at a high concentration, H_2O_2 oxidizes cysteine (Cys) and methionine (Met) residues of amino acids, inactivates Calvin cycle enzymes and SOD enzymes by oxidizing their thiol group, and can even trigger PCD (Das and Roychowdhury 2014).

2.2.4 Hydroxyl Radical ($\cdot$OH)

$\cdot$OH radicals are the most potent form of ROS. They have a single unpaired electron, which enhances their reactivity with triplet ground state oxygen (Sharma et al. 2012). The Haber–Weiss reaction and the Fenton reaction are major pathways/reactions that mediate $\cdot$OH formation inside plant cells under stress. In the presence of transition metals, H_2O_2 and $O_2^{\cdot-}$ are ultimately converted into $\cdot$OH radical via the Haber–Weiss reaction (Richards et al. 2015). In the Fenton reaction, reduced forms of transition metals catalyze the formation of $\cdot$OH radical from H_2O_2 (Richards et al. 2015). Under dark conditions, production of $\cdot$OH radical, possibly via the Fenton reaction, triggered wilting in the epicotyl of pea seedlings (Hideg et al. 2010). Cells do not recruit any enzymatic mechanism to scavenge these radicals; therefore, their excess accumulation leads to cell death (Sharma et al. 2012). $\cdot$OH radicals are short-lived ROS with a half-life time of 1 ns; they have strong positive redox potential and can migrate to a distance of 1 nm (Mitler 2017), and thus often display reactivity near sites of production (Sharma et al. 2012). Due to their highly toxic and reactive nature, $\cdot$OH radicals interact with all biomolecules and subsequently damage cellular proteins, induce lipid peroxidation, and disrupt membranes (Sharma et al. 2012).

2.3 ABIOTIC STRESS AS A PRECURSOR OF ROS OVER-PRODUCTION

Plant growth is determined by their genetic makeup and persisting environmental conditions such as temperature, water, light, radiation, etc., collectively known as abiotic factors. Slight variations in environmental conditions induce mild stress that temporarily restricts plant growth (Gull et al. 2019). However, extreme fluctuations in abiotic factors trigger lasting damage that not only impinges on plants' physiological performance but also alters plants' genomics to an extent that reduces plant growth and development (Raza et al. 2019; Saijo and Loo 2020). Abiotic stress(es) trigger production of ROS within plant cells, which hinders vital processes including transpiration, respiration, and photosynthesis and disrupts enzymatic activities that correspond to ROS over-production (Figure 2.2). Abiotic stress-induced ROS production in important crops, resulting in cellular damage, is summarized in Table 2.1.

2.3.1 Temperature Stress

Variation from the optimum growth temperature induces heat or chilling/freezing stress, resulting in impaired plant architecture, reproduction, and fruit setting (Zaidi et al. 2014; Martinez 2016). An increase in greenhouse gases (GHG), high light intensities, and heat waves have subjected plants to heat stress, which triggers morphological, cellular, and metabolic changes such as elongation of petioles and hypocotyls, diminishes photosynthetic and respiratory performance, reduces enzymatic activity, up-regulates transcription and translation of heat shock proteins (HSP), elevates calcium influx, and increases ROS production (Bita and Gerats 2013). High temperature increases ground-level ozone (O_3) production, which can also trigger oxidative stress in plants (Coates et al. 2016). In contrast to heat stress, chilling stress is characterized by low-temperature events that increase solubility and accumulation of O_2, and promote electron leakage from the photosynthetic and respiratory ETC, aggravating ROS production (Coates et al. 2016). Over-production of ROS under chilling stress affects protein and lipid molecules, resulting in increased membrane fluidity and reduced enzyme activity (Jalil et al. 2017). Due to chilling stress in cucumber seedlings, increased electrolyte leakage along with reduced tissue water and chlorophyll content has been reported by Zhang et al. (2012). Similarly, under low temperature, over-production of H_2O_2 and $O_2^{\cdot -}$ in tomato leaves has been reported with enhanced malondialdehyde (MDA) level and *RBOH1* (*respiratory burst oxidase homolog 1*) expression. Stress has been found to reduce net photosynthesis rate and chlorophyll fluorescence (Liu et al. 2018).

2.3.2 Water Stress

Change in climatic scenarios in the last decades has tremendously affected rainfall patterns, causing erratic precipitation with altered magnitude and seasonal variations (Feng et al. 2013). These extreme changes have resulted in water deficit (drought) or waterlogging (flooding) stress. It has been estimated that the proportion of the world's agricultural land engulfed by drought will be increased two-fold by 2050, which will considerably reduce agricultural productivity (Kumar et al. 2019; Khan et al. 2020a). The effect of water deficit stress greatly varies with the severity and time-length of the stress. Water deficit conditions have been reported to accelerate the rate of stomatal closure, decrease CO_2 fixation, and elevate photoreduction of O_2 in the chloroplast as well as photorespiration, resulting in ROS accumulation triggering oxidative stress in plants (Jalil et al. 2017) Accumulation of ROS in white clover leaves grown under water deficit conditions has been reported along with an increase in MDA content and decrease in dry mass (Lee et al. 2009).

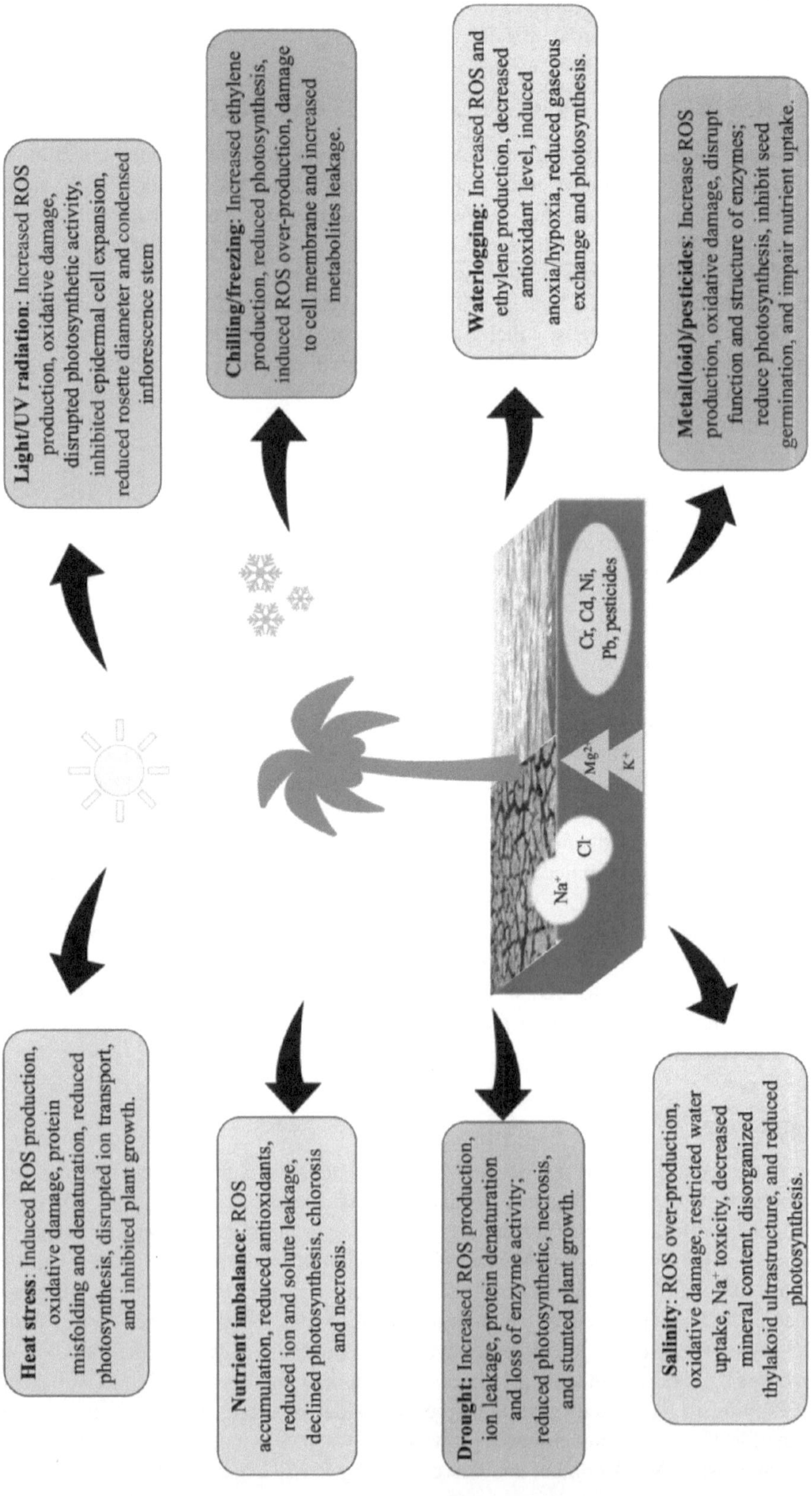

FIGURE 2.2 Abiotic stress-induced over-production of ROS and reduced plant activities.

TABLE 2.1
Production of Reactive Oxygen Species in Major Crops under Abiotic Stress(es)

Stress	Crops	ROS Production and Oxidative Damage	References
Chilling	Tomato (*Lycopersicon esculentum* Mill.)	H_2O_2 content increased in roots and leaves	Diao et al. 2017
	Tomato (*L. esculentum* cv. C.H. Falat)	H_2O_2 content increased two-fold and electrolyte leakage increased by 20%	Ghanbari and Sayyari 2018
Drought	Rice (*Oryza sativa* L.)	Increased production of H_2O_2 and lipid oxidation	Sohag et al. 2020
	Wheat (*Triticum aestivum*)	Increased H_2O_2 production and MDA content	Habib et al. 2020
	Maize (*Zea mays*)	Increased accumulation of ROS, membrane damage, electrolyte leakage. Elevation of lipoxygenase and thiobarbituric acid reactive substance (TBARS)	Anjum et al. 2017
	Tomato (*L. esculentum* Mill.)	Increased MDA content and electrolyte leakage	Malhotra et al. 2017
	Chinese crab apple (*Malus hupehensis*)	Increased lipid peroxidation and accumulation of H_2O_2 and $O_2^{\cdot-}$	Wang et al. 2012
Heat stress	*Arabidopsis thaliana*	Increased ROS level and elevated TBARS content by 68.30%	Kipp and Boyle 2013
	Rice (*Oryza sativa* L.)	Increased H_2O_2 accumulation	Liu et al. 2019
Heavy metal (chromium)	Mustard (*Brassica juncea*)	Increased TBARS content, H_2O_2 production and lipoxygenase activity	Kabala et al. 2019
Heavy metal (cadmium)	Mung bean (*Vigna radiata*)	Increased H_2O_2 and $O_2^{\cdot-}$ production and increased MDA content	Nahar et al. 2017
Ground-level ozone (O_3)	Rice (*Oryza sativa* L.)	Increased production of H_2O_2 and lipid peroxidation	Ueda et al. 2013
	Wheat (*Triticum aestivum* L.)	Increased production of H_2O_2 and $O_2^{\cdot-}$, and enhanced lipid peroxidation	Yadav et al. 2019
Salinity	Sweet pepper (*Capsicum annuum*)	Increased ROS and MDA content, and electrolyte leakage	Abdelaal et al. 2020
UV-B radiation	Cucumber (*Cucumis sativus*)	Increased H_2O_2 accumulation in cotyledons	Rybus-Zajac and Kubis 2010
Water-logging	Barley (*Hordeum vulgare*)	Increased production of $O_2^{\cdot-}$ and MDA content	Luan et al. 2018
	Millet (*Sorghum bicolor*)	Increased MDA content in sensitive cultivar	Zhang et al. 2019

Flooding is another severe abiotic stress that creates a waterlogging condition, characterized by low light, impaired gas exchange, and hypoxia/anoxia, which reduces the diffusion of oxygen, thereby suppressing aerobic activity in the soil such as root respiration (Sasidharan et al. 2018; Khan et al. 2020b). During waterlogging due to deficient O_2 conditions, ROS and volatile gases like ethylene are produced. The anoxic state results in inhibition of photosynthetic and mitochondrial ETC, which consequently ends up in over-production of ROS (Chang et al. 2012; Sasidharan et al. 2018). Waterlogging conditions result in leaching of essential nutrients from the soil, accumulation of salts, and increase in heavy metal availability to the plant due to alteration in soil pH. These changes aggravate stress conditions and trigger nutrient deficiency, salinity, heavy metal, and oxidative stress in plants (Steffen 2014). The onset of oxidative stress in sesame plants subjected to waterlogging conditions has been described by Anee et al. (2019), who reported accumulation of ROS and methylglyoxal content and increase in lipid peroxidation.

2.3.3 Salinity

Soil salinity is one of the most stubborn abiotic stresses affecting agricultural produce (20–50%) and soil fertility worldwide (Yuan et al. 2015; He et al. 2018). It has been estimated that around 20% of the world's total arable land is affected by salinity (Etesami and Noori, 2019), and by the year 2050 more than 50% of cultivable land will be salinized (Etesami and Noori 2019). Climate change has resulted in sea level rise and intrusion of saline water into arable lands, thereby increasing salt content in soil and imposing stress on plant cultivation (Chima et al. 2015). Water scarcity (due to uneven precipitation and low groundwater level) and high temperature resulting in excess water evaporation also aggravate soil saline concentration (Zhang et al. 2018). Salinity stress occurs due to the presence of excessive Na^+ and Cl^- ions, which establishes hypertonic conditions and induces osmotic stress (Ilangumaran and Smith 2017). Induced osmotic stress hinders plant homeostasis and affects physiological processes such as photosynthesis, uptake of nutrients (causing nutrient deficiency) and their translocation, hormonal balance, etc. (Ilangumaran and Smith 2017). Salt stress reduces plants' capacity to absorb water and induces drought-like conditions (Jalil and Ansari 2020). This reduces stomatal conductance, disrupts PS, and affects photosynthetic enzymes, leading to ROS production in plants (Hasanuzzaman et al. 2018). Wheat plants subjected to salinity stress have been reported to accumulate excess ROS, which triggered lipid peroxidation and reduced membrane stability, leading to electrolyte leakage (Kaur et al. 2017).

2.3.4 Heavy Metal Stress

Anthropogenic activities have increased the load of heavy metals, chemicals, and other xenobiotic compounds in the environment, causing a stressful situation. Heavy metals include both essential and non-essential plant nutrients. Accumulation of non-essential metals is known to cause toxicity in plants via ROS generation; however, unrestricted uptake of essential nutrients also induces ROS generation. Heavy metals like iron (Fe), chromium (Cr), and copper (Cu) are major redox-active metals that result in oxidative stress in plants due to their high concentration in soil (Schutzendubel and Polle 2002; Steffens 2014; Khan et al., 2021). Heavy metals trigger ROS production in chloroplasts, mitochondria, apoplast, and peroxisomes (Pandey et al. 2009; Steffens 2014; Trinh et al. 2014). Cadmium (Cd) is a non-essential metal and is known to cause toxicity in plants. It induces over-production in plant cells indirectly by replacing Cu or Fe ions in metalloenzyme antioxidants, thus impairing the respiratory ETC and disturbing redox status (Gupta et al. 2017). Fe is an essential micronutrient, but when it accumulates in excess in plants, it induces over-production of ROS through a chain of reactions (Becana et al. 1998), causing damage to lipid membranes and chlorophyll (Hajiboland 2012). Fe present in reduced form oxidizes and generates H_2O_2 and $O_2^{\cdot -}$ in the cell via a process called auto-oxidation (Schutzendubel and Polle 2002). The H_2O_2 formed oxidizes another reduced Fe molecule to produce a highly toxic ·OH radical (Hajiboland 2012).

2.3.5 Xenobiotic Compound Stress

Similarly to heavy metals, certain xenobiotic compounds, including pesticides, induce oxidative stress by triggering over-production of ROS within plant cells (Sharma et al. 2019). Only 1% of the total applied pesticides reach the target organisms, and the rest contaminate the ecosystem, resulting in pesticide-induced stress (Sachdev and Singh 2018). Pesticides have been well documented to possess the ability to retard growth and photosynthetic efficiency of the plant, cause molecular alterations, induce over-production of ROS, and reduce plants' intrinsic antioxidant-based scavenging activity (Sharma et al. 2019; Yuzbasioglu et al. 2019). Application of pesticide (thiram) to tomato plants has been reported to increase the production of H_2O_2, elevate MDA level, and degrade leaves' chlorophyll content (Yuzbasioglu et al. 2019). Similar effects were reported in *Brassica juncea* treated with the pesticide imidacloprid (Sharma et al. 2019).

2.4 ROS-INDUCED DAMAGE TO CELLULAR BIOMOLECULES

Oxidative bursts caused by over-production of ROS under adverse environmental conditions attack biomolecules, primarily DNA, RNA, proteins, and lipids, cause protein oxidation, enzyme inactivation, lipid peroxidation, disruption of membrane integrity, chlorophyll degradation, and nucleic acid damage, and instigate the apoptosis pathway, leading to PCD under severe conditions (Roychowdhury et al. 2019; Shah et al. 2019). This damage affect growths, development, and ultimately plant survival.

2.4.1 Lipid Membranes

Cell and organelle membranes composed of lipids are a prime target for ROS. The PUFA of lipids are primarily attacked by ROS due to the presence of unsaturated C-C double bonds in fatty acids. The ·OH radical attacks the methylene group of fatty acids and abstracts a hydrogen (H) atom, forming a carbon-centered lipid radical (Anjum et al. 2015). ROS also attack the ester linkage between glycerol and fatty acids and cause membrane phospholipids to disintegrate (Sharma et al. 2012; Das and Roychowdhary 2014). The PUFA of the plasma membrane and mitochondrial membrane-like arachidonic acid, linolenic acid, and linoleic acid are the most susceptible targets of ROS. The ·OH radical initiates a cyclic reaction resulting in peroxidation of PUFA (Das and Roychowdhary 2014). The process of lipid peroxidation involves three stages: Initiation, propagation, and termination (cleavage) (Anjum et al. 2015). Initiation is the first stage, which involves the production of ROS by reduction of O_2. The ROS generated in this way triggers a cascade of reaction leading to the production of lipid radicals (lipid peroxyl radicals, hydroperoxides, etc.), constituting the second or propagation stage. Finally, lipid radicals end up in lipid dimers, resulting in the culmination of the chain reaction and marking the termination of the process (Das and Roychowdhary 2014). The formation of lipid radicals by peroxidation causes membrane destabilization, increases membrane permeability and electrolyte leakage, deactivates membrane-located enzymes and receptors, and enhances the oxidation of other biomolecules such as nucleic acids and proteins (Sharma et al. 2012; Das and Roychowdhary 2014; Anjum et al. 2015).

Oxidative stress induced by stresses like salinity (Katsuhara et al. 2005), drought (Hameed et al. 2011; 2013), high temperature (Ali et al. 2005), metal(loid)s (Singh et al. 2006), pesticides (Majid et al. 2014), etc. has been found to be associated with increased cellular and organelle lipid peroxidation. Arsenic toxicity has been found to induce lipid peroxidation, electrolyte leakage, and oxidative damage in common bean seedlings due to the accumulation of H_2O_2 (Talukdar 2013). MDA, a lipid peroxidation product, indicates the degree of oxidative damage in the cell and is thus considered as a marker for lipid peroxidation (Sharma et al. 2012; Das and Roychowdhary 2014). Increased accumulation of H_2O_2 along with MDA content and lipid peroxidation in tomato plants has been found to be linked with salinity, heat stress, and the combination of these stresses (Martinez et al. 2016).

2.4.2 Proteins

Proteins are important functional and structural components of a plant cell that play a crucial role in facilitating tolerance to abiotic stress by adjusting the physiological characters of plants (Kosova et al. 2018). Protein aggregation or change in conformation modifies their enzymatic, binding, and other functional activities (Oracz et al. 2007; Anjum et al. 2015). Oxidative stress causes protein oxidation by direct and indirect means. Direct oxidation by ROS involves oxidation of side chains of amino acids, such as $^1O_2$- and ·OH-induced oxidation of Cys and Met residues, and degradation of the peptide backbone, which results in carbonylation, nitrosylation, disulfide bond formation, and glutathionylation; whereas indirect oxidation is mediated through products formed during lipid peroxidation (Moller et al. 2011; Das and Roychowdhary 2014). Oxidation also enhances the susceptibility of proteins towards proteolytic digestion (Moller et al. 2007). Protein oxidation under oxidative stress is either irreversible or reversible. ROS, for instance, $O_2^{\cdot -}$, cause irreversible damage to enzymes containing an Fe-S center. Irreversible damage includes carbonylation, protein–protein cross-linking, etc., resulting in functional loss. Reversible changes like glutathionylation and *S*-nitrosylation can facilitate redox regulation (Anjum et al. 2015). Carbonylation of a protein is irreversible and unrepairable damage mediated by oxidative cleavage of proteins and is considered as the best marker for estimation of oxidative damage under stress (Moller et al. 2007). A significant increase in carbonylated protein in cobalt-stressed *B. juncea* leaves has been documented by Karuppanapandian and Kim (2013). Similarly, an increase in protein oxidation under salinity has been reported in cashew plants (Ferreira-Silva et al. 2012). Under drought conditions, also, several-fold higher protein carbonyl has been detected in mitochondria compared with chloroplasts and peroxisomes in wheat leaves (Bartoli et al. 2004).

2.4.3 Nucleic Acids

The plant nucleic acids (DNA and RNA) undergo oxidation by ROS attack generated under abiotic stress, which leads to mutation (Hollosy 2002). Chloroplastic and mitochondrial DNA are more susceptible to oxidation than nuclear DNA due to their closeness to the ROS production site and lack of protective histones and associative proteins (Das and Roychowdhary 2014). ROS oxidizes nucleic acids, resulting in oxidation of sugar residues, alteration of nucleotide bases (insertion or deletion), the abstraction of nucleotides, breaks in DNA strands, and cross-linking of the DNA and protein (Das and Roychowdhary 2014; Ebone et al. 2019). Intersomal nDNA fragmentation has been reported in the sensitive genotype of wheat with PCD in leaves under drought (Hameed et al. 2013). ROS subtracts an H-atom from the C4 position of the deoxyribose sugar backbone, forming a deoxyribose radical, which causes a further break in the DNA strand (Evans et al. 2004). The ·OH radical attacks the double bonds of purine and pyrimidine bases (Halliwell 2006), causing DNA lesions and the formation of 8-hydroquinine with some other products including hydroxyl methyl urea, thymine glycol, etc. (Das and Roychowdhary 2014). The ·OH radical also induces cross-linking of DNA and protein on reaction with either DNA or associated proteins. The repairing of this cross-linkage is a difficult task, and if not repaired before replication or transcription, can cause a lethal effect on plant cells (Das and Roychowdhary 2014). Apart from direct oxidation, the lipid radicals generated during lipid peroxidation result in indirect DNA oxidation (Roldan-Arjona and Ariza 2009). MDA reacts with the guanine (G) residues of DNA and as a result, forms M_1G, i.e., a pyrimidopurinone adduct (Fink and Reddy 1997). Similarly to other biomolecules, RNAs are susceptible to ROS (Liu et al. 2012). RNA oxidation leads to 8-oxo-7,8-dihydroguanosine (8-OHG) formation, which is used as a marker for determining the intensity of RNA oxidation.

2.5 ANTIOXIDANT-BASED DEFENSE SYSTEM

Accumulation of excessive ROS under stress can be toxic and result in oxidative stress, ultimately leading to plant cell death (Gill and Tuteja 2010). To maintain redox equilibrium within the cell, plants contain antioxidant machinery that quenches or detoxifies free radicals with the aid of enzymatic (superoxide dismutase (SOD), catalase (CAT), ascorbate peroxidase (APX), glutathione

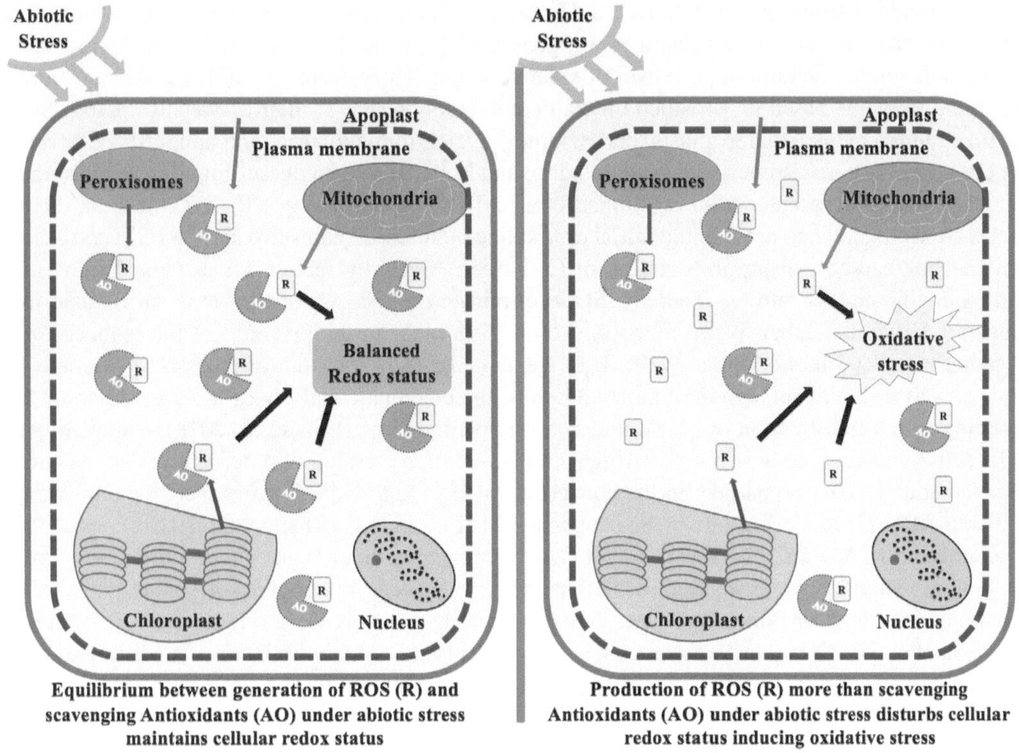

FIGURE 2.3 Imbalance in equilibrium between antioxidants and reactive oxygen species induces oxidative stress.

peroxidase (GPX), glutathione reductase (GR), peroxidase (POX), glutathione *S*-transferase (GST), monodehydroascorbate reductase (MDHAR), dehydroascorbate reductase (DHAR), etc.) and non-enzymatic antioxidants (ascorbate (AsA), glutathione (GSH), phenols, tocopherols, carotenoids, flavonoids, and others) (Dumont and Rivoal 2019; Hasanuzzaman et al. 2020). These antioxidants display their defensive activity at a different level depending upon stage and type of action, i.e., inhibiting ROS production, ROS scavenging, and/or repairing damage caused by ROS (Ighodaro and Akinloye 2018) (Figure 2.3). Based on their activity, antioxidants are categorized into four lines of defense. The first line of defense participates in the prevention of ROS production, such as SOD, CAT, and GPX, which prevent the formation of H_2O_2 and hydroperoxides (Ighodaro and Akinloye 2018). The second line of defense involves ROS scavenging, which includes the activity of AsA, α-tocopherol, etc., whereas the third line of defense undertakes the repair process, and the fourth line of defense participates in the adaption mechanism in response to stress-induced signals (Ighodaro and Akinloye 2018). Plants also display an important metabolic pathway known as the ascorbate-glutathione (AsA-GSH) or Foyer–Halliwell–Asada pathway (Pandey et al. 2017), comprising several enzymatic (APX, MDHAR, DHAR, GPX) and non-enzymatic (AsA, GSH) antioxidants to combat over-production of H_2O_2 (Bartoli et al. 2017). The AsA-GSH pathway occurs in chloroplasts, plastids, mitochondria, and peroxisomes (Pang and Wang 2010).

2.5.1 Enzymatic Antioxidants

Enzymatic antioxidants alleviate ROS levels by breaking them down and removing them from the system through a series of steps, including conversion of ROS to H_2O_2 and then into water (H_2O) molecules in the presence of metallic cofactors (Nimse and Pal 2015).

Superoxide dismutases (SOD; EC: 1.15.1.1) are metalloenzymes present in subcellular loci where free radicals and SOD substrates are generated (Luis et al. 2018). SODs are up-regulated during unfavorable conditions to prevent oxidative stress. Three forms of SOD or SOD isozymes are found in plants based on prosthetic metallic cofactors present at their active sites: CuZn-SOD localized in the cytosol, chloroplasts, peroxisomes, nuclei, mitochondria, and apoplast; Fe-SOD in the chloroplasts, peroxisomes, and mitochondria; and Mn-SOD (manganese) found in peroxisomes, mitochondria, and vascular tissues (Ahmad et al. 2010; Gill and Tuteja 2010; Zafra et al. 2018). SODs are well known to provide the initial or first line of defense against toxic ROS (Gill and Tuteja 2010). They catalyze disproportionation of $O_2^{\cdot -}$ free radicals by reducing one radical into H_2O_2 and oxidizing another into O_2. Thereby, SODs eliminate the risk of production of more toxic free radical ·OH via the Haber–Weiss reaction, which takes place in the presence of the reduced form of redox metals (Gill and Tuteja 2010). H_2O_2 generated as a by-product of enzymatic reactions of SOD acts as a significant signaling molecule; thus, the enzymatic activity of SOD is important for inducing signal transduction in plants under stressful conditions (Luis et al. 2018). Modulation in SOD activity under abiotic stress also triggers a cascade of expression of other antioxidant enzymes (Ahmad et al. 2010) that partake in the conversion of H_2O_2 into H_2O. For instance, the antioxidant enzymes APX, GPX, CAT catalyze the conversion of H_2O_2 to H_2O (Zhang et al. 2016).

Catalases (CAT; EC: 1.11.1.6) are H_2O_2-scavenging tetrameric and heme-prosthetic group-containing metalloenzymes that are located in peroxisomes and mitochondria (Anjum et al. 2016). The function of CAT enzymes is to catalyze the decomposition of H_2O_2 into H_2O and O_2 (Chen et al. 2017; Su et al. 2018), and they thus act as a signal receptor (Anjum et al. 2016). The dismutation of H_2O_2 occurs in two steps in an energy-efficient manner without utilizing any reducing agent (Anjum et al. 2016). In the first step, the heme group of CAT is oxidized by H_2O_2, forming an intermediate Fe peroxide, which is followed by the second step, where the intermediate formed is finally converted into H_2O and O_2 in the presence of H_2O_2 (Anjum et al. 2016).

Ascorbate peroxidases (APX; EC: 1.11.1.11) are H_2O_2-scavenging heme-containing peroxidase enzymes (Anjum et al. 2016; Pandey et al. 2017) having protoporphyrin as a prosthetic group (Pang and Wang 2010). APX are found in the chloroplasts, mitochondria, cytoplasmic matrix or cytosol, microbodies, and peroxisomes/glyoxysomes (Anjum et al. 2016). APX requires the reducing substrate AsA for proper functioning and to maintain its own stability (Pandey et al. 2017). Pang and Wang (2010) mentioned rapid inactivation of APX and loss of stability when the concentration of AsA falls below 20 μM. The catalytic reaction of APX takes place in a few steps. Initially, the heme group of APX is oxidized by H_2O_2 to oxyferryl, which is converted into water molecules by two successive one-electron reactions (Anjum et al. 2016). APX also participates in the AsA-GSH pathway as a key enzyme (Pang and Wang 2010).

Glutathione reductase (GR; EC: 1.6.4.2) is a flavo-protein oxidoreductase that participates as one of the key enzymes in the H_2O_2-scavenging pathway, i.e., the AsA-GSH cycle (Pang and Wang 2010; Yousuf et al. 2012). GR is localized in chloroplasts, mitochondria, cytosol, peroxisomes, and in non-photosynthetic tissues and organelles (Edwards et al. 1990); however, it is predominantly found in chloroplasts, showing 70–80% of its activity (Rao and Reddy 2008; Pang and Wang 2010). GR catalyzes the conversion of glutathione disulfide (GSSG) into the reduced form known as GSH using NADPH as an electron donor (Pang and Wang 2010; Yousuf et al. 2012) and maintains a balanced GSH/GSSG ratio necessary for the detoxification of H_2O_2 (Zandalinas et al. 2017). The functioning of GR is dependent on the combined action of pH at the location where the activity takes place and the concentrations of NADPH and GSSG (Pang and Wang 2010).

2.5.2 Non-Enzymatic Antioxidants and Gamma-Aminobutyric Acid (GABA)

Non-enzymatic antioxidants detoxify ROS by interrupting the free radical chain reaction (Nimse and Pal 2015). Non-enzymatic antioxidants include basic compounds such as vitamins, pigments, secondary metabolites, amino acids, and sugars.

Glutathione (GSH) is a low-molecular-weight non-enzymatic ubiquitous thiol tripeptide antioxidant that comprises glutamate (Glu), Cys, and glycine (Gly) amino acids (Gill et al. 2013). GSH is a powerful antioxidant that participates in the degradation of H_2O_2 in a reaction catalyzed by GPX (Szalai et al. 2009; Bartoli et al. 2017). GSH shows scavenging activity against the production of ROS both directly and indirectly. It takes part in the AsA-GSH pathway as a reductant for DHAR and aids in scavenging H_2O_2 indirectly (Szalai et al. 2009; Bartoli et al. 2017). Directly, GSH participates in the degradation of H_2O_2 and lipid peroxides by forming conjugates and transporting to vacuoles through a reaction catalyzed by the enzymes GPX and GST, respectively (Szalai et al. 2009). GSH also participates in redox signaling and activates defense mechanisms, thereby providing tolerance to abiotic stress (Szalai et al. 2009).

Ascorbate (AsA) or vitamin C (ascorbic acid) is a small, ubiquitous, powerful non-enzymatic antioxidant molecule present in all living cells (Akram et al. 2017; Bartoli et al. 2017). AsA participates in the AsA-GSH pathway as an electron donor for APX (Bartoli et al. 2017) and is a co-factor of POX, thus indirectly or directly participating in the detoxification of ROS (Noctor et al. 2018). AsA regenerates tocopherol and the production of xanthophyll, which take part in the quenching of excitation energy (Noctor et al. 2018). AsA also acts as a stress signal molecule. AsA in conjugation with vitamin E plays a significant role in quenching toxic ROS intermediates and excited reactive forms of O_2 either directly or indirectly through enzymatic catalysis (Akram et al. 2017).

Carotenoids are terpenoids that are present in photosynthetic organisms as light-harvesting pigments (Hix et al. 2004; Uarrato et al. 2018). Carotenoids are photoprotective compounds that alleviate oxidative stress and protect plants from the damaging effects of high light illumination stress by quenching excessive energy and dissipating it in the form of heat (Hix et al. 2004; Uarrato et al. 2018). Carotenoids also avert over-excitation of PS II in the thylakoid membrane by efficiently scavenging $^1Chl^*$ and $^3Chl^*$ and $^1O_2$ (Uarrato et al. 2018). Carotenoids not only protect the cell from ROS but also participate in signaling during stress. During stress, carotenoids scavenge over-produced $^1O_2$, which results in the generation of breakdown products that act as stress-induced signals (Noctor et al. 2018).

Gamma-aminobutyric acid (GABA) is a non-protein amino acid produced within plant cells during stress. It mediates tolerance by quenching free radicals and regulates enzyme activity (Ansari et al. 2014; Jalil and Ansari 2020). GABA either acts as an osmolyte or encourages the production of other osmolytes under stress (Jalil and Ansari 2020). GABA is metabolized by a GABA shunt pathway comprising GABA transaminase (GABA-T), glutamate dehydrogenase (GDH), and succinic semialdehyde dehydrogenase (Ansari et al. 2014; Khan et al. 2021). Jalil et al. (2017) reported reduced GABA and chlorophyll content, photosynthesis, and GDH activity with elevated membrane ion leakage, MDA content, and early leaf senescence in an *A. thaliana* mutant lacking the GABA-T gene under various abiotic stresses. In addition, GABA has been reported to participate in the signal transduction pathway via the increased cytosolic calmodulin-dependent activity of the enzyme glutamate decarboxylase under stress (Jalil and Ansari 2020). Moreover, Khanna et al. (2021) revealed that exogenous application of GABA ameliorates the salt stress-induced adverse effects in wheat by enhancing proline metabolism, the antioxidant system, and N-S assimilation.

It is apparent that the enzymatic and non-enzymatic antioxidants employ crucial direct and indirect pathways for tight regulation of ROS within plant cells and are responsible for the highly efficient amelioration of abiotic stress-induced oxidative stress. The potential role of antioxidants in the alleviation of oxidative stress has been documented by many workers. The increased activity of POX, GSH, and proline in the *B. juncea* tolerant genotype under heat stress has been observed to be responsible for alleviating the stress (Wilson et al. 2014). Similarly, in a study undertaken by Almeselmani et al. (2006), increased activity of SOD, CAT, and APX in tolerant wheat genotypes (HD 2815 and HDR 77) as compared with susceptible genotypes subjected to high-temperature stress prevented chlorophyll loss and membrane damage. Analogously, the production of SOD,

GPX, APX, GR, AsA, GSH, and proline in rice plants has been found to be associated with increased tolerance against Cu-induced oxidative stress (Thounaojam et al. 2012).

2.6 ROS SIGNALING AND DEFENSE MECHANISM

Plants growing in natural conditions are endowed with adaptive tactics that aid survival under severe conditions. Signaling is the main strategy by which plants respond to stress and induce tolerance. Systemic signals under unfavorable situations arise from regions exposed to stress and move toward the unstressed region to alert and activate the defense pathway (Kumar et al. 2017). ROS plays a significant role as a signal transducer under stress. Under stress conditions, cell organelles switch on over-production of ROS (Gilroy et al. 2016; Huang et al. 2016; Saini et al. 2018), which disturb their normal homeostasis and initiate a signal transduction cascade (Foyer and Noctor 2005) involving specific feedback and feedforward responses, facilitating stress tolerance (Kumar et al 2017). The systemic signaling of ROS generates as an auto-propagating wave to the adjacent cells (Mittler et al. 2011) and confers acclimatization toward stress through spatiotemporal communication with hormones and/or amino acid signals particular to stress (Kumar et al. 2017). For instance, a transient increase in ROS level under stress acts as a secondary messenger by triggering signaling via protein oxidation (Kumar et al. 2017). Among various members of the ROS family, H_2O_2 is a relatively stable yet reactive molecule and easily diffuses through membranes via aquaporins, initiating signaling. Therefore, H_2O_2 is considered an important molecule for signal transduction (Sewelam et al. 2016; Kumar et al. 2017). ROS mediate signaling in a highly coordinated manner and activate defense genes, antioxidants, kinases, protein phosphorylation, and the influx of calcium ions, and increase phytohormone (salicylic acid (SA), jasmonic acid (JA), ethylene) synthesis, etc. Controlled production of ROS also ameliorates abiotic stress by inducing changes in developmental stages such as the formation of tracheary elements, lignification, and cross-linking in cell walls, leading to PCD (Jacobson 1996; Schutzendubel and Polle, 2002). For instance, under pathogen attack, a transient increase in ROS in plant cells limits pathogen invasion by initiating an defense response via production of phytoalexin and pathogenesis-related proteins, plant cell wall strengthening, and PCD (Camejo et al. 2016; Kumar et al. 2017; Andersen et al. 2018; Shah et al. 2019).

ROS production in cell organelles under stress induces retrograde signals to the nucleus that propagate at the speed of 8.4 cm min^{-1}. Retrograde signaling plays a crucial role as a messenger and facilitates stress tolerance (Miller et al. 2009; Bhattacharjee 2012; Pucciariello et al. 2012). The retrograde signaling helps the nucleus to modulate anterograde control for acclimatization of plants exposed to abiotic perturbation (Woodson and Chory, 2008; Sachdev et al. 2021). Under stress conditions, ROS burst has been reported to elicit, transcription of stress-responsive genes such as heat shock gene (HSG) (Pucciariello et al. 2012). HSG activates HSP, which act as molecular chaperones, predominantly during heat stress. HSP are known to prevent protein aggregation, misfolding, denaturation, and degradation and also help in protein refolding (Pucciariello et al. 2012). The role of HSP in the regulation of oxidative damage has been reported under various abiotic stresses including light, anoxia, and cold (Heidarvand and Amiri, 2010; Pucciariello et al. 2012). In *Arabidopsis* cell culture under heat stress, H_2O_2 has been reported to modulate HSG expression and trigger production of APX2, HSP17.6, and HSP 18.2 (Volkov et al. 2006). Similarly, over-production of H_2O_2 in *Arabidopsis* has been found to up-regulate HSPs and antioxidant enzyme APX1 production, which quenched excess H_2O_2 and imparted tolerance against high light stress (Pnueli et al. 2003) as well as acclimatizing plants exposed to combined heat and drought stress (Koussevitzky et al. 2008). Analogously, under oxygen deprivation (anoxia/hypoxia stress), H_2O_2 was reported to up-regulate the expression of genes encoding for HSPs; fermentation genes such as *ALCOHOL DEHYDROGENASE* (ADH); ROS-regulated transcription factors including *ZAT10* and *ZAT12*; and proteins, which subsequently facilitates acclimatization to stress (Banti et al. 2010; Pucciariello et al. 2012; Yang and Hong 2015; Sasidharan et al. 2018).

2.7 CONCLUSION AND FUTURE PROSPECTS

In the era of climate change, the frequent occurrence of abiotic stress is a common and inevitable load on plants that hinders their development and productivity. Under the influence of abiotic stress, the normal functioning of plant cells is disrupted, leading to the over-production of ROS. ROS have always been regarded as toxic molecules, showing variable reactivity depending upon their type and location of production/action in a plant cell. To avert ROS-mediated damage, plants establish an antioxidant defense system (with the ability to curb excess ROS); however, under extreme conditions, their scavenging ability fails, leading to oxidative burst. The oxidative burst can result in oxidative stress, which could be fatal for plant sustenance. In addition to damaging properties, recent studies have also highlighted the role of ROS in mediating tolerance against stresses by channeling signaling pathways. In-depth knowledge and a better understanding regarding the toxicity of ROS and their effects, antioxidant-based defense systems, and signaling mechanisms can provide a path to mitigate the adverse effects of ROS on plants and can facilitate the development of improved defense strategies.

REFERENCES

Abdelaal, K. A., EL-Maghraby, L. M., Elansary, H., Hafez, Y. M., Ibrahim, E. I., El-Banna, M., El-Esawi, M., & Elkelish, A. (2020). Treatment of sweet pepper with stress tolerance-inducing compounds alleviates salinity stress oxidative damage by mediating the physio-biochemical activities and antioxidant systems. *Agronomy*, *10*(1), 26.

Ahanger, M. A., Tomar, N. S., Tittal, M., Argal, S., & Agarwal, R. M. (2017). Plant growth under water/salt stress: ROS production; antioxidants and significance of added potassium under such conditions. *Physiology and Molecular Biology of Plants*, *23*(4), 731–744.

Ahmad, P., Jaleel, C. A., Salem, M. A., Nabi, G., & Sharma, S. (2010). Roles of enzymatic and nonenzymatic antioxidants in plants during abiotic stress. *Critical Reviews in Biotechnology*, *30*(3), 161–175.

Akram, N. A., Shafiq, F., & Ashraf, M. (2017). Ascorbic acid-a potential oxidant scavenger and its role in plant development and abiotic stress tolerance. *Frontiers in Plant Science*, *8*, 613.

Ali, M. B., Hahn, E. J., & Paek, K. Y. (2005). Effects of temperature on oxidative stress defense systems, lipid peroxidation and lipoxygenase activity in Phalaenopsis. *Plant Physiology and Biochemistry*, *43*(3), 213–223.

Almeselmani, M., Deshmukh, P. S., Sairam, R. K., Kushwaha, S. R., & Singh, T. P. (2006). Protective role of antioxidant enzymes under high temperature stress. *Plant Science*, *171*(3), 382–388.

Andersen, E. J., Ali, S., Byamukama, E., Yen, Y., & Nepal, M. P. (2018). Disease resistance mechanisms in plants. *Genes*, *9*(7), 339.

Anee, T. I., Nahar, K., Rahman, A., Mahmud, J. A., Bhuiyan, T. F., Alam, M. U., Fujita, M., & Hasanuzzaman, M. (2019). Oxidative damage and antioxidant defense in *Sesamum indicum* after different waterlogging durations. *Plants*, *8*(7), 196.

Anjum, N. A., Sharma, P., Gill, S. S., Hasanuzzaman, M., Khan, E. A., Kachhap, K., Mohamed, A.A., Thangavel, P., Devi, G. D., Vasudhevan, P., Sofo, A., Khan, N. A., Misra, A. N., Lukatin, A. S., Singh, H. P., Pereira, E., & Tuteja, N. (2016). Catalase and ascorbate peroxidase—representative H_2O_2-detoxifying heme enzymes in plants. *Environmental Science and Pollution Research*, *23*(19), 19002–19029.

Anjum, N. A., Sofo, A., Scopa, A., Roychoudhury, A., Gill, S. S., Iqbal, M., Lukatin, A. S., Singh, H. P., Pereira, E., Duarte, A. C., & Ahmad, I. (2015). Lipids and proteins—major targets of oxidative modifications in abiotic stressed plants. *Environmental Science and Pollution Research*, *22*(6), 4099–4121.

Anjum, S. A., Ashraf, U., Tanveer, M., Khan, I., Hussain, S., Shahzad, B., Zohaib, A., Abbas, F., Saleem, M.F., Ali, I., & Wang, L. C. (2017). Drought induced changes in growth, osmolyte accumulation and antioxidant metabolism of three maize hybrids. *Frontiers in Plant Science*, *8*, 69.

Ansari, M. I., Hasan, S., & Jalil, S. U. (2014). Leaf senescence and GABA shunt. *Bioinformation*, *10*(12), 734.

Banti, V., Mafessoni, F., Loreti, E., Alpi, A., & Perata, P. (2010). The heat-inducible transcription factor HsfA2 enhances anoxia tolerance in Arabidopsis. *Plant Physiology*, *152*(3), 1471–1483.

Bartoli, C. G., Buet, A., Grozeff, G. G., Galatro, A., & Simontacchi, M. (2017). Ascorbate-glutathione cycle and abiotic stress tolerance in plants. In: *Ascorbic Acid in Plant Growth, Development and Stress Tolerance* (Hossain, M. A., Munne-Bosch, S., Burritt, D. J., Diaz-Vivancos, P., eds.), Springer, Cham, pp. 177–200.

Bartoli, C. G., Gómez, F., Martínez, D. E., & Guiamet, J. J. (2004). Mitochondria are the main target for oxidative damage in leaves of wheat (*Triticum aestivum* L.). *Journal of Experimental Botany*, *55*(403), 1663–1669.

Becana, M., Moran, J. F., & Iturbe-Ormaetxe, I. (1998). Iron-dependent oxygen free radical generation in plants subjected to environmental stress: toxicity and antioxidant protection. *Plant and Soil*, *201*(1), 137–147.

Bhattacharjee, S. (2012). The language of reactive oxygen species signaling in plants. *Journal of Botany*, 2012, 1–22.

Bhuyan, M. B., Hasanuzzaman, M., Parvin, K., Mohsin, S. M., Al Mahmud, J., Nahar, K., & Fujita, M. (2020). Nitric oxide and hydrogen sulfide: two intimate collaborators regulating plant defense against abiotic stress. *Plant Growth Regulation*, *90*(3), 409–424.

Bita, C., & Gerats, T. (2013). Plant tolerance to high temperature in a changing environment: scientific fundamentals and production of heat stress-tolerant crops. *Frontiers in Plant Science*, *4*, 273.

Bulgari, R., Franzoni, G., & Ferrante, A. (2019). Biostimulants application in horticultural crops under abiotic stress conditions. *Agronomy*, *9*(6), 306.

Camejo, D., Guzmán-Cedeño, Á., & Moreno, A. (2016). Reactive oxygen species, essential molecules, during plant–pathogen interactions. *Plant Physiology and Biochemistry*, *103*, 10–23.

Chang, R., Jang, C. J., Branco-Price, C., Nghiem, P., & Bailey-Serres, J. (2012). Transient MPK6 activation in response to oxygen deprivation and reoxygenation is mediated by mitochondria and aids seedling survival in Arabidopsis. *Plant Molecular Biology*, *78*(1), 109–122.

Chen, N., Teng, X. L., & Xiao, X. G. (2017). Subcellular localization of a plant catalase-phenol oxidase, AcCATPO, from Amaranthus and identification of a non-canonical peroxisome targeting signal. *Frontiers in Plant Science*, *8*, 1345.

Chima, U. D., Ofodile, E. A. U., & Naaluba, M. (2015). Evaluation of NaCl salinity tolerance of *Artocarpus altilis* (Parkinson) Fosberg and *Treculia africana* Decne for climate change adaptation in agroecosystems. *International Journal of Scientific & Engineering Research*, *6*(5), 1516–1523.

Coates, J., Mar, K. A., Ojha, N., & Butler, T. M. (2016). The influence of temperature on ozone production under varying NO x conditions–a modelling study. *Atmospheric Chemistry and Physics*, *16*(18), 11601–11615.

Das, M. K., & Roychoudhury, A. (2014). ROS and responses of antioxidant as ROS-scavengers during environmental stress in plants. *Frontiers in Environmental Science*, *2*, 1–13.

De Grey, A. D. (2002). HO_2•: the forgotten radical. *DNA and Cell Biology*, *21*(4), 251–257.

Diao, Q., Song, Y., Shi, D., & Qi, H. (2017). Interaction of polyamines, abscisic acid, nitric oxide, and hydrogen peroxide under chilling stress in tomato (*Lycopersicon esculentum* Mill.) seedlings. *Frontiers in Plant Science*, *8*, 203.

Dmitrieva, V. A., Tyutereva, E. V., & Voitsekhovskaja, O. V. (2020). Singlet oxygen in plants: generation, detection, and signaling roles. *International Journal of Molecular Sciences*, *21*(9), 3237.

Dumont, S., & Rivoal, J. (2019). Consequences of oxidative stress on plant glycolytic and respiratory metabolism. *Frontiers in Plant Science*, *10*, 166.

Ebone, L. A., Caverzan, A., & Chavarria, G. (2019). Physiologic alterations in orthodox seeds due to deterioration processes. *Plant Physiology and Biochemistry*, *145*, 34–42.

Edwards, E. A., Rawsthorne, S., & Mullineaux, P. M. (1990). Subcellular distribution of multiple forms of glutathione reductase in leaves of pea (*Pisum sativum* L.). *Planta*, *180*(2), 278–284.

Etesami, H., & Noori, F. (2019). Soil salinity as a challenge for sustainable agriculture and bacterial-mediated alleviation of salinity stress in crop plants. In: *Saline Soil-Based Agriculture by Halotolerant Microorganisms* (Kumar, M., Etesami, H., Kumar, V., eds.), Springer, Singapore, pp. 1–22.

Evans, M. D., Dizdaroglu, M., & Cooke, M. S. (2004). Oxidative DNA damage and disease: induction, repair and significance. *Mutation Research/Reviews in Mutation Research*, *567*(1), 1–61.

Feng, X., Porporato, A., & Rodriguez-Iturbe, I. (2013). Changes in rainfall seasonality in the tropics. *Nature Climate Change*, *3*(9), 811–815.

Ferreira-Silva, S. L., Voigt, E. L., Silva, E. N., Maia, J. M., Aragão, T. C. R., & Silveira, J. A. G. (2012). Partial oxidative protection by enzymatic and non-enzymatic components in cashew leaves under high salinity. *Biologia Plantarum*, *56*(1), 172–176.

Fink, S. P., Reddy, G. R., & Marnett, L. J. (1997). Mutagenicity in *Escherichia coli* of the major DNA adduct derived from the endogenous mutagen malondialdehyde. *Proceedings of the National Academy of Sciences*, *94*(16), 8652–8657.

Foyer, C. H., & Noctor, G. (2005). Redox homeostasis and antioxidant signaling: a metabolic interface between stress perception and physiological responses. *The Plant Cell*, *17*(7), 1866–1875.

Ghanbari, F., & Sayyari, M. (2018). Controlled drought stress affects the chilling-hardening capacity of tomato seedlings as indicated by changes in phenol metabolisms, antioxidant enzymes activity, osmolytes concentration and abscisic acid accumulation. *Scientia Horticulturae*, *229*, 167–174.

Gill, S. S., & Tuteja, N. (2010). Reactive oxygen species and antioxidant machinery in abiotic stress tolerance in crop plants. *Plant Physiology and Biochemistry*, *48*(12), 909–930.

Gill, S. S., Anjum, N. A., Hasanuzzaman, M., Gill, R., Trivedi, D. K., Ahmad, I., Pereira, E., & Tuteja, N. (2013). Glutathione and glutathione reductase: a boon in disguise for plant abiotic stress defense operations. *Plant Physiology and Biochemistry*, *70*, 204–212.

Gilroy, S., Białasek, M., Suzuki, N., Górecka, M., Devireddy, A. R., Karpiński, S., & Mittler, R. (2016). ROS, calcium, and electric signals: key mediators of rapid systemic signaling in plants. *Plant Physiology*, *171*(3), 1606–1615.

Gull, A., Lone, A. A., & Wani, N. U. I. (2019). Biotic and abiotic stresses in plants. In: Abiotic and Biotic Stress in Plants (de Oliveira, A. B., ed.), IntechOpen, London, pp. 1–19.

Gupta, D. K., Pena, L. B., Romero-Puertas, M. C., Hernández, A., Inouhe, M., & Sandalio, L. M. (2017). NADPH oxidases differentially regulate ROS metabolism and nutrient uptake under cadmium toxicity. *Plant, Cell & Environment*, *40*(4), 509–526.

Habib, N., Ali, Q., Ali, S., Javed, M. T., Zulqurnain Haider, M., Perveen, R., Shahid, M. R., Rizwan, M., Abdel-Daim, M. M., Elkelish, A., & Bin-Jumah, M. (2020). Use of nitric oxide and hydrogen peroxide for better yield of wheat (*Triticum aestivum* L.) under water deficit conditions: growth, osmoregulation, and antioxidative defense mechanism. *Plants*, *9*(2), 285.

Hajiboland, R. (2012). Effect of micronutrient deficiencies on plants stress responses. In: *Abiotic Stress Responses in Plants* (Ahmad, P., Prasad, M. N. V., eds.), Springer, New York, pp. 283–329.

Halliwell, B. (2006). Reactive species and antioxidants. Redox biology is a fundamental theme of aerobic life. *Plant Physiology*, *141*(2), 312–322.

Hameed, A., Bibi, N., Akhter, J., & Iqbal, N. (2011). Differential changes in antioxidants, proteases, and lipid peroxidation in flag leaves of wheat genotypes under different levels of water deficit conditions. *Plant Physiology and Biochemistry*, *49*(2), 178–185.

Hameed, A., Goher, M., & Iqbal, N. (2013). Drought induced programmed cell death and associated changes in antioxidants, proteases, and lipid peroxidation in wheat leaves. *Biologia Plantarum*, *57*(2), 370–374.

Hasanuzzaman, M., Bhuyan, M. H. M., Zulfiqar, F., Raza, A., Mohsin, S. M., Mahmud, J. A., Fujita, M., & Fotopoulos, V. (2020). Reactive oxygen species and antioxidant defense in plants under abiotic stress: revisiting the crucial role of a universal defense regulator. *Antioxidants*, *9*(8), 681.

Hasanuzzaman, M., Oku, H., Nahar, K., Bhuyan, M. B., Al Mahmud, J., Baluska, F., & Fujita, M. (2018). Nitric oxide-induced salt stress tolerance in plants: ROS metabolism, signaling, and molecular interactions. *Plant Biotechnology Reports*, *12*(2), 77–92.

He, M., He, C. Q., & Ding, N. Z. (2018). Abiotic stresses: general defenses of land plants and chances for engineering multistress tolerance. *Frontiers in Plant Science*, *9*, 1771.

Heidarvand, L., & Amiri, R. M. (2010). What happens in plant molecular responses to cold stress? *Acta Physiologiae Plantarum*, *32*(3), 419–431.

Hideg, É., Vitányi, B., Kósa, A., Solymosi, K., Bóka, K., Won, S., Inoue, Y., Ridge, R. W., & Böddi, B. (2010). Reactive oxygen species from type-I photosensitized reactions contribute to the light-induced wilting of dark-grown pea (*Pisum sativum*) epicotyls. *Physiologia Plantarum*, *138*(4), 485–492.

Hix, L. M., Lockwood, S. F., & Bertram, J. S. (2004). Bioactive carotenoids: potent antioxidants and regulators of gene expression. *Redox Report*, *9*(4), 181–191.

Hollósy, F. (2002). Effects of ultraviolet radiation on plant cells. *Micron*, *33*(2), 179–197.

Huang, S., Van Aken, O., Schwarzländer, M., Belt, K., & Millar, A. H. (2016). The roles of mitochondrial reactive oxygen species in cellular signaling and stress response in plants. *Plant Physiology, 171*(3), 1551–1559.

Ighodaro, O. M., & Akinloye, O. A. (2018). First line defence antioxidants-superoxide dismutase (SOD), catalase (CAT) and glutathione peroxidase (GPX): their fundamental role in the entire antioxidant defence grid. *Alexandria Journal of Medicine, 54*(4), 287–293.

Ilangumaran, G., & Smith, D. L. (2017). Plant growth promoting rhizobacteria in amelioration of salinity stress: a systems biology perspective. *Frontiers in Plant Science, 8*, 1768.

Iqbal, Z., Iqbal, M. S., Hashem, A., Abd Allah, E. F., & Ansari, M. I. (2021). Plant defense responses to biotic stress and its interplay with fluctuating dark/light conditions. *Frontiers in Plant Science, 12*, 297.

Jacobson, M. D. (1996). Reactive oxygen species and programmed cell death. *Trends in Biochemical Sciences, 21*(3), 83–86.

Jalil, S. U., & Ansari, M. I. (2018). Plant microbiome and its functional mechanism in response to environmental stress. *International Journal of Green Pharmacy (IJGP), 12*(01): 81–92.

Jalil, S. U., & Ansari, M. I. (2020). Physiological role of gamma-aminobutyric acid in salt stress tolerance. In: *Salt and Drought Stress Tolerance in Plants* (Hasanuzzaman, M., Tanveer, M., eds.), Springer Nature, Cham, Switzerland, pp. 337–350.

Jalil, S. U., Ahmad, I., & Ansari, M. I. (2017). Functional loss of GABA transaminase (GABA-T) expressed early leaf senescence under various stress conditions in *Arabidopsis thaliana. Current Plant Biology, 9*, 11–22.

Kabała, K., Zboińska, M., Głowiak, D., Reda, M., Jakubowska, D., & Janicka, M. (2019). Interaction between the signaling molecules hydrogen sulfide and hydrogen peroxide and their role in vacuolar H+-ATPase regulation in cadmium-stressed cucumber roots. *Physiologia Plantarum, 166*(2), 688–704.

Karuppanapandian, T., & Kim, W. (2013). Cobalt-induced oxidative stress causes growth inhibition associated with enhanced lipid peroxidation and activates antioxidant responses in Indian mustard (*Brassica juncea* L.) leaves. *Acta Physiologiae Plantarum, 35*(8), 2429–2443.

Katsuhara, M., Otsuka, T., & Ezaki, B. (2005). Salt stress-induced lipid peroxidation is reduced by glutathione S-transferase, but this reduction of lipid peroxides is not enough for a recovery of root growth in Arabidopsis. *Plant Science, 169*(2), 369–373.

Kaur, H., Bhardwaj, R. D., & Grewal, S. K. (2017). Mitigation of salinity-induced oxidative damage in wheat (*Triticum aestivum* L.) seedlings by exogenous application of phenolic acids. *Acta Physiologiae Plantarum, 39*(10), 1–15.

Kerchev, P., van der Meer, T., Sujeeth, N., Verlee, A., Stevens, C. V., Van Breusegem, F., & Gechev, T. (2020). Molecular priming as an approach to induce tolerance against abiotic and oxidative stresses in crop plants. *Biotechnology Advances, 40*, 107503.

Khan, M. I. R., Palakolanu, S. R., Chopra, P., Rajurkar, A. B., Gupta, R., Iqbal, N., & Maheshwari, C. (2020a). Improving drought tolerance in rice: ensuring food security through multi-dimensional approaches. *Physiologia Plantarum, 172*(2), 645–668.

Khan, M. I. R., Trivellini, A., Chhillar, H., Chopra, P., Ferrante, A., Khan, N. A., & Ismail, A. M. (2020b). The significance and functions of ethylene in flooding stress tolerance in plants. *Environmental and Experimental Botany, 179*, 104188.

Khan, M. I. R., Jalil, S. U., Chopra, P., Chhillar, H., Ferrante,A., Khan, N. A., & Ansari, M. I. (2021). Role of GABA in plant growth, development and senescence. *Plant Gene, 26*, 100283.

Khanna, R. R., Jahan, B., Iqbal, N., Khan, N. A., AlAjmi, M. F., Rehman, M. T., & Khan, M. I. R. (2021). GABA reverses salt-inhibited photosynthetic and growth responses through its influence on NO-mediated nitrogen-sulfur assimilation and antioxidant system in wheat, *Journal of Biotechnology 325*, 73–82.

Kipp, E., & Boyle, M. (2013). The effects of heat stress on reactive oxygen species production and chlorophyll concentration in *Arabidopsis thaliana. Research in Plant Sciences, 1*(2), 20–23.

Kohli, S. K., Khanna, K., Bhardwaj, R., Abd Allah, E. F., Ahmad, P., & Corpas, F. J. (2019). Assessment of subcellular ROS and NO metabolism in higher plants: multifunctional signaling molecules. *Antioxidants, 8*(12), 641.

Kosová, K., Vítámvás, P., Urban, M. O., Prášil, I. T., & Renaut, J. (2018). Plant abiotic stress proteomics: the major factors determining alterations in cellular proteome. *Frontiers in Plant Science, 9*, 122.

Koussevitzky, S., Suzuki, N., Huntington, S., Armijo, L., Sha, W., Cortes, D., Shulaev, V., & Mittler, R. (2008). Ascorbate peroxidase 1 plays a key role in the response of *Arabidopsis thaliana* to stress combination. *Journal of Biological Chemistry, 283*(49), 34197–34203.

Kumar, A., Patel, J. S., Meena, V. S., & Srivastava, R. (2019). Recent advances of PGPR based approaches for stress tolerance in plants for sustainable agriculture. *Biocatalysis and Agricultural Biotechnology*, *20*, 101271.

Kumar, V., Khare, T., Sharma, M., & Wani, S. H. (2017). ROS-induced signaling and gene expression in crops under salinity stress. In: *Reactive Oxygen Species and Antioxidant Systems in Plants: Role and Regulation under Abiotic Stress* (Khan, M. I. R., Khan, N. A., eds.), Springer, Singapore, pp. 159–184.

Lee, B. R., Li, L. S., Jung, W. J., Jin, Y. L., Avice, J. C., Ourry, A., & Kim, T. H. (2009). Water deficit-induced oxidative stress and the activation of antioxidant enzymes in white clover leaves. *Biologia Plantarum*, *53*(3), 505–510.

Liu, J., Hasanuzzaman, M., Wen, H., Zhang, J., Peng, T., Sun, H., & Zhao, Q. (2019). High temperature and drought stress cause abscisic acid and reactive oxygen species accumulation and suppress seed germination growth in rice. *Protoplasma*, *256*(5), 1217–1227.

Liu, M., Gong, X., Alluri, R. K., & Wu, J. (2012). Characterization of RNA damage under oxidative stress in *Escherichia coli*. *Biological Chemistry*, *393*(3), 123.

Liu, T., Hu, X., Zhang, J., Zhang, J., Du, Q., & Li, J. (2018). H_2O_2 mediates ALA-induced glutathione and ascorbate accumulation in the perception and resistance to oxidative stress in *Solanum lycopersicum* at low temperatures. *BMC Plant Biology*, *18*(1), 1–10.

Luan, H., Shen, H., Pan, Y., Guo, B., Lv, C., & Xu, R. (2018). Elucidating the hypoxic stress response in barley (*Hordeum vulgare* L.) during waterlogging: a proteomics approach. *Scientific Reports*, *8*(1), 1–13.

Luis, A., Corpas, F. J., López-Huertas, E., & Palma, J. M. (2018). Plant superoxide dismutases: function under abiotic stress conditions. In: *Antioxidants and Antioxidant Enzymes in Higher Plants* (Gupta D. K., Palma, J. M., Corpas, F. J., eds.), Springer Nature, Cham, Switzerland, pp. 1–26.

Majid, U., Siddiqi, T. O., & Iqbal, M. (2014). Antioxidant response of *Cassia angustifolia* Vahl. to oxidative stress caused by Mancozeb, a pyrethroid fungicide. *Acta Physiologiae Plantarum*, *36*(2), 307–314.

Malhotra, C., Kapoor, R. T., Ganjewala, D., & Singh, N. B. (2017). Sodium silicate mediated response of antioxidative defense system in *Lycopersicon esculentum* mill. under water stress. *International Journal of Phytomedicine*, *9*, 364–378.

Martinez, P. A. (2016). Genomics of temperature stress. In: *Plant Genomics and Climate Change* (Edwards, D., Batley J., eds.), Springer, New York, pp. 137–147.

Martinez, V., Mestre, T. C., Rubio, F., Girones-Vilaplana, A., Moreno, D. A., Mittler, R., & Rivero, R. M. (2016). Accumulation of flavonols over hydroxycinnamic acids favors oxidative damage protection under abiotic stress. *Frontiers in Plant Science*, *7*, 838.

Maurya, A. K. (2020). Oxidative stress in crop plants. In: *Agronomic Crops* (Hasanuzzaman, M., ed.), Springer, Singapore, pp. 349–380.

Miller, G., Schlauch, K., Tam, R., Cortes, D., Torres, M. A., Shulaev, V., Dangl, J. L., & Mittler, R. (2009). The plant NADPH oxidase RBOHD mediates rapid systemic signaling in response to diverse stimuli. *Science sSignaling*, *2*(84), ra45–ra45.

Mittler, R. (2017). ROS are good. *Trends in Plant Science*, *22*(1), 11–19.

Mittler, R., Vanderauwera, S., Suzuki, N., Miller, G. A. D., Tognetti, V. B., Vandepoele, K., Gollery, M., Shulae, V., & Van Breusegem, F. (2011). ROS signaling: the new wave? *Trends in Plant Science*, *16*(6), 300–309.

Møller, I. M., Jensen, P. E., & Hansson, A. (2007). Oxidative modifications to cellular components in plants. *Annual Review of Plant Biology*, *58*, 459–481.

Møller, I. M., Rogowska-Wrzesinska, A., & Rao, R. S. P. (2011). Protein carbonylation and metal-catalyzed protein oxidation in a cellular perspective. *Journal of Proteomics*, *74*(11), 2228–2242.

Nahar, K., Hasanuzzaman, M., Alam, M. M., Rahman, A., Mahmud, J. A., Suzuki, T., & Fujita, M. (2017). Insights into spermine-induced combined high temperature and drought tolerance in mung bean: osmoregulation and roles of antioxidant and glyoxalase system. *Protoplasma*, *254*(1), 445–460.

Nimse, S. B., & Pal, D. (2015). Free radicals, natural antioxidants, and their reaction mechanisms. *RSC Advances*, *5*(35), 27986–28006.

Noctor, G., Reichheld, J. P., & Foyer, C. H. (2018). ROS-related redox regulation and signaling in plants. In: *Seminars in Cell & Developmental Biology* (Vol. 80, pp. 3–12). Academic Press, United Kingdom.

Oracz, K., Bouteau, H. E. M., Farrant, J. M., Cooper, K., Belghazi, M., Job, C., Job, D., Corbineau, F., & Bailly, C. (2007). ROS production and protein oxidation as a novel mechanism for seed dormancy alleviation. *The Plant Journal*, *50*(3), 452–465.

Pandey, S., Fartyal, D., Agarwal, A., Shukla, T., James, D., Kaul, T., Negi, Y. K., Arora, S., & Reddy, M. K. (2017). Abiotic stress tolerance in plants: myriad roles of ascorbate peroxidase. *Frontiers in Plant Science*, *8*, 581.

Pandey, V., Dixit, V., & Shyam, R. (2009). Chromium effect on ROS generation and detoxification in pea (*Pisum sativum*) leaf chloroplasts. *Protoplasma*, *236*(1–4), 85–95.

Pang, C. H., & Wang, B. S. (2010). Role of ascorbate peroxidase and glutathione reductase in ascorbate–glutathione cycle and stress tolerance in plants. In: *Ascorbate-Glutathione Pathway and Stress Tolerance in Plants* (Anjum, N. A., Chan, M. T., Umar, S., eds.), Springer Nature, Switzerland, pp. 91–113.

Pnueli, L., Liang, H., Rozenberg, M., & Mittler, R. (2003). Growth suppression, altered stomatal responses, and augmented induction of heat shock proteins in cytosolic ascorbate peroxidase (Apx1)-deficient Arabidopsis plants. *The Plant Journal*, *34*(2), 187–203.

Pucciariello, C., Banti, V., & Perata, P. (2012). ROS signaling as common element in low oxygen and heat stresses. *Plant Physiology and Biochemistry*, *59*, 3–10.

Rao, A. C., & Reddy, A. R. (2008). Glutathione reductase: a putative redox regulatory system in plant cells. In: *Sulfur Assimilation and Abiotic Stress in Plants* (Khan N. A., Singh, S., Umar, S., eds.), Springer, Berlin, Heidelberg, pp. 111–147.

Raza, A., Razzaq, A., Mehmood, S. S., Zou, X., Zhang, X., Lv, Y., & Xu, J. (2019). Impact of climate change on crops adaptation and strategies to tackle its outcome: a review. *Plants*, *8*(2), 34.

Richards, S. L., Wilkins, K. A., Swarbreck, S. M., Anderson, A. A., Habib, N., Smith, A. G., McAinsh, M., & Davies, J. M. (2015). The hydroxyl radical in plants: from seed to seed. *Journal of Experimental Botany*, *66*(1), 37–46.

Roldán-Arjona, T., & Ariza, R. R. (2009). Repair and tolerance of oxidative DNA damage in plants. *Mutation Research/Reviews in Mutation Research*, *681*(2–3), 169–179.

Roychowdhury, R., Khan, M. H., & Choudhury, S. (2019). Physiological and molecular responses for metalloid stress in rice—a comprehensive overview. In: *Advances in Rice Research for Abiotic Stress Tolerance* (Hasanuzzaman, M., Fujita, M., Nahar, K., Biswas, J. K., eds.) Woodhead Publishing, USA, pp. 341–369.

Rybus-Zajac, M., & Kubis, J. (2010). Effect of UV-B radiation on antioxidative enzyme activity in cucumber cotyledons. *Acta Biologica Cracoviensia. Series Botanica*, *52*(2), 97–102.

Sachdev, S., & Singh, R. P. (2018). Isolation, characterisation and screening of native microbial isolates for biocontrol of fungal pathogens of tomato. *Climate Change and Environmental Sustainability*, *6*(1), 46–58.

Sachdev, S., Ansari, S. A., Ansari, M. I., Fujita, M., & Hasanuzzaman, M. (2021). Abiotic stress and reactive oxygen species: generation, signaling, and defense mechanisms. *Antioxidants*, *10*(2), 277.

Saijo, Y., & Loo, E. P. I. (2020). Plant immunity in signal integration between biotic and abiotic stress responses. *New Phytologist*, *225*(1), 87–104.

Saini, P., Gani, M., Kaur, J. J., Godara, L. C., Singh, C., Chauhan, S. S., Francies, R. M., Bhardwaj, A., Kumar, N. B., & Ghosh, M. K. (2018). Reactive oxygen species (ROS): a way to stress survival in plants. In: *Abiotic Stress-Mediated Sensing and Signaling in Plants: an Omics Perspective* (Zargar, S.M., Zargar, M.Y., eds.), Springer, Singapore, pp. 127–153.

Sasidharan, R., Hartman, S., Liu, Z., Martopawiro, S., Sajeev, N., van Veen, H., Yeung, E., & Voesenek, L. A. (2018). Signal dynamics and interactions during flooding stress. *Plant Physiology*, *176*(2), 1106–1117.

Schutzendubel, A., & Polle, A. (2002). Plant responses to abiotic stresses: heavy metal-induced oxidative stress and protection by mycorrhization. *Journal of Experimental Botany*, *53*(372), 1351–1365.

Sewelam, N., Kazan, K., & Schenk, P. M. (2016). Global plant stress signaling: reactive oxygen species at the cross-road. *Frontiers in Plant Science*, *7*, 187.

Shah, K., Chaturvedi, V., & Gupta, S. (2019). Climate change and abiotic stress-induced oxidative burst in rice. In: *Advances in Rice Research for Abiotic Stress Tolerance* (Hasanuzzaman, M., Fujita, M., Nahar, K., Biswas, J. K., eds.), Woodhead Publishing, USA, pp. 505–535.

Sharma, A., Yuan, H., Kumar, V., Ramakrishnan, M., Kohli, S. K., Kaur, R., Thukral, A. K., Bhardwaj, R., & Zheng, B. (2019). Castasterone attenuates insecticide induced phytotoxicity in mustard. *Ecotoxicology and Environmental Safety*, *179*, 50–61.

Sharma, P., Jha, A. B., Dubey, R. S., & Pessarakli, M. (2012). Reactive oxygen species, oxidative damage, and antioxidative defense mechanism in plants under stressful conditions. *Journal of Botany*, *2012*, 1–26.

Singh, S., Eapen, S., & D'souza, S. F. (2006). Cadmium accumulation and its influence on lipid peroxidation and antioxidative system in an aquatic plant, *Bacopa monnieri* L. *Chemosphere*, *62*(2), 233–246.

Smirnoff, N., & Arnaud, D. (2019). Hydrogen peroxide metabolism and functions in plants. *New Phytologist, 221*(3), 1197–1214.
Sohag, A. A. M., Tahjib-Ul-Arif, M., Brestic, M., Afrin, S., Sakil, M. A., Hossain, M. T., Hossain, M. A., & Hossain, M. A. (2020). Exogenous salicylic acid and hydrogen peroxide attenuate drought stress in rice. *Plant, Soil and Environment, 66*(1), 7–13.
Steffens, B. (2014). The role of ethylene and ROS in salinity, heavy metal, and flooding responses in rice. *Frontiers in Plant Science, 5*, 685.
Su, T., Wang, P., Li, H., Zhao, Y., Lu, Y., Dai, P., ... & Ma, C. (2018). The Arabidopsis catalase triple mutant reveals important roles of catalases and peroxisome-derived signaling in plant development. *Journal of Integrative Plant Biology, 60*(7), 591–607.
Szalai, G., Kellős, T., Galiba, G., & Kocsy, G. (2009). Glutathione as an antioxidant and regulatory molecule in plants under abiotic stress conditions. *Journal of Plant Growth Regulation, 28*(1), 66–80.
Talukdar, D. (2013). Arsenic-induced oxidative stress in the common bean legume, *Phaseolus vulgaris* L. seedlings and its amelioration by exogenous nitric oxide. *Physiology and Molecular Biology of Plants, 19*(1), 69–79.
Thounaojam, T. C., Panda, P., Mazumdar, P., Kumar, D., Sharma, G. D., Sahoo, L., & Sanjib, P. (2012). Excess copper induced oxidative stress and response of antioxidants in rice. *Plant Physiology and Biochemistry, 53*, 33–39.
Triantaphylides, C., Krischke, M., Hoeberichts, F. A., Ksas, B., Gresser, G., Havaux, M., Breusegem, F. V., & Mueller, M. J. (2008). Singlet oxygen is the major reactive oxygen species involved in photooxidative damage to plants. *Plant Physiology, 148*(2), 960–968.
Trinh, N. N., Huang, T. L., Chi, W. C., Fu, S. F., Chen, C. C., & Huang, H. J. (2014). Chromium stress response effect on signal transduction and expression of signaling genes in rice. *Physiologia Plantarum, 150*(2), 205–224.
Uarrota, V. G., Stefen, D. L. V., Leolato, L. S., Gindri, D. M., & Nerling, D. (2018). Revisiting carotenoids and their role in plant stress responses: from biosynthesis to plant signaling mechanisms during stress. In: *Antioxidants and Antioxidant Enzymes in Higher Plants* (Gupta, D. K., Palma, J. M., Corpas, F.J., eds.), Springer Nature, Cham, Switzerland, pp. 207–232.
Ueda, Y., Uehara, N., Sasaki, H., Kobayashi, K., & Yamakawa, T. (2013). Impacts of acute ozone stress on superoxide dismutase (SOD) expression and reactive oxygen species (ROS) formation in rice leaves. *Plant Physiology and Biochemistry, 70*, 396–402.
Volkov, R. A., Panchuk, I. I., Mullineaux, P. M., & Schöffl, F. (2006). Heat stress-induced H_2O_2 is required for effective expression of heat shock genes in Arabidopsis. *Plant Molecular Biology, 61*(4), 733–746.
Wang, S., Liang, D., Li, C., Hao, Y., Ma, F., & Shu, H. (2012). Influence of drought stress on the cellular ultrastructure and antioxidant system in leaves of drought-tolerant and drought-sensitive apple rootstocks. *Plant Physiology and Biochemistry, 51*, 81–89.
Wilson, R. A., Sangha, M. K., Banga, S. S., Atwal, A. K., & Gupta, S. (2014). Heat stress tolerance in relation to oxidative stress and antioxidants in *Brassica juncea. Journal of Environmental Biology, 35*(2), 383.
Woodson, J. D., & Chory, J. (2008). Coordination of gene expression between organellar and nuclear genomes. *Nature Reviews Genetics, 9*(5), 383–395.
Yadav, D. S., Rai, R., Mishra, A. K., Chaudhary, N., Mukherjee, A., Agrawal, S. B., & Agrawal, M. (2019). ROS production and its detoxification in early and late sown cultivars of wheat under future O_3 concentration. *Science of the Total Environment, 659*, 200–210.
Yang, C. Y., & Hong, C. P. (2015). The NADPH oxidase Rboh D is involved in primary hypoxia signalling and modulates expression of hypoxia-inducible genes under hypoxic stress. *Environmental and Experimental Botany, 115*, 63–72.
Yousuf, P. Y., Hakeem, K. U. R., Chandna, R., & Ahmad, P. (2012). Role of glutathione reductase in plant abiotic stress. In: *Abiotic Stress Responses in Plants* (Ahmad, P., Prasad, M. N. V., eds.), Springer, New York, pp. 149–158.
Yuan, F., Lyu, M. J. A., Leng, B. Y., Zheng, G. Y., Feng, Z. T., Li, P. H., Zhu, X. G., & Wang, B. S. (2015). Comparative transcriptome analysis of developmental stages of the *Limonium bicolor* leaf generates insights into salt gland differentiation. *Plant, Cell & Environment 38*, 1637–1657. doi: 10.1111/pce.12514
Yüzbaşıoğlu, E., & Dalyan, E. (2019). Salicylic acid alleviates thiram toxicity by modulating antioxidant enzyme capacity and pesticide detoxification systems in the tomato (*Solanum lycopersicum* Mill.). *Plant Physiology and Biochemistry, 135*, 322–330.

Zafra, A., Castro, A. J., & de Dios Alché, J. (2018). Identification of novel superoxide dismutase isoenzymes in the olive (*Olea europaea* L.) pollen. *BMC Plant Biology*, *18*(1), 1–16.

Zaidi, N. W., Dar, M. H., Singh, S., & Singh, U. S. (2014). Trichoderma species as abiotic stress relievers in plants. In: Biotechnology and Biology of Trichoderma (Gupta, V. K., Schmoll, M., Herrera-Estrella, A., Upadhyay, R. S., Druzhinina, I., Tuohy, M. G., eds.), Elsevier, Netherlands, pp. 515–525. doi:10.1016/b978-0-444-59576-8.00038-2

Zandalinas, S. I., Balfagón, D., Arbona, V., & Gómez-Cadenas, A. (2017). Modulation of antioxidant defense system is associated with combined drought and heat stress tolerance in citrus. *Frontiers in Plant Science*, *8*, 953.

Zandalinas, S. I., Fichman, Y., Devireddy, A. R., Sengupta, S., Azad, R. K., & Mittler, R. (2020). Systemic signaling during abiotic stress combination in plants. *Proceedings of the National Academy of Sciences*, *114*(24), 13810–13820.

Zhang, M., Dong, B., Qiao, Y., Yang, H., Wang, Y., & Liu, M. (2018). Effects of sub-soil plastic film mulch on soil water and salt content and water utilization by winter wheat under different soil salinities. *Field Crops Research*, *225*, 130–140.

Zhang, M., Smith, J. A. C., Harberd, N. P., & Jiang, C. (2016). The regulatory roles of ethylene and reactive oxygen species (ROS) in plant salt stress responses. *Plant Molecular Biology*, *91*(6), 651–659.

Zhang, R., Zhou, Y., Yue, Z., Chen, X., Cao, X., Xu, X., Xing, Y., Jiang, B., Ai, X., & Huang, R. (2019). Changes in photosynthesis, chloroplast ultrastructure, and antioxidant metabolism in leaves of sorghum under waterlogging stress. *Photosynthetica*, *57*(4), 1076–1083.

Zhang, Y., Jiang, W., Yu, H., & Yang, X. (2012). Exogenous abscisic acid alleviates low temperature-induced oxidative damage in seedlings of *Cucumis sativus* L. *Transactions of the Chinese Society of Agricultural Engineering*, *28*(1), 221–228.

3 Nitric Oxide-Mediated Salinity Stress Tolerance in Plants
Signaling and Physiological Perspectives

Praveen Gupta[1]*, Dharmendra Kumar*[1]*, and Chandra Shekhar Seth*[2]*
[1]Department of Botany, University of Delhi, India
[2]Associate Professor, Department of Botany
University of Delhi, Delhi, India
*Corresponding author: E-mail: csseth52@gmail.com

CONTENTS

DOI: 10.1201/9781003159636-3

3.1 INTRODUCTION

Soil salinity and associated land degradation are widespread environmental constraints on crop productivity in many regions across the world. The noxious effects of salinity stress on plant growth are generally associated with inhibition of water uptake, specific ions (particularly Na^+ and Cl^-) toxicity, nutritional disequilibrium, and increased generation of reactive oxygen species (ROS) (Munns and Tester, 2008; Hasanuzzaman et al., 2013; Khan et al., 2014). Sodium ion (Na^+) is a primary cause of specific ion toxicity, as it approaches the toxic level before Cl^- does; therefore, most of the studies on salinity stress focus on the transport of Na^+ within a plant and its tolerance mechanisms (Munns and Tester, 2008; Khan et al., 2012). Salinity stress inhibits photosynthesis through structural alterations in thylakoid membranes, inhibition of chlorophyll biosynthesis, reduction of gas exchange parameters, and disruption in the activity of photosynthetic enzymes (Khan et al., 2012; Turan and Tripathy, 2015; Fatma et al., 2016). Excess production of ROS ($O_2^{\cdot-}$, OH·, and H_2O_2) under salinity stress causes oxidation of proteins, lipids, and nucleic acids, which subsequently affects growth, photosynthesis, and nitrogen and antioxidant metabolism in plants (Singh et al., 2016; Gupta et al., 2017; Khan and Khan, 2017; Gupta and Seth, 2020).

Nitric oxide (NO) is a short-lived, redox-active, and gaseous signaling molecule that plays multiple biological roles in the modulation of several physiological and developmental processes throughout the life cycle of plants (Gupta et al., 2017; Khan et al., 2019). In plants, Klepper (1979) demonstrated the generation of NO for the first time in *Glycine max*, and it was titled "Molecule of the Year" in 1992 by the journal *Science*. In 1998, three scientists, Robert F. Furchgott, Louis J. Ignarro, and Ferid Murad, were awarded the Nobel Prize in physiology and medicine for their discovery related to the signaling behavior of NO (Sahay and Gupta, 2017). Since then, several studies have reported the positive effects of NO in regulation of seed germination and seedling growth, root initiation and development, vascular differentiation, flowering initiation, stomatal movements, photosynthesis, and senescence in plants (Simontacchi et al., 2015; Gupta et al., 2017; Khan et al., 2020). At the molecular level, NO can modify enzyme/protein conformation, translocation, and function via specific post-translational modifications (PTMs). Experimental evidence suggests that endogenous and/or exogenous NO plays a significant role in the detoxification of ROS, either directly or indirectly by interacting with a wide range of targets, leading to the stimulation of antioxidant responses through altering gene expression and protein functions in plants (Khan et al., 2020; Alamri et al., 2020). In spite of much evidence showing important roles of NO as a plant growth regulator, the current information about NO biosynthesis and signaling mechanism(s) is still limited. Therefore, this chapter describes the current understanding of NO biosynthesis, signaling, and knowledge of its physiological actions with a particular emphasis on salinity stress conditions in plants.

3.2 BIOSYNTHESIS OF NITRIC OXIDE IN PLANTS

NO production in plants occurs via oxidative and/or reductive routes in mitochondria, chloroplasts, peroxisomes, endoplasmic reticulum, and apoplastic space (Sahay and Gupta, 2017). Oxidative pathways of NO biosynthesis primarily include oxidation of polyamines (spermidine and spermine), hydroxylamine, and L-arginine by nitric oxide synthase (NOS), whereas reductive pathways consist of reduction of nitrite by nitrate reductase (NR), xanthine oxidoreductase (XOR), plasma membrane-bound nitrite-NO reductase (NiNOR), and non-enzymatic (e.g. carotenoids, phenolic compounds, and ascorbic acid) sources (Asgher et al., 2017). In addition, various enzymes, such as cytochrome P450, xanthine oxidase, and copper amine oxidase 1, have also been suggested as potential sources of NO production in plants (Asgher et al., 2017). An increasing amount of evidence indicates that NR and NOS are the two potential enzymes that are involved in the biosynthesis of NO under normal as well as extreme environmental conditions (Gupta and Seth, 2020). NOS catalyzes the oxidation of L-arginine to citrulline and NO in a complex reaction that involves nicotinamide adenine dinucleotide phosphate (NADPH), calmodulin, calcium, tetrahydrobiopterin, flavin adenine dinucleotide (FAD), and flavin mononucleotide (FMN). The idea of NOS-like activity for NO generation in plants originated from the use of mammalian NOS inhibitors and measurements of NOS activity in plant tissues, cellular extracts, and purified cell organelles.

$$\text{L-arginine} + \text{NAD(P)H} + O_2 \xrightarrow{\text{NOS}} \text{L-citrulline} + \text{NO} + \text{NAD(P)}^+ + \text{NO} + H_2O$$

NR-dependent NO production is a well-characterized source of endogenous NO production in plants, as it can be triggered by both biotic and abiotic stress factors (Gupta and Seth, 2020). NR mainly catalyzes the reduction of nitrate (NO_3^-) to nitrite (NO_2^-) in an NAD(P)H-dependent manner; however, it can also reduce NO_2^- to NO in cells.

$$NO_2^- + 2\,H^+ + e^- \xrightarrow{\text{NR}} NO$$

Under standard physiological conditions, nitrite-dependent NO generation is about 1% of total NR activity, suggesting that only a fraction of NR activity contributes to NO production (Rockel et al., 2002). The switch of NR to its accessory NO synthesis function could be due to its reversible PTM by 14-3-3 proteins that regulate its activity by specific phosphorylation and dephosphorylation reactions (Reda et al., 2018). In *Arabidopsis*, NR is encoded by two genes, *NIA1* and *NIA2*; *NIA1* mainly catalyzes NO production, whereas *NIA2* participates in the reduction of nitrate. Genetic-based approaches demonstrated that *A. thaliana* constitutive NR mutants exhibited increased NO_2^- levels and higher emission rates of NO (Lea et al., 2004), whereas double NR mutant *nia1 nia2* failed to produce NO and related responses (Zhao et al., 2019). Moreover, triple *nia1nia2noa1-2* mutant having defective NR and nitric oxide associated 1 activities exhibited a decreased level of NO in the roots (Lozano-Juste and Leon, 2010). Endogenous NO production was found to be sensitive to NR inhibitor (NaN_3/tungstate), whereas NOS inhibitor (L-NNA/L-NAME) did not affect the NO generation in salt-stressed *Phaseolus vulgaris* (Liu et al., 2007) and *Brassica juncea* (Gupta and Seth, 2020) plants. Moreover, Reda et al. (2018) reported that increased NR activity, *CsNR* gene expression, NO_2^- content, and NO_2^-/NO_3^- ratio, with a simultaneous decline in NR activity, promoted NR-dependent NO generation in *Cucumis sativus* under salinity stress. Recently, Gupta and Seth (2020) also demonstrated that inhibition of NR by tungstate application compromised the brassinosteroid-induced salinity stress tolerance in *B. juncea*. Thus, from this discussion, it is clear that the determination of the exact source of NO is a challenging task, and more future work is required to understand the molecular mechanism(s) of NO biosynthesis in plants. Potential mechanisms of biosynthesis, signaling, and physiological actions of NO in plants under salinity stress are depicted in Figure 3.1.

3.3 NITRIC OXIDE-MEDIATED POST-TRANSLATIONAL MODIFICATIONS IN PLANTS

Nitric oxide-based PTMs are involved in the regulation of cell signaling and function of target proteins under normal physiological as well as stress conditions (Begara-Morales et al. 2015). NO-derived molecules such as nitrosonium cation (NO^+), nitroxyl anion (NO), peroxynitrite ($ONOO^-$), peroxynitrate (O_2NOO), peroxynitrous acid (ONOOH), and peroxynitric acid (O_2NOOH) are known to mediate the PTMs of cellular proteins. These NO-PTMs primarily include protein modifications such as S-nitrosylation, metal nitrosylation, and tyrosine nitration. Among the different PTMs, tyrosine nitration and S-nitrosylation are the most frequently occurring NO-mediated protein modifications in plants under salinity stress. However, there is insufficient data about the role of metal nitrosylation in plants under salinity stress (Saddhe et al., 2019). A detailed description of the physiological and signaling roles of NO-mediated PTMs, i.e. S-nitrosylation and tyrosine nitration, is provided in the following section.

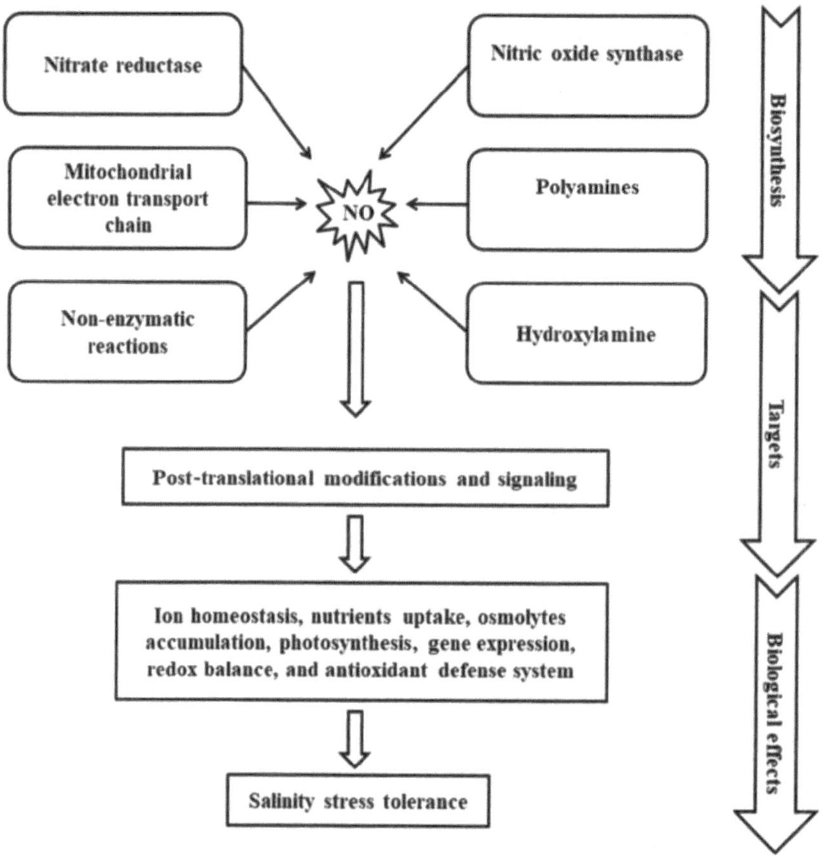

FIGURE 3.1 Overview of nitric oxide production, signaling, and function in plants under salinity stress.

3.3.1 S-Nitrosylation

S-nitrosylation is an enzyme-independent and highly selective post-translational modification that involves covalent attachment of NO to the thiol (-SH) side chain of cysteine (Cys), leading to the formation of S-nitrosothiols (SNOs) in target proteins (Jain and Bhatla, 2018). Experimental evidence has revealed the fundamental role of S-nitrosylation in plant physiology, cell signaling, protein modifications, energy metabolism, photosynthesis, and responses towards various environmental stresses (Camejo et al., 2013; Tanou et al., 2014). It has been demonstrated that NO-induced S-nitrosylation preserved the function of cellular proteins by preventing ROS-induced protein carbonylation under salinity stress (Tanou et al., 2009). Fares et al. (2011) reported that constitutive S-nitrosylation of phosphate transporter (PHT3; 1), vacuolar ATPase subunit, and adenylate translocator at Cys-104, Cys-201, and Cys-130 residues could regulate their activities and enhanced the salinity stress tolerance in *Arabidopsis*. Proteomic analysis of *Pisum sativum* revealed that S-nitrosylation of HSP90, succinate dehydrogenase, ATPase β subunit, and glycine dehydrogenase subunits could regulate the responses to salinity stress by modulating the respiratory and photorespiratory pathways as well as some antioxidant enzymes (Camejo et al., 2013). Monodehydroascorbate reductase (MDHAR) and dehydroascorbate reductase (DHAR) are the important enzymes of the ascorbate-glutathione (AsA-GSH) cycle, responsible for the regeneration of AsA from its oxidative state, i.e. DHA, in an NADPH-dependent manner. Begara-Morales et al. (2015) reported that S-nitrosylation of ascorbate peroxidase (APX) and monodehydroascorbate reductase (MDHAR) differentially regulates their activities in salt-stressed *P. sativum* plants. Importantly, S-nitrosylation of APX at a specific cysteine residue (Cys-32) enhanced its activity, whereas S-nitrosylation of MDHAR resulted in the inhibition of the enzyme activity (Begara-Morales et al., 2015). Proteomic analysis of cotyledon and roots of sunflower seedlings revealed that S-nitrosylation of chaperones, pectinesterase, phospholipase Dα (PLDα), and calmodulin regulates signaling and physiological responses under salinity stress (Jain et al., 2018). Moreover, an increase in NO-mediated protein S-nitrosylation following salinity exposure has been reported in *P. sativum* (Begara-Morales et al., 2015), *Chlamydomonas reinhardtii* (Chen et al., 2016), and *Helianthus annuus* (Jain et al., 2018). Thus, NO-mediated S-nitrosylation plays an important role in modulation of protein function in plants under salinity stress conditions.

3.3.2 Tyrosine Nitration

Protein tyrosine nitration refers to the addition of a nitro group (-NO_2) to tyrosine residues (generally at the ortho position of the phenolic hydroxyl group), resulting in the formation of 3-nitrotyrosine (Kolbert et al., 2017). It is mediated either by peroxynitrite formation ($ONOO^-$) or via production of NO_2^- by heme proteins (the reaction between H_2O_2 and nitrite). This modification converts the tyrosine into a negatively charged residue, which can lead to either activation/inhibition or no change in the function of target proteins. In most cases, tyrosine nitration (NO_2-Tyr) of proteins resulted in a decline in their activities. Therefore, decreased tyrosine nitration of proteins could improve salinity stress tolerance in plants. It has been demonstrated that tyrosine nitration regulated the activity of several proteins associated with photosynthesis, glycolysis, antioxidant metabolism, and protein folding in *Citrus aurantifolia* under salinity stress (Tanou et al., 2012). Confocal laser scanning microscopy (CLSM) and immunoblot analysis revealed the enhancement of NO_2-Tyr content in the columella and peripheral cells in the roots of *Helianthus annuus* under salinity stress (David et al., 2015). Reports have shown that tyrosine nitration could also negatively impact the performance of the AsA-GSH cycle by targeting APX and MDHAR in plants under salinity stress. Begara-Morales et al. (2014, 2015) demonstrated that peroxynitrite-mediated tyrosine nitration of cytosolic APX down-regulates its activity, which could compromise the H_2O_2 scavenging potential of the AsA-GSH cycle. Further, *in silico* and proteomics analysis identified Tyr235 as the primary target, which may disrupt heme-group properties and be responsible for the inactivation of APX (Begara-Morales et al., 2014). Similarly, tyrosine nitration of MDHAR at Tyr345 resulted in the inhibition of its activity, which could disrupt the regeneration of AsA from MDHAR (Begara-Morales et al., 2015). In plants, superoxide dismutase (SOD) and catalase (CAT) constitute a first line of the antioxidant defense system that can effectively neutralize the toxic ROS generated under salt-stress conditions. It has been reported that NO-mediated Tyr-nitration and S-nitrosylation could also modulate the activities of SOD and CAT in plants (Jain and Bhatla, 2018). However, Tyr-nitration does not affect chloroplastic and cytosolic glutathione reductase (GR) activity in *P. sativum* (Begara-Morales et al., 2015). So far, very few studies have been carried out regarding the role of protein tyrosine nitration in higher plants under salt stress. Therefore, identification of key tyrosine-nitrated proteins in the different plant organs will be an exciting challenge for researchers exploring NO signaling in plants. Important studies showing the role of NO-mediated S-nitrosylation and tyrosine nitration in plants under salinity stress are presented in Table 3.1.

3.4 NITRIC OXIDE SIGNALING IN PLANTS UNDER SALINITY STRESS

Nitric oxide actively participates in the modulation of several morpho-physiological responses of plants under salinity stress (Gupta et al., 2017; Khanna et al., 2021; Khan et al., 2021). To date, there is no concluding evidence for the presence of any DNA sequences that can directly interact with NO inside the nucleus. Therefore, NO may activate signaling pathways and regulate

TABLE 3.1
Proteomic Studies on NO-Mediated Post-Translational Modifications in Plants under Salinity Stress

Post-Translational Modification	NaCl Concentration	Plant Species	Target Proteins	Protein Functions	Reference
S-nitrosylation					
	150 mM NaCl	*Citrus aurantium*	Rubisco large subunit, Rubisco activase, glutathione-S-transferase (GST), SOD, APX, peroxiredoxin, tubulin, actin, annexin, translational elongation and initiation factors, ATP synthase, glycolytic enzymes, and heat shock proteins (HSPs)	Energy metabolism, protein targeting, folding and stability, transporters, photosynthesis	Tanou et al., 2009
	100 mM NaCl	*Arabidopsis thaliana*	Photosystem I apoprotein A2, large subunit of Rubisco, UDP-glucose 6-dehydrogenase, PHT3;1, heat shock cognate 70 kDa protein 1, adenosine kinase 1, dehydroascorbate reductase 1, glutathione S-transferase, ascorbate peroxidase 1	Energy metabolism, antioxidant response, molecular chaperones, transporters, photosynthesis, protein folding and stability	Fares et al., 2011
	150 mM NaCl	*Citrus aurantium*	Peroxiredoxin, tubulin, phosphoglycerate kinase, photosystem II proteins, glycine decarboxylase, glyceraldehyde-3-phosphate dehydrogenase, transketolase, and glutathione-S-transferase	Energy metabolism, photosynthesis, defense, signal transduction, etc.	Tanou et al., 2012
	150 mM NaCl	*Pisum sativum*	Peroxiredoxin, HSP90, SOD, thioredoxin, POX, Rubisco small subunit, malate dehydrogenase, and ATP synthase β subunit	Photorespiration, respiration, antioxidant response, translation, and mitochondrial metabolism	Camejo et al., 2013
	150 mM NaCl	*Pisum sativum*	Ascorbate peroxidase	Antioxidant response	Begara-Morales et al., 2014
	100 mM NaCl	*Citrus aurantium*	14-3-3, auxin induced protein, alcohol dehydrogenase, catalase, DHAR, ethylene-induced esterase, isopentenyl pyrophosphate isomerase, GST, SOD, and MDHAR	Photosynthesis, protein synthesis/destination/storage, transport, metabolism, stress response, detoxification, etc.	Tanou et al., 2014

Tyrosine nitration					
	150 mM NaCl	*Citrus aurantium*	Rubisco large subunit, glutathione reductase, 20S proteasome, fructose-1,6-bisphosphate, aldolase, glyceraldehyde 3-phosphate dehydrogenase, and actin	Synthesis, folding, antioxidant response, and protein degradation	Tanou et al., 2012
	150 mM NaCl	*Pisum sativum*	Ascorbate peroxidase, monodehydroascorbate reductase, and glutathione reductase	Antioxidant response	Begara-Morales et al., 2014, 2015
	120 mM NaCl	*Helianthus annuus*	Peroxidase	Antioxidant metabolism	Jain and Bhatla, 2018
	190 mM NaCl	*Arabidopsis thaliana* and *Nicotiana tabacum*	Peroxidase, L-ascorbate oxidase, monodehydroascorbate reductase, phosphoglycerate kinase, and HSP70	Antioxidant defense, synthesis, and protein folding	Szuba et al., 2015

gene expression through intermediate secondary messengers like ROS, reactive nitrogen species (RNS), cGMP, MAPK, Ca^{2+}, phytohormones, lipids, transcription factors, and peptides (Asgher et al., 2017). The important NO signaling mechanism(s) are discussed in detail in the following subsections.

3.4.1 Calcium-Dependent Pathway

Calcium (Ca^{2+}) along with calcium-dependent proteins known as calmodulins (CaMs), participates in the regulation of several NO-associated responses in plants. Garcia-Mata et al. (2003) reported that NO-mediated enhancement of cytoplasmic Ca^{2+} levels regulated the activity of K^+ ion channels that triggered the stomatal closure in *Arabidopsis*. In addition, Ca^{2+} ions also enhanced NOS-dependent NO production in *Arabidopsis* under salinity stress (Corpas et al., 2009). Moreover, Khan et al. (2012) demonstrated that interaction between exogenously supplied calcium (as $CaCl_2$) and NO (as sodium nitroprusside, SNP) alleviated salinity-induced oxidative stress by enhancing the antioxidant defense system in *B. juncea*. Similarly, cross-talk between Ca^{2+} and NO increased the Na^+ efflux and decreased the K^+ efflux by up-regulating the activity/gene expression of Na^+/H^+ antiporters in *Aegiceras corniculatum* under salinity stress (Lang et al., 2014). Protein–protein interaction studies reveal that CaM isoforms AtCaM1 and AtCaM4 can bind with S-nitrosoglutathione reductase (GSNOR) and inhibit its function, which subsequently increased the NO accumulation in *Arabidopsis*. Besides, the analysis of CaM mutants also revealed the role of NO in the maintenance of K^+ and Na^+ homeostasis in plants under salinity stress (Zhou et al., 2016).

3.4.2 G-Protein-Dependent Pathway

The G-protein signaling cascades consist of G-protein coupled receptors (GPCRs), heterotrimeric G-proteins (composed of α, β, and γ subunits), and several downstream secondary signaling components such as protein kinases, phospholipases, guanylate cyclases, phosphodiesterases, and ion channels (Saddhe et al., 2019). Literature reports suggest that G-protein signaling improves salinity stress tolerance through up-regulation of genes related to proline biosynthesis, oxidative stress, ion homeostasis, stress, and abscisic acid (ABA) responses in plants (Ma et al., 2015; Bot et al., 2019). G-protein signaling is reported to regulate extracellular calmodulin (ExtCaM)-mediated NO production and stomatal closure in *Arabidopsis* (Li et al., 2009). Similarly, G-protein signaling mediated the NR and NOS-dependent NO production, which enhanced the activity of antioxidant enzymes in *Zea mays* under salinity stress (Bai et al., 2011).

3.4.3 Mitogen-Activated Protein Kinase Pathway

MAPKs are serine/threonine kinases, which phosphorylate a wide range of targets, such as other kinases and/or transcription factors for induction of stress-responsive genes in plants (Liu et al., 2013). MAPKs such as MPK6 phosphorylated the zinc finger transcriptional regulator (ZAT6), which, in turn, improved the seed germination in *Arabidopsis* under salinity and osmotic stress (Liu et al., 2013). Genetic and biochemical analysis revealed that MPK6 also phosphorylated the NR isoform *NIA2* at Ser-627, which triggered the NR-dependent NO generation in *Arabidopsis* (Wang et al., 2010). Over-expression of *Z. mays* salt-induced MAPK1 (*ZmSIMK1*) and MAPKK (*ZmMKK1*) enhanced the gene expression of stress-responsive marker genes like *CAT*, *POD*, *RAB18*, *RD29A*, and *P5CS1* in *Arabidopsis* under salinity stress (Gu et al., 2010; Cai et al., 2014). Importantly, NO has also been reported to regulate the activity of other MAPKs such as osmotic stress-activated protein kinase (NtOSAK), the glycolytic enzyme glyceraldehyde-3-phosphate dehydrogenase (GAPDH), and phosphoenolpyruvate carboxyl-kinase (PEPCase-K) in plants under salinity stress (Monreal et al., 2013).

3.4.4 Nitro-Fatty Acids

Nitro-fatty acids (NO_2-FAs) are produced by RNS-mediated nitration of free or esterified polyunsaturated fatty acids. The generation and signaling of NO_2-FAs have been extensively studied in animals, whereas few reports are available in plants (Trostchansky et al., 2013). Mata-Pérez et al. (2016) reported that nitro-linolenic acid (NO_2-Ln) initiated a defense mechanism in *Arabidopsis* against salt stress via induction of heat shock proteins (HSPs) and antioxidant enzymes. The signaling mechanism of NO_2-FAs has scarcely been studied in plants. Moreover, it is believed that NO_2-FAs can alleviate oxidative stress by altering the membrane lipid composition, PTMs, and inducing the expression of stress-responsive genes in plants (Mata-Pérez et al., 2018).

3.5 PHYSIOLOGICAL ROLES OF NITRIC OXIDE IN PLANTS UNDER SALINITY STRESS

Most of the studies on NO-related research rely upon the exogenous application of NO donors, scavengers, and inhibitors of its biosynthetic pathways (Thao et al., 2015; Gupta and Seth, 2020). A large number of studies suggest that S-nitroso-N-acetylpenicillamine (SNAP), S-nitrosoglutathione (GSNO), SNP, and diethylamine NONOate sodium (DEANONOate) are the most commonly used NO donors in plants (Gupta et al., 2020). Among the different application methods, direct supplementation in the growth/nutrient media, seed soaking/pretreatment, and foliar application are widely used for NO-related studies in plants. Some of the important physiological responses of endogenously produced and/or exogenously supplied NO in plants under salinity stress are discussed in detail in the following subsections.

3.5.1 Involvement of Nitric Oxide in Seed Germination and Growth Responses in Plants under Salinity Stress

Seed germination and seedling establishment are the most sensitive stages of plant growth, which are severely affected by salinity stress. NO improves seed germination by maintaining the redox and hormonal homeostasis in plants. It has been demonstrated that exogenously applied NO improved seed germination by enhancing the proline accumulation, amylase activity, and antioxidant defense system in plants (Zheng et al., 2009; Silva et al., 2019). The NO-mediated alleviation of salinity stress is well reported in *Cucumis sativus* (Fan et al., 2013), *Triticum aestivum* (Tian et al., 2015), and *Cicer arietinum* (Ahmad et al., 2016) seedlings. In an important study, Kong et al. (2016) reported that exogenously applied NO regulated the expression of *SOS1*, *NHX1*, *NCED2*, *NCED9*, and *IPT1* genes, which delayed salinity-induced leaf senescence by limiting the ABA and Na^+ accumulation and enhancing the K^+ and cytokinin contents in *Gossypium hirsutum*. Exogenous application of the NO donor SNP has been reported to increase the lengths and fresh and dry mass of root and shoot of *B. juncea* (Gupta et al., 2017), *Rubus idaeus* (Ghadakchiasl et al., 2017), and *T. aestivum* (Ali et al., 2017; Sehar et al., 2019) growing under salt stress.

3.5.2 Involvement of Nitric Oxide in Maintenance of Osmotic Balance in Plants under Salinity Stress

Accumulation of compatible osmolytes is an adaptive strategy employed by plants to maintain a favorable gradient for water flow from soil into roots, and to prevent cellular dehydration under salinity stress (Gupta and Seth, 2020; Iqbal et al., 2014). Exogenous application of NO has been reported to enhance the accumulation of compatible osmolytes (proline, glycine betaine, trehalose, γ-aminobutyric acid, and soluble sugars) that maintain optimal cellular functioning by protecting cellular proteins, enzymes, nucleic acids, and other important biomolecules in plants (Gupta et al.,

2017; Ahmad et al., 2018). It has been demonstrated that the NO donor SNP increased proline accumulation by enhancing the activity of Δ′-pyrroline-5-carboxylate synthase (or γ-glutamyl kinase) and decreasing the activity of proline oxidase (or proline dehydrogenase) in *B. juncea* (Gupta et al., 2017) and *T. aestivum* (Sehar et al., 2019) under salt stress. Moreover, Costa-Broseta et al. (2018) demonstrated that *Arabidopsis triple nia1,2noa1-2* mutants with reduced NR activity and NO generation exhibited decreased accumulation of proline metabolites, such as trans-4-hydroxyproline, 5-oxoproline, and N-acetylproline. In a recent study, Kumari et al. (2019) reported that the NO donor diethylenetriamine increased the activity and transcript abundance of betaine aldehyde dehydrogenase (*BADH*, a key gene involved in the biosynthesis of glycine betaine) in *Helianthus annuus* under salinity stress. On the other hand, Phillips et al. (2018) showed that inhibition of NOS-like activity altered the expression of *BADH* in *Z. mays*. NO supplementation has been documented to boost glycine betaine biosynthesis in *Cicer arietinum* (Ahmad et al., 2016), *P. sativum* (Yadu et al., 2017), and *Solanum lycopersicum* (Ahmad et al., 2018), which ultimately enhanced the resistance of plants against salinity stress. In an important study, Gupta and Seth (2020) demonstrated that inhibition of NR-dependent NO production lowered the proline and trehalose contents in *B. juncea* under salinity stress.

3.5.3 Involvement of Nitric Oxide in Maintenance of Photosynthetic Performance in Plants under Salinity Stress

Salinity stress affects the physicochemical properties and functional organization of thylakoid membranes and chloroplast proteins, and inhibits electron flow between PSI and PSII (Fatma et al., 2016). Besides, salinity-induced inhibition of chlorophyll biosynthesis and induction of stomatal closure decrease the partial pressure of CO_2 in the sub-stomatal chamber, which severely affects CO_2 assimilation in plants (Turan and Tripathy, 2015; Khoshbakht et al., 2018). NO plays a crucial role in increasing the light-capturing efficiency by enhancing the chlorophyll content in plants under salinity stress (Fatma et al., 2016; Gupta et al., 2017; Sehar et al., 2019). It has been demonstrated that *Arabidopsis nia1/2* and *noa1* mutants with inhibition in NO biosynthesis exhibited lower chlorophyll accumulation and greater sensitivity to salinity stress (Zhao et al., 2018; Zhao et al., 2019). On the other hand, NO supplementation significantly alleviated the salinity stress-induced decline in photosynthetic pigments and leaf gaseous exchange parameters in several plant species, such as *G. hirsutum* (Kong et al., 2016), *B. juncea* (Fatma et al., 2016; Gupta et al., 2017; Jahan et al., 2020), *T. aestivum* (Sehar et al., 2019), and *Capsicum annuum* (Shams et al., 2019). Importantly, NO supplementation also increased the ribulose-1,5-bisphosphate carboxylase/oxygenase (Rubisco) and Rubisco activase activities to enhance the photosynthetic efficiency of plants (Fatma et al., 2016; Sehar et al., 2019). Moreover, exogenously supplied NO donor SNP has also been reported to improve the maximal PSII photochemical efficiency (F_v/F_m), electron transport rate, intrinsic PSII activity, and photochemical quenching in plants under salinity stress (Khoshbakht et al., 2018; Jahan et al., 2020).

3.5.4 Involvement of Nitric Oxide in Maintenance of Ions and Nutrient Homeostasis in Plants under Salinity Stress

Salinity stress disturbs ion balance by decreasing the availability and uptake of essential nutrients to plants. A large number of studies have demonstrated the inhibitory roles of Na^+ on uptake and accumulation of K^+, Ca^{2+}, and Mg^{2+}, which alter several essential metabolic processes in plants (Tian et al., 2015; Khoshbakht et al., 2018). Besides, salinity-induced disruption of plasma membrane integrity decreased the acquisition of NO_3^-, NO_2^-, NH_4^+, and PO_4^{3-}, ultimately reducing the growth and yield of several crop plants (Miransari, 2011; Singh et al., 2016). It has been reported that *Arabidopsis* mutants (*Atnoa1*) with impaired NOS-like activity and endogenous NO content exhibited decreased seed germination due to higher Na^+/K^+ ratio and H_2O_2 accumulation under salinity stress. However, exogenously supplied NO alleviated the salinity-induced oxidative stress in *Atnoa1* mutants by decreasing the Na^+/K^+ ratio (Zhao et al., 2007). In an important study, Li et al. (2018) demonstrated that the *Arabidopsis SORTING NEXIN 1* (*SNX1*) gene played a crucial role in NOS-dependent NO production in *Arabidopsis* under salinity stress. *SNX1* mutants with decreased NOS-like activity showed reduced proline accumulation, antioxidant defense, and expression of stress-responsive genes under salinity stress. Moreover, exogenous NO donor significantly rescued the hypersensitivity of the *SNX1* mutants to the salinity stress. In plants, regulation of ion homeostasis largely depends on the performance of membrane-localized ion channels and transporters (Evelin et al., 2019). *NHX1*, a tonoplast-localized antiporter, decreases cytosolic Na^+ accumulation by pumping excess Na^+ ions into the vacuole, whereas plasma membrane-located *SOS1* transporter extrudes Na^+ out of the cell into apoplast in an H^+-ATPase-dependent manner (Chen et al., 2015; Zhao et al., 2018). *HKT1* is another K^+ influx channel, responsible for the uptake of K^+ into the cytoplasm (Zhao et al., 2019). Exogenous supplementation of NO has been reported to enhance the uptake and accumulation of K^+, Ca^+, and Mg^{2+} in salt-stressed *T. aestivum* (Tian et al., 2015), *B. juncea* (Gupta et al., 2017), *Citrus* sp. (Khoshbakht et al., 2018), and *Fragaria* sp. (Kaya et al., 2019) plants. Moreover, a large number of studies have demonstrated the role of NO in regulating the gene expression of K^+ and Na^+ transporters, such as *SOS1*, *NHX1*, and *HKT1*, and thereby helping maintain the ion homeostasis in plants under salinity stress (Gong et al., 2014; Chen et al., 2015; Kong et al., 2016; Zhao et al., 2019). However, the NO scavenger cPTIO has been reported to decrease the positive impacts of NO on nutrient homeostasis in plants exposed to saline and sodic alkaline stress conditions (Gong et al., 2014; Zhao et al., 2019). Important studies showing the involvement of NO-donor compounds in the regulation of ion and nutrient homeostasis in plants under salinity stress are presented in Table 3.2.

TABLE 3.2
Nitric Oxide-Mediated Regulation of Ion and Nutrient Homeostasis in Plants under Salinity Stress

Plant species	Salinity Concentration	NO Donor and Dose	Effects	Reference
Kandelia obovata	400 mM NaCl	SNP, 100 μM	Improved K^+/Na^+ balance by activating AKT1-type K^+ channel and Na^+/H^+ antiporter	Chen et al., 2013
Solanum lycopersicum	75 mM $NaHCO_3$	SNP, 0.1 mM	Upregulated gene expression of Na^+/H^+ antiporter (*SlSOS1*), vacuolar Na^+/H^+ exchanger (*SlNHX1* and *SlNHX2*), and Na^+ transporter (*SlHKT1,1* and *SlHKT1,2*)	Gong et al., 2014
Hordeum vulgare	150 mM NaCl	SNP, 100 μM	Enhanced gene expression of H^+-ATPase (*HvHA1*), Na^+/H^+ antiporter (*HvSOS1*), vacuolar Na^+/H^+ antiporter (*HvVNHX2* and *NHE1*), and H^+-ATPase subunit β (*HvVHA-β*)	Chen et al., 2015
Solanum lycopersicum	75 mM $NaHCO_3$	SNP, 100 μM	Enhanced expression of Na extraction gene (*SlSOS1*), Na^+ detoxification genes (*SlNHX1* and *SlNHX2*), and Na^+ transporter genes (*SlHKT1,1* and *SlHKT1,2*)	Liu et al., 2015

(*continued*)

TABLE 3.2 (Continued)
Nitric Oxide-Mediated Regulation of Ion and Nutrient Homeostasis in Plants under Salinity Stress

Plant species	Salinity Concentration	NO Donor and Dose	Effects	Reference
Gossypium hirsutum	150 mM NaCl	SNP, 300 μM	Reduced accumulation of Na^+ by enhancing the gene expression of *SOS1* and *NHX1*	Kong et al., 2016
Brassica napus	200 mM NaCl	SNP, 10 μM	Upregulated gene expression of *SOS2* and *NHX1*, while the Na^+/K^+ ratio is reduced	Zhao et al., 2018
Hylotelephium erythrostictum	200 mM NaCl	SNP, 50 μM	Regulated cytosolic K^+/Na^+ balance via SOS and NHX transporters	Chen et al., 2019
Brassica napus	125 mM NaCl	SNP, 10 μM	Strengthened gene expression of *BnSOS1*, *BnNHX1*, and *BnKT1*	Zhao et al., 2019

3.5.5 Involvement of Nitric Oxide in Regulation of ROS Metabolism and Antioxidant Defense in Plants under Salinity Stress

Salinity stress triggers the over-production of ROS, which causes oxidative damage to cell organelles by enhancing lipid peroxidation, ion leakage, protein oxidation, methylglyoxal production, membrane damage, DNA mutations, and programmed cell death (Evelin et al., 2019; Khan et al., 2020; Gupta and Seth, 2020). Plants are equipped with a well-developed enzymatic as well as non-enzymatic antioxidant defense system that quenches and detoxifies excess ROS in cells (Yadu et al., 2017; Gupta and Seth, 2020; Khan et al., 2021). However, under extreme salinity stress condition, the balance between ROS accumulation and antioxidant defense system becomes disturbed, which negatively affects the essential cellular processes in plants (Evelin et al., 2019). The roles of exogenously supplied NO in augmenting the antioxidant defense system and scavenging excess ROS have been explored in a large number of plant species. The most demonstrated ability of NO in mitigating salinity-induced oxidative stress is the maintenance of cellular redox homeostasis due to its ability to detoxify harmful ROS by activating the antioxidant defense system in plants (Gupta and Seth, 2020). Moreover, as a free-radical molecule, NO can directly react with superoxide or lipid hydroperoxyl radicals and thereby act as a chain breaker to attenuate lipid peroxidation and protein oxidation in plants under salinity stress (Parankusam et al., 2017). Foliar application of NO has been reported to decrease malondialdehyde (MDA) and H_2O_2 content by enhancing the activities of antioxidant enzymes (SOD, CAT, and APX) in *C. arietinum* under salinity stress (Ahmad et al., 2016). Similarly, the NO donor 2,2′-(hydroxynitrosohydrazono) bis-ethanimine (DETA/NO) activated the various isoforms of SOD to scavenge the excess superoxide radicals in *Z. mays* under salinity stress (Klein et al., 2018). The positive effect of exogenously supplied NO on improving the activity of antioxidant enzymes has also been reported in *P. sativum* (Yadu et al., 2017), *S. lycopersicum* (Siddiqui et al., 2017), *Rubus idaeus* (Ghadakchiasl et al., 2017), *T. aestivum* (Sehar et al., 2019), and *C. annuum* (Shams et al., 2019) under salinity stress. NO supplementation also up-regulates the transcription of several antioxidant enzymes that can further enhance the detoxification of harmful ROS in plants under salinity stress (Ahmad et al., 2016; Zhao et al., 2018). Taken together, these studies provide good evidence that exogenous NO can up-regulate the antioxidant defense system and thus stabilize cell membranes and biomolecules, such as lipids and proteins, by quenching free radicals and buffering the cellular redox potential under salinity stress. Important studies showing the involvement of NO in the modulation of growth and physiological responses under salinity stress are depicted in Table 3.3.

TABLE 3.3
Role of Exogenously Supplemented Nitric Oxide in Modulating the Physiological Responses under Salinity Stress

Plant Species	Salinity Concentration	NO Donor and Dose	Physiological Effects	Reference
Cucumis sativus	50, 100, 200, and 400 mM NaCl	50 µM SNP	Decreased MDA production by improving the activities of antioxidant enzymes	Fan et al., 2013
Solanum lycopersicum	120 mM NaCl	300 mM SNP	Increased activities of enzymes involved in antioxidant defense and nitrogen metabolism	Manai et al., 2014
Triticum aestivum	100 mM NaCl	0.10 mM SNP	Decreased production of ROS while enhancing the contents of K, Ca, Mg, Zn, Fe, and Cu	Tian et al., 2015
Cicer arietinum	100 mM NaCl	50 µM SNAP	Lowered electrolyte leakage and H_2O_2 and MDA content by enhancing the activities of SOD, CAT, and APX enzymes	Ahmad et al., 2016
Brassica juncea	100 mM NaCl	100 µM SNP	Enhanced photosynthesis, Rubisco activity, sulfur metabolism, and antioxidant responses	Fatma et al., 2016
Gossypium hirsutum	150 mM NaCl	300 µM SNP	Increased chlorophyll content, net photosynthetic rate, and expression of cytokinin biosynthesis (IPT1 and ZR) genes, while decreasing the expression of ABA biosynthesis genes (NCED2 and NCED9)	Kong et al., 2016
Triticum aestivum	150 mM NaCl	0.1 mM SNP	Improved contents of AsA and total phenols as well as activities of SOD, POD, and CAT enzymes	Ali et al., 2017
Rubus idaeus	50 and 100 mM NaCl	50 and 100 µM SNP	Increased content of proline, total soluble sugars, and activities of antioxidant enzymes	Ghadakchiasl et al., 2017
Brassica juncea	100 mM NaCl	100 µM SNP	Improved photosynthesis, proline and nitrogen metabolism	Gupta et al., 2017
Solanum lycopersicum	100 mM NaCl	100 µM SNP	Improved growth parameters and content of photosynthetic pigments and proline	Siddiqui et al., 2017
Pisum sativum	100 mM NaCl	0.1 mM SNP	Lowered MDA content by increasing SOD, CAT, POX, and APX activities	Yadu et al., 2017
Solanum lycopersicum	200 mM NaCl	50 µM SNAP	Enhanced content of flavonoids, proline, and glycine betaine, as well as antioxidant enzymes	Ahmad et al., 2018
Triticum aestivum	100 mM NaCl	50 µM SNP	Enhanced photosynthesis through upregulation of antioxidant system, proline content, and N assimilation	Sehar et al., 2019
Capsicum annum	50, 100, and 150 mM NaCl	50, 100, and 150 µM SNP	Increased plant growth, photosynthesis, and nutrient (Zn, Fe, B, K, Ca, and Mg) uptake	Shams et al., 2019
Brassica juncea	100 mM NaCl	100 µM SNP	Promoted growth, photosynthesis, and stomatal morphology, and increased nitrogen and sulfur-use efficiency	Jahan et al., 2020

3.6 CONCLUSIONS

Recent literature studies showed the pleiotropic effects of NO in mediating salinity stress responses in plants. The production of endogenous NO and its signaling in plants are supported by NO biosynthetic mutants as well as through the use of exogenously supplied NO donors/scavengers. Exogenous NO donors showed promising effects in enhancing water and nutrient uptake, photosynthesis, and osmolyte accumulation, and up-regulating the antioxidant defense systems under salinity stress. Although NO biosynthetic pathways have been explored in different plant species, the detailed mechanisms of their activation by salinity stress as well as tissue specificity remain ambiguous. NO signaling can also be influenced by other signaling molecules (such as H_2O_2, Ca^{2+}, MAPK, and other plant hormones), which could act synergistically or compete for the control of downstream targets. The cross-talk between NO and other signaling molecules, particularly under salinity stress, needs to be explored. Likewise, the physiological and signaling relevance of NO-mediated PTMs in the modulation of diverse protein functions in plants under salinity stress is still poorly understood. Hence, high-throughput genomic, proteomic, and metabolomic approaches are imperative to understand the role of NO in mediating salinity stress responses in plants.

REFERENCES

Ahmad, Parvaiz, Arafat A. Abdel Latef, Abeer Hashem, Elsayed F. Abd Allah, Salih Gucel, and Lam-Son P. Tran. "Nitric oxide mitigates salt stress by regulating levels of osmolytes and antioxidant enzymes in chickpea." *Frontiers in Plant Science* 7 (2016): 347.

Ahmad, Parvaiz, Mohammad Abass Ahanger, Mohammed Nasser Alyemeni, Leonard Wijaya, Pravej Alam, and Mohammad Ashraf. "Mitigation of sodium chloride toxicity in *Solanum lycopersicum* L. by supplementation of jasmonic acid and nitric oxide." *Journal of Plant Interactions* 13, no. 1 (2018): 64–72.

Alamri, S., Ali, H.M., Khan, M.I.R., Singh, V.P., and Siddiqui, M.H. "Exogenous nitric oxide requires endogenous hydrogen sulfide to induce the resilience through sulfur assimilation in tomato seedlings under hexavalent chromium toxicity." *Plant Physiology and Biochemistry* 155 (2020): 20–34.

Ali, Qasim, M.K. Daud, Muhammad Zulqurnain Haider, Shafaqat Ali, Muhammad Rizwan, Nosheen Aslam, Ali Noman et al. "Seed priming by sodium nitroprusside improves salt tolerance in wheat (*Triticum aestivum* L.) by enhancing physiological and biochemical parameters." *Plant Physiology and Biochemistry* 119 (2017): 50–58.

Asgher, Mohd, Tasir S. Per, Asim Masood, Mehar Fatma, Luciano Freschi, Francisco J. Corpas, and Nafees A. Khan. "Nitric oxide signalling and its crosstalk with other plant growth regulators in plant responses to abiotic stress." *Environmental Science and Pollution Research* 24, no. 3 (2017): 2273–2285.

Bai, Xuegui, Liming Yang, Yunqiang Yang, Parvaiz Ahmad, Yongping Yang, and Xiangyang Hu. "Deciphering the protective role of nitric oxide against salt stress at the physiological and proteomic levels in maize." *Journal of Proteome Research* 10, no. 10 (2011): 4349–4364.

Begara-Morales, Juan C., Beatriz Sánchez-Calvo, Mounira Chaki, Raquel Valderrama, Capilla Mata-Pérez, Javier López-Jaramillo, María N. Padilla, Alfonso Carreras, Francisco J. Corpas, and Juan B. Barroso. "Dual regulation of cytosolic ascorbate peroxidase (APX) by tyrosine nitration and S-nitrosylation." *Journal of Experimental Botany* 65, no. 2 (2014): 527–538.

Begara-Morales, Juan C., Beatriz Sánchez-Calvo, Mounira Chaki, Capilla Mata-Pérez, Raquel Valderrama, María N. Padilla, Javier López-Jaramillo, Francisco Luque, Francisco J. Corpas, and Juan B. Barroso. "Differential molecular response of monodehydroascorbate reductase and glutathione reductase by nitration and S-nitrosylation." *Journal of Experimental Botany* 66, no. 19 (2015): 5983–5996.

Bot, Phearom, Bong-Gyu Mun, Qari Muhammad Imran, Adil Hussain, Sang-Uk Lee, Gary Loake, and Byung-Wook Yun. "Differential expression of AtWAKL10 in response to nitric oxide suggests a putative role in biotic and abiotic stress responses." *PeerJ* 7 (2019): e7383.

Cai, Guohua, Guodong Wang, Li Wang, Yang Liu, Jiaowen Pan, and Dequan Li. "A maize mitogen-activated protein kinase kinase, ZmMKK1, positively regulated the salt and drought tolerance in transgenic *Arabidopsis*." *Journal of Plant Physiology* 171, no. 12 (2014): 1003–1016.

Camejo, Daymi, María del Carmen Romero-Puertas, María Rodríguez-Serrano, Luisa María Sandalio, Juan José Lázaro, Ana Jiménez, and Francisca Sevilla. "Salinity-induced changes in S-nitrosylation of pea mitochondrial proteins." *Journal of Proteomics* 79 (2013): 87–99.

Chen, Juan, Duan-Ye Xiong, Wen-Hua Wang, Wen-Jun Hu, Martin Simon, Qiang Xiao, Ting-Wu Liu, Xiang Liu, and Hai-Lei Zheng. "Nitric oxide mediates root K^+/Na^+ balance in a mangrove plant, *Kandelia obovata*, by enhancing the expression of AKT1-type K^+ channel and Na^+/H^+ antiporter under high salinity." *Plos One* 8, no. 8 (2013): e71543.

Chen, Juan, Wen-Hua Wang, Fei-Hua Wu, En-Ming He, Xiang Liu, Zhou-Ping Shangguan, and Hai-Lei Zheng. "Hydrogen sulfide enhances salt tolerance through nitric oxide-mediated maintenance of ion homeostasis in barley seedling roots." *Scientific Reports* 5 (2015): 12516.

Chen, Xiaodong, Dagang Tian, Xiangxiang Kong, Qian Chen, E. F. Abd Allah, Xiangyang Hu, and Aiqun Jia. "The role of nitric oxide signalling in response to salt stress in *Chlamydomonas reinhardtii*." *Planta* 244, no. 3 (2016): 651–669.

Chen, Zhixin, Xueqi Zhao, Zenghui Hu, and Pingsheng Leng. "Nitric oxide modulating ion balance in *Hylotelephium erythrostictum* roots subjected to NaCl stress based on the analysis of transcriptome, fluorescence, and ion fluxes." *Scientific Reports* 9, no. 1 (2019): 1–12.

Corpas, Francisco J., Makoto Hayashi, Shoji Mano, Mikio Nishimura, and Juan B. Barroso. "Peroxisomes are required for in vivo nitric oxide accumulation in the cytosol following salinity stress of *Arabidopsis* plants." *Plant Physiology* 151, no. 4 (2009): 2083–2094.

Costa-Broseta, Álvaro, Carlos Perea-Resa, Mari-Cruz Castillo, M. Fernanda Ruíz, Julio Salinas, and José León. "Nitric oxide controls constitutive freezing tolerance in *Arabidopsis* by attenuating the levels of osmoprotectants, stress-related hormones and anthocyanins." *Scientific Reports* 8, no. 1 (2018): 1–10.

David, Anisha, Sunita Yadav, František Baluška, and Satish Chander Bhatla. "Nitric oxide accumulation and protein tyrosine nitration as a rapid and long distance signalling response to salt stress in sunflower seedlings." *Nitric Oxide* 50 (2015): 28–37.

Evelin, Heikham, Thokchom Sarda Devi, Samta Gupta, and Rupam Kapoor. "Mitigation of salinity stress in plants by arbuscular mycorrhizal symbiosis: current understanding and new challenges." *Frontiers in Plant Science* 10 (2019): 470.

Fan, Huai-Fu, Chang-Xia Du, Ling Ding, and Yan-Li Xu. "Effects of nitric oxide on the germination of cucumber seeds and antioxidant enzymes under salinity stress." *Acta Physiologiae Plantarum* 35, no. 9 (2013): 2707–2719.

Fares, Abasse, Michel Rossignol, and Jean-Benoît Peltier. "Proteomics investigation of endogenous S-nitrosylation in *Arabidopsis*." *Biochemical and Biophysical Research Communications* 416, no. 3–4 (2011): 331–336.

Fatma, Mehar, Asim Masood, Tasir S. Per, and Nafees A. Khan. "Nitric oxide alleviates salt stress inhibited photosynthetic performance by interacting with sulfur assimilation in mustard." *Frontiers in Plant Science* 7 (2016): 521.

Garcia-Mata, Carlos, Robert Gay, Sergei Sokolovski, Adrian Hills, Lorenzo Lamattina, and Michael R. Blatt. "Nitric oxide regulates K^+ and Cl^- channels in guard cells through a subset of abscisic acid-evoked signalling pathways." *Proceedings of the National Academy of Sciences* 100, no. 19 (2003): 11116–11121.

Ghadakchiasl, Ali, Ali-akbar Mozafari, and Nasser Ghaderi. "Mitigation by sodium nitroprusside of the effects of salinity on the morpho-physiological and biochemical characteristics of *Rubus idaeus* under *in vitro* conditions." *Physiology and Molecular Biology of Plants* 23, no. 1 (2017): 73–83.

Gong, Biao, Xiu Li, Sean Bloszies, Dan Wen, Shasha Sun, Min Wei, Yan Li, Fengjuan Yang, Qinghua Shi, and Xiufeng Wang. "Sodic alkaline stress mitigation by interaction of nitric oxide and polyamines involves antioxidants and physiological strategies in *Solanum lycopersicum*." *Free Radical Biology and Medicine* 71 (2014): 36–48.

Gu, Lingkun, Yukun Liu, Xiaojuan Zong, Lixia Liu, Da-Peng Li, and De-Quan Li. "Overexpression of maize mitogen-activated protein kinase gene, ZmSIMK1 in *Arabidopsis* increases tolerance to salt stress." *Molecular Biology Reports* 37, no. 8 (2010): 4067–4073.

Gupta, Praveen, and Chandra Shekhar Seth. "Interactive role of exogenous 24 Epibrassinolide and endogenous NO in *Brassica juncea* L. under salinity stress: Evidence for NR-dependent NO biosynthesis." *Nitric Oxide* 97 (2020): 33–47.

Gupta, Praveen, Sudhakar Srivastava, and Chandra Shekhar Seth. "24-Epibrassinolide and sodium nitroprusside alleviate the salinity stress in *Brassica juncea* L. cv. Varuna through cross talk among proline, nitrogen metabolism and abscisic acid." *Plant and Soil* 411, no. 1–2 (2017): 483–498.

Hasanuzzaman, Mirza, Kamrun Nahar, and Masayuki Fujita. "Plant response to salt stress and role of exogenous protectants to mitigate salt-induced damages." In *Ecophysiology and Responses of Plants under Salt Stress*, pp. 25–87. Springer, New York, 2013.

Iqbal, N., S. Umar, N.A. Khan, and M.I.R. Khan. "A new perspective of phytohormones in salinity tolerance: Regulation of proline metabolism." *Environmental and Experimental Botany* 100 (2014): 34–42.

Jahan, Badar, Mohamed F. AlAjmi, Md Tabish Rehman, and Nafees A. Khan. "Treatment of nitric oxide supplemented with nitrogen and sulfur regulates photosynthetic performance and stomatal behavior in mustard under salt stress." *Physiologia Plantarum* 168, no. 2 (2020): 490–510.

Jain, Prachi, and Satish C. Bhatla. "Tyrosine nitration of cytosolic peroxidase is probably triggered as a long distance signalling response in sunflower seedling cotyledons subjected to salt stress." *PloS One* 13, no. 5 (2018): e0197132.

Jain, Prachi, Christine von Toerne, Christian Lindermayr, and Satish C. Bhatla. "S-nitrosylation/denitrosylation as a regulatory mechanism of salt stress sensing in sunflower seedlings." *Physiologia Plantarum* 162, no. 1 (2018): 49–72.

Kaya, Cengiz, Nudrat A. Akram, and Muhammad Ashraf. "Influence of exogenously applied nitric oxide on strawberry (*Fragaria*× *ananassa*) plants grown under iron deficiency and/or saline stress." *Physiologia Plantarum* 165, no. 2 (2019): 247–263.

Khan, M. Nasir, Manzer H. Siddiqui, Firoz Mohammad, and M. Naeem. "Interactive role of nitric oxide and calcium chloride in enhancing tolerance to salt stress." *Nitric Oxide* 27, no. 4 (2012): 210–218.

Khan, N.A., M.I.R. Khan, M. Asgher, M. Fatma, A. Masood and S. Shabina Syeed. "Salinity tolerance in plants: Revisiting the role of sulfur metabolites." *Plant Biochemistry and Physiology* 2, no. 1 (2014).

Khan, M.I.R., S. Syeed, R. Nazar, and N.A. Anjum. "An insight into the role of salicylic acid and jasmonic acid in salt stress tolerance". In: Khan, N., Nazar, R., Iqbal, N., Anjum, N. (eds) *Phytohormones and Abiotic Stress Tolerance in Plants*. Springer, Berlin, Heidelberg, 2012.

Khan, M.I.R., and N.A. Khan. *Reactive Oxygen Species and Antioxidant System in Plants: Role and Regulation under Abiotic Stress*. Springer, Singapore, 2017.

Khan, M.I.R., N.A. Khan, B. Jahan, V. Goyal, J. Hamid, S. Khan, N. Iqbal, S. Alamri, and M.H. Siddiqui. "Phosphorus supplementation modulates nitric oxide biosynthesis and stabilizes the defence system to improve arsenic stress tolerance in mustard." *Plant Journal*, 23 (Suppl 1) (2021): 152–161.

Khan, M.I.R., A. Singh and P. Poor. *Improving Abiotic Stress Tolerance in Plants*. CRC Press, Boca Raton, Florida, 2020.

Khan, M.I.R., F Ashfaque, H. Chhillar, M. Irfan, and N.A. Khan. "The intricacy of silicon, plant growth regulators and other signaling molecules for abiotic stress tolerance: An entrancing crosstalk between stress alleviators." *Plant Physiology and Biochemistry* 162 (2021): 36–47.

Khan, M.I.R., P.S. Reddy, A. Ferrante and N.A. Khan. *Plant Signaling Molecules: Role and Regulation under Stressful Environments*. Woodhead Publishing, Cambridge, UK, 2019.

Khanna, R.R., Iqbal N. Badar Jahan, N.A. Khan, M.F. AlAjmi, M.T. Rehman, and M.I.R. Khan. "GABA reverses salt-inhibited photosynthetic and growth responses through its influence on NO-mediated nitrogen-sulfur assimilation and antioxidant system in wheat." *Journal of Biotechnology* 325 (2021): 73–82.

Khoshbakht, D., M. R. Asghari, and M. Haghighi. "Effects of foliar applications of nitric oxide and spermidine on chlorophyll fluorescence, photosynthesis and antioxidant enzyme activities of citrus seedlings under salinity stress." *Photosynthetica* 56, no. 4 (2018): 1313–1325.

Klein, A., L. Hüsselmann, M. Keyster, and N. Ludidi. "Exogenous nitric oxide limits salt-induced oxidative damage in maize by altering superoxide dismutase activity." *South African Journal of Botany* 115 (2018): 44–49.

Klepper, Lowell. "Nitric oxide (NO) and nitrogen dioxide (NO_2) emissions from herbicide-treated soybean plants." *Atmospheric Environment (1967)* 13, no. 4 (1979): 537–542.

Kolbert, Zsuzsanna, Gábor Feigl, Ádám Bordé, Árpád Molnár, and László Erdei. "Protein tyrosine nitration in plants: Present knowledge, computational prediction and future perspectives." *Plant Physiology and Biochemistry* 113 (2017): 56–63.

Kong, Xiangqiang, Tao Wang, Weijiang Li, Wei Tang, Dongmei Zhang, and Hezhong Dong. "Exogenous nitric oxide delays salt-induced leaf senescence in cotton (*Gossypium hirsutum* L.)." *Acta Physiologiae Plantarum* 38, no. 3 (2016): 61.

Kumari, Archana, Rupam Kapoor, and Satish C. Bhatla. "Nitric oxide and light co-regulate glycine betaine homeostasis in sunflower seedling cotyledons by modulating betaine aldehyde dehydrogenase transcript levels and activity." *Plant Signalling and Behavior* 14, no. 11 (2019): 1666656.

Lang, Tao, Huimin Sun, Niya Li, Yanjun Lu, Zedan Shen, Xiaoshu Jing, Min Xiang, Xin Shen, and Shaoliang Chen. "Multiple signalling networks of extracellular ATP, hydrogen peroxide, calcium, and nitric oxide in the mediation of root ion fluxes in secretor and non-secretor mangroves under salt stress." *Aquatic Botany* 119 (2014): 33–43.

Lea, Unni S., Floor Ten Hoopen, Fiona Provan, Werner M. Kaiser, Christian Meyer, and Cathrine Lillo. "Mutation of the regulatory phosphorylation site of tobacco nitrate reductase results in high nitrite excretion and NO emission from leaf and root tissue." *Planta* 219, no. 1 (2004): 59–65.

Li, Jian-Hua, Yin-Qian Liu, Pin Lü, Hai-Fei Lin, Yang Bai, Xue-Chen Wang, and Yu-Ling Chen. "A signalling pathway linking nitric oxide production to heterotrimeric G protein and hydrogen peroxide regulates extracellular calmodulin induction of stomatal closure in *Arabidopsis*." *Plant Physiology* 150, no. 1 (2009): 114–124.

Li, Ting-Ting, Wen-Cheng Liu, Fang-Fang Wang, Qi-Bin Ma, Ying-Tang Lu, and Ting-Ting Yuan. "SORTING NEXIN 1 functions in plant salt stress tolerance through changes of NO accumulation by regulating NO synthase-like activity." *Frontiers in Plant Science* 9 (2018): 1634.

Liu, Yinggao, Ruru Wu, Qi Wan, Gengqiang Xie, and Yurong Bi. "Glucose-6-phosphate dehydrogenase plays a pivotal role in nitric oxide-involved defense against oxidative stress under salt stress in red kidney bean roots." *Plant and Cell Physiology* 48, no. 3 (2007): 511–522.

Liu, Xiao-Min, Xuan Canh Nguyen, Kyung Eun Kim, Hay Ju Han, Jaehyeong Yoo, Kyunghee Lee, Min Chul Kim, Dae-Jin Yun, and Woo Sik Chung. "Phosphorylation of the zinc finger transcriptional regulator ZAT6 by MPK6 regulates *Arabidopsis* seed germination under salt and osmotic stress." *Biochemical and Biophysical Research Communications* 430, no. 3 (2013): 1054–1059.

Liu, Na, Biao Gong, Zhiyong Jin, Xiufeng Wang, Min Wei, Fengjuan Yang, Yan Li, and Qinghua Shi. "Sodic alkaline stress mitigation by exogenous melatonin in tomato needs nitric oxide as a downstream signal." *Journal of Plant Physiology* 186 (2015): 68–77.

Lozano-Juste, Jorge, and José León. "Enhanced abscisic acid-mediated responses in nia1nia2noa1-2 triple mutant impaired in NIA/NR-and AtNOA1-dependent nitric oxide biosynthesis in *Arabidopsis*." *Plant Physiology* 152, no. 2 (2010): 891–903.

Ma, Ya-nan, Ming Chen, Dong-bei Xu, Guang-ning Fang, Er-hui Wang, Shi-qing Gao, Zhao-shi Xu, Lian-cheng Li, Xiao-hong Zhang, and Dong-hong Min. "G-protein β subunit AGB1 positively regulates salt stress tolerance in *Arabidopsis*." *Journal of Integrative Agriculture* 14, no. 2 (2015): 314–325.

Manai, Jamel, Houda Gouia, and Francisco J. Corpas. "Redox and nitric oxide homeostasis are affected in tomato (*Solanum lycopersicum*) roots under salinity-induced oxidative stress." *Journal of Plant Physiology* 171, no. 12 (2014): 1028–1035.

Mata-Pérez, Capilla, Beatriz Sánchez-Calvo, María N. Padilla, Juan C. Begara-Morales, Francisco Luque, Manuel Melguizo, Jaime Jiménez-Ruiz et al. "Nitro-fatty acids in plant signalling: Nitro-linolenic acid induces the molecular chaperone network in *Arabidopsis*." *Plant Physiology* 170, no. 2 (2016): 686–701.

Mata-Pérez, Capilla, María N. Padilla, Beatriz Sánchez-Calvo, Juan C. Begara-Morales, Raquel Valderrama, Mounira Chaki, and Juan B. Barroso. "Biological properties of nitro-fatty acids in plants." *Nitric Oxide* 78 (2018): 176–179.

Miransari, Mohammad. "Arbuscular mycorrhizal fungi and nitrogen uptake." *Archives of Microbiology* 193, no. 2 (2011): 77–81.

Monreal, José A., Cirenia Arias-Baldrich, Vanesa Tossi, Ana B. Feria, Alfredo Rubio-Casal, Carlos García-Mata, Lorenzo Lamattina, and Sofía García-Mauriño. "Nitric oxide regulation of leaf phosphoenolpyruvate carboxylase-kinase activity: Implication in sorghum responses to salinity." *Planta* 238, no. 5 (2013): 859–869.

Munns, Rana, and Mark Tester. "Mechanisms of salinity tolerance." *Annual Review of Plant Biology* 59 (2008): 651–681.

Parankusam, Santisree, Srivani S. Adimulam, Pooja Bhatnagar-Mathur, and Kiran K. Sharma. "Nitric oxide (NO) in plant heat stress tolerance: Current knowledge and perspectives." *Frontiers in Plant Science* 8 (2017): 1582.

Phillips, Kyle, Anelisa Majola, Arun Gokul, Marshall Keyster, Ndiko Ludidi, and Ifeanyi Egbichi. "Inhibition of NOS-like activity in maize alters the expression of genes involved in H2O2 scavenging and glycine betaine biosynthesis." *Scientific Reports* 8, no. 1 (2018): 1–9.

Reda, Małgorzata, Agnieszka Golicka, Katarzyna Kabała, and Małgorzata Janicka. "Involvement of NR and PM-NR in NO biosynthesis in cucumber plants subjected to salt stress." *Plant Science* 267 (2018): 55–64.

Rockel, Peter, Frank Strube, Andra Rockel, Juergen Wildt, and Werner M. Kaiser. "Regulation of nitric oxide (NO) production by plant nitrate reductase *in vivo* and *in vitro*." *Journal of Experimental Botany* 53, no. 366 (2002): 103–110.

Saddhe, Ankush Ashok, Manali Ramakant Malvankar, Suhas Balasaheb Karle, and Kundan Kumar. "Reactive nitrogen species: Paradigms of cellular signalling and regulation of salt stress in plants." *Environmental and Experimental Botany* 161 (2019): 86–97.

Sahay, Seema, and Meetu Gupta. "An update on nitric oxide and its benign role in plant responses under metal stress." *Nitric Oxide* 67 (2017): 39–52.

Sehar, Zebus, Asim Masood, and Nafees A. Khan. "Nitric oxide reverses glucose-mediated photosynthetic repression in wheat (*Triticum aestivum* L.) under salt stress." *Environmental and Experimental Botany* 161 (2019): 277–289.

Shams, Mostafakamal, Melek Ekinci, Selda Ors, Metin Turan, Guleray Agar, Raziye Kul, and Ertan Yildirim. "Nitric oxide mitigates salt stress effects of pepper seedlings by altering nutrient uptake, enzyme activity and osmolyte accumulation." *Physiology and Molecular Biology of Plants* 25, no. 5 (2019): 1149–1161.

Siddiqui, Manzer H., Saud A. Alamri, Mutahhar Y. Al-Khaishany, Mohammed A. Al-Qutami, Hayssam M. Ali, Al-Rabiah Hala, and Hazem M. Kalaji. "Exogenous application of nitric oxide and spermidine reduces the negative effects of salt stress on tomato." *Horticulture, Environment, and Biotechnology* 58, no. 6 (2017): 537–547.

Silva, Aparecida Leonir da, Daniel Teixeira Pinheiro, Laércio Junio da Silva, and Denise Cunha Fernandes dos Santos Dias. "Effect of cyanide by sodium nitroprusside (SNP) application on germination, antioxidative system and lipid peroxidation of *Senna macranthera* seeds under saline stress." *Journal of Seed Science* 41, no. 1 (2019): 86–96.

Simontacchi, Marcela, Andrea Galatro, Facundo Ramos-Artuso, and Guillermo E. Santa-María. "Plant survival in a changing environment: The role of nitric oxide in plant responses to abiotic stress." *Frontiers in Plant Science* 6 (2015): 977.

Singh, Madhulika, Vijay Pratap Singh, and Sheo Mohan Prasad. "Responses of photosynthesis, nitrogen and proline metabolism to salinity stress in *Solanum lycopersicum* under different levels of nitrogen supplementation." *Plant Physiology and Biochemistry* 109 (2016): 72–83.

Szuba, Agnieszka, Anna Kasprowicz-Maluśki, and Przemysław Wojtaszek. "Nitration of plant apoplastic proteins from cell suspension cultures." *Journal of Proteomics* 120 (2015): 158–168.

Tanou, Georgia, Claudette Job, Loïc Rajjou, Erwann Arc, Maya Belghazi, Grigorios Diamantidis, Athannasios Molassiotis, and Dominique Job. "Proteomics reveals the overlapping roles of hydrogen peroxide and nitric oxide in the acclimation of citrus plants to salinity." *The Plant Journal* 60, no. 5 (2009): 795–804.

Tanou, Georgia, Panagiota Filippou, Maya Belghazi, Dominique Job, Grigorios Diamantidis, Vasileios Fotopoulos, and Athanassios Molassiotis. "Oxidative and nitrosative-based signalling and associated post-translational modifications orchestrate the acclimation of citrus plants to salinity stress." *The Plant Journal* 72, no. 4 (2012): 585–599.

Tanou, Georgia, Vasileios Ziogas, Maya Belghazi, Anastasis Christou, Panagiota Filippou, Dominique Job, Vasileios Fotopoulos, and Athanassios Molassiotis. "Polyamines reprogram oxidative and nitrosative status and the proteome of citrus plants exposed to salinity stress." *Plant, Cell and Environment* 37, no. 4 (2014): 864–885.

Thao, N.P., M.I.R. Khan, N.B.A. Thu, X.L.T. Hoang, M. Asgher, N.A. Khan, and L-S. P. Tran. "Role of ethylene and its cross talk with other signaling molecules in plant responses to heavy metal stress". *Plant Physiology* 169 (2015): 73–84.

Tian, Xianyi, Mingrong He, Zhenlin Wang, Jiwang Zhang, Yiling Song, Zhenli He, and Yuanjie Dong. "Application of nitric oxide and calcium nitrate enhances tolerance of wheat seedlings to salt stress." *Plant Growth Regulation* 77, no. 3 (2015): 343–356.

Trostchansky, Andrés, Lucía Bonilla, Lucía González-Perilli, and Homero Rubbo. "Nitro-fatty acids: formation, redox signalling, and therapeutic potential." *Antioxidants and Redox Signalling* 19, no. 11 (2013): 1257–1265.

Turan, S., and B. C. Tripathy. Salt-stress induced modulation of chlorophyll biosynthesis during de-etiolation of rice seedlings. *Physiologia Plantarum* 153, no. 3 (2015): 477–491.

Wang, Pengcheng, Yanyan Du, Yuan Li, Dongtao Ren, and Chun-Peng Song. "Hydrogen peroxide-mediated activation of MAP kinase 6 modulates nitric oxide biosynthesis and signal transduction in *Arabidopsis*." *The Plant Cell* 22, no. 9 (2010): 2981–2998.

Yadu, Shrishti, Teman Lal Dewangan, Vibhuti Chandrakar, and S. Keshavkant. "Imperative roles of salicylic acid and nitric oxide in improving salinity tolerance in *Pisum sativum* L." *Physiology and Molecular Biology of Plants* 23, no. 1 (2017): 43–58.

Zhao, Min-Gui, Qiu-Ying Tian, and Wen-Hao Zhang. "Nitric oxide synthase-dependent nitric oxide production is associated with salt tolerance in *Arabidopsis*." *Plant Physiology* 144, no. 1 (2007): 206–217.

Zhao, Gan, Yingying Zhao, Xiuli Yu, Felix Kiprotich, Han Han, Rongzhan Guan, Ren Wang, and Wenbiao Shen. "Nitric oxide is required for melatonin-enhanced tolerance against salinity stress in rapeseed (*Brassica napus* L.) seedlings." *International Journal of Molecular Sciences* 19, no. 7 (2018): 1912.

Zhao, Gan, Yingying Zhao, Wang Lou, Jiuchang Su, Siqi Wei, Xuemei Yang, Ren Wang, Rongzhan Guan, Huiming Pu, and Wenbiao Shen. "Nitrate reductase-dependent nitric oxide is crucial for multi-walled carbon nanotube-induced plant tolerance against salinity." *Nanoscale* 11, no. 21 (2019): 10511–10523.

Zheng, Chunfang, Dong Jiang, Fulai Liu, Tingbo Dai, Weicheng Liu, Qi Jing, and Weixing Cao. "Exogenous nitric oxide improves seed germination in wheat against mitochondrial oxidative damage induced by high salinity." *Environmental and Experimental Botany* 67, no. 1 (2009): 222–227.

Zhou, Shuo, Lixiu Jia, Hongye Chu, Dan Wu, Xuan Peng, Xu Liu, Jiaojiao Zhang, Junfeng Zhao, Kunming Chen, and Liqun Zhao. "*Arabidopsis* CaM1 and CaM4 promote nitric oxide production and salt resistance by inhibiting S-nitrosoglutathione reductase via direct binding." *PLoS Genetics* 12, no. 9 (2016): e1006255.

4 *S*-Nitrosylation and Denitrosylation

A Regulatory Mechanism during Abiotic Stress Tolerance in Crops

*Priyanka Babuta and Renu Deswal**
Molecular Physiology and Proteomics Laboratory,
Department of Botany, University of Delhi, Delhi
*Corresponding author: E-mail: rdeswal@botany.du.ac.in

CONTENTS

4.1 INTRODUCTION

Plants are exposed to a plethora of environmental stressors, mainly due to climate change, which limits their growth and development, thereby negatively impacting productivity and crop yield. Abiotic stresses like drought, salinity, extreme temperatures (high or low), light intensity, oxidative stress, and heavy metal exposure are major contributors to the decline in both quantity and quality of crops (Zhu, 2016; Zhang et al., 2020). Plants have evolved various mechanisms at both gene and protein level to adapt to these environmental changes. Proteins are crucial for the maintenance of cellular homeostasis and are key players in the stress tolerance mechanism by regulating physiological characteristics in plants under abiotic stresses (Liu et al., 2019). An important characteristic of eukaryotes is that the function of proteins depends not only on their molecular structure but also on their subcellular localization and post-translational modifications (PTMs) (Kosova et al., 2018). PTMs are known to diversify protein functions, which is crucial to maintain cellular homeostasis by regulating almost all molecular events, including gene expression and signal transduction (Deribe

DOI: 10.1201/9781003159636-4

et al., 2010). Currently, more than 400 PTMs are known to dynamically regulate cellular homeostasis (Uniprot Consortium), of which those well studied in plants include nitrosylation, phosphorylation, glycosylation, and ubiquitination. *S*-nitrosylation and denitrosylation is an important NO-based PTM switch that regulates the functions of a wide variety of proteins involved in many cellular processes, including abiotic stress tolerance.

"Omics" technologies, covering metabolomics, transcriptomics, proteomics, and genomics, are being widely applied in plant sciences to decipher the mechanism underlying plant stress tolerance (Van Emon, 2016; Mosa et al., 2017). These omics approaches facilitate a detailed understanding of the mechanism and identification of key players involved in complex networks associated with plant development and stress tolerance mechanisms. Proteomics is one important branch that provides multiple platforms (gel-based, gel-free, labeled, label-free, or targeted proteomics) to deal with the investigation of proteins, including identification, PTMs, protein–protein interactions, and differential protein profiles under different conditions (Perez-Clemente et al., 2013; Pappireddi et al., 2019). Proteomics in relation to various biotic and abiotic plant stress responses and tolerance mainly aims to elucidate the functional aspects of the biological complexes and biochemistry in plants exposed to stress (Kosova et al., 2018). Proteomics approaches have been extensively used to study plants under abiotic stresses and are well reviewed by many researchers (Tanou et al., 2012; Tan et al., 2017; Kim et al., 2017b; Kosova et al., 2018; Komatsu, 2019).

PTMs are an important aspect of proteins that is crucial for maintaining the stability, localization conformation, and function of the proteins. PTMs provide functional diversity and versatility to proteins. Major PTMs, such as phosphorylation glycosylation, ubiquitination, nitrosylation, sumoylation, carbonylation, and lysine acetylation, are highly involved in plant responses to unfavorable conditions (Hashiguchi and Komatsu, 2017; Gupta et al., 2018a; Gupta et al., 2018b). Advancements in proteomic techniques provide powerful tools that allow extensive research in PTMs in plant biology (Spoel, 2018). Nitric oxide (NO)-based redox modifications are of great interest, as NO is involved in a plethora of physiological and biochemical processes in plants under environmental stress. In this chapter, we are assessing the regulatory mechanism of NO and NO-based PTM switch (*S*-nitrosylation/denitrosylation) in plants under abiotic stress conditions. We will also focus on the role of denitrosylases under various abiotic stresses.

4.2 NITRIC OXIDE IN PLANTS

Over the past 40 years, NO has been proved to be a major player in plant cell signaling (Kolbert et al., 2019). NO, being a highly reactive gaseous molecule, is extensively involved in a multitude of physiological processes, such as seed dormancy and germination; plant growth and development; root development; stomatal closure; fruit ripening,; and adaptive responses to various biotic and abiotic stresses (Abat et al., 2008; Besson Bard et al., 2008; Hu et al., 2017; Palma et al., 2019). NO production is attributed to two main pathways: the oxidative (arginine-dependent) pathway via nitric oxide synthase (NOS) and the reductive (nitrite-dependent) pathway involving nitrate reductase (Corpas and Barroso, 2017; Astier et al., 2018). The arginine-dependent pathway is well characterized as the main route in animals, but the occurrence of NOS in photosynthetic organisms is still controversial (Zaffagnini et al., 2016; Hancock and Neill, 2019). However, several studies revealed the presence of NOS in algae (Foresi et al., 2010; Astier et al., 2021). A few studies reported the detection of NOS-like activity in plants, which could be because of the synergistic effect of oxygenase and reductase domains of a new protein to produce NO (Talwar et al., 2012; Barroso et al., 1999; Corpas et al., 2006, 2008; Santolini et al., 2017; Corpas and Barroso, 2017). However, higher accumulation of NO in arginine-deficient mutant Arabidopsis seedlings supplied with exogenous polyamines provided an indirect clue to the presence of an arginine-dependent pathway for NO production in photosynthetic plants (Flores et al, 2008). Nitrate reductase (NR) is considered as the main enzyme involved in NO biosynthesis by providing an electron source for the

NO-forming nitrite reductase (NOFNiR) (Lindermayr, 2017; Chamizo-Ampudia et al., 2017). Also, root cell-specific nitrite:NO reductase (NiNOR) catalyzes the reduction of nitrite, resulting in NO formation (Stohr et al., 2001).

The regulation of numerous physiological processes through NO can be done either by changing gene transcription or by PTM of proteins (Kolbert et al., 2021; Matamoros et al., 2021). NO and some of its adducts can interact with reactive oxygen species (ROS) to form reactive nitrogen species (RNS) such as nitrogen oxides (NO_2, N_2O_3) and peroxynitrite ($ONOO^-$), which are more reactive and sometimes toxic as compared with NO (Stamler et al., 2001; Corpas and Barroso, 2017). However, these RNS, if present in optimal concentrations, act as signaling molecules in various physiological processes and stress-related responses in higher organisms (Neill et al., 2003). RNS interacts with protein or non-protein sulfhydryl-containing compounds to exert the effect of NO via various redox-based PTMs, including *S*-nitrosylation of thiol groups, nitration of tyrosine, and binding of NO to metal centers (metal nitrosylation) in cell signaling events (Frungillo et al., 2013). However, NO exerts its ubiquitous influence on signal transduction and other aspects of cellular function mainly through *S*-nitrosylation of Cys residues in proteins (Hess and Stamler, 2012).

4.3 *S*-NITROSYLATION IN PLANTS: INTRODUCTION AND RELEVANCE

NO is a small biological messenger that regulates plant growth and developmental processes and stress responses mainly through *S*-nitrosylation and denitrosylation (Zhang et al., 2020). *S*-nitrosylation is a highly conserved NO-dependent PTM that regulates protein activity and involves a covalent coupling of an NO moiety to sulfur groups of cysteine residues to form S-nitrosothiol (SNO), which represents a relatively stable NO reserve (Lamotte et al., 2015; Feng et al., 2019). S-nitrosoglutathione (GSNO), the most abundant low-molecular-weight SNO and mobile reservoir of NO bioactivity, is formed by reversible *S*-nitrosylation reaction of NO with reduced glutathione (GSH) tripeptide (γ-Glu-Cys-Gly) under aerobic conditions (Diaz et al., 2003). GSNO can release NO or transnitrosylate proteins to modulate the transduction of a myriad of cellular signals affecting major biological parameters including enzymatic activity, subcellular localization, protein–protein interactions, and protein stability (Derakhshan et al., 2007; Lindermayr and Durner, 2009; Marino and Gladyshev, 2010; Anand et al., 2012). Structural analysis of *S*-nitrosylated proteins showed that Cys residues prone to nitrosylation are flanked by acidic or basic residues. A hydrophobic environment, low pKa value, and solvent exposure of Cys residues are thought to favor *S*-nitrosylation (Marino and Gladyshev, 2010).

Various studies have indicated that NO can also be a double-edged sword. The beneficial or harmful effects of NO are dependent upon its concentration (Rinalducci et al., 2008). NO can act as a chemical messenger regulating various physiological processes and a defense mechanism to improve stress tolerance at optimum concentration, but at high concentrations, over-accumulation of RNS leads to severe nitrosative stress, which subsequently causes damage at the subcellular level (Neill et al., 2003; Neill et al., 2008; Ziogas et al., 2013). Therefore, steady-state *S*-nitrosylation is regulated by denitrosylase activity to maintain dynamic equilibria between SNO-proteins and low-molecular-weight SNOs, which is crucial for cellular NO homeostasis (Stomberski et al., 2019).

4.4 DENITROSYLATION: A MECHANISM TO MAINTAIN NO HOMEOSTASIS

Cells have devised certain ways to scavenge excess NO and thus to keep a check on nitrosative stress. Denitrosylation, a mechanism of SNO catabolism, is important for maintaining the level of SNO and confers cellular protection by alleviating the symptoms of pathological conditions linked to nitrosative stress (Benhar et al., 2010). Both enzymatic and non-enzymatic systems are known to regulate the cellular NO levels.

Non-enzymatic catabolism of SNOs occurs via heat, light, or nucleophilic compounds that cleave the SNO bond, resulting in the release of NO radical or ionic species (NO^-). Various studies have also reported an increase in *S*-nitrosoglutathione reductase (GSNOR) activity under high light exposure conditions (Corpas et al., 2008; Kubienova et al., 2013; Lopes-Oliveira et al., 2021). For instance, NO scavenging capacity was detected to be four times higher in light-exposed tomato seedlings (Zuccarelli et al., 2017), suggesting an indirect involvement of light in causing denitrosylation by mediating factors affecting direct NO catabolism in cells. Hemoglobin, another potent NO scavenger, was reported to be extensively involved in oxidative denitrosylation (Helms et al., 2018). Oxyleghemoglobin (hemoglobin present in leguminous plants) was also shown to scavenge NO and peroxynitrite in root nodules (Herold and Rock, 2005). Glutathione (GSH) is a low-molecular-weight thiol in the cell that scavenges NO from SNO when present in excess (Zaffagnini et al., 2013). It was observed that the cleavage of GSNO by GSH is efficiently catalyzed in the presence of copper, which forms the Cu^+-GSH complex (Gorren et al., 1996).

Enzymatic denitrosylation requires the activity of denitrosylases, the enzymes responsible for the removal of the NO moiety from the Cys thiol side chain of proteins or low-molecular-weight thiols. Denitrosylases play a crucial role in maintaining cellular NO homeostasis by abstracting NO from proteins, ameliorating nitrosative stress, and regulating signal transduction (Anand et al., 2012). Cellular enzymatic denitrosylation is mainly categorized into two major classes: low-molecular-weight-SNO denitrosylases (including the glutathione/*S*-nitrosoglutathione reductase (GSH/GSNOR) system and the SNO-CoA reductase system) and protein–SNO denitrosylases (including thioredoxin/thioredoxin reductase (Trx/TrxR) systems) (Stomberski et al., 2019). GSNOR reduces GSNO, which is in equilibrium with the thiol pool, causing indirect protein denitrosylation (SNO reduction), whereas Trx causes a direct denitrosylation reaction via transnitrosylation of NO from proteins to Trx.

Aldo-keto reductase (AKR1) is characterized in yeast and mammalian cells and is involved in the reduction of SNO-CoA and GSNO; however, no such activity has been reported in plants so far (Anand et al., 2014; Stomberski et al., 2019). Several other enzymes have also been identified as potential denitrosylases for small-molecular-weight SNOs, such as copper, zinc superoxide dismutase (Cu/Zn-SOD), glutathione peroxidase (Hou et al., 1996), xanthine oxidase, protein disulfide isomerase (Sliskovic et al., 2005), and carbonyl reductase (Bateman et al., 2008). Xanthine oxidase catalyzes the decomposition of low-molecular-weight SNOs (such as GSNO and Cys-NO) under aerobic conditions (Trujillo et al., 1998). It was reported that Cu/Zn-SOD, but not Mn-SOD, catalyzes the decomposition of GSNO in the presence of GSH in a concentration gradient manner (Jourd'heuil et al., 1999). Also, dihydrolipoic acid has been shown to catalyze the denitrosation of GSNO, S-nitrosocaspase 3, S-nitrosoalbumin, and S-nitroso-metallothionein *in vitro* (Stoyanovsky et al., 2005). The physiological relevance and mechanism of action of all candidate denitrosylases and their specific substrates remain to be studied and validated in detail.

4.4.1 S-Nitrosoglutathione Reductase

GSNOR (also known as alcohol dehydrogenase 5 (ADH5)) is a zinc-dependent dehydrogenase that belongs to the Class III alcohol dehydrogenase (ADH3; EC 1.1.1.1) family. A unique property of ADH5 is that it does not use primary short-chain alcohols (like ethanol) as its substrate (Barnett et al., 2017). GSNOR was initially identified as glutathione-dependent formaldehyde dehydrogenase (GS-FDH; EC 1.2.1.1) due to its role in the formaldehyde detoxification pathway, where it catalyzes the oxidation of S-hydroxymethylglutathione (an intermediate of spontaneous reaction between glutathione and formaldehyde) in the presence of NADP as a cofactor. Later, it was renamed GSNOR due to its role in the metabolism of SNOs, specifically GSNO (Jensen et al., 1998). GSNOR has a greater affinity for GSNO than S-hydroxymethyl glutathione (Liu et al., 2001). The GSNOR-catalyzed reaction involves trans-nitrosylation of NO from GSNO to protein thiols, thereby releasing GSH to regulate protein nitrosylation. Therefore, GSNOR is considered a key enzyme in NO metabolism and the modulation of NO-dependent signaling pathways (Frungillo et al., 2013).

GSNOR, a highly conserved enzyme, acts as a regulator for protein de-nitrosation through the reduction of GSNO, subsequently reducing nitrosylated proteins (Liu et al., 2001). It is homodimeric, with each monomer consisting of two domains: a major catalytic domain containing two zinc atoms and a minor non-catalytic domain containing a binding site for coenzyme NAD^+ (Lindermayr, 2017). In the catalytic domain of GSNOR, Zn^{2+} plays a role in the catalysis and is coordinated by two cysteines that are sensitive to oxidation: Cys47 and Cys177. On the other hand, the other Zn^{2+} only has a structural role, is coordinated by four cysteines, Cys94, Cys99, Cys102, and Cys105, and is not involved in enzyme activity (Kubienova et al., 2013; Kovacs et al., 2016). Interestingly, *S*-nitrosylation of conserved non-zinc coordinating cysteines (Cys-10, Cys-271, and Cys-370) takes place through a cooperative mechanism, where nitrosylation at the first site favors subsequent nitrosylation and inhibition of *At*GSNOR via a feedback mechanism (Guerra et al., 2016). The modification of GSNOR at Cys-10 is specifically mediated by transnitrosylase activity of ROG1 (repressor of gsnor1) (Chen et al., 2020). The transnitrosylation facilitates its binding to "Autophagy-related" 8 (ATG8) proteins of the autophagy machinery by inducing local conformational changes, finally degrading GSNOR via autophagy (Zhan et al., 2018). Similarly, human GSNOR and *Saccharomyces cerevisiae* (yeast) GSNOR have three (Cys8, Cys195, and Cys268 in human GSNOR; Cys12, Cys276, and Cys375 in yeast GSNOR) predicted sites for *S*-nitrosylation, resulting in the reversible inhibition of the enzyme (Guerra et al., 2016).

Genes encoding for the GSNOR enzymes are mainly single copy, as reported in Arabidopsis, tomato, and tobacco,; however, gene duplication may result in multiple gene copies in some plant species, such as *Phaseolus vulgaris* (bean) and *Glycine max* (soybean) (Martinez et al., 1996; Diaz et al., 2003; Xu et al., 2013; Kubienova et al., 2013). Recently, GSNOR was reported to have two isoforms in *Lotus japonica*, with similar kinetics but different expression profiles in tissues (Matamoros et al., 2020). Subcellular localization of GSNOR is restricted to cytosol and nucleus, with dramatic exclusion from the nucleolus. The presence of a putative mitochondrial targeting peptide is reported in Arabidopsis, rice, and *Selaginella* GSNOR (Xu et al., 2013). Reumann et al. (2007) also suggested the localization of the enzyme in Arabidopsis peroxisomes.

4.4.2 Thioredoxin–Thioredoxin Reductase System

Thioredoxins are ubiquitous redox proteins that are known to play a key role in many important biological processes, such as redox signaling, and also act as antioxidants by reducing other proteins via cysteine thiol–disulfide exchange. Animals and yeasts contain two types of Trx: Trx1 (present in cytosol and nucleus) and Trx2 (mitochondrial form) (Watson et al., 2004), whereas the plant Trx system is particularly complex due to the presence of a large diversity of Trx isoforms, organized in different multigenic families in plants, which are located in the cytosol (type h), chloroplast (types f, m, z, y, and x), and mitochondria (type o) (Kang et al., 2019). Trxh represents the largest multigenic family, with three different subgroups. The specificity of Trxh is probably linked to tissue and subcellular localization and specificity of reduction pathway (Gelhaye et al., 2005). Trx is thought to play a role in numerous physiological functions through both its transnitrosylase and denitrosylase activities depending on its oxidation state, mainly in the context of nitrosative stress (nitroso-redox imbalance). Interestingly, various studies have reported *S*-nitrosylation of different Trx isoforms. In plants, Trxm5 and Trxh were found to be the target of *S*-nitrosylation in *Arabidopsis thaliana* and *Brassica juncea*, respectively (Lindermayr et al., 2005; Sehrawat and Deswal, 2014). The redox state of conserved cysteine residues Cys-Gly-Pro-Cys (CGPC) in the active site is essential for its oxidoreduction and denitrosylation activity (Meyer et al., 2002; Lafaye et al., 2016). When Cys32 and Cys35 are reduced into dithiols, Trx1 serves as a denitrosylase, and when a disulfide is formed between the two (oxidized state), it acts as a trans-nitrosylase (Wu et al., 2011). Trx1 is nitrosylated at Cys73 only after Cys32–Cys35 disulfide linkage, which subsequently results in transnitrosylation of target proteins, including peroxiredoxin 1 and caspase 3 (Mitchell et al., 2005; Wu et al., 2011). On the other hand, reduced Trx2-mediated denitrosylation (via Cys32 and Cys35) of procaspase-3 is reported after Fas stimulation, resulting in apoptosis (Sengupta and Holgren, 2013; Sun et al., 2013).

Trx proteins are mainly reduced by NADPH-dependent TrxR (NTR). NTR is well characterized as a pyridine nucleotide-disulfide oxidoreductase containing a central catalytic domain, NADPH binding site, and FAD binding site that maintains cellular redox equilibrium (Obiero et al., 2010). Uncoupling of Trx from NTR allows *S*-nitrosylation of (oxidized) Trx, which further promotes the function of SNO-Trx as a transnitrosylase. The importance of CGPC motif and cis-Pro residues in determining denitrosylation activity is demonstrated in engineered DsbG (Trx family proteins) (Lafaye et al., 2016) The TrxR inhibitor auranofin blocks reduction of oxidized Trx, resulting in the inhibition of denitrosylation (Gromer et al., 1998; Sun et al., 2013).

Two variants of TrxR have been proposed to have evolved independently: high-molecular-weight TrxR, which is found exclusively in higher eukaryotes, including animals, and low-molecular-weight TrxR, which is found in prokaryotes and some eukaryotes, like fungi and plants (Hirt et al., 2002, Gelhaye et al., 2005). High-molecular-weight TrxR is a homodimer of a 55 kDa subunit, containing selenocysteine at the C-terminal active site, while low-molecular-weight TrxR is a homodimer of a 35 kDa subunit without selenocysteine. Plant NADPH-dependent TrxRs (NTRs) are further classified as universal NTRs (NTRA and NTRB) and plastid NTRs (NTRC). Of these, NTRA proteins are mainly involved in the reduction of cytosolic Trx, with meager efficiency even for mitochondrial Trx, whereas NTRB is mainly responsible for the reduction of mitochondrial Trx (Laloi et al., 2001; Hirt et al., 2002). Both NTRA and NTRB in plants contain only TrxR motif with 82% sequence similarity and are functionally redundant (Jacquot et al., 1994; Laloi et al., 2001; Reichheld et al., 2007). NTRC has been identified in plastids of various photosynthetic plants, such as *A. thaliana*, *Oryza sativa* (rice), *Medicago truncatula*, *Hordeum vulgare* (barley), and some cyanobacteria (Serrato et al., 2004; Pérez-Ruiz et al., 2006; Alkhalfioui et al., 2007; Machida et al., 2007; Pascual et al., 2011; Wulff et al., 2011). This enzyme contains both Trx and TrxR at the C-terminus and N-terminus, respectively, forming a heterodimer where the Trx domain of one subunit is reduced by the TrxR domain of the other subunits (Serrato et al., 2004; Perez-Ruiz et al., 2009). Unlike the cytoplasmic and mitochondrial Trx systems, some chloroplastic Trx is also reduced by a ferredoxin–Trx (FTR), which is a unique protein that transfers the electrons from photochemically reduced ferredoxin to Trx in the chloroplasts (Schürmann et al., 2008; Okegawa et al., 2016). Protein denitrosylation by cytosolic and mitochondrial Trx–TrxR has been reported in plants (Benhar et al., 2008, 2009).

4.5 *S*-NITROSYLATION/DENITROSYLATION DURING ABIOTIC STRESS IN PLANTS

NO is a well-known signaling molecule that is involved in mitigating the negative effects of stress conditions in plants (Sahay et al., 2017; Sami et al., 2018). *S*-nitrosylation, mediated by GSNOR, plays a key role in the regulation of multiple developmental processes and stress responses in plants (Gong et al., 2019). Denitrosation by GSNOR and the Trx–TrxR system ameliorates the effects of nitrosative stress caused by increased NO level and ROS production in the early phase of both biotic and abiotic stress conditions in plants (Ticha et al., 2018; Zhang et al., 2020; Jedelska et al., 2020). Modulation of GSNOR activity under abiotic stress conditions has been studied in several plant systems (Table 4.1), whereas less is known about the role of the Trx–TrxR system. Extensive research is being done to comprehend the role of GSNOR and Trx–TrxR role in protecting plants against multiple abiotic conditions.

TABLE 4.1
Effect of Different Abiotic Stresses on GSNOR

Type of Stress	Plant System	Duration of Stress	GSNOR Activity	Fold Change	References
Heat stress	*Arabidopsis thaliana*	55 °C – 8h	Increase then decrease	2.0	De Pinto et al., 2013
	Poplar	35 °C up to 72 h	Increase	2.0	Cheng et al., 2018
	Pisum sativum	30 °C – 1 h/ 35 °C – 1 h/ 38 °C – 4 h	Increase	1.5	Corpas et al., 2008
	Citrus aurantium	42 °C – 1 h	Increase	1.25	Ziogas et al., 2013
	Helianthus annuus	30 °C – 1 h/ 35 °C – 1 h/ 38 °C – 4 h	Decrease	–	Chaki et al., 2011
Cold stress	*Pisum sativum*	8 °C–48 h	Increase	1.67	Corpas et al., 2008
	Capsicum annuum	8 °C – 1–3 h	Increase	1.32	Airaki et al., 2012
	Citrus aurantium	–8 °C – 5 h	Decrease	1.7	Ziogas et al., 2013
	Solanum lycopersicum	Chilling – 4 °C for 6 days	Increase	2.5	Lv et al., 2017
Salinity stress	*Solanum lycopersicum*	100 mmol/l $NaHCO_3$ – up to 12 d	Increase	2.0	Gong et al., 2015
	Citrus aurantium	100 mM NaCl – 12 d	Increase	1.3	Ziogas et al., 2013
	Pisum sativum (mitochondria)	150 mM NaCl – 5 d/14 d	Increase	1.5	Camejo et al., 2013
	Helianthus annuus	120 mM NaCl – 48 h	Decrease	0.5	Jain et al., 2018
	Arabidopsis thaliana	100 mM NaCl – 7 days	Decrease	–	Zhou et al., 2016
Drought stress	*Citrus aurantium*	13% PEG –12 d	Increase	1.1	Ziogas et al., 2013
	Lotus japonicus	5 days	Decrease	0.75	Signorelli et al., 2013
Heavy metal stress	*Oryza sativa* (aluminum)	75 µM	Increase	1.25	Yang et al., 2013
	Arabidopsis thaliana (arsenic)	100–1000 µM – 7 d	Decrease	–	Leterrier et al., 2012
	Solanum tuberosum (aluminum)	250 µM – 48 h	Increase	1.7	Arasimowicz-Jelonek et al., 2013
	Pisum sativum (cadmium)	50 µm – 14 d	Decrease	1.3	Barroso et al., 2006
	Pisum sativum (cadmium)	50 µm – 14 d	Decrease	–	Ortega-Galisteo et al., 2012
High light	*Pisum sativum*	1,189 $\mu Es^{-1}m^{-2}$ – 4 h	Increase	–	Corpas et al., 2008
Mechanical damage / wounding	*Arabidopsis thaliana*	–	Decrease	–	Diaz et al., 2003
	Nicotiana tabacum	–	Decrease	–	Diaz et al., 2003
	Pisum sativum	4 h	Increase	1.25	Corpas et al., 2008
	Helianthus annuus	–	Decrease	2.0	Chaki et al., 2011

4.5.1 Cold Stress

Cold stress can affect plant growth and productivity by modulating essential processes like gene expression, photosynthesis, cellular homeostasis, and induction of antioxidant systems (Sehrawat et al., 2013b; Selvarajan et al., 2018). NO metabolism is negatively impacted by cold stress (Corpas et al., 2008; Airaki et al., 2012; Sehrawat et al., 2013a). Cold stress-induced changes in the *S*-nitrosoproteome have been reported in *B. juncea* and *A. thaliana.* Proteins involved in major pathways like photosynthetic, glycolytic, defense-related, and signaling-associated were differentially regulated during cold stress (Abat and Deswal, 2009; Puyaubert et al., 2014). Cold stress increased NOS-like mediated NO production, which in turn increased the total SNO content and nitrated proteins, causing nitrosative stress (Corpas et al., 2008). Accumulation of ROS-related molecules was observed under cold stress-induced oxidative stress in *Chorispora bungeana* (Liu et al., 2010). Moreover, cold stress induced an imbalance in ROS and RNS metabolism in pepper leaves, resulting in increased oxidative and nitrosative stress (Airaki et al., 2012). Further, an increase in *S*-nitrosylation of antioxidant enzymes like Cu/Zn SOD and dehydroascorbate reductase (DHAR) was observed, which contributes to ROS detoxification and acts as the first line of defense against cold stress (Sehrawat et al., 2013a; Sehrawat and Deswal, 2014). In wheat seedlings, cold stress-induced negative impacts were decreased by exogenous SNP treatment that stimulated the antioxidant systems (Esim et al., 2014a, b).

Acclimatization of pepper plants to cold stress has been linked to the enhancement of GSNOR activity and decrease in the NO content (Airaki et al., 2011). Cold stress-induced accumulation of GSNOR and NO has been reported in *Pisum sativum*, *Helianthus annus*, and *Capsicum annuum* (Corpas et al., 2008; Chaki et al., 2011; Airaki et al., 2012). Cheng et al. (2015) also reported an accumulation of GSNOR during cold stress in *Populus trichocarpa*, which was later suppressed after prolonged cold stress. Similar results were also observed in pepper, where GSNOR activity increased by 32% after 1 day of cold stress followed by a decline on the second day of the cold stress, suggesting the existence of a feedback mechanism (Airaki et al., 2012). In contrast to these results, a downregulation of GSNOR was reported at both transcript and protein level in citrus during cold stress (Ziogas et al., 2013). Together with NR, GSNOR was found to be involved in cold acclimation-induced chilling tolerance in tomatoes. GSNOR activity increased to approx. 2.5-fold as compared with control in tomato plants subjected to chilling stress for 6 days (Lv et al., 2017).

The Trx–TrxR system is also linked with cold and freezing stress in plants. Expressions of *MaTrx6* and *MaTrx12* (homolog of Arabidopsis Trx-*h*3) in banana peels were upregulated by low-temperature stress. Also, harvested banana fruit treated with ethylene exhibited chilling tolerance, suggesting the role of redox homeostasis in Trx-mediated stress tolerance (Wu et al., 2016). Another study by Moon et al. (2015) showed that overexpression of *AtNTRC* in *A. thaliana* also conferred cold and freezing stress tolerance on the transgenic lines as compared with the wild-type. The recombinant AtNTRC proteins so produced showed cryoprotective activity for malate dehydrogenase and lactic dehydrogenase *in vitro*, thereby possibly increasing the cold and freezing stress tolerance of transgenic lines.

4.5.2 Heat Stress

High-temperature stress induces both oxidative and nitrosative stress in plants (Chaki et al., 2011; Poór et al. 2021). Activation of RNS metabolism and increase in total SNO content were reported during heat stress in many plant systems, such as pea and brassica (Corpas et al., 2008; Abat and Deswal, 2009). The levels of NO, SNO, and O_2 were significantly increased in citrus leaves exposed to heat stress (Ziogas et al., 2013). Further, Tobacco Bright Yellow-2 (BY-2) cells subjected to heat shock showed increased total protein SNO and S-nitrosylated cytosolic ascorbate peroxidase (cAPx) with a significant reduction in APx activity during programmed cell death (PCD) (de Pinto et al., 2013).

A well-known GSNOR mutant of Arabidopsis, hot5, was named because of its high sensitivity to heat stress. The mutant showed an increase in the production of NO, suggesting the role of NO homeostasis in the plant tolerance to temperature stress (Lee et al., 2008). High temperature resulted in the increment of GSNOR activity by up to two-fold, as reported in *A. thaliana*, *P. sativum*, *Citrus aurantium*, and poplar (Corpas et al., 2008; Ziogas et al., 2013; Cheng et al., 2018). In addition, heat shock (55 °C) to tobacco Bright Yellow-2 cells showed an increase in the GSNOR activity and GSNO content, which resulted in PCD (de Pinto et al., 2013).

Trx-*h*3, isolated from heat-treated cytosolic extracts, conferred improved tolerance to heat shock in Arabidopsis (Park et al., 2009). An overexpression study revealed that NTRC confers increased thermotolerance in Arabidopsis, while its knockdown mutant (hot1) was found to be sensitive to heat, drought salinity, and oxidative stress conditions (Chae et al., 2013).

4.5.3 Salinity Stress

Salinity stress is another abiotic stress that is responsible for enormous losses of crop yield. A rise in L-arginine-dependent production of NO in response to salt treatments has been reported in tobacco cells and olive trees, which increases the SNO content, causing nitrosative stress (Gould et al., 2003; Valderrama et al., 2007; Riyazuddin et al., 2020). Besides, exogenous application of NO prevented oxidative damage caused by salt stress, thus ameliorating the negative effects provoked by salinity (Manai et al., 2014). Advance proteomics approaches identified *S*-nitrosylated proteins in citrus plant and Arabidopsis cell suspension subjected to salinity stress (Tanou et al., 2009; Fares et al., 2011). In pea plants under salt stress, APx, mono dehydroascorbate reductase (MDHAR), recombinant mitochondrial PsPrxII F, and proteins involved in photorespiration were *S*-nitrosylated, contributing to stress tolerance (Begara-Morales et al., 2014). GSNOR is involved in the regulation of tomato plant tolerance to alkaline stress by the increase in transcript and protein levels to confer the RNS scavenging effect (Gong et al., 2015). Mitochondrial GSNOR activity in pea was enhanced in both short and long-term treatment with NaCl, with the detection of more nitrated proteins (Camejo et al., 2013). On the contrary, downregulation of GSNOR was observed in several plant systems. In salt-treated *Nicotiana tabacum* plants, second messengers (AtCaM1 and AtCaM4) directly bind to GSNOR to inhibit its activity and to increase the endogenous NO level for maintaining cellular ion homeostasis (Zhou et al., 2016). In addition, a decrease in GSNOR activity was also observed in citrus and sunflower under salt stress (Ziogas et al., 2013; Jain et al., 2018). Salinity stress suppressed the GSNOR activity through polyamines, suggesting a role of polyamines in nitrosative signaling under salinity stress (Tanou et al., 2014).

Along with the GSNOR, the Trx–NTR system is also modulated by salinity stress. Expression of Arabidopsis *Trxh-2* enhanced tolerance to salt stress in *Brassica napus* (Ji et al., 2020). Expression of *Solanum lycopersicum* NTR was increased in response to oxidative and salinity stress but not in cold stress (Dai et al., 2011). an NTRC-deficient mutant showed severe growth inhibition in *O. sativa*, suggesting the role of NTRC in protection against abiotic stress (Serrato et al., 2004).

4.5.4 Heavy Metal Stress

Increasing contamination of heavy metals in the environment is impairing plant growth and inhibiting photosynthesis, which results in loss of crop yield. In the past few decades, researchers have begun focusing on the role of NO-based PTMs in plants under heavy metal stress. In pea seedling leaves, both cadmium and 2,4-D caused a decrease in NO production and modulated *S*-nitrosylation of peroxisomal proteins like glycolate oxidase (GOX), catalase (CAT), and malate dehydrogenase

(MDH) as determined by mass spectrometry (Ortega-Galisteo et al., 2012). GSNOR activity was increased by 2,4-D and inhibited by cadmium (Barroso et al., 2006; Ortega-Galisteo et al., 2012). The *S*-nitrosylation level and SNO content were decreased in cadmium-treated *Boehmeria nivea* (L.) Gaud seedlings (Wang et al., 2015). Phytochelatins, produced in response to heavy metals in plants, were found to be regulated by *S*-nitrosylation in cadmium-treated plants (Asgher et al., 2015). The toxicity of heavy metals like cadmium was found to affect GSNOR expression in plants. GSNOR expression and activity was reduced to about 30% in pea plants treated with cadmium (Barroso et al., 2006). Leterrier et al (2012) also reported the reduction of plant growth with a significant decrease in the enzyme activity of GSNOR and GSNO content in Arabidopsis plants grown in 0.5 mM arsenic. Moreover, a high level of NO in a loss of function mutant of AtGSNOR exhibited higher tolerance to selenite (Lehotai et al., 2012). Potato plants exposed to aluminum showed altered GSNOR activity in leaves and stems but not in roots (Arasimowicz-Jelonek et al., 2013). An increase in GSNOR activity was observed in rice seedlings subjected to aluminum stress (Yang et al., 2013). This suggests the crucial role of NO-based PTM switch in the systemic response to toxic heavy metals.

4.5.5 Drought Stress

Roles of NO in drought stress have been extensively investigated, but much less is known about the role of NO-associated *S*-nitrosylation under drought conditions. Drought stress induced by 13% polyethylene glycol (PEG) in *C. aurantium* for 12 days increased the activity of GSNOR (Ziogas et al., 2013). In *Antiaris toxicaria* Lesch. seeds, drought stress decreased the *S*-nitrosylation of enzymes involved in the ascorbate–GSH cycle, like APx, glutathione reductase (GR), MDHAR, and DHAR (Bai et al., 2011). A drought-tolerant genotype of sugarcane showed higher GSNOR activity than the drought-sensitive genotype in both hydrated and water deficit conditions. Further, the drought-tolerant genotype exhibited a decrease in root GSNOR and an increase in leaf GSNOR activity under water deficit conditions (Silveira et al., 2017). Moreover, water stress in *Lotus japonicus* resulted in the reduction of GSNOR activity and an increase in NO content (Signorelli et al., 2013).

In addition, the Trx–TrxR system is also involved in the stress tolerance mechanism, as overexpression of NTRA conferred elevated drought and oxidative stress tolerance in *A. thaliana* by regulating the level of ROS (Cha et al., 2014). Similar results were presented for NTRC in *A. thaliana*, where overexpression of NTRC conferred elevated drought and oxidative stress tolerance (Kim et al., 2017a).

4.6 CONCLUSION

Abiotic stresses have greatly impacted the growth and productivity of crops and have become a serious threat to food security across the globe. ROS and RNS are considered to be fundamental signaling molecules that govern the signaling events during stress conditions in plants. ROS scavenging and osmolyte accumulation are major protective mechanisms developed by plants to combat environmental stress conditions. NO is known as a key bioactive molecule that mediates a multitude of physiological processes and stress responses in plants. Out of various NO-based PTMs, the *S*-nitrosylation and denitrosylation switch regulates NO-dependent signaling and its function in abiotic stress tolerance in plants. It is becoming more apparent that not only *S*-nitrosylation but also denitrosylation plays an important role in regulating plant tolerance to both biotic and abiotic stresses by modulating multiple physiological and biochemical aspects. An increase in *S*-nitrosylation by various stresses is mainly due to the increase in total SNO content and feedback inhibition of GSNOR via repressor of GSNOR (ROG1) protein. This in turn is responsible for the activation of antioxidant machinery and ROS detoxification. Regulation of ROS by NO-based PTMs

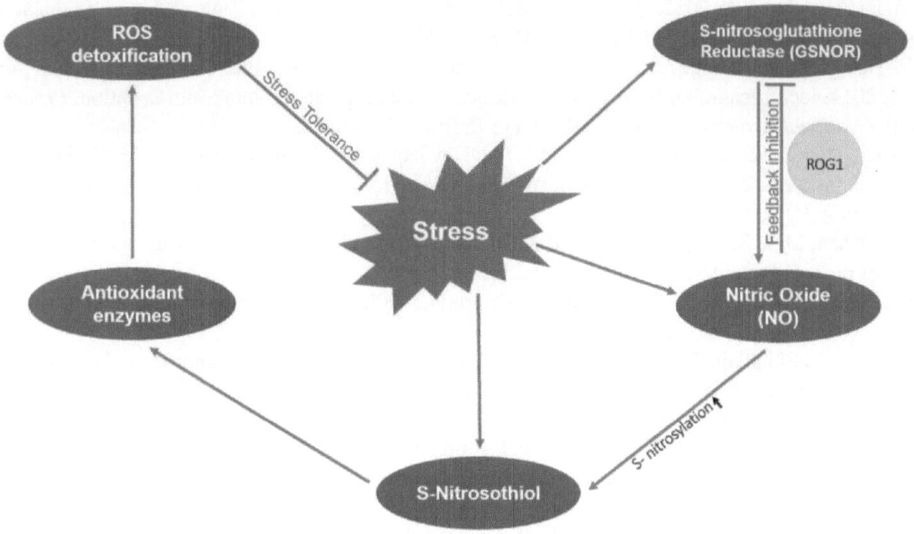

FIGURE 4.1 Mechanism of NO-based post-translational modification dependent stress tolerance in plants.

plays a crucial role in conferring tolerance to abiotic stresses (Figure 4.1). Various stresses impact the level of endogenous NO content and alter the *S*-nitrosylation and denitrosylation mechanism by altering the enzymes involved in NO homeostasis. Several reports have suggested the role of GSNOR under abiotic stress, whereas very little is known about Trx–TrxR. Further research on all denitrosylating enzymes will be necessary to decipher the exact mechanism of the *S*-nitrosylation/denitrosylation switch in responses to adverse stress conditions. However, extensive characterization of *S*-nitrosylation/denitrosylation targets is indispensable to understand the physiological relevance of NO signaling leading to plant tolerance to abiotic stresses.

ACKNOWLEDGEMENT

This work was supported by Faculty Research Programme Grant – Institute of Eminence, University of Delhi (Ref no. IoE/2021/12/FRP).

REFERENCES

Abat, J. K., and Deswal, R. 2009. Differential modulation of S-nitrosoproteome of *Brassica juncea* by low temperature: change in *S*-nitrosylation of Rubisco is responsible for the inactivation of its carboxylase activity. *Proteomics* 9 (18): 4368–4380.

Abat, J. K., Mattoo, A. K., and Deswal, R. 2008. S-nitrosylated proteins of a medicinal CAM plant *Kalanchoe pinnata* – ribulose-1,5-bisphosphate carboxylase/oxygenase activity targeted for inhibition. *The FEBS Journal* 275 (11): 2862–2872.

Airaki, M., Leterrier, M., Mateos, R. M., Valderrama, R., Chaki, M., Barroso, J. B., Del Rio, L. A., Palma, J. M., and Corpas, F. J. 2012. Metabolism of reactive oxygen species and reactive nitrogen species in pepper (*Capsicum annuum* L.) plants under low temperature stress. *Plant, Cell and Environment* 35(2): 281–295.

Airaki, M., Sanchez-Moreno, L., Leterrier, M., Barroso, J. B., Palma, J. M., and Corpas, F. J. 2011. Detection and quantification of S-nitrosoglutathione (GSNO) in pepper (*Capsicum annuum L.*) plant organs by LC-ES/MS. *Plant and Cell Physiology* 52: 2006–2015.

Alkhalfioui, F., Renard, M., and Montrichard, F. 2007. Unique properties of NADP-thioredoxin reductase C in legumes. *Journal of Experimental Botany* 58: 969–978.

Anand, P., and Stamler, J. S. 2012. Enzymatic mechanisms regulating protein S-nitrosylation: implications in health and disease. *Journal of Molecular Medicine* 90: 233–244.

Anand, P., Hausladen, A., Wang, Y. J., Zhang, G. F., Stomberski, C., Brunengraber, H., Hess, D. T., and Stamler, J. S. 2014. Identification of S-nitroso-CoA reductases that regulate protein S-nitrosylation. *Proceedings of the National Academy of Sciences USA* 111 (52): 18572–18577.

Arasimowicz-Jelonek, M., Floryszak-Wieczorek, J., Drzewiecka, K., Chmielowska-Bąk, J., Abramowski, D., and Izbianska, K. 2013. Aluminum induces cross-resistance of potato to *Phytophthora infestans*. *Planta* 239: 679–694.

Asgher, M., Khan, M. I. R., Anjum, N. A., and Khan, N. A. 2015. Minimising toxicity of cadmium in plants – role of plant growth regulators. *Protoplasma* 252 (2): 399–413.

Astier, J., Gross, I., and Durner, J. 2018. Nitric oxide production in plants: an update. *Journal of Experimental Botany* 69 (14): 3401–3411.

Astier, J., Rossi, J., Chatelain, P., Klinguer, A., Besson-Bard, A., Rosnoblet, C., Jeandroz, S., Nicolas-Frances, V., and Wendehenne, D. 2021. Nitric oxide production and signalling in algae. *Journal of Experimental Botany* 72(3): 781–792.

Bai, X., Yang, L., Tian, M., Chen, J., Shi, J., Yang, Y., and Hu, X. 2011. Nitric oxide enhances desiccation tolerance of recalcitrant *Antiaris toxicaria* seeds via protein *S*-nitrosylation and carbonylation. *PLoS One* 6 (6): e20714.

Barnett, S. D., and Buxton, I. L. 2017. The role of S-nitrosoglutathione reductase (GSNOR) in human disease and therapy. *Critical Reviews in Biochemistry and Molecular Biology* 52 (3): 340–354.

Barroso, J. B., Corpas, F. J., Carreras, A., Sandalio, L. M., Valderrama, R., Palma, J., and del Río, L. A. 1999. Localization of nitric-oxide synthase in plant peroxisomes. *Journal of Biological Chemistry* 274 (51): 36729–36733.

Barroso, J. B., Corpas, F. J., Carreras, A., Rodriguez-Serrano, M., Esteban, F. J., Fernandez-Ocana, A., Chaki, M., Romero-Puertas, M. C., Valderrama, R., Sandalio, L. M., and del Rio, L. A. 2006. Localization of S-nitrosoglutathione and expression of S-nitrosoglutathione reductase in pea plants under cadmium stress. *Journal of Experimental Botany* 57: 1785–1793.

Bateman, R. L., Rauh, D., Tavshanjian, B., and Shokat, K. M. 2008. Human carbonyl reductase 1 is an S-nitrosoglutathione reductase. *Journal of Biology and Chemistry* 283: 35756–35762.

Begara-Morales, J. C., Sánchez-Calvo, B., Chaki, M., Valderrama, R., Mata-Pérez, C., López-Jaramillo, J., and Barroso, J. B. 2014. Dual regulation of cytosolic ascorbate peroxidase (APX) by tyrosine nitration and S-nitrosylation. *Journal of Experimental Botany* 65(2): 527–538.

Benhar, M., Forrester, M. T., and Stamler, J. S. 2009. Protein denitrosylation: enzymatic mechanisms and cellular functions. *Nature Reviews Molecular Cell Biology* 10: 721–732.

Benhar, M., Forrester, M. T., Hess, D. T., and Stamler, J. S. 2008. Regulated protein denitrosylation by cytosolic and mitochondrial thioredoxins. *Science* 320: 1050–1054.

Benhar, M., Thompson, J. W., Moseley, M. A., and Stamler, J. S. 2010. Identification of S-nitrosylated targets of thioredoxin using a quantitative proteomic approach. *Biochemistry* 49: 6963–6969.

Besson-Bard, A., Pugin, A., and Wendehenne, D. 2008. New insights into nitric oxide signaling in plants. *Annual Review of Plant Biology* 59: 21–39.

Camejo, D., Romero-Puertas, M., del.C, Rodriguez-Serrano, M., Sandalio, L. M., Lazaro, J. J., Jimenez, A., and Sevilla, F. 2013. Salinity-induced changes in *S*-nitrosylation of pea mitochondrial proteins. *Journal of Proteomics* 79: 87–99.

Cha, J. Y., Kim, J. Y., Jung, I. J., Kim, M. R., Melencion, A., Alam, S. S., Yun, D. J., Lee, S. Y., Kim, M. G., and Kim, W. Y. 2014. NADPH-dependent thioredoxin reductase A (NTRA) confers elevated tolerance to oxidative stress and drought. *Plant Physiology and Biochemistry* 80: 184–191.

Chae, H. B., Moon, J. C., Shin, M. R., Chi, Y. H., Jung, Y. J., Lee, S. Y., Nawkar, G. M., Jung, H. S., Hyun, J. K., Kim, W. Y., Kang, C. H., Yun, D. J., Lee, K. O., and Lee, S. Y. 2013. Thioredoxin reductase type C (NTRC) orchestrates enhanced thermotolerance to Arabidopsis by its redox-dependent holdase chaperone function. *Molecular Plant* 6: 323–336.

Chen, L., Wu, R., Feng, J., Feng, T., Wang, C., Hu, J., Zhan, N., Li, Y., Ma, X., Ren, B., and Zuo, J. 2020. Transnitrosylation mediated by the non-canonical catalase ROG1 regulates nitric oxide signaling in plants. *Developmental Cell* 53(4): 444–457.

Chaki, M., Valderrama, R., Fernandez-Ocana, A. M., Carreras, A., Gomez-Rodriguez, M. V., Pedrajas, J. R., Begara-Morales, J. C., Sanchez-Calvo, B., Luque, F., Leterrier, M., Corpas, F. J., and Barroso, J. B.

2011 Mechanical wounding induces a nitrosative stress by downregulation of GSNO reductase and an increase in S-nitrosothiols in sunflower (*Helianthus annuus*) seedlings. *Journal of Experimental Botany* 62: 1803–1813.

Chamizo-Ampudia, A., Sanz-Luque, E., Llamas, A., Galvan, A., and Fernandez, E. 2017. Nitrate reductase regulates plant nitric oxide homeostasis. *Trends in Plant Science* 22(2): 163–174.

Cheng, T., Chen, J., Abd Allah, E. F., Wang, P., Wang, G., Hu, X., and Shi, J. 2015. Quantitative proteomics analysis reveals that S-nitrosoglutathione reductase (GSNOR) and nitric oxide signaling enhance poplar defense against chilling stress. *Planta* 242: 1361–1390.

Cheng, T., Shi, J., Dong, Y., Ma, Y., Peng, Y., Hu, X., and Chen, J. 2018. Hydrogen sulfide enhances poplar tolerance to high-temperature stress by increasing S-nitrosoglutathione reductase (GSNOR) activity and reducing reactive oxygen/nitrogen damage. *Plant Growth Regulation* 84(1): 11–23.

Corpas, F. J., and Barroso, J. B. 2017. Nitric oxide synthase-like activity in higher plants. *Nitric Oxide: Biology and Chemistry* 68: 5–6.

Corpas, F. J., Barroso, J. B., Carreras, A., Valderrama, R., Palma, J. M., León, A. M., Sandalio, L. M., and del Río, L. A. 2006. Constitutive arginine-dependent nitric oxide synthase activity in different organs of pea seedlings during plant development. *Planta* 224: 246–254.

Corpas, F. J., Chaki, M., Fernandez-Ocana, A., Valderrama, R., Palma, J. M., Carreras, A., Begara-Morales, J. C., Airaki, M., del Rio, L. A., and Barroso, J. B. 2008. Metabolism of reactive nitrogen species in pea plants under abiotic stress conditions. *Plant Cell Physiology* 49: 1711–1722.

Dai, C., and Wang, M. H. 2011. Isolation and characterization of thioredoxin and NADPH-dependent thioredoxin reductase from tomato (Solanum lycopersicum). *BMB Reports* 44(10): 692–697.

De Pinto, M. C., Locato, V., Sgobba, A., Romero-Puertas, M. C., Gadaleta, C., Delledonne, M., and De Gara, L. 2013. *S*-Nitrosylation of ascorbate peroxidase is part of programmed cell death signaling in Tobacco Bright Yellow-2 Cells. *Plant Physiology* 163(4): 1766–1775.

Derakhshan, B., Hao, G., and Gross, S. S. 2007. Balancing reactivity against selectivity: the evolution of protein *S*-nitrosylation as an effector of cell signaling by nitric oxide. *Cardiovascular Research* 75: 210–219.

Deribe, Y. L., Pawson, T., and Dikic, I. 2010. Post-translational modifications in signal integration. *Nature Structural and Molecular Biology* 17: 666–672.

Dıaz, M., Achkor, H., Titarenko, E., and Martınez, M. C. 2003. The gene encoding glutathione-dependent formaldehyde dehydrogenase/GSNO reductase is responsive to wounding, jasmonic acid and salicylic acid. *FEBS Letters* 543: 136–139.

Esim, N., and Atici, O. 2014a. Nitric oxide improves chilling tolerance of maize by affecting apoplastic antioxidative enzymes in leaves. *Plant Growth Regulation* 72(1): 29–38.

Esim, N., Atici, O., and Mutlu, S. 2014b. Effects of exogenous nitric oxide in wheat seedlings under chilling stress. *Toxicology and Industrial Health* 30(3): 268–274.

Fares, A., Rossignol, M., and Peltier, J. B. 2011. Proteomics investigation of endogenous S-nitrosylation in Arabidopsis. *Biochemical and Biophysical Research Communications* 416(3–4): 331–336.

Feng, J., Chen, L., and Zuo, J. 2019. Protein S-nitrosylation in plants: current progresses and challenges. *Journal of Integrative Plant Biology* 61(12): 1206–1223.

Flores, T., Todd, C. D., Tovar-Mendez, A., Dhanoa, P. K., Correa-Aragunde, N., Hoyos, M. E., and Polacco, J. C. 2008. Arginase-negative mutants of Arabidopsis exhibit increased nitric oxide signaling in root development. *Plant Physiology* 147(4): 1936–1946.

Foresi, N., Correa-Aragunde, N., Parisi, G., Caló, G., Salerno, G., and Lamattina, L. 2010. Characterization of a nitric oxide synthase from the plant kingdom: NO generation from the green alga *Ostreococcus tauri* is light irradiance and growth phase dependent. *The Plant Cell* 22(11): 3816–3830.

Frungillo, L., de Oliveira, J. F., Saviani, E. E., Oliveira, H. C., Martınez, M. C., and Salgado, I. 2013. Modulation of mitochondrial activity by S-nitrosoglutathione reductase in *Arabidopsis thaliana* transgenic cell lines. *Biochimica et Biophysica Acta* 1827: 239–247.

Gelhaye, E., Rouhier, N., Navrot, N., and Jacquot, J. P. 2005. The plant thioredoxin system. *Cellular and Molecular Life Sciences* 62: 24–35.

Gong, B, Wen, D., Wang, X., Wei, M., Yang, F., Li, Y., and Shi, Q. 2015. S-nitrosoglutathione reductase-modulated redox signaling controls sodic alkaline stress responses in *Solanum lycopersicum* L. *Plant Cell Physiology* 56(4): 790–802.

Gong, B., Yan, Y., Zhang, L., Cheng, F., Liu, Z., and Shi, Q. 2019. Unravelling GSNOR-mediated S-nitrosylation and multiple developmental programs in tomato plants. *Plant and Cell Physiology* 60(11): 2523–2537.

Gorren, A. C., Schrammel, A., Schmidt, K., and Mayer, B. 1996. Decomposition of S-nitrosoglutathione in the presence of copper ions and glutathione. *Archives of Biochemistry and Biophysics* 330(2): 219–228.

Gould, K. S., Lamotte, O., Klinguer, A., Pugin, A., and Wendehenne, D. 2003. Nitric oxide production in tobacco leaf cells: a generalized stress response?. *Plant, Cell and Environment* 26(11): 1851–1862.

Gromer, S., Arscott, L. D., Williams Jr, C. H., Schirmer, R. H., and Becker, K. 1998. Human placenta thioredoxin reductase: Iisolation of the selenoenzyme, steady state kinetics, and inhibition by therapeutic gold compounds. *Journal of Biological Chemistry* 273(32): 20096–20101.

Guerra, D., Ballard, K., Truebridge, I., and Vierling, E. 2016. S-nitrosation of conserved cysteines modulates activity and stability of S-nitrosoglutathione reductase (GSNOR). *Biochemistry* 55(17): 2452–2464.

Gupta, R., Min, C. W., Meng, Q., Agrawal, G. K., Rakwal, R., and Kim, S. T. 2018a. Comparative phosphoproteome analysis upon ethylene and abscisic acid treatment in *Glycine max* leaves. *Plant Physiology and Biochemistry* 130: 173–180.

Gupta, R., Min, C. W., Meng, Q., Jun, T. H., Agrawal, G. K., Rakwal, R. and Kim, S. T. 2018b. Phosphoproteome data from abscisic acid and ethylene treated *Glycine max* leaves. *Data in Brief* 20: 516–520.

Hancock, J. T., and Neill, S. J. 2019. Nitric oxide: its generation and interactions with other reactive signaling compounds. *Plants* 8(2): 41.

Hashiguchi, A., and Komatsu, S. 2017. Posttranslational modifications and plant–environment interaction. *Methods in Enzymology* 586: 97–113.

Helms, C. C., Gladwin, M. T., and Kim-Shapiro, D. B. 2018. Erythrocytes and vascular function: oxygen and nitric oxide. *Frontiers in Physiology* 9: 125.

Herold, S., and Röck, G. 2005. Mechanistic studies of S-nitrosothiol formation by NO/O_2 and by NO/methemoglobin. *Archives of Biochemistry and Biophysics* 436(2): 386–396.

Hess, D. T., and Stamler, J. S. 2012. Regulation by S-nitrosylation of protein post-translational modification. *Journal of Biology and Chemistry* 287: 4411–4418.

Hirt, R. P., Müller, S., Embley, T. M., and Coombs, G. H. 2002. The diversity and evolution of thioredoxin reductase: new perspectives. *Trends in Parasitology* 18: 302–308.

Hou, Y., Guo, Z., Li, J., and Wang, P. G. 1996. Seleno compounds and glutathione peroxidase catalyzed decomposition of s-nitrosothiols. *Biochemical and Biophysical Research Communications* 228: 88–93.

Hu, J., Yang, H., Mu, J., Lu, T., Peng, J., Deng, X., Kong, Z., Bao, S., Cao, X., and Zuo, J. 2017. Nitric oxide regulates protein methylation during stress responses in plants. *Molecular Cell* 67(4): 702–710.

Jacquot, J. P., Rivera-Madrid, R., Marinho, P., Kollarova, M., Le Marechal, P., Miginiac-Maslow, M., and Meyer, Y. 1994. *Arabidopsis thaliana* NADHP thioredoxin reductase. cDNA characterization and expression of the recombinant protein in *Escherichia coli. Journal of Molecular Biology* 235: 1357–1363.

Jain, P., von Toerne, C., Lindermayr, C., and Bhatla, S. C. 2018. S-nitrosylation/denitrosylation as a regulatory mechanism of salt stress sensing in sunflower seedlings. *Physiologia Plantarum* 162(1): 49–72.

Jedelská, T., Luhová, L., and Petřivalský, M. 2020. Thioredoxins: emerging players in the regulation of protein s-nitrosation in plants. *Plants* 9(11): 1426.

Jensen, D. E., Belka, G. K., and Du Bois, G. C. 1998. S-nitrosoglutathione is a substrate for rat alcohol dehydrogenase class III isoenzyme. *The Biochemical Journal* 331: 659–668.

Ji, M. G., Park, H. J., Cha, J. Y., Kim, J. A., Shin, G. I., Jeong, S. Y., Lee, E. S., Yun, D. J., Lee, S. Y., and Kim, W. Y. 2020. Expression of *Arabidopsis thaliana* thioredoxin-h2 in *Brassica napus* enhances antioxidant defenses and improves salt tolerance. *Plant Physiology and Biochemistry* 147: 313–321.

Jourd'heuil, D., Laroux, F. S., Miles, A. M., Wink, D. A., and Grisham, M. B. 1999. Effect of superoxide dismutase on the stability of *S*-nitrosothiols. *Archives of Biochemistry and Biophysics* 361(2): 323–330.

Kang, Z., Qin, T., and Zhao, Z. 2019. Thioredoxins and thioredoxin reductase in chloroplasts: a review. *Gene* 706: 32–42.

Kim, M. R., Khaleda, L., Jung, I. J., Kim, J. Y., Lee, S. Y., Cha, J. Y., and Kim, W. Y. 2017a. Overexpression of chloroplast-localized NADPH-dependent thioredoxin reductase C (NTRC) enhances tolerance to photo-oxidative and drought stresses in *Arabidopsis thaliana. Journal of Plant Biology* 60(2): 175–180.

Kim, S. W., Lee, S. H., Min, C. W., Jo, I. H., Bang, K. H., Hyun, D. Y., Agrawal, G. K., et al. 2017b. Ginseng (*Panax* Sp.) Proteomics: an update. *Applied Biological Chemistry* 60: 311–320.

Kolbert, Z., Barroso, J. B., Brouquisse, R., Corpas, F. J., Gupta, K. J., Lindermayr, C., and Hancock, J. T. 2019. A forty year journey: the generation and roles of NO in plants. *Nitric Oxide* 93: 53–70.

Kolbert, Z., Lindermayr, C., and Loake, G. J. 2021. The role of nitric oxide in plant biology: current insights and future perspectives. *Journal of Experimental Botany* 72(3): 777–780.

Komatsu, S. 2019. Plant Proteomic Research 2.0: trends and perspectives. International *Journal* of *Molecular Sciences* 20(10): 2495.

Kosová, K., Vítámvás, P., Urban, M. O., Prášil, I. T., and Renaut, J. 2018. Plant abiotic stress proteomics: the major factors determining alterations in cellular proteome. *Frontiers in Plant Science* 9: 122.

Kovacs, I., Holzmeister, C., Wirtz, M., Geerlof, A., Fröhlich, T., Römling, G., and Lindermayr, C. 2016. ROS-mediated inhibition of S-nitrosoglutathione reductase contributes to the activation of anti-oxidative mechanisms. *Frontiers in Plant Science* 7: 1669.

Kubienova, L., Kopecny, D., Tylichova, M., Briozzo, P., Skopalová, J., Šebela, M., Navrátil, M., Tâche, R., Luhová, L., Barroso, J. B., and Petřivalský, M. 2013. Structural and functional characterization of a plant S-nitrosoglutathione reductase from *Solanum lycopersicum*. *Biochimie* 95: 889–902.

Lafaye, C., Van Molle, I., Dufe, V. T., Wahni, K., Boudier, A., Leroy, P., Collet, J. F., and Messens, J. 2016. Sulfur denitrosylation by an engineered Trx-like DsbG enzyme identifies nucleophilic cysteine hydrogen bonds as key functional determinant. *Journal of Biological Chemistry* 291(29): 15020–15028.

Laloi, C., Rayapuram, N., Chartier, Y., Grienenbeger, J. M. , Bonnard, G., and Meyer, Y. 2001. Identification and characterization of a mitochondrial thioredoxin system in plants. *Proceedings of the National Academy of Sciences USA* 98: 14144–14149.

Lamotte, O., Bertoldo, J. B., Besson-Bard, A., Rosnoblet, C., Aimé, S., Hichami, S., and Wendehenne, D. 2015. Protein S-nitrosylation: specificity and identification strategies in plants. *Frontiers in Chemistry* 2: 114.

Lee, U., Wie, C., Fernandez, B. O., Feelisch, M., and Vierling, E. 2008. Modulation of nitrosative stress by S-nitrosoglutathione reductase is critical for thermotolerance and plant growth in Arabidopsis. *Plant Cell* 20: 786–802.

Lehotai, N., Kolbert, Z., Pető, A., Feigl, G., Ördög, A., Kumar, D., Tari, I., and Erdei, L. 2012. Selenite-induced hormonal and signalling mechanisms during root growth of *Arabidopsis thaliana* L. *Journal of Experimental Botany* 63(15): 5677–5687.

Leterrier, M., Airaki, M., Palma, J. M., Chaki, M., Barroso, J. B., and Corpas, F. J. 2012. Arsenic triggers the nitric oxide (NO) and S-nitrosoglutathione (GSNO) metabolism in Arabidopsis. *Environmental Pollution* 166: 136–143.

Lindermayr, C., and Durner, J. 2009. S-Nitrosylation in plants: pattern and function. *Journal of Proteomics* 73: 1–9.

Lindermayr, C., Saalbach, G., and Durner, J. 2005. Proteomic identification of S-nitrosylated proteins in Arabidopsis. *Plant Physiology* 137(3): 921–930.

Lindermayr, C. 2017. Crosstalk between reactive oxygen species and nitric oxide in plants: key role of S-nitrosoglutathione reductase. *Free Radical Biology and Medicine* 122: 110–115.

Liu, L., Hausladen, A., Zeng, M., Que, L., Heitman, J., and Stamler, J. S. 2001. A metabolic enzyme for S-nitrosothiol conserved from bacteria to humans. *Nature* 410: 490–494.

Liu, Y., Jiang, H., Zhao, Z., and An, L. 2010. Nitric oxide synthase like activity-dependent nitric oxide production protects against chilling-induced oxidative damage in *Chorispora bungeana* suspension cultured cells. *Plant Physiology and Biochemistry* 48(12): 936–944.

Liu, Y., Lu, S., Liu, K., Wang, S, Huang, L., and Guo, L. 2019. Proteomics: a powerful tool to study plant responses to biotic stress. *Plant Methods* 15: 135.

Lopes-Oliveira, P. J., Oliveira, H. C., Kolbert, Z., and Freschi, L. 2021. The light and dark sides of nitric oxide: multifaceted roles of nitric oxide in plant responses to light. *Journal of Experimental Botany* 72(3): 885–903.

Lv, X., Ge, S., Jalal Ahammed, G., Xiang, X., Guo, Z., Yu, J., and Zhou, Y. 2017. Crosstalk between nitric oxide and MPK1/2 mediates cold acclimation-induced chilling tolerance in tomato. *Plant and Cell Physiology* 58(11): 1963–1975.

Machida, T., Kato, E., Ishibashi, A., Ohashi, N., Honjoh, K., and Miyamoto, T. 2007. Molecular characterization of low-temperature-inducible NTR-C in *Chlorella vulgaris*. In *Nucleic Acids Symposium Series* 51(1): 463–464. Oxford University Press.

Manai, J., Kalai, T., Gouia, H., and Corpas, F. J. 2014. Exogenous nitric oxide (NO) ameliorates salinity-induced oxidative stress in tomato (*Solanum lycopersicum*) plants. *Journal of Soil Science and Plant Nutrition* 14(2): 433–446.

Marino, S. M., and Gladyshev, V. N. 2010. Structural analysis of cysteine S-nitrosylation: a modified acid-based motif and the emerging role of trans-nitrosylation. *Journal of Molecular Biology* 395 (4): 844–859.

Martinez, M. C., Achkor, H., Persson, B., Fernández, M. R., Shafqat, J., Farrés, J., Jörnvall, H., and Parés, X. 1996. Arabidopsis formaldehyde dehydrogenase. Molecular properties of plant class III alcohol dehydrogenase provide further insights into the origins, structure and function of plant class p and liver class I alcohol dehydrogenases. *European Journal of Biochemistry* 241: 849–857.

Matamoros, M. A., and Becana, M. 2021. Molecular responses of legumes to abiotic stress: protein post-translational modifications and redox signaling. *Journal of Experimental Botany* 72 (16): 5876–5892.

Matamoros, M. A., Cutrona, M. C., Wienkoop, S., Begara-Morales, J. C., Sandal, N., Orera, I., Barroso, J. B., Stougaard, J., and Becana, M. 2020. Altered plant and nodule development and protein S-nitrosylation in *Lotus japonicus* mutants deficient in S-nitrosoglutathione reductases. *Plant and Cell Physiology* 61(1): 105–117.

Meyer, A. J., and Fricker, M. D. 2002. Control of demand-driven biosynthesis of glutathione in green Arabidopsis suspension culture cell. *Plant Physiology* 130: 1927–1937.

Mitchell, D. A., and Marletta, M. A. 2005. Thioredoxin catalyzes the S-nitrosation of the caspase-3 active site cysteine. *Nature Chemical Biology* 1: 154–158.

Moon, J. C., Lee, S., Shin, S. Y., Chae, H. B., Jung, Y. J., Jung, H. S., Lee, K.O., Lee, J. R., and Lee, S. Y. 2015. Overexpression of Arabidopsis NADPH-dependent thioredoxin reductase C (AtNTRC) confers freezing and cold shock tolerance to plants. *Biochemical and Biophysical Research Communications* 463(4): 1225–1229.

Mosa, K. A., Ismail, A., and Helmy, M. 2017. Omics and system biology approaches in plant stress research. In *Plant Stress Tolerance* (pp. 21–34). Springer, Cham.

Neill, S., Barros, R., Bright, J., Desikan, R., Hancock, J., Harrison, J., Morris, P., Ribeiro, D., and Wilson, I. 2008. Nitric oxide, stomatal closure, and abiotic stress. *Journal of Experimental Botany* 59: 165–176.

Neill, S. J., Desikan, R., and Hancock, J. T. 2003. Nitric oxide signalling in plants. *New Phytologist* 159: 11–35.

Obiero, J., Pittet, V., Bonderoff, S. A., and Sanders, D. A. 2010. Thioredoxin system from *Deinococcus radiodurans*. *Journal of Bacteriology* 192(2): 494–501.

Okegawa, Y., and Motohashi, K. 2016. Expression of spinach ferredoxin-thioredoxin reductase using tandem T7 promoters and application of the purified protein for in vitro light dependent thioredoxin-reduction system. *Protein Expression and Purification* 121: 46–51.

Ortega-Galisteo, A. P., Rodríguez-Serrano, M., Pazmiño, D. M., Gupta, D. K., Sandalio, L. M., and Romero-Puertas, M. C. 2012. S-Nitrosylated proteins in pea *(Pisum sativum* L.) leaf peroxisomes: changes under abiotic stress. *Journal of Experimental Botany* 63(5): 2089–2103.

Palma, J. M., Freschi, L., Rodríguez-Ruiz, M., González-Gordo, S., and Corpas, F. J. 2019. Nitric oxide in the physiology and quality of fleshy fruits. *Journal of Experimental Botany* 70(17): 4405–4417.

Pappireddi, N., Martin, L., and Wühr, M. 2019. A review on quantitative multiplexed proteomics. *ChemBioChem* 20(10): 1210–1224.

Park, S. K., Jung, Y. J., Lee, J. R., Lee, Y. M., Jang, H. H., Lee, S. S., Park, J. H., Kim, S. Y., Moon, J. C., Lee, S. Y., and Lee, S. Y. 2009. Heat-shock and redox-dependent functional switching of an h-type Arabidopsis thioredoxin from a disulfide reductase to a molecular chaperone. *Plant Physiology* 150(2): 552–561.

Pascual, M. B., Mata-Cabana, A., Florencio, F. J., Lindahl, M., and Cejudo, F. J. 2011. A comparative analysis of the NADPH thioredoxin reductase C-2-Cys peroxiredoxin system from plants and cyanobacteria. *Plant Physiology* 155: 1806–1816.

Pérez-Clemente, R. M., Vives, V., Zandalinas, S. I., López-Climent, M. F., Muñoz, V., and Gómez-Cadenas, A. 2013. Biotechnological approaches to study plant responses to stress. *BioMed Research International*: 654120.

Pérez-Ruiz, J. M., Gonzalez, M., Spinola, M. C., Sandalio, L. M., and Cejudo, F. J. 2009. The quaternary structure of NADPH thioredoxin reductase C is redox-sensitive. *Molecular Plant* 2: 457–467.

Pérez-Ruiz, J. M., Spinola, M. C., Kirchsteiger, K., Moreno, J., Sahrawy, M., and Cejudo, F. J. 2006. Rice NTRC is a high-efficiency redox system for chloroplast protection against oxidative damage. *Plant Cell* 18: 2356–2368.

Poór, P., Nawaz, K., Gupta, R., Ashfaque, F., and Khan, M. I. R. 2021. Ethylene involvement in the regulation of heat stress tolerance in plants. *Plant Cell Reports* 1–24.

Puyaubert, J., Fares, A., Rézé, N., Peltier, J. B., and Baudouin, E. 2014. Identification of endogenously S-nitrosylated proteins in Arabidopsis plantlets: effect of cold stress on cysteine nitrosylation level. *Plant Science* 215: 150–156.

Reichheld, J. P., Khafif, M., Riondet, C., Droux, M., Bonnard, G., and Meyer, Y. 2007. Inactivation of thioredoxin reductases reveals a complex interplay between thioredoxin and glutathione pathways in Arabidopsis development. *Plant Cell* 19: 1851–1865.

Reumann, S., Babujee, L., Ma, C., Wienkoop, S., Siemsen, T., Antonicelli, G. E., Rasche, N., Lüder, F., Weckwerth, W., and Jahn, O. 2007. Proteome analysis of Arabidopsis leaf peroxisomes reveals novel targeting peptides, metabolic pathways, and defense mechanisms. *Plant Cell* 19(10): 3170–3193.

Rinalducci, S., Murgiano, L., and Zolla, L. 2008. Redox proteomics: basic principles and future perspectives for the detection of protein oxidation in plants. *Journal of Experimental Botany* 59: 3781–3801.

Riyazuddin, R., Verma, R., Singh, K., Nisha, N., Keisham, M., Bhati, K. K., Kim, S. T., and Gupta, R. 2020. Ethylene: a master regulator of salinity stress tolerance in plants. *Biomolecules* 10: 1–22.

Sahay, S., and Gupta, M. 2017. An update on nitric oxide and its benign role in plant responses under metal stress. *Nitric Oxide* 67: 39–52.

Sami, F., Faizan, M., Faraz, A., Siddiqui, H., Yusuf, M., and Hayat, S. 2018. Nitric oxide-mediated integrative alterations in plant metabolism to confer abiotic stress tolerance, NO crosstalk with phytohormones and NO-mediated post translational modifications in modulating diverse plant stress. *Nitric Oxide* 73: 22–38.

Santolini, J., André, F., Jeandroz, S., and Wendehenne, D. 2017. Nitric oxide synthase in plants: where do we stand?. *Nitric Oxide* 63: 30–38.

Sehrawat, A., Abat, J. K., and Deswal, R. 2013a. RuBisCO depletion improved proteome coverage of cold responsive S-nitrosylated targets in *Brassica juncea*. *Frontiers in Plant Science* 4: 342.

Sehrawat, A., and Deswal, R. 2014. S-nitrosylation analysis in *Brassica juncea* apoplast highlights the importance of nitric oxide in cold-stress signaling. *Journal of Proteome Research* 13(5): 2599–2619.

Sehrawat, A., Gupta, R., and Deswal, R. 2013b. Nitric oxide-cold stress signalling cross-talk, evolution of a novel regulatory mechanism. *Proteomics* 13: 1816–1835.

Selvarajan, D., Mohan, C., Dhandapani, V., Nerkar, G., Jayanarayanan, A. N., Mohanan, M. V., Murugan, N., Kaur, L., Chennappa, M., Kumar, R., and Chinnaswamy, A. 2018. Differential gene expression profiling through transcriptome approach of *Saccharum spontaneum* L. under low temperature stress reveals genes potentially involved in cold acclimation. *3 Biotech* 8(4): 1–18.

Sengupta, R., and Holmgren, A. 2013. Thioredoxin and thioredoxin reductase in relation to reversible S-nitrosylation. *Antioxidant and Redox Signalling* 18: 259–269.

Serrato, A. J., Perez-Ruiz, J. M., Spinola, M. C., and Cejudo, F. J. 2004. A novel NADPH thioredoxin reductase, localized in the chloroplast, which deficiency causes hypersensitivity to abiotic stress in *Arabidopsis thaliana*. *Journal of Biology and Chemistry* 279: 43821–43827.

Schürmann, P., and Buchanan, B. B. 2008. The ferredoxin/thioredoxin system of oxygenic photosynthesis. *Antioxidants & Redox Signaling* 10(7): 1235–1274.

Signorelli, S., Corpas, F. J., Omar Borsani, O., Barroso, J. B., and Monza, J. 2013. Water stress induces a differential and spatially distributed nitro-oxidative stress response in roots and leaves of *Lotus japonicus*. *Plant Science* 201–202: 137–146.

Silveira, N. M., Hancock, J. T., Frungillo, L., Siasou, E., Marcos, F. C., Salgado, I., Machado, E. C., and Ribeiro, R. V. 2017. Evidence towards the involvement of nitric oxide in drought tolerance of sugarcane. *Plant Physiology and Biochemistry* 115: 354–359.

Sliskovic, I., Raturi, A., and Mutus, B. 2005. Characterization of the S-denitrosation activity of protein disulfide isomerase. *Journal of Biology and Chemistry* 280: 8733–8741.

Spoel, S. H. 2018. Orchestrating the proteome with post-translational modifications. *Journal of Experimental Botany* 69(19): 4499–4503.

Stamler, J. S., Lamas, S., and Fang, F. C. 2001. Nitrosylation: the prototypic redox-based signaling mechanism. *Cell* 106: 675–683.

Stöhr, C., Strube, F., Marx, G., Ullrich, W. R., and Rockel, P. 2001. A plasma membrane-bound enzyme of tobacco roots catalyses the formation of nitric oxide from nitrite. *Planta* 212(5): 835–841.

Stomberski, C. T., Hess, D. T., and Stamler, J. S. 2019. Protein S-nitrosylation: determinants of specificity and enzymatic regulation of S-nitrosothiol-based signaling. *Antioxidants and Redox Signaling* 30(10): 1331–1351.

Stoyanovsky, D. A., Tyurina, Y. Y., Tyurin, V. A., Anand, D., Mandavia, D. N., Gius, D., Ivanova, J., Pitt, B., Billiar, T. R., and Kagan, V. E. 2005. Thioredoxin and lipoic acid catalyze the denitrosation of low molecular weight and protein S-nitrosothiols. *Journal of the American Chemical Society* 127(45): 15815–15823.

Sun, N., Hao, J. R., Li, X. Y., Yin, X. H., Zong, Y. Y., Zhang, G. Y., and Gao, C. 2013. GluR6-FasL-Trx2 mediates denitrosylation and activation of procaspase-3 in cerebral ischemia/reperfusion in rats. *Cell Death and Disease* 4(8): e771–e771.

Talwar, P. S., Gupta, R., Maurya, A. K., and Deswal, R. 2012.*Brassica juncea* nitric oxide synthase like activity is stimulated by PKC activators and calcium suggesting modulation by PKC-like kinase. *Plant Physiology and Biochemistry* 60: 157–64.

Tan, B. C., Lim, Y. S., and Lau, S. E. 2017. Proteomics in commercial crops: an overview. *Journal of Proteomics* 169: 176–188.

Tanou, G., Fotopoulos, V., and Molassiotis, A. 2012. Priming against environmental challenges and proteomics in plants: update and agricultural perspectives. *Frontiers in Plant Science* 3: 216.

Tanou, G., Ziogas, V., Belghazi, M., Christou, A., Filippou, P., Job, D., Fotopoulos, V., and Molassiotis, A. 2014. Polyamines reprogram oxidative and nitrosative status and the proteome of citrus plants exposed to salinity stress. *Plant, Cell and Environment* 37(4): 864–885.

Tanou, G., Job, C., Rajjou, L., Arc, E., Belghazi, M., Diamantidis, G., Molassiotis, A., and Job, D. 2009. Proteomics reveals the overlapping roles of hydrogen peroxide and nitric oxide in the acclimation of citrus plants to salinity. *Plant Journal* 60: 795–804.

Tichá, T., Sedlářová, M., Činčalová, L., Trojanová, Z. D., Mieslerová, B., Lebeda, A., Luhová, L., and Petřivalský, M. 2018. Involvement of S-nitrosothiols modulation by S-nitrosoglutathione reductase in defence responses of lettuce and wild *Lactuca spp.* to biotrophic mildews. *Planta* 247(5): 1203–1215.

Trujillo, M., Alvarez, M. N., Peluffo, G., Freeman, B. A., and Radi, R. 1998. Xanthine oxidase-mediated decomposition of s-nitrosothiols. *Journal of Biology and Chemistry* 273: 7828–7834.

Valderrama, R., Corpas, F. J., Carreras, A., Fernández-Ocaña, A., Chaki, M., Luque, F., Gómez-Rodríguez, M.V., Colmenero-Varea, P., Luis, A., and Barroso, J. B. 2007. Nitrosative stress in plants. *FEBS Letters* 581(3): 453–461.

Van Emon, J. M. 2016. The omics revolution in agricultural research. *Journal of Agricultural and Food Chemistry* 64(1): 36–44.

Wang, D., Liu, Y., Tan, X., Liu, H., Zeng, G., Hu, X., Jian, H., and Gu, Y. 2015. Effect of exogenous nitric oxide on antioxidative system and S-nitrosylation in leaves *of Boehmeria nivea* (L.) Gaud under cadmium stress. *Environmental Science and Pollution Research* 22(5): 3489–3497.

Watson, W. H., Yang, X., Choi, Y. E., Jones, D. P., and Kehrer, J. P. 2004. Thioredoxin and its role in toxicology. *Toxicological Sciences* 78: 3–14.

Wu, C., Parrott, A. M., Fu, C., Liu, T., Marino, S. M., Gladyshev, W. N., Jain, M. R., Baykal, A. T., Li, Q., Oka, S., Sadoshima, J., Beuve, A., Simmons, W. J., and Li, H. 2011. Thioredoxin 1-mediated post-translational modifications: reduction, transnitrosylation, denitrosylation and related proteomics methodologies. *Antioxidants and Redox Signaling* 15(9): 2565–2604.

Wu, F., Li, Q., Yan, H., Zhang, D., Jiang, G., Jiang, Y., and Duan, X. 2016. Characteristics of three thioredoxin genes and their role in chilling tolerance of harvested banana fruit. *International Journal of Molecular Sciences* 17(9): 1526.

Wulff, R. P., Lundqvist, J., Rutsdottir, G., Hansson, A., Stenbaek, A., Elmlund, D., Elmlund, H., Jensen, P. E., and Hansson, M. 2011. The activity of barley NADPH-dependent thioredoxin reductase C is independent of the oligomeric state of the protein: tetrameric structure determined by cryo-electron microscopy. *Biochemistry* 50(18): 3713–3723.

Xu, S., Guerra, D., Lee, U., and Vierling, E. 2013. S-nitrosoglutathione reductases are low-copy number, cysteine-rich proteins in plants that control multiple developmental and defense responses in Arabidopsis. *Frontiers in Plant Science* 4(430): 1–13.

Yang, L., Tian, D., Todd, C. D., Luo, Y., and Hu, X. 2013. Comparative proteome analyses reveal that nitric oxide is an important signal molecule in the response of rice to aluminum toxicity. *Journal of Proteome Research* 12(3): 1316–1330.

Zaffagnini, M., De Mia, M., Morisse, S., Di Giacinto, N., Marchand, C. H., Maes, A., Lemaire, D., and Trost, P. 2016. Protein S-nitrosylation in photosynthetic organisms: a comprehensive overview with future perspectives. *Biochimica et Biophysica Acta (BBA)-Proteins and Proteomics* 1864(8): 952–966.

Zaffagnini, M., Morisse, S., Bedhomme, M., Marchand, C. H., Festa, M., Rouhier, N., Lemaire, S. D., and Trost, P. 2013. Mechanisms of nitrosylation and denitrosylation of cytoplasmic glyceraldehyde-3-phosphate dehydrogenase from *Arabidopsis thaliana*. *Journal of Biological Chemistry* 288: 22777–22789.

Zhan, N., Wang, C., Chen, L., Yang, H., Feng, J., Gong, X., Ren, B., Wu, R., Mu, J., Li, Y., and Zuo, J. 2018. S-nitrosylation targets GSNO reductase for selective autophagy during hypoxia responses in plants. *Molecular Cell* 71(1): 142–154.

Zhang, J., and Liao, W. 2020. Protein S-nitrosylation in plant abiotic stresses. *Functional Plant Biology* 47(1): 1–10.

Zhou, S., Jia, L., Chu, H., Wu, D., Peng, X., Liu, X., Zhang, J., Zhao, J., Chen, K., and Zhao, L. 2016. Arabidopsis CaM1 and CaM4 promote nitric oxide production and salt resistance by inhibiting S-nitrosoglutathione reductase via direct binding. *PLoS Genetics* 12(9): e1006255.

Zhu, J. K. 2016. Abiotic stress signaling and responses in plants. *Cell* 167: 313–324.

Ziogas, V., Tanou, G., Filippou, P., Diamantidis, G., Vasilakakis, M., Fotopoulos, V., and Molassiotis, A. 2013. Nitrosative responses in citrus plants exposed to six abiotic stress conditions. *Plant Physiology and Biochemistry* 68: 118–126.

Zuccarelli, R., Coelho, A. C., Peres, L. E., and Freschi, L. 2017. Shedding light on NO homeostasis: light as a key regulator of glutathione and nitric oxide metabolisms during seedling deetiolation. *Nitric Oxide* 68: 77–90.

5 Calcium Signaling Is a Hub of the Signaling Network in Response and Adaptation of Plants to Heat Stress

Zhong-Guang Li[1,2,3]
[1]School of Life Sciences, Yunnan Normal University, Kunming, P.R. China
[2]Engineering Research Center of Sustainable Development and Utilization of Biomass Energy, Ministry of Education, Kunming, P.R. China
[3]Key Laboratory of Biomass Energy and Environmental Biotechnology, Yunnan Province, Yunnan Normal University, Kunming, P.R. China
E-mail: zhongguang_li@163.com

CONTENTS

5.1 INTRODUCTION

Calcium is an essential macroelement for plants, animals, and microbes, which is involved in the functioning of different life forms on our planet. Calcium has multiple functions: it is involved in cellular structure, is a potential extracellular cofactor and an intracellular regulator, and so on (Dodd et al. 2010; Ohama et al. 2017; Campbell 2018; Demidchik et al. 2018; Feijó and Wudick 2018; Poovaiah and Du 2018). In addition, calcium as a signaling molecule has attracted much attention in plant biology (Khan et al. 2015). Calcium commonly exerts its role in the form of calcium ion (Ca^{2+}). Calcium signaling centers on Ca^{2+} and its receptor calmodulin (CaM) in plants, which take part in the entire life cycle of plants, from seed germination to plant senescence, and even in the adaptation of plants to stressful environments, including abiotic and biotic stress (Dodd et al. 2010; Ohama et al. 2017; Campbell 2018; Demidchik et al. 2018; Kudla et al. 2018; Poovaiah and Du 2018). Ca^{2+} at a high level (above 0.1 mM) binds easily to inorganic phosphate (Pi) and produces insoluble calcium phosphate ($Ca_3(PO_4)_2$), which disturbs energy metabolism and leads to damage to membranes and organelles, ultimately exhibiting a toxic effect (Dodd et al. 2010; Ohama et al. 2017; Campbell

DOI: 10.1201/9781003159636-5

2018; Demidchik et al. 2018; Liu et al. 2018; Poovaiah and Du 2018). Therefore, cytosolic Ca^{2+} homeostasis is necessary for plant cells.

Calcium signaling, a rapid increase in cytosolic Ca^{2+} concentration, can be triggered by different types of environmental stresses activating extracellular and/or intracellular Ca^{2+} stores, which is known as Ca^{2+} signature (also known as Ca^{2+} pattern or Ca^{2+} shaping). A given environmental stress can induce a specific Ca^{2+} signature, such as Ca^{2+} transients, plateaus, oscillations, waves, tides, puffs, spikes, and sparks (Bouche et al. 2005; Dodd et al. 2010; Ohama et al. 2017; Campbell 2018; Costa et al. 2018; Demidchik et al. 2018; Poovaiah and Du 2018). These signatures can be decoded by different Ca^{2+} sensors (e.g. Ca^{2+} spikes are decoded by catalase), which include calmodulins (CaMs), calmodulin-like proteins (CMLs), calcium-dependent protein kinases (CDPKs), calcium/calmodulin-dependent protein kinases (CCaMKs), calcineurin B–like proteins (CBLs), CBL-interacting protein kinases (CIPKs), calcium-dependent transcription factors (CDTFs, such as CaM-binding transcription factors (CAMTs)), GT-element-binding-like proteins (GTL, MYBs, WRKYs, and NACs), antioxidant enzymes (superoxide dismutase (SOD), catalase (CAT), guaiacol peroxidase (GPX), ascorbate peroxidase (APX), glutathione reductase (GR)), and the methylglyoxal detoxification system (glyoxalase I (Gly I), glyoxalase II (Gly II)) (Figure 5.1) (Xu and Huystee 1993; Yang and Poovaiah 2003; Bouche et al. 2005; Mura et al. 2005; Dodd et al. 2010; Virdi et al. 2015; Ohama et al. 2017; Campbell 2018; Demidchik et al. 2018; Poovaiah and Du 2018). Ca^{2+} signaling is now a young, blooming field of plant biology research (Kudla et al. 2018).

Plants, as sessile organisms, are simultaneously or sequentially subjected to abiotic (heat, low temperature, salt, drought, flooding, and heavy metal stress) and biotic (bacteria, fungi, virus, herbivore, and parasite) stresses (Neill et al. 2002; Dodd et al. 2010; Gill et al. 2010; Li 2016; Li et al. 2016; Zhu 2016; da-Silva and Modolo 2018; Sami et al. 2018). Among all these, heat stress is one of the major environmental stresses that affect cellular metabolism, seed germination, plant growth,

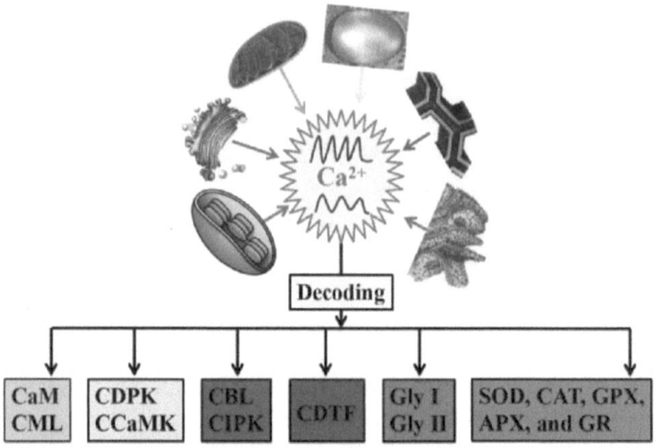

FIGURE 5.1 Triggering and decoding of calcium signaling in plant cells. Calcium signaling can be triggered by different environment stresses or clues (including heat stress) by activating Ca^{2+} channels located on plasma membrane, tonoplast, mitochondrial membrane, chloroplast membrane, and endoplasmic reticulum. And then, calcium signaling can be decoded by calmodulins (CaMs), calmodulin-like proteins (CMLs), calcium-dependent protein kinases (CDPKs), calcium/calmodulin-dependent protein kinases (CCaMKs), calcineurin B-like proteins (CBLs), CBL-interacting protein kinases (CIPKs), calcium-dependent transcription factors (CDTFs), superoxide dismutase (SOD), catalase (CAT), guaiacol peroxidase (GPX), ascorbate peroxidase (APX), glutathione reductase (GR), glyoxalase I (Gly I), and glyoxalase II (Gly II), thus responding and adapting to the changing environment.

development, geographical distribution, and productivity (Poór et al. 2021). Heat stress usually results in direct damage (loss of membrane integrity and protein denaturation) and indirect damage (oxidative and osmotic stress) to the molecular, cellular, physiological, and biochemical hierarchy (Wahid et al. 2007; Dodd et al. 2010; Asthir 2015; Hemmati et al. 2015). In addition, heat stress can trigger a series of signaling cascades and forms a complex signaling network consisting of Ca^{2+}, reactive oxygen species (ROS, mainly H_2O_2), reactive nitrogen species (nitric oxide (NO)), reactive sulfur species (hydrogen sulfide (H_2S)), and reactive carbonyl species (methylglyoxal (MG)) (Neill et al. 2002; Dodd et al. 2010; Gill et al. 2010; Li 2016; Li et al. 2016; Zhu 2016; da-Silva and Modolo 2018; Sami et al. 2018). In this signaling network, calcium signaling (centering on Ca^{2+} and CaM) plays a central role in the response and adaptation of plants to abiotic stress, including heat tolerance. In this chapter, we discuss the crosstalk between calcium signaling and other signaling molecules such as H_2O_2, NO, H_2S, MG, and phytohormones for regulating responses and adaptations of plants to heat stress. The objective of this chapter is to further understand the Ca^{2+}-mediated signaling networks and their crosstalks for elucidating possible mechanisms in the acquisition of heat tolerance in plants.

5.2 INTERACTION OF CALCIUM SIGNALING WITH OTHER SIGNALING MOLECULES IN THE RESPONSE AND ADAPTATION TO HEAT STRESS

As mentioned earlier, the formation of heat tolerance is involved in the interaction of signaling molecules such H_2O_2, NO, H_2S, MG, and plant hormones in plants (Figure 5.2) (Zhu 2016; Campbell 2018; Demidchik et al. 2018; Mhamdi and Breusegem 2018; Mohanta et al. 2018; Sami et al. 2018; Sajid et al. 2018). The detailed signaling crosstalk mechanisms are discussed in the following subsections.

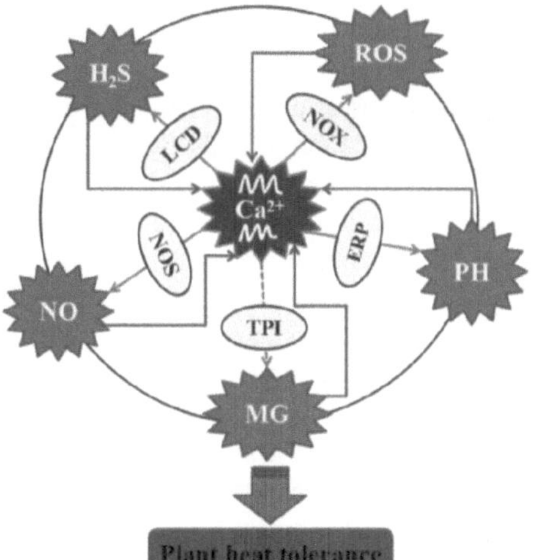

FIGURE 5.2 Calcium signaling-centered signaling crosstalk in the acquisition of heat tolerance in plants. Calcium can separately activate NADPH oxidase (NOX), nitric oxide synthase (NOS), L-cysteine desulfhydrase (LCD), triosephosphate isomerase (TPI), and enzymes related to plant hormone synthesis (ERP), which in turn induce reactive oxygen species (ROS), nitric oxide (NO), hydrogen sulfide (H_2S), methylglyoxal (MG), and plant hormones (PH, such as abscisic acid (ABA), salicylic acid (SA), and ethylene (ETH)) signaling and their interactions. Conversely, ROS, NO, H_2S, MG, and PH also can activate Ca^{2+} channels, followed by triggering calcium signaling in plants. The solid arrows indicate identified pathways; the imaginary arrows represent supposed pathways.

5.2.1 Interplay of Calcium Signaling and H_2O_2 Signaling

In plant cells, there are many Ca^{2+} stores, including intracellular and extracellular Ca^{2+} pools. The intracellular Ca^{2+} pools in the plant cells include vacuole, endoplasmic reticulum, Golgi apparatus, mitochondrion, and chloroplast. The extracellular Ca^{2+} pool refers to cell wall (Figure 5.1). The concentrations of Ca^{2+} are different in the intracellular and extracellular Ca^{2+} pools. The Ca^{2+} concentrations in the intracellular and extracellular Ca^{2+} pools (~1–10 mM Ca^{2+}) are 10000 times greater than that of cytosol (~100 μM Ca^{2+}) (Dodd et al. 2010; Campbell 2018). In general, Ca^{2+} concentration in the cytosol can maintain homeostasis or fluctuation in a certain range in plant cells by the coordinated role of Ca^{2+} channels and Ca^{2+} pumps. Ca^{2+} signaling, that is, the change in Ca^{2+} concentrations with different amplification, duration, and density, can be triggered by different environmental stresses, including heat stress, which in turn induces heat tolerance in various plant species (Dodd et al. 2010; Campbell 2018; Costa et al. 2018; Konrad et al. 2018). In transgenic tobacco seedlings transformed with the Ca^{2+}-sensitive, luminescent protein aquaporin, heat shock (at 39, 43, and 47 °C) triggered a rapid increase in cytoplasmic Ca^{2+} concentration within several minutes, which in turn induced thermo-tolerance (Gong et al. 1998a). These effects were enhanced by exogenous Ca^{2+} and diminished by ethylene glycol-bis(b-aminoethylether)-N,N,N′,N′-tetraacetic acid (EGTA, Ca^{2+} chelator), $LaCl_3$ (plasma membrane Ca^{2+}-channel blocker), ruthenium red (intracellular Ca^{2+}-channel blocker), and neomycin solution (phospholipase C inhibitor) (Gong et al. 1998a). These results showed that heat shock-induced cytoplasmic Ca^{2+} signaling and resulting heat tolerance are a result of the combination of extracellular and intracellular Ca^{2+} stores in plants. In maize seedlings, heat shock could generate an endogenous H_2O_2 peak at 3 hours, which in turn induced adaptation to heat, chilling, drought, and salt stress (Gong et al. 2001). These indicate that heat shock can induce Ca^{2+} and H_2O_2 signaling, and the latter (several hours) lags behind the former (several minutes). In addition, irrigation with H_2O_2 also could enhance multi-resistance to heat, chilling, drought, and salt stress (Gong et al. 2001; Ashfaque et al. 2014; Khan et al. 2016). These results suggest that H_2O_2 plays an important role in the downstream processing of Ca^{2+} signaling in plant stress tolerance (Khan et al. 2019).

In *Zea mays* (maize) seedlings, both abscisic acid (ABA) and H_2O_2 increased the concentration of cytosolic Ca^{2+} in the protoplasts of mesophyll cells, which in turn induced the gene expression of CaM and the content of CaM in leaves of maize plants (Hu et al. 2007). In addition, both treatments also activated the gene expression of antioxidant enzymes (SOD, APX, and GR) and the corresponding enzyme activity. The action of antioxidant enzymes was completely inhibited by two CaM antagonists (trifluoperazine (TFP), *N*-(6-aminohexyl)-1-naphthalene sulfonamide hydrochloride (W5), and *N*-(6-aminohexyl)-5-chloro-1-naphthalene sulfonamide hydrochloride (W7)) (Hu et al. 2007). Also, CaM antagonists blocked ABA-induced H_2O_2 generation, while an inhibitor (diphenyleneiodonium chloride (DPI)) and a scavenger (dimethylthiourea (DMTU)) of ROS had no effect on the early increase in the content of CaM induced by ABA (Hu et al. 2007). These results showed that Ca^{2+}/CaM took part in the activation of the antioxidant defense system induced by ABA through the interaction with H_2O_2 in maize seedlings. Similarly, ABA up-regulated the gene expression and activity of maize Ca^{2+}/CaM-dependent protein kinase (ZmCCaMK) located in the nucleus, cytoplasm, and plasma membrane, which in turn activated the antioxidant defense system (Ma et al. 2012). In addition, NO also induced gene expression and activity of ZmCCaMK, which was blocked by the NO scavenger 2-4-carboxyphenyl-4,4,5,5- tetramethylimidazoline-1-oxyl-3-oxide (cPTIO), while RNAi silencing of ZmCCaMK had no significant effect on ABA-induced NO generation, indicating that NO was acquired for ABA-induced ZmCCaMK (Ma et al. 2012). Moreover, H_2O_2 increased the transcription and activity of ZmCCaMK, and this increase was inhibited by an NO scavenger (cPTIO) and inhibitor (N^G-nitro-l-Arg methyl ester (l-NAME)), suggesting that H_2O_2 activated ZmCCaMK in the upstream of NO in ABA-induced ZmCCaMK (Ma et al. 2012).

Therefore, ABA could induce ZmCCaMK and the subsequent antioxidant defense system by the crosstalk between NO and H_2O_2 in maize seedlings, and this crosstalk was closely associated with the acquisition of heat tolerance (Gong et al. 1998a, 2001).

Also, Ca^{2+} channels can be activated by H_2O_2, which in turn increases Ca^{2+} levels in cytosol (Price et al. 1994; Pei et al. 2000). In parallel, NADPH oxidase located on the plasma membrane is a key enzyme in H_2O_2 synthesis in plants, which can also be activated by Ca^{2+}, and this activation needs a continuous Ca^{2+} supply (Keller et al. 1998). These studies show a feedback mechanism between calcium and ROS signaling during heat stress in plants.

5.2.2 Interplay of Calcium Signaling and NO Signaling

In higher plant cells, which have many Ca^{2+} pools, NO also can be synthesized via several biosynthetic pathways. The interplay between calcium signaling and NO signaling can be found in many environmental stress responses and adaptations, including heat stress, in plants. The calcium channels can be opened by NO at a certain concentration. In parallel, the key enzymes in NO biosynthesis, especially NO synthetase or NO synthetase-like, can be activated by Ca^{2+} and its receptor (CaM). Therefore, both Ca^{2+} signaling and NO signaling can be found to closely and mutually regulate cellular metabolism, seed germination, plant growth and development, and the response and tolerance to environmental stresses, including heat stress. Similarly to calcium signaling, heat stress also can trigger NO signaling in *Medicago sativa* (alfalfa) (Leshem 2000) and *Symbiodinium microadriaticum* (Bouchard and Yamasaki 2008, 2009). In callus of *Phragmites communis* (reed), sodium nitroprusside and S-nitroso-N-acetylpenicillamine (SNAP) (NO donors) alleviated the increase in ion leakage, H_2O_2, and malondialdehyde (MDA) content, which in turn reduced the decrease in growth suppression and cell viability under heat stress at 45 °C (Song et al. 2006). In addition, NO donors increased the activities of antioxidant enzymes (SOD, CAT, APX, and GPX) in calluses under heat stress, while an NO scavenger (cPTIO) blocked the NO donor-mediated protective effects (Song et al. 2006). This indicates that NO as a signaling molecule could confer thermo-tolerance by stimulating ROS-scavenging enzymes to effectively overcome oxidative stress induced by heat stress. In both wild-type *Arabidopsis thaliana* and mutant-type lacking a functional gene encoding GLB3 (ΔAtGLB3), a homologue of bacterial truncated hemoglobin (trHb), the seeds of the wild-type and ΔAtGLB3 can normally and similarly germinate under optimal temperature conditions at 22 °C. Under high temperature at 32 °C, the ΔAtGLB3 seeds did not germinate, while wild-type seeds maintained a high germination rate. The former were partially restored by the supplementation of NO scavengers (cPTIO, 3-(3,4-dihydroxycinnamoyl) quinic acid, bovine serum Hb, or isoprene) (Hossain et al. 2010). The results suggest that the inhibition of seed germination by high temperature might be closely associated with the excessive accumulation of NO induced by high temperature in *A. thaliana*, further supporting the signaling role of NO in plant response to high-temperature stress.

In fava bean (*Vicia faba* L.), heat stress led to many injury symptoms, as reflected in the increase in the levels of ROS (H_2O_2, $O_2^{\bullet -}$), NO, proline (Pro), MDA, electrolyte leakage, DNA damage, and chlorophyll (Chl) degradation, and the decrease in leaf relative water content (RWC) and Chl content. In addition, the production of NO and Pro, the activities of Pro metabolizing enzymes (Δ^1-pyrroline-5-carboxylate synthetase (P5CS), ornithine-δ-aminotransferase (OAT), and proline dehydrogenase (ProDH) and the transcript levels of heat shock proteins (Hsp17.6, Hsp70, Hsp90-1, and Hsp101) induced by heat stress were enhanced by increasing temperature (32, 37, and 42 °C) (Alamri et al. 2018). Afterward, induced biosynthesis of NO and Pro under heat stress further activated the antioxidant defense system, as shown in a high activity of antioxidant enzymes (SOD, CAT, GR, and APX) and a high content of non-enzymatic antioxidants (reduced glutathione (GSH), ascorbic acid (AsA), and oxidized glutathione (GSSG)). However, the increase in GSSG was lower than that of GSH under heat stress. On the contrary, the application of cPTIO reduced the accumulation of leaf

RWC, Pro, and NO (Alamri et al. 2018). Thus, NO might act as both signaling molecule and promoter of Pro biosynthesis to induce the acquisition of heat tolerance in plants.

In addition, not only can NO directly stimulate the activity of target proteins, Ca^{2+} channels located on the plasma membrane, through nitrosylation (-SNO); it also indirectly activates Ca^{2+} channels located in the intracellular Ca^{2+} stores via the mediation of cyclic ADP-ribose (cADPR), cyclic GMP (cGMP), and phosphorylation-dependent processes, which in turn increase the content of Ca^{2+} in cytosol, indicating that NO acts as a Ca^{2+}-mobilizing intracellular messenger (Courtois et al. 2008).

5.2.3 Interplay of Calcium Signaling and H_2S Signaling

In higher plants, H_2S can be biosynthesized at least by five biosynthetic pathways. Among these pathways, L-cysteine desulfhydrase (LCD, EC 4.4.1.1), D-cysteine desulfhydrase (DCD, EC 4.4.1.15), sulfite reductase (SiR, EC 1.8.7.1), cyanoalanine synthase (CAS, EC 4.4.1.9), and cysteine synthase (CS, EC 4.2.99.8) are the key enzymes in H_2S synthesis in plants (Li 2013; Li et al. 2016). Under environmental stress, including heat stress, H_2S signaling can be triggered in plants by mainly activating LCD and then induces the acquisition of abiotic stress tolerance, including heat tolerance, which can be induced by exogenously applied NaHS (an H_2S donor) (Zhou et al. 2018). In tobacco seedlings, high temperature at 35 °C activated the activity of LCD and then induced the accumulation of endogenous H_2S, maintaining a high level for 3 days (Chen et al. 2016). Similarly, in *Fragaria ananassa* (strawberry) seedlings, heat stress at 42 °C led to a significant increase in endogenous H_2S content after 1, 4, and 8 h compared with control plants without heat stress treatment (Christou et al. 2014). In addition, a similar increase in H_2S also was recorded in plants pretreated with NaHS at the early phase (1 h) during heat stress exposure and then gradually recovered to the level of the control (Christou et al. 2014).

In tobacco suspension cell cultures, Ca^{2+} and CaM stimulated the activity of LCD, which in turn increased the accumulation of endogenous H_2S. This stimulation and accumulation were weakened by EGTA (Ca^{2+} chelator) and chlorpromazine (CPZ, CaM antagonist) (Li et al. 2015). In addition, the increased LCD activity and endogenous H_2S level by Ca^{2+} and CaM were abolished by H_2S synthesis inhibitors (DL-propargylglycine (PAG), aminooxy acetic acid (AOA), potassium pyruvate (PP), and hydroxylamine (HA)) and its scavenger (hypotaurine (HT)) (Li et al. 2015). These results imply that H_2S homeostasis in plant cells can be closely regulated by Ca^{2+} and CaM by stimulating LCD activity. Interestingly, heat tolerance in tobacco cells could be separately improved by NaHS, Ca^{2+}, and CaM, while improved heat tolerance by Ca^{2+} and CaM was markedly reinforced by the exogenous supplementation of NaHS, but blocked by PAG, AOA, PP, HA, or HT (Li et al. 2015). Thus, H_2S derived from LCD partially regulated the heat tolerance of tobacco suspension cell cultures by activating Ca^{2+} and CaM. Similarly, in *Setaria italica* (foxtail millet), chromium stress triggered H_2S and Ca^{2+} signaling (Fang et al. 2014). In addition, chromium tolerance could be strongly induced by H_2S and Ca^{2+} and alleviated by H_2S synthesis inhibitor and Ca^{2+} chelators, indicating that the H_2S–Ca^{2+} interaction modulated chromium stress by activating the antioxidant system (Fang et al. 2014), further supporting the interaction of calcium signaling and H_2S signaling in plant abiotic stress tolerance.

In wheat (*Triticum aestivum* L.) coleoptile cells, NaHS treatment led to a transient increase in ROS ($O_2^{\bullet-}$ and H_2O_2), which in turn stimulated the activities of antioxidant enzymes (SOD, CAT, GPX), followed by inducing the resistance of coleoptile to heat stress (Kolupaev et al. 2017). In addition, the protective effects of H_2S were blocked by treatment with dimethylthiourea (H_2O_2 scavenger), imidazole (NADPH oxidase inhibitor), EGTA (extracellular calcium chelator), and neomycin (phosphatidylinositol-specific phospholipase C inhibitor) (Kolupaev et al. 2017). These results indicate that H_2S could increase the heat tolerance of plants by activating the ROS generation deriving from NADPH oxidase (NOX) and depending on Ca^{2+} homeostasis to induce antioxidant enzymes.

5.2.4 Interplay of Calcium Signaling and Methylglyoxal Signaling

In general, MG homeostasis in plant cells can be tightly regulated by biosynthesis and degradation. Plant cells can generate MG via enzymatic and non-enzymatic pathways, but it can be scavenged by the glyoxalase system (including Gly I and Gly II) and the non-glyoxalase system (Khan et al. 2020). It has been found that glyoxalase system activity is controlled by calcium signaling, especially Ca^{2+} and CaM, in many plant species. In *Brassica juncea*, Gly I, also known as *S*-lactoylglutathione-lyase (EC 4.4.1.5), a heterodimeric protein with 56 kDa, can be activated by Ca^{2+}/CaM, exhibiting a 2.6-fold increase in its activity (Deswal and Sopory 1999), indicating that Gly I may be a Ca^{2+}/CaM-binding protein. Afterwards, this hypothesis was further confirmed by electrophoresis, Chelex-100 assay, and gel overlays using $^{45}CaCl_2$. Conversely, Gly I activated by Ca^{2+}/CaM was inhibited by CaM antibodies and a CaM inhibitor (TFP) (Deswal and Sopory 1999). In addition, a CaM-Sepharose column bound Gly I, which was eluted by EGTA, and the eluted protein fractions were also activated by CaM (Deswal and Sopory 1999). These results confirm that Gly I can be activated by Ca^{2+}/CaM, thus regulating MG homeostasis in plant cells.

In rice (*Oryza sativa* L. cv. BRRI dhan29) seedlings, cadmium stress resulted in higher cadmium accumulation, which in turn induced oxidative stress, as reflected in the overproduction of ROS and MG by inhibiting the antioxidant defense system and the glyoxalase system. Thus, cadmium stress led to chlorosis, leaf roll, and growth inhibition (Rahman et al. 2016). In addition, exogenous supplementation of Ca^{2+} reduced cadmium uptake and significantly increased the activities of antioxidant enzymes (SOD, CAT, glutathione *S*-transferase (GST), monodehydroascorbate reductase (MDHAR), and dehydroascorbate reductase (DHAR)) and the MG-detoxification system (Gly I and Gly II) as well as the content of non-enzymatic antioxidant (AsA) (Rahman et al. 2016). Therefore, Ca^{2+} application reduced the toxicity of cadmium by regulating the antioxidant defense and glyoxalase systems to reverse overproduced ROS and MG, further supporting the interaction of calcium signaling and MG signaling in plant abiotic stress response and adaptation (Rahman et al. 2016).

5.2.5 Interplay of Calcium Signaling and Phytohormone Signaling

Plant hormones such as auxin (AUX), gibberellin (GA), cytokinin (CTK), ABA, and ethylene (ETH) play a very important role in plant growth, development, and response and adaptation to environmental stress by their interactions or crosstalk with other signaling molecules, including calcium signaling (Peleg and Blumwald 2011; Wani et al. 2016; Zhu 2016; Ciura and Kruk 2018; Khan et al. 2021a). More recently, the protective role of plant hormones has attracted more attention in the response and adaptation of plants to environmental stress, including heat stress (Peleg and Blumwald 2011; Khan et al. 2013; Wani et al. 2016; Zhu 2016; Ciura and Kruk 2018).

ABA is a stress hormone that has long been studied in plant response and adaptation to environmental stress, including heat stress. In maize seedlings, ABA treatment remarkably enhanced heat tolerance; this enhancement was reinforced by the addition of $CaCl_2$ solution by raising Ca^{2+} content, while it was weakened by EGTA (the Ca^{2+} chelator) by lowering the Ca^{2+} content of maize seedlings (Gong et al. 1998b). In addition, pretreatment of maize seedlings with La^{3+} and verapamil (the plasma membrane Ca^{2+} channel blockers) blocked the heat tolerance induced by ABA, indicating that the ABA-mediated heat tolerance needs the entry of extracellular Ca^{2+} into cells (Gong et al. 1998b). However, W7 and chlorpromazine (CPZ) (the calmodulin antagonists) had no significant effect on the ABA-induced heat tolerance in maize seedlings (Gong et al. 1998b). To explore the mechanisms of ABA–Ca^{2+} interaction-induced heat tolerance, the activities of antioxidant enzymes (SOD, CAT, APX, and GPX) and the extent of lipid peroxidation (MDA) were measured in maize seedlings under heat stress. The results showed that the ABA-treated maize seedlings maintained a higher activity of antioxidant enzymes and a lower level of MDA under both normal culture temperature and heat stress (Gong et al. 1998b). In addition, the ABA-activated SOD and APX were enhanced by the exogenous supplement of Ca^{2+}, which in turn reduced the accumulation of MDA induced by heat stress in the maize seedlings treated with ABA, while this was reversed by EGTA (Gong et al. 1998b). These data show that ABA can induce the heat tolerance of maize seedlings by activating the Ca^{2+} channels located on the plasma membrane and triggering calcium signaling, and the antioxidant enzyme system is involved in the acquisition of ABA-induced heat tolerance (Gong et al. 1998b).

In *Arabidopsis*, heat stress resulted in an increase in MDA level (an indicator of oxidative damage to membranes) and a decrease in survival percentage, but the exposure of plants to a mild heat stress resulted in the acquisition of heat tolerance (Larkindale and Knight 2002). These negative effects were worsened by Ca^{2+} channel blockers (La^{3+}, nifedipine, and verapamil,) and CaM inhibitors (W7 and TFP), while they were reversed by exogenous Ca^{2+}, suggesting that Ca^{2+}/CaM plays a protective role in the heat-induced oxidative damage in *Arabidopsis* (Larkindale and Knight 2002). Similarly, pretreatment with salicylic acid (SA), 1-aminocyclopropane-1-carboxylic acid (an ETH precursor), and ABA also could protect plants from oxidative damage induced by heat stress (Larkindale and Knight 2002). Also, the ethylene-insensitive mutant etr-1, the ABA-insensitive mutant abi-1, and a transgenic line expressing *nahG* (inhibiting SA synthesis) exhibited an increase in the susceptibility to heat stress, suggesting that ETH, ABA, and SA are involved in the protection against oxidative damage induced by heat stress in *Arabidopsis* (Larkindale and Knight 2002). In addition, during heat stress, a significant increase in cytosolic Ca^{2+} levels was not detected in *Arabidopsis*, but during recovery from heat stress, Ca^{2+} showed transient elevations, and the Ca^{2+} peaks were greater in heat-tolerant plants. Furthermore, during the recovery phase, treatment with the aforementioned Ca^{2+} channel blockers and CaM inhibitors prevents heat tolerance in *Arabidopsis* (Larkindale and Knight 2002), indicating that Ca^{2+} signaling plays the role of a hub in the acquisition of plant heat tolerance triggered by plant hormones.

5.3 CONCLUSION AND PERSPECTIVE

As stated earlier, the acquisition of plant heat tolerance is implicated in a complicated signaling network from heat perception to alteration in gene expression, which in turn triggers change in the metabolic, physiological, and biochemical parameters, ultimately exhibiting a response (damage or adaptation) to heat stress. The signaling network is involved in a variety of interactions between signaling molecules, such as Ca^{2+}, ROS, NO, H_2S, and MG, and plant hormones (ABA, SA, and ETH) (Wahid et al. 2007; Saidi et al. 2011; Hasanuzzaman et al. 2013; Asthir 2015; Hemmati et al. 2015; Ohama et al. 2017; Mohanta et al. 2018; Khan et al. 2013; 2021b). For example, in creeping bentgrass (*Agrostis stolonifera* var. *palustris*), foliar application of H_2O_2, Ca^{2+}, ABA, SA, and ETH increased heat tolerance by activating the antioxidant system (APX, GPX, CAT, and SOD) to reduce oxidative damage (Larkindale and Huang 2004). Similarly, in wheat, treatment with Ca^{2+}, H_2O_2, ABA, and SA increased the activities of antioxidant enzymes (SOD, APX, and CAT) and NADPH oxidase (NOX), which were decreased by EGTA (a calcium chelator) and diphenylene iodonium chloride (DPI, a specific inhibitor of NOX), followed by inducing plant abiotic stress tolerance, indicating that crosstalk among abiotic stress signaling molecules (Ca^{2+}, H_2O_2, ABA, and SA) stimulated transcription factors related to antioxidant enzymes (Agarwal et al. 2005). In addition, crosstalk among Ca^{2+}, H_2O_2, and NO in *Ulva compressa* was observed (Gonzalez et al. 2012).

Among the metabolic, physiological, and biochemical changes induced by signaling crosstalk (Ca^{2+}, H_2O_2, NO, H_2S, MG, ABA, SA, and ETH) triggered by heat stress, the activation of the antioxidant system (including antioxidant enzymes and non-enzymatic antioxidants) plays a crucial role in the acquisition of heat tolerance in plants (Wahid et al. 2007; Hasanuzzaman et al. 2013; Asthir 2015; Hemmati et al. 2015; Ohama et al. 2017). However, heat tolerance acquisition in plants is a complex molecular, cellular, physiological, and biochemical event, which is involved in the repair and rebuilding of biomembranes (including plasma membrane and organelle membrane), synthesis of HSP (including sHSP, HSP60, HSP70, HSP90, and HSP100) and osmolytes (Pro, GB, Tre, and soluble sugar), enhancement of the ROS-scavenging system (SOD, CAT, APX, GR, GPX, AsA, and GSH) and the MG-detoxification system (Gly I, Gly II, and GSH), and so on (Wahid et al. 2007; Hasanuzzaman et al. 2013; Qu et al. 2013; Asthir 2015; Hemmati et al. 2015; Ohama et al. 2017). Therefore, with the finding of novel signaling molecules, the heat stress-induced signaling network requires to be further updated and supplemented in plants. In addition, with the development of "omics" (genomics, transcriptomics, proteomics, metabolomics, and phenomics), the detailed mechanisms of plant responses to heat stress by signaling molecules and their interactions also need to be uncovered in the future.

REFERENCES

Agarwal, S., R.K. Sairam, G.C. Srivastava, A. Tyagi, and R.C. Meena. 2005. Role of ABA, salicylic acid, calcium and hydrogen peroxide on antioxidant enzymes induction in wheat seedlings. *Plant Science* 169:559–570.

Alamri, S.A., M.H. Siddiqui, M.Y. Al-Khaishany, M.N. Khan, M.H. Ali, and K.A. Alakeel. 2018. Nitric oxide-mediated cross-talk of proline and heat shock proteins induce thermotolerance in *Vicia faba* L. *Environmental and Experimental Botany* 23: in press.

Ashfaque, F., M.I.R. , and N. Khan. 2014. Exogenously applied H_2O_2 promotes proline accumulation, water relations, photosynthetic efficiency and growth of wheat (*Triticum aestivum L.*) under salt stress. *Annual Research & Review in Biolog* 4(1):105–120.

Asthir, B. 2015. Mechanisms of heat tolerance in crop plants. *Biologica Plantarum* 59:620–628.

Bouche, N., A. Yellin, W.A. Snedden, and H. Fromm. 2005. Plant-specific calmodulin-binding proteins. *Annual Review of Plant Biology* 56:435–66.

Bouchard, J.N., and H. Yamasaki. 2008. Heat stress stimulates nitric oxide production in *Symbiodinium microadriaticum*: a possible linkage between nitric oxide and the coral bleaching phenomenon. *Plant & Cell Physiology* 49:641–652.

Bouchard, J.N., and H. Yamasaki. 2009. Implication of nitric oxide in the heat-stress-induced cell death of the symbiotic alga *Symbiodinium microadriaticum*. *Marine Biology* 156:2209–2220.

Campbell, A.K. 2018. *Fundamentals of Intracellular Calcium*. Chichester: Wiley.

Chen, X., Q. Chen, X. Zhang, et al. 2016. Hydrogen sulfide mediates nicotine biosynthesis in tobacco (*Nicotiana tabacum*) under high temperature conditions. *Plant Physiology and Biochemistry* 104:174–179.

Christou, A., P. Filippou, G.A. Manganaris, and V. Fotopoulos. 2014. Sodium hydrosulfide induces systemic thermotolerance to strawberry plants through transcriptional regulation of heat shock proteins and aquaporin. *BMC Plant Biology* 14:42.

Ciura, J., and J. Kruk. 2018. Phytohormones as targets for improving plant productivity and stress tolerance. *Journal of Plant Physiology* 229:32–40.

Costa, A., L. Navazio, and I. Szabo. 2018. The contribution of organelles to plant intracellular calcium signalling. *Journal of Experimental Botany* 69:4175–4193.

Courtois, C., A. Besson, J. Dahan, et al. 2008. Nitric oxide signalling in plants: interplays with Ca^{2+} and protein kinases. *Journal of Experimental Botany* 59:155–163.

da-Silva, C.J., and L.V. Modolo. 2018. Hydrogen sulfide: a new endogenous player in an old mechanism of plant tolerance to high salinity. *Acta Botanica Brasilica* 32:150–160.

Deswal, R., and S.K. Sopory. 1999. Glyoxalase I from *Brassica juncea* is a calmodulin stimulated protein. *Biochimica Biophysica Acta-Molecular Cell Research* 1450:460–467.

Demidchik, V., F. Maathuis, and O. Voitsekhovskaja. 2018. Unravelling the plant signalling machinery: an update on the cellular and genetic basis of plant signal transduction. *Functional Plant Biology* 45:1–8.

Dodd, A.N., J. Kudla, and D. Sanders. 2010. The language of calcium signaling. *Annual Review of Plant Biology* 61:593–620.

Fang, H., T. Jing, Z. Liu, L. Zhang, Z. Jin, and Y. Pei. 2014. Hydrogen sulfide interacts with calcium signaling to enhance the chromium tolerance in *Setaria italica*. *Cell Calcium* 56:472–481.

Feijó, J.A., and M.M. Wudick. 2018. 'Calcium is life'. *Journal of Experimental Botany* 69:4147–4150.

Gill, S.S., and N. Tuteja. 2010. Reactive oxygen species and antioxidant machinery in abiotic stress tolerance in crop plants. *Plant Physiology and Biochemistry* 48:909–930.

Gong, M., B. Chen, Z.G. Li, and L.H. Guo. 2001. Heat-shock-induced cross adaptation to heat, chilling, drought and salt stress in maize seedlings and involvement of H_2O_2. *Journal of Plant Physiology* 158:1125–1130.

Gong, M., A.H. van der Luit, M.R. Knight, and A.J. Trewavas. 1998a. Heat-shock-induced changes in intracellular Ca^{2+} level in tobacco seedlings in relation to thermotolerance. *Plant Physiology* 116:429–437.

Gong, M., Y.J. Li, and S.N. Chen. 1998b. Abscisic acid-induced thermotolerance in maize seedlings is mediated by calcium and associated with antioxidant systems. *Journal of Plant Physiology* 153:488–496.

Gonzalez, A., M.D. Cabrera, M.J. Henriquez, R.A. Contreras, B. Morales, and A. Moenne. 2012. Cross talk among calcium, hydrogen peroxide, and nitric oxide and activation of gene expression involving calmodulins and calcium-dependent protein kinases in *Ulva compressa* exposed to copper excess. *Plant Physiology* 158:1451–1462.

Hasanuzzaman, M., K. Nahar, M.M. Alam, R. Roychowdhury, and M. Fujita. 2013. Physiological, biochemical, and molecular mechanisms of heat stress tolerance in plants. *International Journal of Molecular Science* 14:9643–9684.

Hemmati, H., D. Gupta, and C. Basu. 2015. Molecular physiology of heat stress responses in plants. In *Elucidation of Abiotic Stress Signaling in Plants: Functional Genomics Perspectives*, ed. G.K. Pandey, 109–142. New York: Springer.

Hossain, K.K., R.D. Itoh, G. Yoshimura, et al. 2010. Effects of nitric oxide scavengers on thermoinhibition of seed germination in *Arabidopsis thaliana. Russian Journal of Plant Physiology* 57:222–232.

Hu, X., M. Jiang, J. Zhang, A. Zhang, F. Lin, and M. Tan. 2007. Calcium–calmodulin is required for abscisic acid-induced antioxidant defense and functions both upstream and downstream of H_2O_2 production in leaves of maize (*Zea mays*) plants. *New Phytologist* 173:27–38.

Keller, T., H.G. Damude, D. Werner, P. Doerner, R.A. Dixon, and C. Lamb. 1998. A plant homolog of the neutrophil NADPH oxidase gp91phox subunit gene encodes a plasma membrane protein with Ca21 binding motifs. *Plant Cell* 10:255–266.

Khan, M.I.R., N. Iqbal, A. Masood, T.S. Per, and N.A. Khan. 2013. Salicylic acid alleviates adverse effects of heat stress on photosynthesis through changes in proline production and ethylene formation. *Plant Signaling & Behavior*, 8:11.

Khan, M.I.R., P. Chopra, H. Chhillar, M.A. Ahanger, S.J. Hussain, and C. Maheshwari. 2021a. Regulatory hubs and strategies for improving heavy metal tolerance in plants: chemical messengers, omics and genetic engineering. *Plant Physiology and Biochemistry* 164:260–278.

Khan, M.I.R., F. Ashfaque, H. Chhillar, M. Irfan, and N.A. Khan. 2021b. The intricacy of silicon, plant growth regulators and other signaling molecules for abiotic stress tolerance: an entrancing crosstalk between stress alleviators. *Plant Physiology and Biochemistry* 162:36–47.

Khan, M.I.R., M. Fatma, T.S. Per, N.A. Anjum, and N.A. Khan. 2015. Salicylic acid-induced abiotic stress tolerance and underlying mechanisms in plants. *Frontiers in Plant Science* 6:462.

Khan, M.I.R., N.A. Khan, A. Masood, T.S. Per, and M. Asgher. 2016. Hydrogen peroxide alleviates nickel-inhibited photosynthetic responses through increase in use-efficiency of nitrogen and sulfur, and glutathione production in mustard. *Frontiers in Plant Science* 7:44.

Khan, M.I.R., P.S. Reddy, A. Ferrante, and N.A. Khan (Eds). 2019. *Plant Signaling Molecules: Role and Regulation under Stressful Environments*. Woodhead Publishing.

Khan, M.I.R., B. Jahan, M.F. AlAjmi, M.T. Rehman, and N.A. Khan. 2020. Ethephon mitigates nickel stress by modulating antioxidant system, glyoxalase system and proline metabolism in Indian mustard. *Physiology and Molecular Biology of Plants* 26:1201–1213.

Kolupaev, Y.E., E.N. Firsova, T.O. Yastreb, and A.A. Lugovaya. 2017. The participation of calcium ions and reactive oxygen species in the induction of antioxidant enzymes and heat resistance in plant cells by hydrogen sulfide donor. *Applied Biochemistry and Microbiology* 53:573–579.

Konrad, K.R, T. Maierhofer, and R. Hedrich. 2018. Spatio-temporal aspects of Ca^{2+} signalling: lessons from guard cells and pollen tubes. *Journal of Experimental Botany* 69:4195–4214.

Kudla, J., D. Becker, E. Grill, et al. 2018. Advances and current challenges in calcium signaling. *New Phytologist* 218:414–431.

Larkindale, J., and B. Huang. 2004. Thermotolerance and antioxidant systems in *Agrostis stolonifera*: involvement of salicylic acid, abscisic acid, calcium, hydrogen peroxide, and ethylene. *Journal of Plant Physiology* 161:405–413.

Larkindale, J., and M.R. Knight. 2002. Protection against heat stress-induced oxidative damage in Arabidopsis involves calcium, abscisic acid, ethylene, and salicylic acid. *Plant Physiology* 128:682–695.

Leshem, Y.Y. 2000. *Nitric Oxide in Plants: Occurrence, Function and Use*. Boston: Kluwer Academic Publishers.

Li, Z.G. 2013. Hydrogen sulfide: a multifunctional gaseous molecule in plants. *Russian Journal of Plant Physiology* 60:733–740.

Li, Z.G. 2016. Methylglyoxal and glyoxalase system in plants: old players, new concepts. *Botanical Review* 82:183–203.

Li, Z.G., W.B. Long, S.Z. Yang, et al. 2015. Endogenous hydrogen sulfide regulated by calcium is involved in thermotolerance in tobacco *Nicotiana tabacum* L. suspension cell cultures. *Acta Physiologiae Plantarum* 37:219.

Li, Z.G., X. Min, and Z.H. Zhou. 2016. Hydrogen sulfide: a signal molecule in plant cross-adaptation. *Frontiers in Plant Science* 7:1621.

Liu, J., Y. Niu, J. Zhang, Y. Zhou, Z. Ma, and X. Huang. 2018. Ca^{2+} channels and Ca^{2+} signals involved in abiotic stress responses in plant cells: recent advances. *Plant Cell, Tissue and Organ Culture* 132:413–424.

Ma, F., R. Lu, H. Liu, et al. 2012. Nitric oxide-activated calcium/calmodulin- dependent protein kinase regulates the abscisic acid-induced antioxidant defence in maize. *Journal of Experimental Botany* 63:4835–4847.

Mhamdi, A., and F.V. Breusegem. 2018. Reactive oxygen species in plant development. *Development* 145: dev164376.

Mohanta, T.K., T. Bashir, A. Hashem, E.F. Abd Allah, A.L. Khan, and A.S. Al-Harrasi. 2018. Early events in plant abiotic stress signaling: interplay between calcium, reactive oxygen species and phytohormones. *Journal of Plant Growth Regulation*. in press.

Mura, A., R. Medda, S. Longu, G. Floris, and A.C. Rinaldi. 2005. A Ca^{2+}/calmodulin-binding peroxidase from *Euphorbia* latex: novel aspects of calcium hydrogen peroxide cross-talk in the regulation of plant defenses. *Biochemistry* 44:14120–14130.

Neill, N., R. Desikan, and J. Hancock. 2002. Hydrogen peroxide signaling. *Current Opinion in Plant Biology* 5:388–395.

Ohama, N., H. Sato, K. Shinozaki, and K. Yamaguchi-Shinozaki. 2017. Transcriptional regulatory network of plant heat stress response. *Trends in Plant Science* 22:53–65.

Peleg, Z., and E. Blumwald. 2011. Hormone balance and abiotic stress tolerance in crop plants. *Current Opinion in Plant Biology* 14:290–295.

Pei, Z.-M., Y. Murata, G. Benning, et al. 2000. Calcium channels activated by hydrogen peroxide mediate abscisic acid signalling in guard cells. *Nature* 406:731–734.

Poór, P., Nawaz, K., Gupta, R. et al. Ethylene involvement in the regulation of heat stress tolerance in plants. *Plant Cell Reports* (2021).

Poovaiah, B.W., and L. Du. 2018. Calcium signaling: decoding mechanism of calcium signatures. *New Phytologist* 217:1394–1396.

Price, A.H., A. Taylor, S.J. Ripley, A. Griffiths, A.J. Trewavas, and M.R. Knight. 1994. Oxidative signals in tobacco increase cytosolic calcium. *The Plant Cell* 6:1301–1310.

Qu, A.L., Y.F. Ding, Q. Jiang, and C. Zhu. 2013. Molecular mechanisms of the plant heat stress response. *Biochemical and Biophysical Research Communications* 432:203–207.

Rahman, A., M.G. Mostofa, K. Nahar, M. Hasanuzzaman, and M. Fujita. 2016. Exogenous calcium alleviates cadmium-induced oxidative stress in rice (*Oryza sativa* L.) seedlings by regulating the antioxidant defense and glyoxalase systems. *Brazilian Journal of Botany* 39:393–407.

Saidi, Y., A. Finka, and P. Goloubinoff. 2011. Heat perception and signalling in plants: a tortuous path to thermotolerance. *New Phytologist* 190:556–565.

Sajid, M., B. Rashid, Q. Ali, and T. Husnain. 2018. Mechanisms of heat sensing and responses in plants. It is not all about Ca^{2+} ions. *Biologia Plantarum* 62:409–420.

Sami, F., M. Faizan, A. Faraz, H. Siddiqui, M. Yusuf, and S. Hayat. 2018. Nitric oxide-mediated integrative alterations in plant metabolism to confer abiotic stress tolerance, NO crosstalk with phytohormones and NO-mediated post translational modifications in modulating diverse plant stress. *Nitric Oxide* 73:22–38.

Song, L., W. Ding, M. Zhao, B. Sun, and L. Zhang. 2006. Nitric oxide protects against oxidative stress under heat stress in the calluses from two ecotypes of reed. *Plant Science* 171:449–458.

Virdi, A.S., S. Singh, and P. Singh. 2015. Abiotic stress responses in plants: roles of calmodulin-regulated proteins. *Frontiers in Plant Science* 6:809.

Wahid, A., S. Gelani, M. Ashraf, and M.R. Foolad. 2007. Heat tolerance in plants: an overview. *Environmental and Experimental Botany* 61:199–223.

Wani, S.H., V. Kumar, V. Shriram, and S.K. Sah. 2016. Phytohormones and their metabolic engineering for abiotic stress tolerance in crop plants. *Crop Journal* 4:162–176.

Xu, Y., and R.B. Huystee. 1993. Association of calcium and calmodulin to peroxidase secretion and activation. *Journal of Plant Physiology* 141:141–146.

Yang, T., and B.W. Poovaiah. 2003. Calcium/calmodulin-mediated signal network in plants. *Trends in Plant Science* 8:505–512.

Zhou, Z.H., Y. Wang, X.Y. Ye, and Z.G. Li. 2018. Signaling molecules hydrogen sulfide improves seed germination and seedling growth of maize (*Zea mays* L.) under high temperature by inducing antioxidant system and osmolyte biosynthesis. *Frontiers in Plant Science* 9:1288.

Zhu, J.K. 2016. Abiotic stress signaling and responses in plants. *Cell* 167:313–324.

6 Functions of Polyamines in Abiotic Stress Tolerance in Plants

*Péter Pálfi, Riyazuddin Riyazuddin, László Bakacsy, and Ágnes Szepesi**
Department of Plant Biology, Institute of Biology, Faculty of Science and Informatics, University of Szeged, Szeged, Közép fasor 52, Hungary
*Corresponding author: Ágnes Szepesi
E-mail: szepesia@bio.u-szeged.hu

CONTENTS

6.1 INTRODUCTION

Polyamines (PAs) are N-containing essential polycationic molecules with regulatory and quality control functions not only in plant growth and development but also in abiotic stress tolerance (Tiburcio et al., 2014; Gill and Tuteja, 2010; Gupta et al., 2013). There is enormous evidence that PAs in different forms, like free, conjugated, or bound, can induce abiotic stress tolerance in plants (Minocha et al., 2014; Shi and Chan, 2014; Alcázar et al., 2020) and also in the engineered enhancement of tolerance; however, their interactions are complex, and our knowledge needs to deepen to use them

DOI: 10.1201/9781003159636-6

precisely (Hussain et al., 2011, Khajuria et al., 2018). There is a complex interplay between PAs and nitrogen during stress responses (Paschalidis et al., 2019; Pál et al., 2015). Moreover, there is also crosstalk between reactive oxygen species (ROS) and PAs in regulating ion transport at the plasma membrane that contributes to the adaptation responses (Pottosin et al., 2014). Among the different PAs, spermine (Spm), in particular, exhibits a critical role in both biotic and abiotic stress responses in a different way from other PAs (Seifi and Shelp, 2019). Here, we provide evidence that exogenous applications of PAs could be effective in alleviating stress injuries and enhancing abiotic stress tolerance in plants. Several studies have suggested that PAs could be prominent plant growth regulators in the service of sustainable agriculture, inducing abiotic stress tolerance in plants (Szepesi, 2019).

6.1.1 Biosynthesis of PAs in Plants

PAs can be synthesized from the amino acids L-arginine or L-ornithine to produce the diamine putrescine (Put), which may be a substrate for biosynthesis of higher PAs such as triamine spermidine (Spd) and tetramine spermine (Spm). These PAs are under the most intensive examination, but other PAs such as cadaverine (Cad) and thermospermine (T-Spm) can also play important roles in the development and stress responses in plants (Jancewicz et al., 2016). Recently, excellent reviews have been published describing the precise mode of action and the biosynthetic pathways of PAs, and different modes of studying these processes, including the use of pharmacological inhibitors or transgenic modifications (Alcázar et al., 2010; Chen et al., 2019). As decarboxylation of S-adenosylmethionine (dcSAM) by S-adenosylmethionine decarboxylase (SAMDC) is essential to generate Spd, some studies suggest that the biosynthetic pathways of PAs and ethylene (ET) are competitive (Majumdar et al., 2017). More recently, some relevant parts of the crosstalk between PAs and ET have been elucidated, which are described later in this chapter. The enzyme Spd synthase (SPDS) is involved in the synthesis of Spd, and Spm synthase (SPMS) catalyzes Spm formation. Moreover, there is evidence for the involvement of SPDS in some senescence-related processes in plants, as overexpression of this enzyme from yeast altered the ripening, senescence, and decay of tomato fruits (Nambeesan et al., 2010). Additional spraying with a solution of the polyamine biosynthesis inhibitor dl-α-difluoromethylarginine (DFMA) increased the abiotic stress tolerance of yellow lupin, indicating that PAs could be involved in drought stress tolerance (Juzon et al., 2017).

6.1.2 Catabolism of PAs

Similarly to the biosynthesis, catabolism of PAs has critical functions in plant development and defense responses (Cona et al., 2006; Moschou et al., 2008). Catabolism of PAs is mediated by two different enzymes: diamine oxidases or copper-containing amine oxidases (DAOs or CuAOs) and FAD-containing polyamine oxidases (PAOs). During catabolic processes, reactive aldehydes and the signaling molecule hydrogen peroxide (H_2O_2) can be synthesized as by-products contributing to fine-tuning of stress tolerance (Yu et al., 2019). For example, PA catabolism could induce leaf senescence (Sobieszczuk-Nowicka, 2017). There is a growing body of evidence suggesting that some PAOs are able to convert higher PAs into Spd or Put, while others only take part in terminal catabolism. Interestingly, catabolism of PAs contributes to the biosynthesis of reactive aldehydes, H_2O_2, and gamma-amino-butyric acid (GABA), which can have different effects on plant abiotic stress responses (Gupta et al., 2016). In addition to these, Takács et al. (2017) provided evidence that PAO is responsible for nitric oxide (NO) production together with H_2O_2 under salt stress in tomatoes, suggesting that even NO synthesis can be triggered by the catabolism of PAs. Evidence has also now emerged to reveal the importance of the molecule GABA in plant stress responses (Li et al., 2021). In pepper, Kim et al. (2013) showed that arginine decarboxylase (ADC) is essential for proper PAs and GABA signaling pathways in cell death processes and defense. Xie et al. (2019) found that flavonoid accumulation was paralleled by GABA synthesis, as some flavonoid biosynthesis genes were also affected as GABA levels varied; however, a GABA receptor has not yet been identified in plants. Yue et al. (2018)

detected the role of GABA shunt in poplar adventitious root growth by inhibition of α-ketoglutarate dehydrogenase. GABA, as a priming agent, protected against *Botrytis cinerea* via ROS production (van Rensburg and van Ende, 2020), showing that GABA can be a useful molecule to induce biotic stress tolerance in crops. Enzymatic conversion of glutamate and GABA could be an indicator of the plant stress response (Eprintsev et al., 2020). Exogenously added GABA could rapidly metabolize to succinic acid, feeding the TCA cycle (Hijaz and Killiny, 2019). Further, it has been reported that inhibition of GABA transaminases enhanced GABA accumulation and resulted in the development of dwarf and infertile mutant tomato plants, demonstrating the importance of GABA homeostasis in plant development (Koike et al., 2013). GABA was capable of inducing heat and drought tolerance in creeping bentgrass (*Agrostis stolonifera*) by altering the transcript levels of stress-responsive genes and transcriptional factors (Li et al., 2018). Also, GABA was proved to alter endogenous PAs and organic metabolites in creeping bentgrass (Li et al., 2020). Podlesakova et al. (2019) revealed a connection between phytohormones and PAs regulating plant stress responses via an altered GABA pathway. Moreover, GABA and related amino acids can act in plant immune responses, as demonstrated by Tarkowski et al. (2020). CuAOs and PAOs show different roles in plant tissue differentiation and organ development (Tavladoraki et al., 2016). For example, recent studies have shown the importance of PA catabolism during lateral root initiation in Arabidopsis (Kaszler et al., 2021).

6.1.3 Fruit and Ripening-Specific PA Metabolism

PAs are among the prominent molecules in improving the shelf life of fruits, enhancing the yield and thereby preventing food waste (Valero et al., 2002). Higher PAs could be the main compounds involved in restoring and enhancing a potential metabolic memory in fruit ripening (Mattoo and Handa, 2008; Mattoo et al., 2010). Put was efficient in alleviating chilling injury, contributing to the enhanced shelf life and fruit quality of pomegranate fruit under cold storage conditions (Barman et al., 2011). Moreover, Put also affected the ethylene emissions and flesh firmness of peach (*Prunus persica*) fruits (Bregoli et al., 2002). In tomatoes, Hazarika et al. (2011) suggested that constitutively expressed SAMDC gene caused enhanced stress tolerance of plants. Besides, PA catabolism was demonstrated to be involved in grape ripening (Agudelo-Romero et al., 2013, 2014). PA modulation could occur at the transcriptional level for abiotic and biotic stress responses in fruit species (Fortes et al., 2019). There are some fruit-specific differences between the climacteric and non-climacteric fruit ripening processes in PA metabolism (Fortes et al., 2018). It is well known that PAs are effective in defense against biotic and abiotic stress by connecting with other stress-induced metabolites (Nehela and Killiny, 2020) such as in citrus plants. In kiwi fruit, PAs effectively affect ET production, respiration, and ripening (Petkou et al., 2004). Exogenous Put was effective in increasing the shelf life of strawberry (*Fragaria ananassa* Duch.) fruit (Khosroshahi et al., 2007) by decreasing ET levels. Postharvest PA application is promising for reducing chilling injury in apricot (Koushesh Saba et al., 2012). Extending shelf life and reducing mechanical damage by the application of PAs are important in fruit species, e.g. in the case of mango (Malik et al., 2005) or apricot (Martínez-Romero et al., 2002). Nambeesan et al. (2019), studying tomato floral organ identity and fruit/seed set, suggested a nexus between PA ratios and developmental programs of plants.

6.2 EXOGENOUS APPLICATION OF PAS TO ENHANCE ABIOTIC STRESS TOLERANCE

6.2.1 Water Balance Improvement by PAs

6.2.1.1 Osmotic Stress

Osmotic stress is induced as secondary stress during drought and salinity stress, threatening the crop yield. It has been reported that both exogenously added PAs and transgenic modifications of PA metabolism could enhance osmotic stress tolerance in agriculturally important crops (reviewed in Gupta et al., 2013). Pál et al. (2018) revealed an interaction between PAs, abscisic acid (ABA), and proline (Pro) in osmotic stress-induced wheat leaves. Spm could promote acclimation to osmotic stress, as it could modify antioxidant level, ABA, and jasmonic acid (JA) in soybean (Radhakrishnan et al., 2013) (Table 6.1).

6.2.1.2 Drought Stress

Drought stress, together with other stress types such as salt stress, causes huge yield losses, threatening food safety and food production worldwide (Khan et al., 2021; Ghatak et al., 2017). PAs are used as plant growth regulators to fight against drought stress-induced injuries (Li et al., 2015). Different types of PA application by seed priming, irrigation, or foliar spray could lead to increased photosynthesis, accumulation of osmolytes, or enhanced antioxidant defense responses during drought stress conditions (Table 6.1).

6.2.1.3 Salt Stress

Salt stress tolerance in plants is a focus of many research groups in order to improve plant tolerance (Wani et al., 2020). PAs have diverse effects on plants exposed to salt stress. They can alleviate photosynthetic inhibition or induce antioxidant defense mechanisms. Sometimes they can elevate the osmoprotectant levels, e.g. Pro, to protect plants from salt stress (Szepesi and Szőllősi, 2018). GABA homeostasis is one of the targets in plants during salt stress conditions, as GABA metabolism is strongly controlled by ROS accumulation (Bao et al., 2015). GABA was able to alleviate the salt-induced damage in tomatoes in connection with ROS metabolism modulating the Na^+ uptake (Wu et al., 2020). At the transcriptomic level, GABA could affect salinity-induced genes via ABA and ET signaling pathways in poplar (Ji et al., 2018) and modifying GABA transaminase (GABA-T) enzyme activities (Ji et al., 2020). There is growing evidence that shows the interaction between NO and PAs during salt stress (Napieraj et al., 2020). Also, a possible interaction between PAs and salicylic acid (SA) was revealed in some studies, suggesting their implication in regulating salt stress tolerance mechanisms in plants (Szepesi et al., 2009, 2011). PAs are also capable of reprogramming oxidative and nitrosative status and proteome of salt-stressed citrus plants (Tanou et al., 2014).

6.2.2 Thermotolerance Induction by PAs

6.2.2.1 Cold Stress

Cold stress is one of the major stresses affecting crop yield in the winter season (Gupta and Deswal, 2014; Sehrawat et al., 2013). Exogenously added PAs are rarely investigated in the literature, except for Put, which could alleviate freezing and cold responses, inducing tolerance by ABA regulation (Cuevas et al., 2008). Sun et al. (2020) provided evidence that Put was efficient in enhancing the chilling tolerance of the ornamental plant *Anthurium andreanum*, reducing the yield loss of this plant in the winter season. Moreover, an interaction was deciphered between PAs and ABA, NO, and H_2O_2 in tomato chilling responses (Diao et al., 2017), which suggests complex crosstalk between these phytohormones. In the future, the combination of cold and drought stress should be investigated to reveal the common and distinct events in PA homeostasis in order to use PAs in enhancing defense mechanisms against these two types of stresses.

6.2.2.2 Heat Stress

Heat stress induces complex responses in plants, involving the destruction of cellular homeostasis, deregulated plant hormone pathways, and ROS production (Kim et al., 2019; Poór et al., 2021). In extreme climate conditions, this stress type can cause elevated yield loss in agriculture. PAs applied exogenously could be a way to alleviate injuries caused by heat stress; however, it is worth mentioning that Spm is an exception (Table 6.1).

TABLE 6.1
Exogenous Application of PAs to Enhance Abiotic Stress Tolerance in Plants

Stress Type	Applied PA	Plant Species	Mode of Application	Applied Concentration	Responses	Reference
Osmotic stress	Put	*Triticum aestivum*	Irrigation	0.5 mM	Higher photosynthetic and antioxidant activity	Doneva et al., 2021
		Triticum aestivum	Irrigation	1 mmol/L	Increased antioxidants	Marcińska et al., 2021
		Triticum aestivum	Irrigation	0.5 mM	Altered fatty acid composition	Pál et al., 2018
		Nicotiana tabacum	Leaf disc treatment	1 mM	Inhibited water loss	Kotakis et al., 2014
	Spd	*Triticum aestivum*	Irrigation	1 mmol/L	Increased antioxidants	Marcińska et al., 2021
	Spm	*Trifolium repens*	Irrigation	20 μM	Enhanced osmotic adjustment and water use efficiency	Li et al., 2020
		Triticum aestivum	Irrigation	1 mmol/L	Increased antioxidants	Marcińska et al., 2021
Drought stress	Put	*Triticum aestivum*	Seed priming	1–100 μM	Osmolyte accumulation	Ebeed et al., 2017
		Triticum aestivum	Seed priming, foliar spray	100 μM	Membrane protection	Hassan et al., 2020
		Eremochloa ophiuroides	Irrigation	0.1–0.5 mM	Increased antioxidants, photosynthesis	Liu et al., 2017
		Cynodon dactylon	Irrigation	5 mM	Accumulation of osmolytes	Shi et al., 2013
	Spd	*Eremochloa ophiuroides*	Irrigation	0.1–0.5 mM	increased antioxidants, photosynthesis	Liu et al., 2017
		Zea mays	Irrigation	0.05–0.1 mM	improved photosynthesis, increased level of phytohormones	Li et al., 2018
		Agrostis stolonifera	Irrigation	500 μM/L	Transcript analysis	Ma et al., 2017
		Cynodon dactylon	Irrigation	5 mM	Accumulation of osmolytes	Shi et al., 2013
		Trifolium repens	Irrigation	20 μM	Activation of antioxidant system	Peng et al., 2016
	Spm	*Triticum aestivum*	Seed priming	1–100 μM	Osmolyte accumulation	Ebeed et al., 2017
		Triticum aestivum	Seed priming, foliar spray	100 μM	Membrane protection	Hassan et al., 2020
		Eremochloa ophiuroides	Irrigation	0.1–0.5 mM	Increased antioxidants, photosynthesis	Liu et al., 2017
		Trifolium repens	Irrigation	0.5 mM	Improved antioxidants, higher level of phytohormones	Zhang et al., 2018
		Cynodon dactylon	Irrigation	5 mM	Accumulation of osmolytes	Shi et al., 2013

(*continued*)

TABLE 6.1 (Continued)
Exogenous Application of PAs to Enhance Abiotic Stress Tolerance in Plants

Stress Type	Applied PA	Plant Species	Mode of Application	Applied Concentration	Responses	Reference
		Trifolium repens	Irrigation	0.5 mM	Dehydrin biosynthesis, carbohydrate metabolism	Li et al., 2015
Salt stress	Put	*Camellia sinensis*	Irrigation	5 mM	Improvement of photosynthetic efficiency	Xiong et al., 2018
		Cucumis sativus	Foliar spray	8 mM	Leaf starch accumulation	Shen et al., 2019
		Cucumis sativus	Foliar spray	8 mM	Alleviated photoinhibition	Wu et al., 2019
		Psidium guajava	Irrigation	250 and 500 ppm	Increased antioxidant system	Ghalati et al., 2020
		Cucumis sativus	Foliar spray	8 mM	Enhanced energy from glycolysis and Krebs cycle	Zhong et al., 2016
		Luffa acutangula	Irrigation	100 µM	Synergistic modulation of antioxidant system with kinetin	Kapoor and Hasanuzzaman, 2020
		Cucumis sativus	Foliar spray	8 mM	Positive effect on Chl metabolism and xanthophyll cycle	Yuan et al., 2018
	Spd	*Zoyzia japonica Steud*	Irrigation	0.15–0.60 mM	PA metabolism	Li et al., 2016
		Zoyzia japonica Steud	Irrigation	0.15 mM	Increased PA metabolism	Li et al., 2017
		Oryza sativa	Irrigation	1 mM	Metabolic adjustment	Saha and Giri, 2017
		Cucumis sativus	Irrigation	0.1 mM	Enhanced antioxidant capacity	Wu et al., 2018
		Cucumis sativus	Irrigation	0.1 mM	Promoted root metabolism	Liu et al., 2020
		Carya illinoinensis	Foliar spray	1 mM	Antioxidant system activation	Wu et al., 2020
		Oryza sativa	Irrigation	0.2–1.5 mM	Stabilizing chloroplast and thylakoids	Jiang et al., 2020
		Poa pratensis	Foliar spray	1 mM	Protecting ion metabolism	Puyang et al., 2016
		Gladiolus gandavensis	Foliar spray	0.1 mmol/L	Osmotic adjustment, antioxidant activity	Qian et al., 2020
		Sorghum bicolor	Irrigation	0.25 mM	Enhanced Calvin cycle and photosynthetic efficiency	El Sayed et al., 2019
		Cucumis sativus	Irrigation	0.1 mM	Photosynthesis improvement	Sang et al., 2016
	Spm	*Solanum lycopersicum*	Irrigation	0.5 mM	Ion homeostasis	Xu et al., 2020
		Solanum lycopersicum	Foliar spray	100 µM	Modulation of osmolytes	Ahanger et al., 2019
		Triticum aestivum	Irrigation	0.25 mM	Enhanced antioxidant system	Elsayed et al., 2018
		Oryza sativa	Priming	2.5 mM Spm or 5 mM Spd	Abiotic stress-responsive gene expression	Paul and Roychoudhury, 2017

Heat stress	Put	*Brassica oleracea L.*	Foliar spray	2.5 mM	Lower lipid peroxidation	Collado-González et al., 2021
		Triticum aestivum	Irrigation	2.5 mM	Increased antioxidant capacity	Kumar et al., 2014
	Spd	*Trifolium repens*	Irrigation	70 μM	Altered GABA metabolism, enhanced antioxidant system	Luo et al., 2020
		Triticum aestivum	Foliar spray	1 mM/L	Enhanced antioxidants, increased Pro levels	Jing et al., 2020
		Pinella ternata	Foliar spray	100 μM	Increased Chl, decreased MDA level	Ma et al., 2020
		Cucumis sativus	Foliar spray	1 mmol/L	miRNA identification	Wang et al., 2018
	Spm	–	–	–	–	–
Heavy metal stress	Put	–	–	–	–	–
	Spd	*Raphanus sativus*	Irrigation	1 mM	Impact on indole acetic acid and ABA	Choudhary et al., 2012
		Boehmeria nivea (L.) Gaudich.	Foliar spray	0.1 mM	Growth promotion, reduced oxidative stress	Gong et al., 2016
		Triticum aestivum	Seed presoaking	2 mM	Alleviate Cd harmful effects	Rady and Hemida, 2015
		Salix matsudana	Irrigation	0.25 mmol/L	Promoted antioxidant system	Tang et al., 2019
		Vigna angularis	Foliar spray	200 μM	Upregulated antioxidant and N metabolism	Ahanger et al., 2020
		Typha latifolia	Foliar spray	0–0.5 mmol/L	Altered Cd ion compartmentation	Tang et al., 2009
		Zea mays	Foliar spray	0.05–0.1 mM	Promoted antioxidant system, increased osmolyte levels	Naz et al., 2021
		Nymphoides peltatum	Foliar spray	0.1 mM	Reduced lipid peroxidation, increased ratio of higher PAs	Wang et al., 2007
	Spm	*Triticum aestivum*	Seed presoaking	2 mM	Alleviate Cd harmful effects	Rady and Hemida, 2015
		Vigna radiata	Irrigation	0.25 mM	Improved antioxidant capacity and osmoregulation	Nahar et al., 2016
		Nymphoides peltatum	Foliar spray	0.1 mM	Reduced lipid peroxidation, increased ratio of higher PAs	Wang et al., 2007

6.2.2.3 Heavy Metal Stress

Environmental pollution and anthropogenic activities produce increased heavy metal stress, causing significant problems not just in plants but in other living organisms. PAs act as radical scavengers to inhibit membrane injuries or promote antioxidant activities. Successful applications of PAs have been reported to diminish the harmful effects of heavy metal stress in plants, and it can be concluded that most results come from the application of higher PAs such as Spd or Spm in the case of heavy metal stress (Table 6.1).

6.2.3 Cross Talk between PAs and ET, NO, GABA, and H_2S during Abiotic Stresses in Crop Plants

Plants adapt to survive under various abiotic stress conditions, including salinity, drought, ionic, nutrient, heavy metals, and cold stresses, through numerous mechanisms. Moreover, PAs, including Put, Spd, and Spm, are ubiquitous biological organic amines involved in plant growth, development, and the response to the above-mentioned abiotic stresses (Takahashi and Kakehi, 2010). NO, GABA, ET, and hydrogen sulfide (H_2S) could be important regulators of plant growth and adaptation to abiotic stresses.

6.2.3.1 PAs and Ethylene (ET)

Abiotic stress tolerance in plants is the cumulative result of coordinated gene expression and metabolic processes along with crosstalk between phytohormones. Of the various phytohormones, ET is known to show coordinated signaling with other signaling factors such as NO, H_2O_2, and ROS and to regulate the plant's response to external stressors. Its signal transduction pathway and key modulators under salinity stress, along with crosstalk with other phytohormones, are discussed by Riyazuddin et al. (2020). Ali et al. (2020) discussed the role of PAs in mitigating abiotic stress tolerance, as they are crucial in maintaining the structure and function of various cellular components during abiotic stress.

ET promotes senescence, whereas PAs are considered to be anti-senescence growth regulators. As ET promotes leaf senescence, flower fading, fruit ripening, and abscission, PAs act antagonistically to ET (Anwar et al., 2015). SAM is the precursor for both ET and PAs; therefore, the concentration of one affects the other as well. However, it is notable that the availability of SAM *in vivo* is not the rate-limiting step in the biosynthesis of either ET or Pas, and both the pathways can run simultaneously (Anwar et al., 2015). For instance, Mehta et al. (2002) demonstrated that overproduction of Spd and Spm during ripening in tomato by overexpressing yeast *SAMDC* gene (*ySAMdc*; *Spe2*) fused with ripening inducible E8 promoter, which resulted in an enhanced level of lycopene and therefore, prolonged the vine life of tomato fruit. Along with this, it was also observed that the ET level was comparably higher in transgenic than in non-transgenic fruits, suggesting simultaneous operation of both the biosynthesis pathways. During stress, ET and PAs show a coordinated response and help plants to withstand unfavorable environmental conditions. For example, under salt stress in salt-tolerant maize plants, an ET peak was observed at 5.5 hours, along with elevated content of H_2O_2 and upregulation of the PA-catabolizing enzyme DAO activity and *Zea mays* diamine oxidase 3 (*ZmDAO3*) gene expression. On the other hand, biphasic ET production was observed in salt-sensitive maize plants at 5.5 and 12.5 hours, and the plants failed to activate PA-dependent H_2O_2 generation during the first ET peak. Thus, ET signaling controls the balance of PA enzyme-catalyzed H_2O_2 production, which in turn acts as a second messenger activating downstream targets to confer salt tolerance on maize plants, a genotype-dependent response at least during short-term salinity stress (Freitas et al., 2018). Overexpression of a cold-responsive ethylene-responsive factor (ERF), *MfERF1* from *Medicago falcata*, led to higher expression of *SAMDC1*, *SAMDC2*, *SPDS1*, *SPDS2*, and *SPMS* genes involved in Spd and Spm synthesis, which increased tolerance to freezing and chilling conditions (Zhuo et al., 2018). As PAs are known to stabilize membranes and minimize water stress (Li et al., 2020), they are vital for plant growth and development during stresses such as cold or drought. For example, PAs and ET are involved in mitigating the effect of severe drought on grain filling in wheat. External application of Spd markedly enhanced grain weight and grain filling rates, whereas application of ethephon, an ET releaser, showed opposite results. Moreover, when ET emission was studied in grains, it was noted that under severe drought conditions, ET concentration was remarkably increased in both superior and inferior grain types, more prominently in a drought-sensitive variety (JM22). The opposite results were obtained for PA concentrations, as Spd and Spm concentration dropped in both types of grains, with a more pronounced drop in the drought-sensitive variety. Based on these results, it can be concluded that a balance between ET and PA concentration determines plant growth and development and their antagonistic relationship (Yang et al., 2014). Chen et al. (2013) shared similar results in rice and suggested a potential metabolic interaction between PAs and ET biosynthesis that mediates the grain filling of inferior spikelets in rice under drought stress. In addition, a study by Wang et al. (2012) on six rice cultivars presented results in accordance with Chen et al. (2013) and Yang et al. (2014), showing a significant negative correlation between ET evolution rates and grain filling rates and positive correlation with free Spd and Spm concentrations.

Plants under heavy metal stress initiate a cascade of mechanisms that help in mitigating negative effects. The crosstalk between PAs and ET helps in reversing heavy metal toxicity. For instance, aluminum (Al) increases ET concentration, which inhibits root growth in *Triticum aestivum* seedlings, and Put-mediated inhibition in ACC synthase (ACS) activity was observed, which lowered ethylene biosynthesis and ultimately mitigated the negative effect of Al-induced root inhibition (Yu et al., 2016). In *Glycine max*, cadmium (Cd) induced the expression of genes encoding *ACS*, *SAMDC*, *MAPK*, and *MAPKK2* and the DOF1, MYBZ2, and bZIP62 transcription factors involved in ET and PA metabolism, NO generation, and MAPK cascades (Chmielowska-Bąk et al., 2013).

6.2.3.2 PAs and Nitric Oxide

NO is a highly diffusible gaseous signaling molecule that acts as a cellular messenger to regulate defense responses against different abiotic stresses such as drought, salinity, temperature extremes, mineral deficiency, and wounding (Neill et al., 2003; Shapiro et al., 2005; Cona et al., 2006; Neill et al., 2008; Qiao and Fan, 2008; Kusano et al., 2008; Moschou et al., 2008). It is well known that NO acts as an intermediate signaling molecule in PA signaling (Tun et al., 2006). A number of studies have shown that crosstalk of PA and NO signaling with other signaling pathways plays an important role in abiotic stress management (Wimalasekera et al., 2011). Further studies have reported that PAs and NO are interlinked through the L-arginine-dependent pathway because L-arginine acts as a precursor for both PAs and NO. Also, under anoxic conditions, NO reacts with PAs to produce 1-substituted diazen-1-ium-1,2-diolates (NONOates), which act as NO donors and show a direct chemical link between PAs and NO (Keefer et al., 1996). Moreover, many studies have also revealed that PAs and NO form a variety of conjugates under different stress conditions, and their interaction increased tolerance to heavy metal stress by activating antioxidants with ROS and reactive nitrogen-scavenging functions, by protection of biomolecules and biomembranes, as well as by enhancing metal chelation (Müller et al., 2003; Sakihama et al., 2003; Hao and Zhang, 2009; Chen et al., 2013). Nahar et al. (2016) reported that the combined effect of PAs and NO improved tolerance to Cd toxicity in *Vigna radiata* L. seedlings via upregulating antioxidant defense, detoxification of metals, and methylglyoxal systems, and improved growth, photosynthetic pigment, and plant water status by regulating osmolytes. Further, it has also been reported that Cd stress increased the root and shoot Cd content, leading to decreased growth, reduced chlorophyll, modulated Pro, reduced relative water content, and increased oxidative damage toxicity. Likewise, the combined interaction of sodium nitroprusside (SNP; an NO donor) and Spd was more effective in enhancing the tolerance to chilling stress in ginger seedlings (*Zingiber officinale* Roscoe) via detoxification of ROS and maintaining the high level of unsaturated fatty acids. Furthermore, co-application of SNP and Spd showed increased NO generation under normal and stress conditions compared with the use of the compounds individually (Li et al., 2014). Moreover, it has been shown that plant species and cultivars that are tolerant to chilling stress accumulate Pas, especially Spd and Spm. In addition, exogenously applied PAs play an important role in the protection of photosynthetic apparatus from unfavorable effects of chilling stress via entering the intact chloroplasts (He et al., 2002). The crosstalk between PAs and NO has also been found to elevate tolerance against drought stress in cucumber (*Cucumis sativus* cv. Dar). The application of exogenous SNP showed significant enhancement of endogenous Put, Spd, and Spm levels under drought conditions as compared with the control (Arasimowicz-Jelonek et al., 2009). It was reported that exogenous NO can reduce membrane damage under water deficit conditions, and it was concluded that NO acts as a downstream signal of PAs under these conditions (Arasimowicz-Jelonek et al., 2009). A study by Diao et al. (2017) revealed that the exogenous application of Spd and Spm increased the production of NO and H_2O_2, as well as enhanced nitrite reductase (NR) activity, NO synthase (NOS)-like activity, and PA oxidase activities, under chilling stress in tomato seedlings (*Lycopersicon esculentum* Mill.). Likewise, application of NO elevated endogenous Put and Spd levels via upregulation of the PA biosynthetic genes during chilling stress (Diao et al., 2017). In short, the combined effects of PAs and NO, rather than applications of the individual compounds, induced significant tolerance to different abiotic stresses by increasing antioxidant defense compared with normal conditions.

6.2.3.3 PAs and GABA

The biosynthesis of GABA is mainly dependent on two pathways: one is the decarboxylation of glutamate from the catalytic action of glutamate decarboxylase in the cytosol, and the other is the degradation of PAs through CuAO and PAO activities (Bhatnagar et al., 2001; Feehily and Karatzas, 2013). A detailed mechanism of GABA biosynthesis and the function of GABA in abiotic tolerance has been discussed in recent reviews (Podlešáková et al., 2019; Khan et al., 2021). PA metabolism is closely linked to other compounds such as GABA and Pro, all known to be involved in stress tolerance (Gupta et al., 2013). Emerging reports also indicate the accumulation of GABA along with ROS in response to abiotic stresses (Kumar et al., 2019; Li et al., 2020; Wu et al., 2020). To examine the effect of exogenous GABA in abiotic stresses, Li et al. (2020) reported that GABA induced changes in the level of endogenous PAs in creeping bentgrass under heat, drought, and salt stresses. Surprisingly, GABA treatment significantly elevated the levels of Spd, glutamic acid, alanine, phenylalanine, aspartic acid, and glycine, which could play important roles as osmolytes, during heat stress, salt stress, and drought stress (Li et al., 2020). Moreover, it has also been suggested that high levels of GABA induced by salt stress could be caused by PA degradation, indicating that PAs can perform their functions via GABA formation under salt stress (Su et al., 2003). Similarly, priming of white clover (*Trifolium repens* cv. Haifa) seeds with GABA reduced the negative impact of salt stress during seed germination by improving the antioxidant defense and promoting the accumulation of proteins known as dehydrins (DHNs) under salt stress (Cheng et al., 2018; Riyazuddin et al., 2021). In 2015, Li and co-workers reported that exogenous application of Spm enhanced drought tolerance in white clover via accumulation of carbohydrates and synthesis of DHNs (Li et al., 2015). Therefore, it can be concluded that exogenous application of PAs could significantly promote salt and drought tolerance in white clover associated with DHN accumulation. In yet another similar study, exogenously applied GABA enhanced the levels of glutamate, PAs, and Pro, associated with providing drought stress tolerance via positive regulation in the GABA shunt, PAs, and Pro metabolism (Yong et al., 2017). Kumar et al. (2019) examined the underlying role of GABA in the amelioration of arsenic (As(III)) toxicity and found that a low concentration of GABA application provided stress tolerance against arsenic toxicity in rice seedlings. Moreover, exogenous application of GABA enhanced the level of PAs, followed by the upregulation of genes that participated in the PA synthesis pathway, namely, ADC, SPMS, and SPDS, against As(III) treatments, resulting in enhanced root PA level. Higher levels of PAs act as antioxidants and reduce the lipid peroxidation and H_2O_2 level in rice seedlings compared with control conditions (Kumar et al., 2019). A recent study found that the application of GABA enhanced Cd tolerance associated with the regulation of antioxidant defense and endogenous PA levels in maize (*Zea mays*) (Seifikalhor et al., 2020). Based on these findings, it can be concluded that GABA acts as a signaling molecule and can induce an antioxidant defense mechanism to provide adaptation to crop plants against abiotic stresses. However, further analysis of changes in different types of PAs commonly or differentially affected by GABA under abiotic stresses will help to better understand the GABA-regulated stress tolerance in plants. The detailed knowledge of the GABA crosstalk will bring new possibilities in the utilization of simple technologies to improve the quality of food products, with added benefits such as nutraceuticals and functional components (Takayama and Ezura, 2015).

6.2.3.4 PAs and H_2S

H_2S is a relatively newly discovered gaseous phytohormone, with a resemblance to NO gasotransmitters, and is involved in the regulation of various physiological processes in both animals and plants. Sodium hydrosulfide (NaHS), a donor of H_2S, has been applied to examine the physiological, morphological, and biochemical functions of H_2S in plants. A growing body of evidence suggests an interaction of H_2S with other phytohormones, including ABA, ET, and SA, as well as PAs in plant development of abiotic stress tolerance. For instance, Chen et al. (2016) reported that exogenous application of NaHS elevates the biosynthesis of free and conjugated PAs under drought conditions to improve drought stress tolerance through accumulating the antioxidants, taking effect as stress-signaling regulators, inducing stomatal closure, and enhancing leaf water content (Shi et al., 2010; Yin et al., 2014; Muilu-Makela et al., 2015). In addition, NaHS treatment enhanced the leaf water potential, increased water uptake and photosynthesis, decreased the level of MDA, and reduced the water loss and accumulation of ROS compared with control conditions, suggesting that H_2S can protect *Spinacia oleracea* seedlings against drought through the inhibition of oxidative stress (Chen et al., 2016). Moreover, Chen and co-workers showed that the H_2S-induced PAs were involved in enhancing drought tolerance in *S. oleracea* seedlings through upregulating the transcription levels of arginine decarboxylase *SoADC*, N-carbamoyl putrescine amidohydrolase *SoCPA*, ornithine decarboxylase *SoODC*, and spermidine synthase *SoSPDS*, leading to the increased biosynthesis and accumulation of PAs (Chen et al., 2016). Therefore, H_2S can enhance plant tolerance to abiotic stresses through interacting with phytohormones such as ABA, JA, ET, SA, and PAs. However, the detailed mechanisms of the effects of H_2S on abiotic stresses remain unclear, especially with respect to physiological processes in plants. H_2S could be a candidate to be involved in plant abiotic stress tolerance; however, its connection to PAs has rarely been studied in plants (Huang et al., 2021). Therefore, further studies are required to clarify the complex molecular networks regulated by NO, ET, PAs, H_2S, and GABA, especially its signal pathway, in response to abiotic stresses.

6.3 CONCLUSIONS AND FUTURE PERSPECTIVES

The combination of stresses and their impacts on growth and development leads to a significant loss of crop yield. Multiple reports have deciphered the role of PAs in plant abiotic stress tolerance; however, the picture is still complex, as PAs could be key central molecules participating in plant stress signaling. There is growing evidence that PA-specific responses depend on plant developmental age, studied organs, and abiotic stress types. Both exogenously applied PAs and transgenic modifications prove that PAs could significantly enhance the abiotic stress tolerance of plants as osmoprotectants or by adjusting antioxidant defense or inducing signal pathways to synthesize other important molecules. A crucial step will be to study the connections of PAs further, as these pathways could

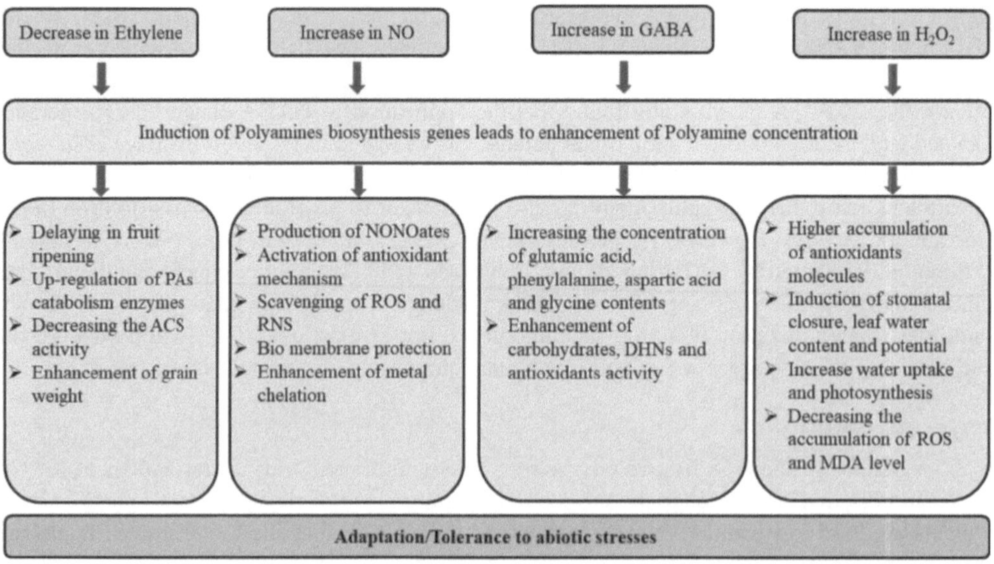

FIGURE 6.1 PA-related events contributing to adaptation or abiotic stress tolerance in plants.

be promising targets to modify plant abiotic stress tolerance. A summary of PA-related events that could induce adaptation or stress tolerance in plants is shown in Figure 6.1.

There are new directions in sustainably using PAs to induce abiotic stress tolerance. A growing body of evidence shows that PAs are essential molecules in all living organisms and could be a common link between plants and microbes, indicating that PA metabolism could play an important role in plant–microbe interactions and recognition or defense mechanisms. Deciphering these interactions could uncover new applications of PAs derived from microbes to enhance plant abiotic stress tolerance; however, much research will be needed in both plants and microbes.

Nanotechnology could be another promising application to enhance crop abiotic stress tolerance in sustainable agriculture (do Espirito Santo Pereira et al., 2021). Nanomaterials can be applied as nanofertilizers, nanopesticides, or nanopriming and could efficiently contribute to efforts to achieve food safety and to increase plant nutrition in a sustainable manner (Fraceto et al., 2016). A combination of Put and carbon dots creating a nano-structure "Put-CQD" was shown to be potentially an efficient priming agent to induce defense against salt stress in grapevine (Gohari et al., 2021). Therefore, it can be concluded that PAs in combination with nanomaterials could be a new technique to improve crop protection against abiotic stress.

The most recent evidence has deciphered the role of PAs in halotropism, a salt avoidance or salt foraging mechanism (Szepesi, 2020), indicating another important target to improve the PA homeostasis of plants, enhancing their capacity to grow in suboptimal growth conditions such as salt stress. Further, metabolomic and omics-based tools could be used in the future to explore the novel connections between PA homeostasis and secondary metabolites, contributing to the development of new biotechnological or agrochemical methods to improve plant abiotic stress tolerance.

ACKNOWLEDGMENTS

The authors are thankful for financial support to NRDI (National Research, Development, and Innovation) Office by Hungarian Ministry under the grant numbers FK129061 and NKFI-6K 125265, Postdoctoral Hungarian State Scholarship 2020/2021 (AK-00205-004/2020), and Bilateral State Scholarship 2019/2020 (AK-00210-002/2019), Tempus Public Foundation.

REFERENCES

Agudelo-Romero, P., Ali, K., Choi, Y. H., Sousa, L., et al. 2014. Perturbation of polyamine catabolism affects grape ripening of *Vitis vinifera* cv. Trincadeira. *Plant Physiology and Biochemistry* 74:141–155. doi: 10.1016/j.plaphy.2013.11.002.

Agudelo-Romero, P., Bortolloti, C., Pais, M. S., Tiburcio, A. F., and Fortes, A. M. 2013. Study of polyamines during grape ripening indicate an important role of polyamine catabolism. *Plant Physiology and Biochemistry* 67:105–119. doi: 10.1016/j.plaphy.2013.02.024.

Ahanger, M. A., Aziz, U., Sahli, A. A., Alyemeni, M. N., and Ahmad, P. 2020. Combined kinetin and spermidine treatments ameliorate growth and photosynthetic inhibition in *Vigna angularis* by up-regulating antioxidant and nitrogen metabolism under cadmium stress. *Biomolecules* 10(1):147. doi:10.3390/biom10010147.

Ahanger, M. A., Qin, C., Maodong, Q., Dong, X. X., Ahmad, P., Abd Allah, E. F., and Zhang, L. 2019. Spermine application alleviates salinity induced growth and photosynthetic inhibition in *Solanum lycopersicum* by modulating osmolyte and secondary metabolite accumulation and differentially regulating antioxidant metabolism. *Plant Physiology and Biochemistry* 144:1–13.

Alcázar, R., Altabella, T., Marco, F., Bortolotti, C., Reymond, M., Koncz, C., Carrasco, P., and Tiburcio, A. F. 2010. Polyamines: molecules with regulatory functions in plant abiotic stress tolerance. *Planta* 231(6):1237–1249. doi:10.1007/s00425-010-1130-0.

Alcázar, R., Bueno, M., and Tiburcio, A. F. 2020. Polyamines: small amines with large effects on plant abiotic stress tolerance. *Cells* 9(11):2373. doi:10.3390/cells9112373.

Ali, R., Hassan, S., Shah, D., Sajjad, N., and Bhat, E. A. 2020. Role of polyamines in mitigating abiotic stress. In *Protective Chemical Agents in the Amelioration of Plant Abiotic Stress: Biochemical and Molecular Perspectives*, eds. A. Roychoudhury and D. K. Tripathi, 291–305. Wiley Blackwell, New York, New York, USA.

Anwar, R., Mattoo, A. K., and Handa, A. K. 2015. Polyamine interactions with plant hormones: crosstalk at several levels. In *Polyamines*, eds. T. Kusano and H. Suzuki, 267–302. Springer, Tokyo.

Arasimowicz-Jelonek, M., Floryszak-Wieczorek, J., and Kubiś, J. 2009. Interaction between polyamine and nitric oxide signaling in adaptive responses to drought in cucumber. *Journal of Plant Growth Regulation* 28(2):177–186.

Bao, H., Chen, X., Lv, S., Jiang, P., et al. 2015. Virus-induced gene silencing reveals control of reactive oxygen species accumulation and salt tolerance in tomato by γ-aminobutyric acid metabolic pathway. *Plant, Cell & Environment* 38(3):600–613. doi:10.1111/pce.12419.

Barman, K., Asrey, R., and Pal, R. K. 2011. Putrescine and carnauba wax pretreatments alleviate chilling injury, enhance shelf life and preserve pomegranate fruit quality during cold storage. *Scientia Horticulturae* 130:795–800. doi: 10.1016/j.scienta.2011.09.005.

Bhatnagar, P., Glasheen, B. M., Bains, S. K., Long, S. L., et al. 2001. Transgenic manipulation of the metabolism of polyamines in poplar cells. *Plant Physiology* 125(4):2139–2153.

Bregoli, A. M., Scaramagli, S., Costa, G., Sabatini, E., et al. 2002. Peach (*Prunus persica*) fruit ripening: aminoethoxyvinylglycine (AVG) and exogenous polyamines affect ethylene emission and flesh firmness. *Physiologia Plantarum* 114:472–481. doi: 10.1034/j.1399-3054.2002.1140317.x.

Chen, D., Shao, Q., Yin, L., Younis, A., et al. 2019. Polyamine function in plants: metabolism, regulation on development, and roles in abiotic stress responses. Frontiers in Plant Science 9:1945.

Chen, J., Shang, Y. T., Wang, W. H., Chen, X. Y., He, E. M., Zheng, H. L., and Shangguan, Z. 2016. Hydrogen sulfide-mediated polyamines and sugar changes are involved in hydrogen sulfide-induced drought tolerance in *Spinacia oleracea* seedlings. *Frontiers in Plant Science* 7:1173.

Chen, L., Wang, L., Chen, F., Korpelainen, H., & Li, C. 2013. The effects of exogenous putrescine on sex-specific responses of *Populus cathayana* to copper stress. *Ecotoxicology and Environmental Safety* 97:94–102.

Chen, T., Xu, Y., Wang, J., Wang, Z., Yang, J., and Zhang, J. 2013. Polyamines and ethylene interact in rice grains in response to soil drying during grain filling. *Journal of Experimental Botany* 64(8):2523–2538.

Cheng, B., Li, Z., Liang, L., Cao, Y., et al. 2018. The γ-aminobutyric acid (GABA) alleviates salt stress damage during seeds germination of white clover associated with Na+/K+ transportation, dehydrins accumulation, and stress-related genes expression in white clover. *International Journal of Molecular Sciences* 19(9):2520.

Chmielowska-Bąk, J., Lefèvre, I., Lutts, S., & Deckert, J. 2013. Short term signaling responses in roots of young soybean seedlings exposed to cadmium stress. *Journal of Plant Physiology* 170:1585–1594.
Choudhary, S. P., Oral, H. V., Bhardwaj, R., Yu, J. Q., and Tran, L. S. (2012). Interaction of brassinosteroids and polyamines enhances copper stress tolerance in *Raphanus sativus*. *Journal of Experimental Botany* 63(15):5659–5675. doi:10.1093/jxb/ers219.
Collado-González, J., Piñero, M. C., Otálora, G., López-Marín, J., and Del Amor, F. M. 2021. Effects of different nitrogen forms and exogenous application of putrescine on heat stress of cauliflower: photosynthetic gas exchange, mineral concentration and lipid peroxidation. *Plants (Basel, Switzerland)* 10(1):152. doi:10.3390/plants10010152.
Cona, A., Rea, G., Angelini, R., Federico, R., and Tavladoraki, P. 2006. Functions of amine oxidases in plant development and defence. *Trends in Plant Science* 11:80–88. doi: 10.1016/j.tplants.2005.12.009.
Cuevas, J. C., López-Cobollo, R., Alcázar, R., Zarza, X., Koncz, C., Altabella, T., et al. 2008. Putrescine is involved in Arabidopsis freezing tolerance and cold acclimation by regulating abscisic acid levels in response to low temperature. *Plant Physiology* 148:1094–1105. doi: 10.1104/pp.108.122945.
Diao, Q., Song, Y., Shi, D., and Qi, H. 2017. Interaction of polyamines, abscisic acid, nitric oxide, and hydrogen peroxide under chilling stress in tomato (*Lycopersicon esculentum* Mill.) seedlings. *Frontiers in Plant Science* 8:203.
do Espirito Santo Pereira, A., Caixeta Oliveira, H., Fernandes Fraceto, L., and Santaella, C. 2021. Nanotechnology potential in seed priming for sustainable agriculture. *Nanomaterials (Basel, Switzerland)* 11(2):267. doi:10.3390/nano11020267.
Doneva, D., Pál, M., Brankova, L., Szalai, G., et al. 2021. The effects of putrescine pre-treatment on osmotic stress responses in drought-tolerant and drought-sensitive wheat seedlings. *Physiologia Plantarum* 171(2):200–216. doi:10.1111/ppl.13150.
Ebeed, H. T., Hassan, N. M., and Aljarani, A. M. 2017. Exogenous applications of polyamines modulate drought responses in wheat through osmolytes accumulation, increasing free polyamine levels and regulation of polyamine biosynthetic genes. *Plant Physiology and Biochemistry* 118:438–448. doi:10.1016/j.plaphy.2017.07.014.
El Sayed, A. I., El-Hamahmy, M. A. M., Rafudeen, M. S., and Ebrahim, M. K. H. 2019. Exogenous spermidine enhances expression of Calvin cycle genes and photosynthetic efficiency in sweet sorghum seedlings under salt stress. *Biologia Plantarum* 63:511–518.
El Sayed, A. I., Rafudeen, M. S., El-Hamahmy, M. A., Odero, D. C., and Hossain, M. S. 2018. Enhancing antioxidant systems by exogenous spermine and spermidine in wheat (*Triticum aestivum*) seedlings exposed to salt stress. *Functional Plant Biology* 45(7):745–759.
Eprintsev, A.T., Selivanova, N.V., Igamberdiev, A.U. 2020. Enzymatic conversions of glutamate and γ-aminobutyric acid as indicators of plant stress response. In *Nitrogen Metabolism in Plants. Methods in Molecular Biology*, ed. K. Gupta, 2057. Humana, New York, New York, USA.
Feehily, C., and Karatzas, K. A. 2013. Role of glutamate metabolism in bacterial responses towards acid and other stresses. *Journal of Applied Microbiology* 114(1):11–24. doi:10.1111/j.1365-2672.2012.05434.x.
Fortes, A. M., and Agudelo-Romero, P. 2018. Polyamine metabolism in climacteric and non-climacteric fruit ripening. In *Polyamines*, eds. R. Alcázar and A. F. Tiburcio, 433–447. Springer, New York.
Fortes, A. M., Agudelo-Romero, P., Pimentel, D., and Alkan, N. 2019. Transcriptional modulation of polyamine metabolism in fruit species under abiotic and biotic stress. *Frontiers in Plant Science* 10:816. doi: 10.3389/fpls.2019.00816.
Fraceto, L. F., Grillo, R., de Medeiros, G. A., Scognamiglio, V., Rea, G., & Bartolucci, C. 2016. Nanotechnology in agriculture: which innovation potential does it have? *Frontiers in Environmental Sciences* 4:20.
Freitas, V. S., de Souza Miranda, R., Costa, J. H., de Oliveira, D. F., et al. 2018. Ethylene triggers salt tolerance in maize genotypes by modulating polyamine catabolism enzymes associated with H_2O_2 production. *Environmental and Experimental Botany* 145:75–86.
Ghalati, R. E., Shamili, M., & Homaei, A. 2020. Effect of putrescine on biochemical and physiological characteristics of guava (*Psidium guajava* L.) seedlings under salt stress. *Scientia Horticulturae* 261:108961.
Ghatak, A., Chaturvedi, P., Paul, P., Agrawal, G. K., et al. 2017. Proteomics survey of Solanaceae family: current status and challenges ahead. *Journal of Proteomics* 169: 41–57. doi: 10.1016/j.jprot.2017.05.016.
Gill, S. S., and Tuteja, N. 2010. Polyamines and abiotic stress tolerance in plants. *Plant Signaling and Behavior* 5(1):26–33. doi:10.4161/psb.5.1.10291.

Gohari, G., Panahirad, S., Sadeghi, M., Akbari, A., et al. 2021. Putrescine-functionalized carbon quantum dot (put-CQD) nanoparticles effectively prime grapevine (*Vitis vinifera* cv. "Sultana") against salt stress. *BMC Plant Biology* 21:120.

Gong, X., Liu, Y., Huang, D., Zeng, G., et al. 2016. Effects of exogenous calcium and spermidine on cadmium stress moderation and metal accumulation in *Boehmeria nivea* (L.) Gaudich. *Environmental Science and Pollution Research International* 23(9):8699–8708. doi:10.1007/s11356-016-6122-6.

Gupta, K., Dey, A., and Gupta, B. 2013. Plant polyamines in abiotic stress responses. *Acta Physiologiae Plantarum* 35(7):2015–2036.

Gupta, K., Dey, A., and Gupta, B. 2013. Polyamines and their role in plant osmotic stress tolerance. In *Climate Change and Plant Abiotic Stress Tolerance*, eds. N. Tuteja and S. S. Gill,1053–1072. Wiley Blackwell, New York, New York, USA.

Gupta, K., Sengupta, A., Chakraborty, M., and Gupta, B. 2016. Hydrogen peroxide and polyamines act as double edged swords in plant abiotic stress responses. *Frontiers in Plant Science* 7:1343. doi:10.3389/fpls.2016.01343.

Gupta, R., and Deswal, R. 2014. Antifreeze proteins enable plants to survive in freezing conditions. *Journal of Biosciences* 39: 931–44. doi: 10.1007/s12038-014-9468-2.

Hao, G., and Zhang, J. 2009. The role of nitric oxide as a bioactive signaling molecule in plants under abiotic stress. In *Nitric Oxide in Plant Physiology*, eds. S. Hayat, M. Mori, J. Pichtel, and A. Ahmad, 115–138. Wiley-Blackwell, New York, New York, USA.

Hassan, N., Ebeed, H., and Aljaarany, A. 2020. Exogenous application of spermine and putrescine mitigate adversities of drought stress in wheat by protecting membranes and chloroplast ultra-structure. *Physiology and Molecular Biology of Plants* 26(2):233–245. doi:10.1007/s12298-019-00744-7.

Hazarika, P., and Rajam, M. V. 2011. Biotic and abiotic stress tolerance in transgenic tomatoes by constitutive expression of S-adenosylmethionine decarboxylase gene. *Physiology and Molecular Biology of Plants* 17:115–128. doi: 10.1007/s12298-011-0053-y.

He, L. X., Nada, K., Kasukabe, Y., and Tachibana, S., 2002. Enhanced susceptibility of photosynthesis to low-temperature photoinhibition due to interruption of chilling induced increase of S-adenosylmethionine decarboxylase activity in leaves of spinach (*Spinacia oleracea* L.). *Plant and Cell Physiology* 43:196–206.

Hijaz, F., and Killiny, N. 2019. Exogenous GABA is quickly metabolized to succinic acid and fed into the plant TCA cycle. *Plant Signaling and Behavior* 14(3):e1573096. doi:10.1080/15592324.2019.1573096.

Huang, D., Huo, J., and Liao, W. 2021. Hydrogen sulfide: roles in plant abiotic stress response and crosstalk with other signals. *Plant Science* 302:110733. doi:10.1016/j.plantsci.2020.110733.

Hussain, S. S., Ali, M., Ahmad, M., and Siddique, K. H. 2011. Polyamines: natural and engineered abiotic and biotic stress tolerance in plants. *Biotechnology Advances* 29(3):300–311. doi:10.1016/j.biotechadv.2011.01.003.

Jancewicz, A. L., Gibbs, N. M., and Masson, P. H. 2016. Cadaverine's functional role in plant development and environmental response. *Frontiers in Plant Science* 7:870. doi: 10.3389/fpls.2016.00870.

Janse van Rensburg, H. C., and Van den Ende, W. 2020. Priming with γ-aminobutyric acid against *Botrytis cinerea* reshuffles metabolism and reactive oxygen species: dissecting signalling and metabolism. *Antioxidants (Basel, Switzerland)* 9(12):1174. doi:10.3390/antiox9121174.

Ji, J., Shi, Z., Xie, T., Zhang, X., et al. 2020. Responses of GABA shunt coupled with carbon and nitrogen metabolism in poplar under NaCl and $CdCl_2$ stresses. *Ecotoxicology and Environmental Safety* 193:110322. doi:10.1016/j.ecoenv.2020.110322.

Ji, J., Yue, J., Xie, T., Chen, W., et al. 2018. Roles of γ-aminobutyric acid on salinity-responsive genes at transcriptomic level in poplar: involving in abscisic acid and ethylene-signalling pathways. *Planta* 248(3):675–690. doi:10.1007/s00425-018-2915-9.

Jiang, D. X., Chu, X., Li, M., Hou, J. J., et al. 2020. Exogenous spermidine enhances salt-stressed rice photosynthetic performance by stabilizing structure and function of chloroplast and thylakoid membranes. *Photosynthetica* 58(1):61–71.

Jing, J., Guo, S., Li, Y., and Li, W. 2020. The alleviating effect of exogenous polyamines on heat stress susceptibility of different heat resistant wheat (*Triticum aestivum* L.) varieties. *Scientific Reports* 10(1):7467. doi:10.1038/s41598-020-64468-5.

Juzoń, K., Czyczyło-Mysza, I., Marcińska, I., et al. 2017. Polyamines in yellow lupin (*Lupinus luteus* L.) tolerance to soil drought. *Acta Physiologiae Plantarum* 39:202. doi:10.1007/s11738-017-2500-z.

Kapoor, R. T., and Hasanuzzaman, M. 2020. Exogenous kinetin and putrescine synergistically mitigate salt stress in *Luffa acutangula* by modulating physiology and antioxidant defence. *Physiology and Molecular Biology of Plants* 26(11):2125–2137.

Kaszler, N., Benkő, P., Bernula, D., Szepesi, Á., Fehér, A., and Gémes, K. 2021. Polyamine metabolism is involved in the direct regeneration of shoots from Arabidopsis lateral root primordia. *Plants (Basel, Switzerland)* 10(2). doi:10.3390/plants10020305.

Keefer, L. K., Nims, R. W., Davies, K. M., and Wink, D. A., 1996. "NONOates" (1-substituted diazen-1-ium-1,2-diolates) as nitric oxide donors: convenient nitric oxide dosage forms. *Methods of Enzymology* 268:281–293.

Khajuria, A., Sharma, N., Bhardwaj, R., and Ohri, P. 2018. Emerging role of polyamines in plant stress tolerance. *Current Protein and Peptide Science* 19(11):1114–1123. doi:10.2174/13892037196661807181 24211.

Khan, M. I. R., Jalil, S. U., Chopra, P., Chillar, H., Ferrante, A., Khan, N. A., and Ansari, M. I. 2021. Role of GABA in plant growth, development and senescence. *Plant Gene* 26:100283.

Khan, M. I. R., Palakolanu, S. R., Chopra, P., Rajurkar, A. B., et al. 2021. Improving drought tolerance in rice: ensuring food security through multi-dimensional approaches. *Physiologia Plantarum* 172:645–668.

Khosroshahi, M. R. Z., Esna-Ashari, M., and Ershadi, A. 2007. Effect of exogenous putrescine on post-harvest life of strawberry (*Fragaria ananassa* Duch.) fruit, cultivar Selva. *Scientia Horticulturae* 114:27–32. doi:10.1016/j.scienta.2007.05.006.

Kim, N. H., Kim, B. S., and Hwang, B. K. 2013. Pepper arginine decarboxylase is required for polyamine and γ-aminobutyric acid signaling in cell death and defence response. *Plant Physiology* 162:2067–2083. doi: 10.1104/pp.113.217372.

Kim, S. W., Gupta, R., Min, C. W., Lee, S. H., et al. 2019. Label-free quantitative proteomic analysis of *Panax ginseng* leaves upon exposure to heat stress. *Journal of Ginseng Research* 43:143–153.

Koike, S., Matsukura, C., Takayama, M., Asamizu, E., and Ezura, H. 2013. Suppression of γ-aminobutyric acid (GABA) transaminases induces prominent GABA accumulation, dwarfism and infertility in the tomato (*Solanum lycopersicum* L.). *Plant and Cell Physiology* 54(5):793–807. doi:10.1093/pcp/pct035.

Kotakis, C., Theodoropoulou, E., Tassis, K., Oustamanolakis, C., Ioannidis, N. E., and Kotzabasis, K. (2014). Putrescine, a fast-acting switch for tolerance against osmotic stress. *Journal of Plant Physiology* 171(2):48–51. doi:10.1016/j.jplph.2013.09.015.

Koushesh Saba, M., Arzani, K., and Barzegar, M. 2012. Postharvest polyamine application alleviates chilling injury and affects apricot storage ability. *Journal of Agricultural and Food Chemistry* 60:8947–8953. doi:10.1021/jf302088e.

Kumar, N., Gautam, A., Dubey, A. K., Ranjan, R., et al. 2019. GABA mediated reduction of arsenite toxicity in rice seedling through modulation of fatty acids, stress responsive amino acids and polyamines biosynthesis. *Ecotoxicology and Environmental Safety* 173:15–27.

Kumar, R. R., Sharma, S. K., Rai, G. K., Singh, K., et al. 2014. Exogenous application of putrescine at pre-anthesis enhances the thermotolerance of wheat (*Triticum aestivum* L.). *Indian Journal of Biochemistry and Biophysics* 51(5):396–406.

Kusano, T., Berberich, T., Tateda, C., and Takahashi, Y. 2008. Polyamines: essential factors for growth and survival. *Planta* 228:367–381.

Li, L., Dou, N., Zhang, H., and Wu, C. 2021. The versatile GABA in plants. *Plant Signaling and Behavior* 16(3):1862565. doi:10.1080/15592324.2020.1862565.

Li, L., Gu, W., Li, J., Li, C., et al. 2018. Exogenously applied spermidine alleviates photosynthetic inhibition under drought stress in maize (*Zea mays* L.) seedlings associated with changes in endogenous polyamines and phytohormones. *Plant Physiology and Biochemistry* 129:35–55. doi:10.1016/j.plaphy.2018.05.017.

Li, S., Cui, L., Zhang, Y., Wang, Y., and Mao, P. 2017. The variation tendency of polyamines forms and components of polyamine metabolism in Zoysiagrass (*Zoysia japonica* Steud.) to salt stress with exogenous spermidine application. *Frontiers in Plant Physiology* 8:208.

Li, S., Jin, H., and Zhang, Q. 2016. The effect of exogenous spermidine concentration on polyamine metabolism and salt tolerance in Zoysiagrass (*Zoysia japonica* Steud) subjected to short-term salinity stress. *Frontiers in Plant Science* 7:1221.

Li, X., Gong, B., and Xu, K. 2014. Interaction of nitric oxide and polyamines involves antioxidants and physiological strategies against chilling-induced oxidative damage in *Zingiber officinale* Roscoe. *Scientia Horticulturae* 170:237–248.

Li, Z., Cheng, B., Peng, Y., and Zhang, Y. 2020. Adaptability to abiotic stress regulated by γ-aminobutyric acid in relation to alterations of endogenous polyamines and organic metabolites in creeping bentgrass. *Plant Physiology and Biochemistry* 157:185–194. doi:10.1016/j.plaphy.2020.10.025.

Li, Z., Hou, J., Zhang, Y., Zeng, W., et al. 2020. Spermine regulates water balance associated with Ca^{2+}-dependent aquaporin (TrTIP2-1, TrTIP2-2 and TrPIP2-7) expression in plants under water stress. *Plant and Cell Physiology* 61(9):1576–1589. doi:10.1093/pcp/pcaa080.

Li, Z., Jing, W., Peng, Y., Zhang, X. Q., et al. 2015. Spermine alleviates drought stress in white clover with different resistance by influencing carbohydrate metabolism and dehydrins synthesis. *PloS One* 10(3):e0120708. doi:10.1371/journal.pone.0120708.

Li, Z., Peng, Y., and Huang, B. 2018. Alteration of transcripts of stress-protective genes and transcriptional factors by γ-aminobutyric acid (GABA) associated with improved heat and drought tolerance in creeping bentgrass (*Agrostis stolonifera*). *International Journal of Molecular Sciences* 19(6):1623. doi:10.3390/ijms19061623.

Liu, B., Peng, X., Han, L., Hou, L., and Li, B. 2020. Effects of exogenous spermidine on root metabolism of cucumber seedlings under salt stress by GC-MS. *Agronomy (Basel, Switzerland)*: 10(4):459.

Liu, M., Chen, J., Guo, Z., and Lu, S. 2017. Differential responses of polyamines and antioxidants to drought in a centipedegrass mutant in comparison to its wild type plants. *Frontiers in Plant Science* 8:792. doi:10.3389/fpls.2017.00792.

Luo, L., Li, Z., Tang, M. Y., Cheng, B. Z., et al. 2020. Metabolic regulation of polyamines and γ-aminobutyric acid in relation to spermidine-induced heat tolerance in white clover. *Plant Biology (Stuttgart, Germany)* 22(5):794–804. doi:10.1111/plb.13139.

Ma, G., Zhang, M., Xu, J., Zhou, W., and Cao, L. 2020. Transcriptomic analysis of short-term heat stress response in *Pinellia ternata* provided novel insights into the improved thermotolerance by spermidine and melatonin. *Ecotoxicology and Environmental Safety* 202:110877. doi:10.1016/j.ecoenv.2020.110877.

Ma, Y., Shukla, V., and Merewitz, E. B. 2017. Transcriptome analysis of creeping bentgrass exposed to drought stress and polyamine treatment. *PLoS One* 12(4):e0175848. doi: 10.1371/journal.pone.0175848.

Majumdar, R., Shao, L., Turlapati, S. A., and Minocha, S. C. 2017. Polyamines in the life of Arabidopsis: profiling the expression of S-adenosylmethionine decarboxylase (SAMDC) gene family during its life cycle. *BMC Plant Biology* 17:264. doi:10.1186/s12870-017-1208-y.

Malik, A. U., and Zora, S. 2005. Pre-storage application of polyamines improves shelf-life and fruit quality of mango. *Journal of Horticultural Science and Biotechnology* 80:363–369. doi: 10.1080/14620316.2005.11511945.

Marcińska, I., Dziurka, K., Waligórski, P., Janowiak, F., et al. 2020. Exogenous polyamines only indirectly induce stress tolerance in wheat growing in hydroponic culture under polyethylene glycol-induced osmotic stress. *Life (Basel, Switzerland)* 10(8):151. doi:10.3390/life10080151.

Martínez-Romero, D., Serrano, M., Carbonell, A., Burgos, L., Riquelme, F., and Valero, D. 2002. Effects of postharvest putrescine treatment on extending shelf life and reducing mechanical damage in apricot. *Journal of Food Science* 67:1706–1712. doi:10.1111/j.1365-2621.2002.tb08710.x.

Mattoo, A. K., and Handa, A. K. 2008. Higher polyamines restore and enhance metabolic memory in ripening fruit. *Plant Science* 174:386–393. doi:10.1016/j.plantsci.2008.01.011.

Mattoo, A. K., Minocha, S. C., Minocha, R., and Handa, A. K. 2010. Polyamines and cellular metabolism in plants: transgenic approaches reveal different responses to diamine putrescine versus higher polyamines spermidine and spermine. *Amino Acids* 38:405–413. doi: 10.1007/s00726-009-0399-4.

Mehta, R., Cassol, T., Li, N., Ali, N., et al. 2002. Engineered polyamine accumulation in tomato enhances phytonutrient content, juice quality, and vine life. *Nature Biotechnology* 20:613–618. doi:10.1038/nbt0602-613.

Minocha, R., Majumdar, R., and Minocha, S. C. 2014. Polyamines and abiotic stress in plants: a complex relationship. *Frontiers in Plant Science* 5:175. doi:10.3389/fpls.2014.00175.

Moschou, P. N., Paschalidis, K. A., and Roubelakis-Angelakis, K. A. 2008. Plant polyamine catabolism: the state of the art. *Plant Signaling and Behavior* 3:1061–1066. doi: 10.4161/psb.3.12.7172.

Moschou, P. N., Paschalidis, K. A., Delis I. D., Andriopoulou, A. H., et al. 2008. Spermidine exodus and oxidation in the apoplast induced by abiotic stress is responsible for H_2O_2 signatures that direct tolerance responses in tobacco. *Plant Cell* 20:1708–1724.

Muilu-Mäkelä, R., Vuosku, J., Läärä, E., Saarinen, M., et al. 2015. Water availability influences morphology, mycorrhizal associations, PSII efficiency and polyamine metabolism at early growth phase of Scots pine seedlings. *Plant Physiology and Biochemistry* 88:70–81.

Müller, S., Liebau, E., Walter, R. D., and Krauth-Siegel, R. L. 2003. Thiol-based redox metabolism of protozoan parasites. *Trends in Parasitology* 19:320–328.

Nahar, K., Hasanuzzaman, M., Alam, M. M., Rahman, A., et al. 2016. Polyamine and nitric oxide crosstalk: antagonistic effects on cadmium toxicity in mung bean plants through upregulating the metal detoxification, antioxidant defense and methylglyoxal detoxification systems. *Ecotoxicology and Environmental Safety* 126:245–255.

Nahar, K., Rahman, M., Hasanuzzaman, M., Alam, M. M., et al. 2016. Physiological and biochemical mechanisms of spermine-induced cadmium stress tolerance in mung bean (*Vigna radiata* L.) seedlings. *Environmental Science and Pollution Research International* 23(21):21206–21218. doi:10.1007/s11356-016-7295-8.

Nambeesan, S., Datsenka, T., Ferruzzi, M. G., Malladi, A., et al. 2010. Overexpression of yeast spermidine synthase impacts ripening, senescence and decay symptoms in tomato. *The Plant Journal* 63:836–847. doi:10.1111/j.1365-313X.2010.04286.x.

Nambeesan, S., Mattoo, A. K., and Handa, A. K. 2019. Nexus between spermidine and floral organ identity and fruit/seed set in tomato. *Frontiers in Plant Science* 10:1033. doi:10.3389/fpls.2019.01033.

Napieraj, N., Reda, M. G., and Janicka, M. G. 2020. The role of NO in plant response to salt stress: interactions with polyamines. *Functional Plant Biology* 47(10):865–879. doi:/10.1071/FP19047.

Naz, R., Sarfraz, A., Anwar, Z., Yasmin, H., et al. 2021. Combined ability of salicylic acid and spermidine to mitigate the individual and interactive effects of drought and chromium stress in maize (*Zea mays* L.). *Plant Physiology and Biochemistry* 159:285–300. doi:10.1016/j.plaphy.2020.12.022.

Nehela, Y., and Killiny, N. 2020. The unknown soldier in citrus plants: polyamines-based defensive mechanisms against biotic and abiotic stresses and their relationship with other stress-associated metabolites. *Plant Signaling and Behavior* 15(6):1761080. doi:10.1080/15592324.2020.1761080.

Neill, S. J., Desikan, R., and Hancock, J. T. 2003. Nitric oxide signalling in plants. *New Phytologist* 159:11–35.

Neill, S., Barros, R., Bright, J., Desikan, R., et al. 2008. Nitric oxide, stomatal closure, and abiotic stress. *Journal of Experimental Botany* 59:165–176.

Pál, M., Szalai, G., and Janda, T. 2015. Speculation: polyamines are important in abiotic stress signaling. *Plant Science* 237:16–23. doi:10.1016/j.plantsci.2015.05.003.

Pál, M., Tajti, J., Szalai, G., et al. 2018. Interaction of polyamines, abscisic acid and proline under osmotic stress in the leaves of wheat plants. *Scientific Reports* 8:12839. doi:10.1038/s41598-018-31297-6.

Paschalidis, K., Tsaniklidis, G., Wang, B.-Q., Delis, C., et al. 2019. The interplay among polyamines and nitrogen in plant stress responses. *Plants* 8:315. doi:10.3390/plants8090315.

Paul, S., and Roychoudhury, A. 2017. Seed priming with spermine and spermidine regulates the expression of diverse groups of abiotic stress-responsive genes during salinity stress in the seedlings of indica rice varieties. *Plant Gene* 11:124–132.

Peng, D., Wang, X., Li, Z., Zhang, Y., et al. 2016. NO is involved in spermidine-induced drought tolerance in white clover via activation of antioxidant enzymes and genes. *Protoplasma* 253(5):1243–1254. doi:10.1007/s00709-015-0880-8.

Petkou, I. T., Pritsa, T. S., and Sfakiotakis, E. M. 2004. Effects of polyamines on ethylene production, respiration and ripening of kiwifruit. *Journal of Horticultural Science and Biotechnology* 79:977–980. doi:10.1080/14620316.2004.11511876.

Podlešáková, K., Ugena, L., Spíchal, L., Doležal, K., and De Diego, N. 2019. Phytohormones and polyamines regulate plant stress responses by altering GABA pathway. *New Biotechnology* 48:53–65. doi:10.1016/j.nbt.2018.07.003.

Poór, P., Nawaz, K., Gupta, R., Ashfaque, F., and Khan, M. I. R. 2021. Ethylene involvement in the regulation of heat stress tolerance in plants. *Plant Cell Reports* 1–24. doi: 10.1007/s00299-021-02675-8.

Pottosin, I., Velarde-Buendía, A. M., Bose, J., Zepeda-Jazo, I., et al. 2014. Cross-talk between reactive oxygen species and polyamines in regulation of ion transport across the plasma membrane: implications for plant adaptive responses. *Journal of Experimental Botany* 65: 1271–1283. doi:10.1093/jxb/ert423.

Puyang, X., An, M., Xu, L., Han, L., and Zhang, X. 2016. Protective effect of exogenous spermidine on ion and polyamine metabolism in Kentucky bluegrass under salinity stress. *Horticulture, Environment, and Biotechnology* 57(1):11–19.

Qian, R., Ma, X., Zhang, X., Hu, Q., Liu, H., and Zheng, J. 2020. Effect of exogenous spermidine on osmotic adjustment, antioxidant enzymes activity, and gene expression of *Gladiolus gandavensis* seedlings under salt stress. *Journal of Plant Growth Regulation* 1:15.

Qiao, W., and Fan, L. M. 2008. Nitric oxide signaling in plant responses to abiotic stresses. *Journal of Integrative Plant Biology* 50:1238–1246.

Radhakrishnan, R., and Lee, I.-J. 2013. Spermine promotes acclimation to osmotic stress by modifying antioxidant, abscisic acid, and jasmonic acid signals in soybean. *Journal of Plant Growth Regulation* 32:22–30. doi:10.1007/s00344-012-9274-8.

Rady, M. M., and Hemida, K. A. 2015. Modulation of cadmium toxicity and enhancing cadmium-tolerance in wheat seedlings by exogenous application of polyamines. *Ecotoxicology and Environmental Safety* 119:178–185. doi:10.1016/j.ecoenv.2015.05.008.

Riyazuddin, R., Nisha, N., Singh, K., et al. 2021. Involvement of dehydrin proteins in mitigating the negative effects of drought stress in plants. *Plant Cell Reports.* https://doi.org/10.1007/s00299-021-02720-6.

Riyazuddin, R., Verma, R., Singh, K., Nisha, N., Keisham, M., Bhati, K. K., Kim, S. T., and Gupta, R. 2020. Ethylene: a master regulator of salinity stress tolerance in plants. *Biomolecules (Basel, Switzerland)* 10:959. doi:10.3390/biom10060959.

Saha, J., and Giri, K. 2017. Molecular phylogenomic study and the role of exogenous spermidine in the metabolic adjustment of endogenous polyamine in two rice cultivars under salt stress. *Gene (Basel, Switzerland)* 609:88–103.

Sakihama, Y., Tamaki, R., Shimoji, H., Ichiba, T., et al. 2003. Enzymatic nitration of phytophenolics: evidence for peroxynitrite-independent nitration of plant secondary metabolites. *FEBS Letters* 553:377–380.

Sang, T., Shan, X., Li, B., and Shu, S. 2016. Comparative proteomic analysis reveals the positive effect of exogenous spermidine on photosynthesis and salinity tolerance in cucumber seedlings. *Plant Cell Reports* 35:1769–1782.

Sehrawat, A., Gupta, R., and Deswal, R. 2013. Nitric oxide-cold stress signalling cross-talk, evolution of a novel regulatory mechanism. *Proteomics* 13:1816–1835.

Seifi, H. S., and Shelp, B. J. 2019. Spermine differentially refines plant defense responses against biotic and abiotic stresses. *Frontiers in Plant Science* 10:117. doi:10.3389/fpls.2019.00117.

Seifikalhor, M., Aliniaeifard, S., Bernard, F., Seif, M., et al. 2020. γ-Aminobutyric acid confers cadmium tolerance in maize plants by concerted regulation of polyamine metabolism and antioxidant defense systems. *Scientific Reports* 10(1):1–18.

Shapiro, A. D. 2005. Nitric oxide signaling in plants. *Vitamins and Hormones* 72:339–398.

Shen, J. L., Wang, Y., Shu, S., Jahan, M. S., et al. 2019. Exogenous putrescine regulates leaf starch overaccumulation in cucumber under salt stress. *Scientia Horticulturae* 253:99–110.

Shi, H., and Chan, Z. 2014. Improvement of plant abiotic stress tolerance through modulation of the polyamine pathway. *Journal of Integrative Plant Biology* 56(2):114–121. doi:10.1111/jipb.12128.

Shi, H., Ye, T., and Chan, Z. 2013. Comparative proteomic and physiological analyses reveal the protective effect of exogenous polyamines in the bermudagrass (*Cynodon dactylon*) response to salt and drought stresses. *Journal of Proteome Research* 12(11):4951–4964. doi:10.1021/pr400479k.

Shi, J., Fu, X. Z., Peng, T., Huang, X. S., et al. 2010. Spermine pretreatment confers dehydration tolerance of citrus in vitro plants via modulation of antioxidative capacity and stomatal response. *Tree Physiology* 30(7):914–922.

Sobieszczuk-Nowicka, E. 2017. Polyamine catabolism adds fuel to leaf senescence. *Amino Acids* 49:49–56. doi:10.1007/s00726-016-2377-y.

Su, Y. C., Wang, J. J., Lin, T. T., and Pan, T. M. 2003. Production of the secondary metabolites γ-aminobutyric acid and monacolin K by Monascus. *Journal of Industrial Microbiology and Biotechnology* 30(1):41–46.

Sun, X., Yuan, Z., Wang, B., Zheng, L., Tan, J., and Chen, F. 2020. Physiological and transcriptome changes induced by exogenous putrescine in anthurium under chilling stress. *Botanical Studies* 61(1):28. doi:10.1186/s40529-020-00305-2.

Szepesi, Á. 2019. Molecular mechanisms of polyamines-induced abiotic stress tolerance in plants. In *Approaches for Enhancing Abiotic Stress Tolerance in Plants*, eds. M. Hasanuzzaman, K. Nahar, M. Fujita, H. Oku, and T. Islam, 387–403. CRC Press, Taylor and Francis, Boca Raton, Florida, USA.

Szepesi, Á. 2020. Halotropism: phytohormonal aspects and potential applications. *Frontiers in Plant Science* 11:571025. doi:10.3389/fpls.2020.571025.

Szepesi, Á., Csiszár, J., Gémes, K., Horváth, E., Horváth, F., Simon, M. L., and Tari, I. 2009. Salicylic acid improves acclimation to salt stress by stimulating abscisic aldehyde oxidase activity and abscisic acid accumulation, and increases Na+ content in leaves without toxicity symptoms in *Solanum lycopersicum* L. *Journal of Plant Physiology* 166(9):914–925. doi:10.1016/j.jplph.2008.11.012.

Szepesi, Á., Gémes, K., Orosz, G., Pető, A., Takács, Z., Vorák, M., and Tari, I. 2011. Interaction between salicylic acid and polyamines and their possible roles in tomato hardening processes. *Acta Biologica Szegediensis* 55(1):165–166.

Szepesi, Á., and Szőllősi, R. 2018. Mechanism of proline biosynthesis and role of proline metabolism enzymes under environmental stress in plants. In *Plant Metabolites and Regulation under Environmental Stress*, eds. P. Ahmad, M. A. Ahanger, V. P. Singh, D. K. Tripathi, P. Alam, and M. N. Alyemeni, 337–353. Academic Press, Cambridge, Massachusetts, USA. doi:10.1016/B978-0-12-812689-9.00017-0.

Takács, Z., Poór, P., Szepesi, Á., and Tari, I. 2017. In vivo inhibition of polyamine oxidase by a spermine analogue, MDL-72527, in tomato exposed to sub-lethal and lethal salt stress. *Functional Plant Biology* 44:480–492.

Takahashi, T., and Kakehi, J. I. 2010. Polyamines: ubiquitous polycations with unique roles in growth and stress responses. *Annals of Botany* 105:1–6.

Takayama, M., and Ezura, H. 2015. How and why does tomato accumulate a large amount of GABA in the fruit? *Frontiers in Plant Science* 6:612. doi:10.3389/fpls.2015.00612.

Tang, C. F., Zhang, R. Q., Wen, S. Z., Li, C. F., et al. 2009. Effects of exogenous spermidine on subcellular distribution and chemical forms of cadmium in *Typha latifolia* L. under cadmium stress. *Water Science and Technology* 59(8):1487–1493. doi:10.2166/wst.2009.170.

Tang, C., Zhang, R., Hu, X., Song, J., et al. 2019. Exogenous spermidine elevating cadmium tolerance in *Salix matsudana* involves cadmium detoxification and antioxidant defense. *International Journal of Phytoremediation* 21(4):305–315. doi.10.1080/15226514.2018.1524829.

Tanou, G., Ziogas, V., Belghazi, M., Christou, A., et al. 2014. Polyamines reprogram oxidative and nitrosative status and the proteome of citrus plants exposed to salinity stress. *Plant Cell and Environment* 37:864–885. doi:10.1111/pce.12204.

Tarkowski, Ł. P., Signorelli, S., and Höfte, M. 2020. γ-Aminobutyric acid and related amino acids in plant immune responses: emerging mechanisms of action. *Plant, Cell and Environment* 43(5):1103–1116. doi:10.1111/pce.13734.

Tavladoraki, P., Cona, A., and Angelini, R. 2016. Copper-containing amine oxidases and FAD-dependent polyamine oxidases are key players in plant tissue differentiation and organ development. *Frontiers in Plant Science* 7:824. doi:10.3389/fpls.2016.00824.

Tiburcio, A. F., Altabella, T., Bitrián, M., and Alcázar, R. 2014. The roles of polyamines during the lifespan of plants: from development to stress. *Planta* 240:1–18. doi:10.1007/s00425-014-2055-9.

Tun, N. N., Santa-Catarina, C., Begum, T., Silveira, V., et al. 2006. Polyamines induce rapid biosynthesis of nitric oxide (NO) in *Arabidopsis thaliana* seedlings. *Plant and Cell Physiology* 47:346–354.

Valero, D., Martínez-Romero, D., and Serrano, M. 2002. The role of polyamines in the improvement of the shelf life of fruit. *Trends in Food Science and Technology* 13:228–234. doi:10.1016/S0924-2244(02)00134-6.

Wang, X., Shi, G., Xu, Q., and Hu, J. 2007. Exogenous polyamines enhance copper tolerance of *Nymphoides peltatum*. *Journal of Plant Physiology* 164(8):1062–1070. doi:10.1016/j.jplph.2006.06.003.

Wang, Y., Guo, S., Wang, L., Wang, L., et al. 2018. Identification of microRNAs associated with the exogenous spermidine-mediated improvement of high-temperature tolerance in cucumber seedlings (*Cucumis sativus* L.). *BMC Genomics* 19(1):285. doi:10.1186/s12864-018-4678-x.

Wang, Z., Xu, Y., Wang, J., Yang, J., et al. 2012. Polyamine and ethylene interactions in grain filling of superior and inferior spikelets of rice. *Plant Growth Regulation* 66:215–228.

Wani, S. H., Kumar, V., Khare, T., Guddimalli, R., et al. 2020. Engineering salinity tolerance in plants: progress and prospects. *Planta* 251(4):76. doi:10.1007/s00425-020-03366-6.

Wimalasekera, R., Tebartz, F., and Scherer, G. F. 2011. Polyamines, polyamine oxidases and nitric oxide in development, abiotic and biotic stresses. *Plant Science* 181(5):593–603.

Wu, J., Shu, S., Li, C., Sun, J., and Guo, S. 2018. Spermidine-mediated hydrogen peroxide signaling enhances the antioxidant capacity of salt-stressed cucumber roots. *Plant Physiology and Biochemistry* 128:152–162.

Wu, X., Jia, Q., Ji, S., Gong, B., et al. 2020. Gamma-aminobutyric acid (GABA) alleviates salt damage in tomato by modulating Na^+ uptake, the GAD gene, amino acid synthesis and reactive oxygen species metabolism. *BMC Plant Biology* 20(1):465. doi:10.1186/s12870-020-02669-w.

Wu, X., Shu, S., Wang, Y., Yuan, R., and Guo, S. 2019. Exogenous putrescine alleviates photoinhibition caused by salt stress through cooperation with cyclic electron flow in cucumber. *Photosynthesis Research* 141(3):303–314.

Wu, Z., Wang, J., Yan, D., Yuan, H., et al. 2020. Exogenous spermidine improves salt tolerance of pecan-grafted seedlings via activating antioxidant system and inhibiting the enhancement of Na+/K+ ratio. *Acta Physiologiae Plantarum* 42(5):1–13.

Xie, T., Ji, J., Chen, W., Yue, J., et al. 2019. γ-Aminobutyric acid is closely associated with accumulation of flavonoids. *Plant Signaling and Behavior* 14(7):1604015. doi:10.1080/15592324.2019.1604015.

Xiong, F., Liao, J., Ma, Y., Wang, Y., et al. 2018. The protective effect of exogenous putrescine in the response of tea plants (*Camellia sinensis*) to salt stress. *HortScience* 53(11):1640–1646.

Xu, J., Yang, J., Xu, Z., Zhao, D., and Hu, X. 2020. Exogenous spermine-induced expression of SlSPMS gene improves salinity–alkalinity stress tolerance by regulating the antioxidant enzyme system and ion homeostasis in tomato. *Plant Physiology and Biochemistry* 157:79–92.

Yang, W., Yin, Y., Li, Y., et al. 2014. Interactions between polyamines and ethylene during grain filling in wheat grown under water deficit conditions. *Plant Growth and Regulation* 72: 189–201. doi:10.1007/s10725-013-9851-2.

Yin, L., Wang, S., Liu, P., Wang, W., et al. 2014. Silicon-mediated changes in polyamine and 1-aminocyclopropane-1-carboxylic acid are involved in silicon-induced drought resistance in *Sorghum bicolor* L. *Plant Physiology and Biochemistry* 80:268–277.

Yong, B., Xie, H., Li, Z., Li, Y. P., et al. 2017. Exogenous application of GABA improves PEG-induced drought tolerance positively associated with GABA-shunt, polyamines, and proline metabolism in white clover. *Frontiers in Plant Physiology* 8:1107.

Yu, Y., Jin, C., Sun, C., Wang, J., et al. 2016. Inhibition of ethylene production by putrescine alleviates aluminium induced root inhibition in wheat plants. *Scientific Reports* 6:18888.

Yu, Z., Jia, D., and Liu, T. 2019. Polyamine oxidases play various roles in plant development and abiotic stress tolerance. *Plants (Basel, Switzerland)* 8(6):184. doi:10.3390/plants8060184.

Yuan, R. N., Shu, S., Guo, S. R., Sun, J., and Wu, J. Q. 2018. The positive roles of exogenous putrescine on chlorophyll metabolism and xanthophyll cycle in salt-stressed cucumber seedlings. *Photosynthetica* 56(2):557–566.

Yue, J., Du, C., Ji, J., Xie, T., et al. 2018. Inhibition of α-ketoglutarate dehydrogenase activity affects adventitious root growth in poplar via changes in GABA shunt. *Planta* 248(4):963–979. doi:10.1007/s00425-018-2929-3.

Zhang, Y., Li, Z., Li, Y. P., Zhang, X. Q., et al. 2018. Chitosan and spermine enhance drought resistance in white clover, associated with changes in endogenous phytohormones and polyamines, and antioxidant metabolism. *Functional Plant Biology* 45(12):1205–1222. doi.10.1071/FP18012.

Zhong, M., Yuan, Y., Shu, S., Sun, J., et al. 2016. Effects of exogenous putrescine on glycolysis and Krebs cycle metabolism in cucumber leaves subjected to salt stress. *Plant Growth Regulation* 79(3):319–330.

Zhuo, C., Liang, L., Zhao, Y., Guo, Z., and Lu, S. 2018. A cold responsive ethylene responsive factor from *Medicago falcata* confers cold tolerance by up-regulation of polyamine turnover, antioxidant protection, and proline accumulation. *Plant, Cell and Environment* 41(9):2021–2032.

7 Decoding the Multifaceted Role of Glycine Betaine in Heavy Metal Stress Regulation

Harsimran Kaur[1], Sukhmeen Kaur Kohli[2], Sakshi Sharma[3], and Renu Bhardwaj[2]*

[1]Post Graduate Department of Botany, Khalsa College, Amritsar, Punjab, India

[2]Plant Stress Physiology Laboratory, Department of Botanical and Environment Sciences, Guru Nanak Dev University, Amritsar, Punjab, India

[3]Department of Botany, DAV College, Amritsar, Punjab, India

Corresponding Author: Assistant Professor, Dr. Harsimran Kaur, PG Department of Botany, Khalsa College, Amritsar, Punjab, India

*Corresponding author: e-mail: harsimran.dhingra@gmail.com

CONTENTS

7.1 INTRODUCTION

Heavy metals (HMs) are dangerous environmental pollutants that are toxic to organisms even at extremely low concentrations and are well documented for their persistence, toxicity and non-biodegradability in the environment (Ali et al. 2020). HMs such as copper (Cu), lead (Pb), arsenic (As), nickel (Ni), cadmium (Cd), cobalt (Co) and zinc (Zn) are phytotoxic in nature when present above a certain threshold level (Anjum et al. 2014). Not only can the concentrations at which these HMs occur in the environment be hazardous to the biosphere, but their biomagnification at each level

DOI: 10.1201/9781003159636-7

of the food chain is also of prime concern. They disturb the environment by adversely affecting soil properties such as fertility, crop yields and eventually human health (Fariduddin et al. 2013). Depending upon the plant species and the concentration of HMs present in the contiguous environment, plants can take up, translocate and accumulate these toxic metals in different parts and at different levels (Ali et al 2020). This not only negatively affects the plant growth attributes but also has toxic effects on consumers. HM toxicity in plants leads to excessive production of reactive oxygen species (ROS), which leads to oxidative damage to the cellular constituents (Fariduddin et al. 2013). This manifests in the form of altered morphological and physiological characteristics of the plants (Ali et al 2020). Toxicity due to HMs depends largely on the concentration of the metal in the environment, and it also varies from one plant species to another. Certain HMs, such as Cd and Cr, reduce nutrient availability to plants, which negatively affects several plant metabolic processes, eventually resulting in reduced plant growth and yield (Hossain et al. 2010; Jabeen et al. 2015; He et al. 2019; Ahmad et al. 2020; Khan et al. 2021).

Plants, being sessile organisms, have evolved an array of strategies to resist or escape different types of stresses, both abiotic and biotic, to which they are continuously exposed during their lifetime (Carillo-Campos et al. 2018). On exposure to HMs, plants accumulate certain organic metabolites, commonly known as osmolytes or compatible solutes, in their cells for protection from metal toxicity. Osmolytes are water soluble and exhibit non-toxicity even at high concentrations (Ali et al. 2015; Dikilitaks et al. 2020; Ashfaque et al. 2020). These molecules allow plant cells to preserve water and avoid disturbances in their normal functioning on exposure to HM stress. Polyols, amino acids, sugars, proline and glycine betaine (GB) are the most common osmolytes produced by plants under stress. GB, a quaternary amine, is the most well-studied and documented compatible solute and is synthesized in large amounts in plants under stress (Rasheed et al, 2014; Gupta and Thind 2019). Enhanced levels of GB are often correlated with increased tolerance (Hasanuzzaman et al. 2019). GB can directly quench the ROS that are over-produced in stressed plants, improving photosynthesis, nutrient uptake and transpiration rate.

Several reports suggest the diverse roles of GB in higher plants, which include stabilization of complex proteins and enzyme quaternary structures, protection of the photosynthetic machinery, and stabilizing membrane structures against the damage resulting from oxidative stress (Masood et al. 2016). Besides, it also modulates the expression of antioxidative genes and acts as both a molecular chaperone and an osmoregulator (Yadu et al. 2017). Understanding the detoxification strategies, specifically the biochemical aspects, adopted by plants against the oxidative stress prompted by abiotic stress is important for manipulating the metal tolerance in different plants (Hossain et al. 2010). A number of plant species, such as *Oryza sativa*, *Solanum lycopersicum* and *S. tuberosum*, lack the ability to synthesize GB naturally. An exogenous supply of GB to these plants is considered to be an important strategy for overcoming stress arising from abiotic factors. Exogenous GB is reported to diminish HM toxicity in several plant species (Ali et al. 2015; Farooq et al. 2016; Kumar et al. 2019). To increase the resistance of these plants towards HM stress, introduction of the GB synthesis pathway via genetic engineering offers an alternative to enhance plant tolerance to such abiotic stresses (Fariduddin et al. 2013). With advancement in genomic and proteomic knowledge and genetic engineering techniques, several plants have been genetically modified with the genes of the GB biosynthesis pathway and have acquired tolerance to HM stress. In this chapter, we summarize the effect of HM toxicity on GB accumulation, the ameliorative effect of both endogenous and exogenous GB on plants, and the recent progress in genetic engineering for development of GB-biosynthesizing transgenic plants exhibiting tolerance towards HM stress.

7.2 HEAVY METAL STRESS INDUCED SIGNALING

Environmental stresses such as HMs, drought, temperature and salinity lead to physiological, morphological and biochemical alterations in plants. Both biotic and abiotic stresses create oxidative stress within the plants, which readily damages cell components, resulting in their dysfunction. Nicotinamide adenine dinucleotide phosphate (NADPH) oxidases are involved in the generation of ROS, which tend to accumulate in cellular organelles, mainly the nucleus, mitochondria and the cytoplasm. ROS over-production leads to lipid peroxidation, protein denaturation, DNA and RNA oxidation, and carbohydrate oxidation in the cellular compartments besides affecting the enzymatic functioning in plants (Ali et al. 2020).

In most crops, the initial signs of metal toxicity are similar to those of other abiotic stresses. An imbalance in the generation and scavenging of ROS is a common feature under HM stress in plants that initiates their antioxidant system. At toxic concentrations, these HMs impede metabolic processes by causing a disturbance in the structure and functioning of proteins, altering cytoplasmic membrane integrity, causing nutritional imbalance, and repressing photosynthesis, respiration and enzymatic activities (Rucinska-Sobkowiak 2016). These similar responses reflect an interconnection between the complex signaling networks transduced under abiotic stress conditions. The interplay and conjunction of these pathways finally leads to regulation of several transcription factors, which further activates stress responsive genes (Jalmi et al. 2018). Against the metal overdose, plants boost defense responses such as metal sequestration into vacuoles, metal chelation, and regulating transporters for metal intake as well as intensifying antioxidative mechanisms. These plant responses are attributable to an intricate signaling network that functions within the cell and transmits extracellular stimuli as an intracellular response. Numerous signal transduction cascades activate in response to HM toxicity, and their induction/activation depends upon the plant species and the concentration of the metal (Jalmi et al. 2018).

Diverse signaling pathways, including mitogen activated protein kinases (MAPKs), ROS signaling, calcium dependent signaling and hormone signaling, are initiated in plants in response to HM stress (Figure 7.1). These signal transduction pathways enhance the expression of stress responsive genes, which leads to adaptive cell responses comprising an improved antioxidant system, rapid ROS scavenging, and accumulation of stress proteins and compatible solutes (Ali et al. 2020). Plants generally synthesize different kinds of solutes, commonly known as osmoprotectants, as an acclimatory response to stress, and these include polyols, sucrose, trehalose, proline, alanine betaine and GB, which provide plant protection against dehydration injuries.

7.3 GLYCINE BETAINE SYNTHESIS IN PLANTS

GB occurs naturally as a quaternary ammonium compound (N, N, N-trimethylglycine) that behaves as a dipolar zwitterion. At physiological pH, it is electrically neutral and is synthesized in the chloroplast to protect the thylakoid membranes and maintain the photosynthetic efficiency of the plant cells under stress (Fariduddin et al. 2013; Gupta and Thind 2019; Hasanuzzaman et al. 2019). Choline and glycine are the two major substrates involved in biosynthesis of GB in most living organisms, including microorganisms, plants and animals. In plants, choline, the substrate for GB synthesis, is produced in the cytoplasm by a reaction catalyzed by phosphoethanolamine N-methyltransferase (PEAMT). Three-step methylation of phosphoethanolamine catalyzed by cytosolic PEAMT converts it into phosphocholine, the choline synthesis precursor. From the cytosol, choline is transported to the chloroplast, where it is further oxidized in two steps (Figure 7.2). Firstly, choline is oxidized to betaine aldehyde, which is a toxic intermediate. Choline monooxygenase (CMO), a ferredoxin-dependent soluble protein possessing a specific Rieske-type iron–sulfur protein motif, catalyzes this oxidation step. The next oxidation is catalyzed by nicotinamide adenine dinucleotide $(NAD)^+$ -dependent betaine aldehyde dehydrogenase (BADH), which oxidizes the toxic betaine aldehyde to GB. Both the enzymes, CMO and BADH, encoded by nuclear genes, have transit sequences that target them into the chloroplasts, where they become localized in the stroma. The activity of these enzymes and the concentration of GB tend to be enhanced during abiotic stress (Surabhi and Rout 2020). Certain microorganisms, such as *Escherichia coli*, synthesize betaine from choline

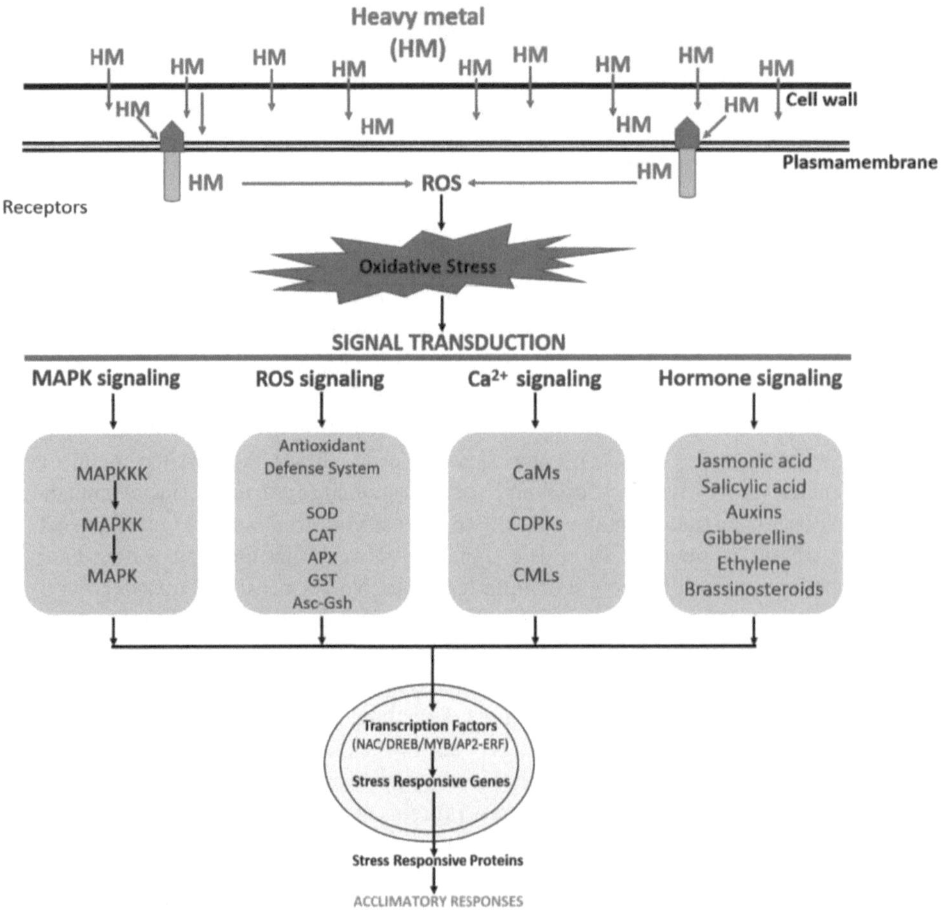

FIGURE 7.1 Crosstalk between signaling cascades in plants under the influence of heavy metal stress. Passive diffusion of HMs occurs from the soil spaces into the root via the cell wall, where they either bind to certain receptors or pass into the symplast through metal transporters present in the plasma membrane and trigger a signal. The signal perpetuates in the form of excess reactive oxygen species (ROS) in the cell, which leads to oxidative stress. The signal response further activates MAPKs, ROS, Ca^{2+} and hormone signaling pathways that induce nuclear responses.

employing an oxygen-dependent and membrane-bound choline dehydrogenase (CDH) enzyme along with BADH. In contrast to this, *Arthrobacter globiformis* (a soil bacterium) biosynthesizes betaine *via* a reaction catalyzed by choline oxidase (COD), a single flavoenzyme, instead of the two enzymes otherwise involved, and releases hydrogen peroxide as a by-product (Sakamoto and Murata 2000, Ahmad et al. 2013).

Another biosynthetic pathway of GB has been described in the halophytic cyanobacteria *Ectothiorhodospira halochloris* and *Actinopolyspora halophila*, where the bacteria use glycine instead of choline for GB synthesis. The pathway involves three-step methylation of glycine, with sarcosine and dimethylglycine as intermediate molecules, eventually producing GB. Glycinesarcosine methyltransferase (GSMT) and sarcosine dimethylglycine methyltransferase (SDMT) catalyze these methylation reactions. GSMT catalyzes the first methylation step, both GSMT and SDMT catalyze the second, and SDMT alone catalyzes the third methylation reaction (Ahmad et al. 2013; Surabhi and Rout 2020).

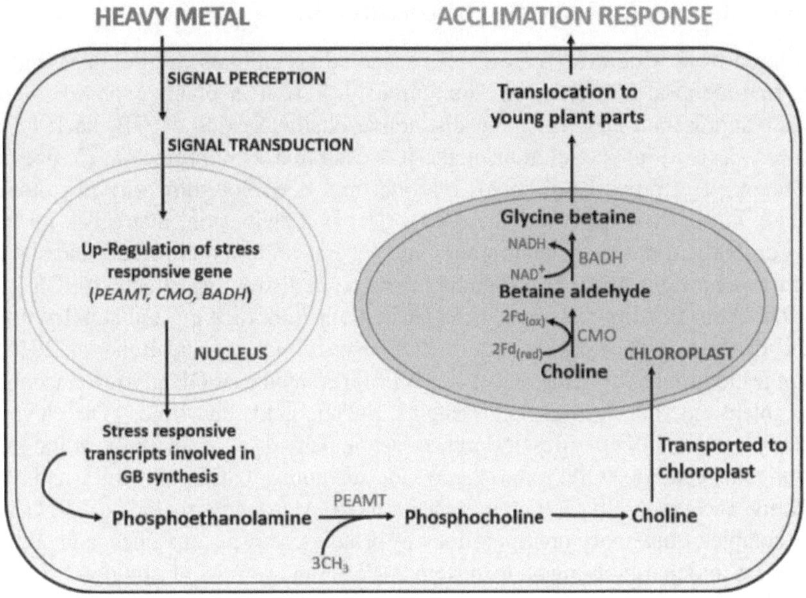

FIGURE 7.2 Biosynthesis of glycine betaine in plants.

7.3.1 Glycine Betaine Accumulation in Plants under Stress

Young leaves of plants tend to accumulate maximum amounts of GB in their chloroplasts, since they are the site of GB biosynthesis. From here, GB is translocated to different plant parts via phloem (Sharma et al. 2019). In a study conducted by Ladyman et al. (1980) in barley plants, GB was identified as an inactive end product, which accumulated in the young shoots of the stressed plants. When the stress cues were removed, GB degradation did not occur. Radiotracers revealed that GB distribution among plant tissues changes as time proceeds. In a well-watered environment, GB is translocated from mature leaves to young expanding leaves (Hasanuzzaman et al. 2019). In transgenic *Arabidopsis* and tomato plants accumulating GB, high levels of GB are present in actively growing tissues like shoot apices and flowers, thus indicating efficient translocation of GB from the source to the sink *via* phloem (Chen and Murata 2011; Masood et al. 2016). Thus, the stress tolerance capability of GB is credited to chloroplast-produced GB, which helps in regulating not only the photosynthetic machinery but also the lipids and enzymes crucial for the flow of electrons in and out of the thylakoid membrane (Sharma et al. 2019).

GB levels in plants increase under abiotic stresses, as observed in *Beta vulgaris*, *Triticum aestivum* and several other crops. Under normal conditions, these plants tend to accumulate low concentrations of GB. However, on exposure to stressful conditions, they amass higher amounts of GB. But not all plant species produce GB either under normal or under stressful conditions. Several plant species, such as *A. thaliana*, *O. sativa*, *Nicotiana tabacum* and Brassica spp., do not amass GB under normal or stress conditions (Masood et al. 2016). In tobacco, a GB-deficient plant, absence of CMO is the key limitation on GB synthesis. Similarly, GB-deficient sorghum and maize plants under abiotic stress had enhanced concentrations of phosphocholine and choline but not GB. Their inability to amass GB is not due to lack of CMO but is attributed to increased levels of phosphocholine, which act as an inhibitor of PEAMT (Chen and Murata 2011). However, natural accretion of GB is not adequate to shield plants from abiotic stressors. Thus, exogenous GB application is proposed to be a beneficial strategy for overcoming such stresses. In several plants exposed to HM toxicity, exogenous GB ameliorates the response. Nonetheless, the response depends largely on the plant species and the genotypes (Bhatti et al. 2013; Ali et al. 2015; Farooq et al. 2016).

7.3.2 Glycine Betaine: An Essential Osmoprotectant

GB is sequestered by a wide array of organisms, including animals, fungi, algae and cyanobacteria (Turkan and Demiral 2009), but are ubiquitously present in plants exposed to abiotic cues (Lokhande and Suprasanna 2012). A few distinctive characteristics of GB, such as high solubility and lower viscosity of its solution, make it a desirable osmoprotectant (Yancey 2005). As an osmoprotectant, it effortlessly dissolves in water and does not show any phytotoxicity, even at comparatively high concentrations (Giri 2011). GB, as a zwitterion, interacts with both hydrophobic and hydrophilic domains of membranes and several protein complexes, and stabilizes their functional and structural integrity, thus shielding them from distressing effects of ROS (Annunziata et al. 2019). The main functions attributed to GB in plants embrace protein stabilization, osmotic regulation and decreasing the toxic effects of ROS (Roychoudhury and Banerjee 2016). Another significant suggestion, made by Umezwa et al. (2006), revealed that GB at low concentrations had the ability to shield macromolecules, i.e., proteins, nucleic acids and lipids. The elevation in GB levels is directly correlated with stress tolerance, accompanied by an increase in ion homeostasis and the activities of catalase (CAT) and superoxide dismutase (SOD) enzymes and a decline in lipid peroxidation and eventually, cell membrane damage (Alasvandyari et al. 2017). GB efficiently stabilizes the complex quaternary configurations of proteins, various enzymes, and components of photosynthesis, i.e., oxygen evolving photosystem II (PS II) and activity of ribulose-1,5-bisphosphate carboxylase-oxygenase (RuBisCo) enzyme (Nayar et al. 2005; Shahbaz et al. 2011). However, the exact role of GB in environmental stress endurance is unknown, but two major functions ascribed to it are cellular compatibility and osmotic adjustment. In the former mechanism, GB substitutes for water in biochemical reactions, sustaining normal metabolism under stress, whereas in the latter, it affects the osmotic pressure in a concentration-dependent manner and absorbs extra water from the surroundings (Giri 2011).

7.4 POTENTIAL ROLE OF GLYCINE BETAINE AS HEAVY METAL STRESS MANAGER

HMs and metalloids are crucial environmental pollutants, which are phytotoxic even at low concentrations. HMs, i.e., Cu, Cd, As, Co, Zn and Ni, are toxic to plants if present above optimal levels. Excessive generation of ROS in retaliation to HM stress leads to oxidative damage to the cellular components (Banu et al. 2009). A plethora of experimental evidence affirms that increased GB sequestration stimulates the response of a plant to HM toxicity, its mitigation and regulation of signaling cascades (Gratao et al. 2019), although a few studies argue that the elevation in GB under stress cues is not an acclimatizing approach but a result of stress (Paradisone et al. 2015). However, it has a commanding role in maintaining osmotic balance, shielding the sub-cellular structures in plant cells exposed to stress, and participates as a molecular chaperone in the rearrangement or altered folding of enzymes (Hussain Wani et al. 2013). Moreover, in addition to working as an osmoprotectant, GB also restrains the production of ROS and other free radicals (Fariduddin et al. 2013). GB appears to play a significant role in modulation of growth and crop productivity, photosynthesis, ROS homeostasis, metabolic implications and nutrients, as well as HM uptake, which will be elaborated later. Figure 7.3 explains the underlying mechanism of HM tolerance in plants upon GB supplementation. As well, there are ample reports suggesting the effect of GB on plants under HM stress (Table 7.1).

7.4.1 Glycine Betaine and Growth/Crop Productivity

GB has momentous significance in the modulation of morphological and biological phenomena in plants. Growth improvement in response to augmented concentration of GB is in terms of elevation in number of seeds produced, fruit set and size of the flowers (Ahanger et al. 2018). A wide variety

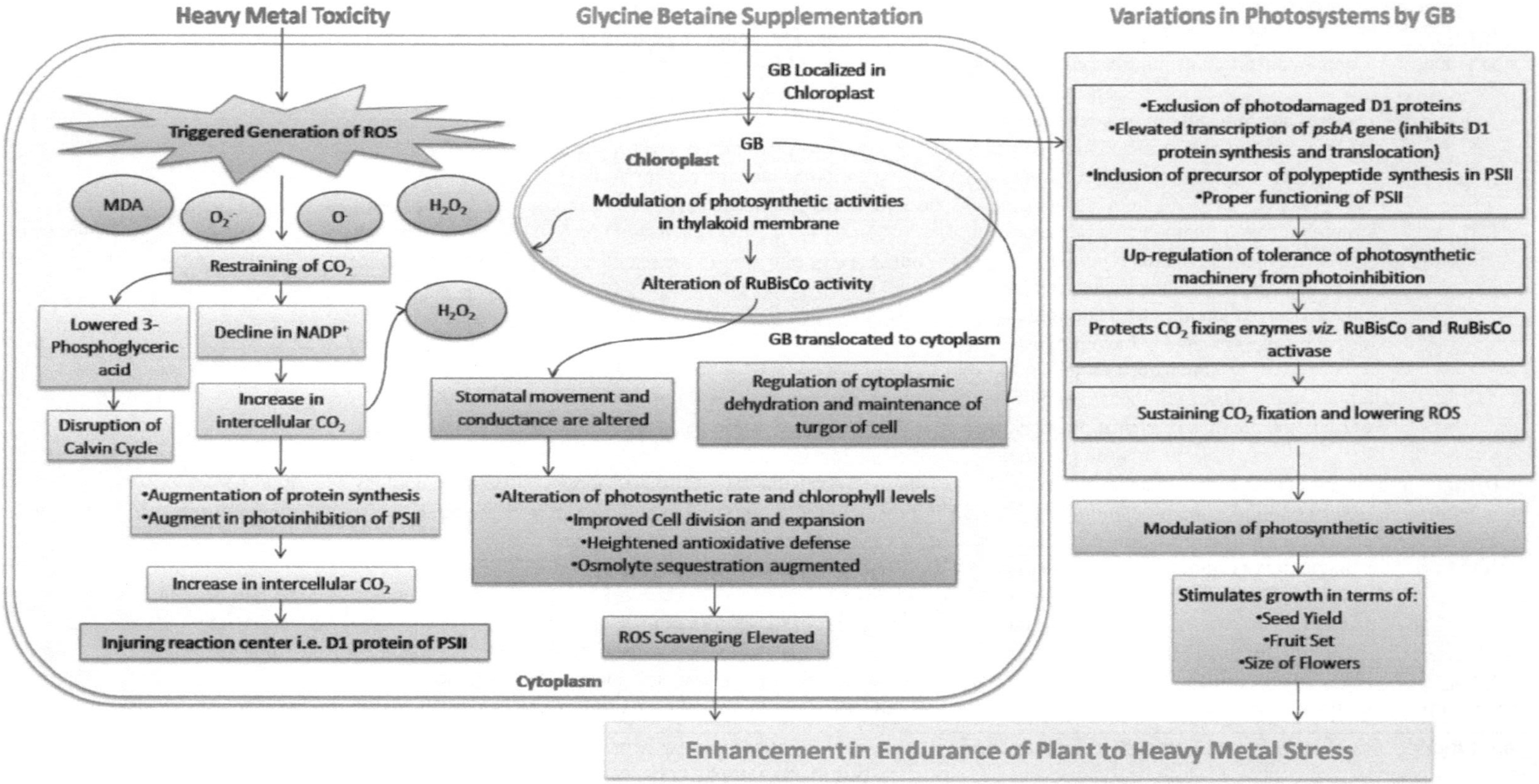

FIGURE 7.3 Schematic representation of underlying mechanism of enhanced endurance of plants to heavy metal stress in response to glycine betaine application.

TABLE 7.1
Effect of Exogenous and Endogenous Glycine Betaine Levels on Heavy Metal Stress Tolerance in Plants and Underlying Mechanism

Plant Species	Heavy Metal Treatment	Glycine Betaine Levels	Regulation of Growth, Photosynthetic Attributes, ROS Homeostasis and Biomolecule Metabolism	Mechanism of Action	Reference
Brassica oleracea	Chromium (10, 100, and 200 μM)	Exogenous foliar application (1 mM)	Plant growth in plant height, root length, number of leaves and leaf area were improved EL and accumulation of MDA and H_2O_2 were lowered Antioxidative enzyme activities (SOD, CAT and POD) were up-regulated	Enhanced photosynthetic rate accompanied by improved stomatal conductance and RuBisCo activity resulting in improved growth GB reduced ROS production in response to increased activity of antioxidant enzymes	Ahmad et al. (2020)
Sorghum bicolor L.	Chromium (2 and 4 ppm)	Exogenous soil supplementation (50 and 100 mM)	Fresh weight, dry weight, root length and shoot length were replenished Chlorophyll levels and grain yield enhanced APOX, CAT, PPO, SOD, GR and POD enzyme activity were further augmented	Better development in nutrient uptake and gas exchange is a probable reason for improved growth RuBisCo and RuBisCo activase activity are augmented, leading to augmented yield and photosynthetic attributes Increase in the enzymatic activities due to the decrease in chromium uptake or reduction in electrolyte leakage	Kumar et al. (2019)
Nicotiana tabacum L.	Cadmium (5 μM)	Exogenous foliar application (500 μM)	Plant height, root growth and SPAD value were enhanced Reduction in MDA levels and up-regulation in antioxidative enzyme activity, i.e., SOD, POD, CAT and APOX were elevated Gas exchange attributes were up-regulated	GB application recovered shoot/root iron, copper and zinc content, hence regulating nutrient metabolism GB can harmonize the activities of antioxidant enzymes and protect cells and tissues against oxidative damage caused by oxidative stress	He et al. (2019)
Spinacia oleracea	Cadmium (25 and 50 mg kg^{-1})	Exogenous foliar application (25 mM)	Growth attributes, i.e., root growth, shoot growth, fresh biomass, dry biomass Total chlorophyll, chl a and chl b levels were replenished MDA content was decremented	GB might increase nutrient uptake, which results in more biomass production under the stress conditions Foliar GB also improved the activities of antioxidant enzymes and reduced the uptake of Cd	Aamer et al. (2018)

Olea europaea	Lead (150 and 450 mg kg^{-1} soil)	Exogenous irrigation water supplementation (20 mM)	Impairment in fresh and dry weights of roots and leaves was lowered Increased the SOD, CAT, APX and GPX activities in roots and leaves, which indicated an improvement of their enzymatic antioxidant defense system Enhanced accumulation of secondary metabolites	Increased uptake of nutrient ions reduces the accumulation of lead ions and hence improves growth Augmentation in net CO_2 assimilation rate and therefore the sub-stomatal CO_2 concentration replenishes growth GB may act directly as an antioxidant and/or indirectly through activation of antioxidant responses, and then limits the oxidative stress and further stimulates secondary metabolite synthesis	Zouari et al. (2018)
Cajanus cajan L.	Fluoride (75 ppm)	Exogenous seed soaking (50 μM)	GB application exhibited improved growth, MSI, genomic template stability, protein and proline accumulation Stress markers such as MDA, 4-HNE, LOX and DNA polymorphism were lowered Augmentation of protein levels and proline content Alterations in the activities/isoforms of SOD, CAT, POD and APOX	Improved growth is attributed to GB-influenced enhancement in rate of cell division and elongation, nutrient uptake and activity of antioxidant enzymes GB is possibly involved in quenching of free radicals by up-regulating the activities of AOS scavenging system GB lessens the permeability of membranes and peroxidation of membrane-bound lipids, and maintains the integrity of membranes GB stabilizes the quaternary structure of proteins Proline level stabilizes the detoxifying enzymes and protein turnover machinery and stimulates the accumulation of stress protective proteins	Yadu et al. (2017)

(*continued*)

TABLE 7.1 (Continued)
Effect of Exogenous and Endogenous Glycine Betaine Levels on Heavy Metal Stress Tolerance in Plants and Underlying Mechanism

Plant Species	Heavy Metal Treatment	Glycine Betaine Levels	Regulation of Growth, Photosynthetic Attributes, ROS Homeostasis and Biomolecule Metabolism	Mechanism of Action	Reference
Gossypium hirsutum L.	Cadmium (1 and 5 µm)	Exogenous foliar application (1 mM)	Growth in terms of plant height, root length, number of leaves per plant, and fresh and dry weights of leaf, stem and root was replenished in response to GB application Photosynthetic pigment content, i.e., total chlorophyll, chl a, chl b and carotenoid levels, was augmented The gas exchange attributes such as Pn, gs, E, A/E of a fully expanded youngest leaf were also enhanced SOD, POD, CAT and APOX activities were augmented Decrement in oxidative damage as evidenced by the decreased production of EL, and H_2O_2 and MDA content	Crop productivity was elevated in response to improved growth, augmentation in level of photosynthetic pigments, protein content and antioxidant enzymes GB maintains RuBisCo activity and chloroplast ultrastructure Enhanced photochemical activity of PS II GB protects the PSII complex by stabilizing the association of the extrinsic PSII complex proteins under stress conditions GB increased antioxidant enzyme activities and led to a decrease in oxidative stress, which eventually lowered production of MDA, H_2O_2 content and EL in both roots and leaves	Farooq et al. (2016)
Triticum aestivum L.	Chromium (0.25 and 0.50 mM)	Exogenous foliar application (0.1 M)	Plant height, root length, kernel length and number of tillers Chlorophylls and carotenoid levels were replenished Enhancement in antioxidative activity in terms of enzyme and antioxidant content	Increase in biomass with GB might be due to enhanced uptake of nutrients, such as nitrogen, phosphorus and potassium Net CO_2 assimilation rate, transpiration rate and sub-stomatal CO_2 concentration are replenished and elevate growth Alleviation of damage to chloroplast ultrastructure, thylakoid lamellae and improved photosynthetic capacity by increasing stomatal conductance GB could contribute to the detoxification of ROS by enhancing APOX and CAT activities	Ali et al. (2015)

Vigna radiata	Chromium (250 and 500 μM)	Exogenous foliar application (50 and 100 mM)	Plant growth, biomass and chlorophyll contents were augmented Chromium accumulation and EL reduced Antioxidative enzyme activities, i.e., SOD, POD and CAT, were up-regulated	Enhanced nutrient uptake by plants and increase in gas exchange attributes of plants Expression of genes of ROS scavenging enzymes, which protects the photosynthetic machinery from damage Protective role of GB in the cell membranes and as a result, less chromium enters the cytoplasm	Jabeen et al. (2015)
Lolium perenne	Cadmium (0.5 mM)	Exogenous seed soaking (20 and 50 mM)	Turf quality and VSGR were increased NRT and chlorophyll content were replenished Enzyme SOD, CAT and POD activity as well as levels of organic acids, including oxalic and tartaric acid, were elevated EL and MDA levels were decremented	- Elevation in gene expression level of *MnSOD*, *FeSOD*, *CytCu/ZnSOD*, *ChlCu/ZnSOD* and *POD* genes was reflected in the enzyme activity Lower translocation factor along with Cd accumulation in roots Lowered EL and MDA suggests membrane stability	Lou et al. (2015)
Triticum aestivum L.	Cadmium (200 μM)	Exogenous foliar spray (20 mM)	Shoot and root fresh masses were enhanced Less degradation of chlorophylls, carotenoids, anthocyanins and leaf phenolics Reduced sequestration of MDA and H_2O_2	GB readily detoxifies ROS and protects chlorophyll Improvement in the efficacy of the photosystems was recorded	Rasheed et al. (2014)
Gossypium hirsutum	Lead (50 and 100 μM)	Exogenous seed soaking (1 mM)	Augmentation in growth, leaf area and biomass of plants Pn, gs, E and A/E were improved SPAD Value, total chlorophyll, chl a, chl b and carotenoid levels were elevated Antioxidative enzyme activity, e.g., SOD, POD, CAT and APOX, was improved MDA and H_2O_2 levels were decreased	GB regulates photosynthetic capacity by increasing stomatal conductance and maintaining RuBisCo activity and chloroplast ultrastructure GB addition could markedly enhance the capacity of defense against oxidative damage induced by lead toxicity	Bharwana et al. (2014)
Lemna gibba	Cadmium (0.5, 1 and 3 mM)	Exogenous seed soaking (0.5, 1, 2 and 5 mM)	Reduction in MDA levels Chlorophyll and proline content was raised Antioxidative enzymes, i.e., SOD, POD and APOX, were up-regulated	GB application increased the intracellular GB levels, which improved inhibition of ion accumulation and osmotic regulation GB could contribute to the detoxification of ROS by increasing the activity of antioxidative enzymes and detoxification and reduced oxidative damage	Duman et al. (2011)

(*continued*)

TABLE 7.1 (Continued)
Effect of Exogenous and Endogenous Glycine Betaine Levels on Heavy Metal Stress Tolerance in Plants and Underlying Mechanism

Plant Species	Heavy Metal Treatment	Glycine Betaine Levels	Regulation of Growth, Photosynthetic Attributes, ROS Homeostasis and Biomolecule Metabolism	Mechanism of Action	Reference
Vigna radiata	Cadmium (1 mM)	Exogenous seed soaking (5 mM)	Elevation in ascorbic acid, GSH and GSSG Enzymatic antioxidants, i.e., APOX, MDHAR, DHAR, GR and GST, were augmented GST enzyme activity was not significantly influenced Gly 1 and Gly2 enzyme activities were up-regulated GB application significantly lowered MDA and H_2O_2 levels	Ascorbic acid regeneration affects the efficient functioning of various antioxidative enzymes specifically GSH level increases Gly I and Gly II activities were able to enhance GSH regeneration and glutathione redox state via the glyoxalase system	Hossain et al. (2010)
Nicotiana tabacum L.	Cadmium (100 mM)	Exogenous seed soaking (1 mM and 10 mM)	Cell viability was elevated by GB supplementation Betaine suppressed HM-induced ROS production Up-regulation of the ASC-GSH cycle and GST induced by GB	Efficient recycling of GSH from GSSG by the activity of NADPH-dependent GR Enhanced antioxidative enzyme activity results in detoxification of H_2O_2	Islam et al. (2009)

Note: HM, heavy metal; GB, glycine betaine; ROS, reactive oxygen species; EL, electrolyte leakage; MDA, malondialdehyde; H_2O_2, hydrogen peroxide; RuBisCo, ribulose-1,5-bisphosphatecarboxylase-oxygenase; SOD, superoxide dismutase; CAT, catalase; Pod, guaiacol peroxidase; APOX, ascorbate peroxidase; PPO, polyphenol oxidase; GR, glutathione reductase; SPAD, soil plant analysis development; chl, chlorophyll; GPX, glutathione peroxidase; CO_2, carbon dioxide; MSI, membrane stability index; 4-HNE, 4-hydroxy-2-nonenal; LOX, lipooxygenase; DNA, deoxyribonucleic acid; AOS, alternate oxidases; Pn, photosynthetic rate; gs, stomatal conductance; E, transpiration rate; A/E, water use efficiency; PSII, photosystem II; VSGR, vertical shoot growth rate; NRT, normalized relative transpiration; GSH, glutathione; GSSG, glutathione disulfide; MDHAR, malate dehydrogenase; DHAR, dehydroascorbate reductase; GR, glutathione reductase; GST, glutathione-S-transferase; ASC, ascorbate.

of experimental evidence has confirmed the growth-enhancing ability of GB by altering the photosynthetic rate and content of chlorophyll in plants (Wang et al. 2010; Farooq et al. 2016; Tian et al. 2017). Exploitation of endogenous GB noticeably boosts plant growth by improving antioxidative defense ability and osmolyte sequestration (Malekzadeh 2015). Antioxidative defense components (enzymatic and non-enzymatic) actively participate in ROS scavenging (Khan et al. 2009). GB, under both stress and non-stress cues, restores the growth of various plant species by modulating over-production of ROS (Akram et al. 2012). Moreover, it is apparent that GB aids in augmenting chlorophyll levels and photosynthetic rates in stressed plants (Farooq et al. 2016; Lou et al. 2019). Also, exogenous application of GB efficiently stimulates the growth, biomass and resilience of plants (Molla et al. 2014; Raza et al. 2014).

In response to exposure to HM toxicity, plants experience faltering and retardation in growth. HM stress results in a considerable decrement in growth, number of leaves, height of plant, and weight of roots and stems (Rasheed et al. 2014; Stepien et al. 2016). Furthermore, HMs hinder processes such as replication and transpiration from the leaves, which eventually affects the growth of plants due to constrained cell division (Malar et al. 2016). GB defends plant's photosynthetic machinery, which results in replenishment of growth in plants (Kurepin et al. 2017; Ahmad et al. 2020). Foliar supplementation of GB aids in alleviation of Cd toxicity in plants, consequently resulting in improvement in cell division and growth (Farooq et al. 2016). A similar observation of amelioration of chromium toxicity in *Triticum aestivum* plants was made by Ali et al. (2015). They further confirmed an improvement in growth, root and shoot weights upon exogenous GB application. Furthermore, it was suggested by Bhatti et al. (2013) that the increase in the growing power of plants in response to augmented GB levels can be attributed to simultaneous increase in root and shoot lengths as well as biomass. Exogenous supplementation of GB replenishes growth by invigorating leaf expansion, maintaining turgidity and elevating the synthesis of photosynthetic pigments in sorghum plants (Ibrahim et al. 2006). Elevated Cd toxicity decreases both fresh and dry weights of tobacco plants, whereas exogenous supplementation of GB ameliorates the deleterious impact of Cd on growth and biomass of the stressed plant (Islam et al. 2009). GB treatment considerably alleviates the inhibitory effects of Cd toxicity in wheat crops (Rasheed et al. 2014). Corroborative observations were made by Mahdavian et al. (2016) in *Peganum harmala* plants.

Cr toxicity in *Brassica oleracea* reduced growth and biomass, which is attributable to the interaction of the HM with vital nutrients and reduction in their uptake. GB application to growing cauliflower seedlings ameliorated Cr stress by enhancing the uptake of nutrients in the stressed plants (Ahmad et al. 2020). Similar results of GB-mediated improvement in plant growth were obtained by Jabeen et al. (2015) in mung bean due to boosted nutrient uptake and escalation in gas exchange features of the plants. Accumulation of flavonoids and phenolics after supplementation of GB to plants under the influence of HM stress also contributes to plant growth and development (Zouari et al. 2018).

7.4.2 Glycine Betaine and Photosynthetic Attributes

HM toxicity is confirmed to alter plant water relations, which consequently affects the rate of photosynthesis, movement of sap, nutrient and water uptake, stomatal opening and disruption of chlorophyll synthesis, and eventually results in a deleterious impact on photosynthesis (Cuypers et al. 2011; Sheetal et al. 2016). Photosynthesis is one of the most vulnerable processes affected by HM stress conditions (Cuypers et al. 2011). Toxic levels of HM ions augment the generation of ROS in stressed plants, subsequently leading to their interaction with varied electron transport activities localized in mitochondria and chloroplasts, membranes and cellular compartments (Chen et al. 2014). A wide array of experimental evidence has confirmed that the toxicity caused by the HM is attributed to ROS-triggered lipid peroxidation and its intermediates, acrolein, linolenic acid-13-ketotriene and 12-oxo-phytodienoic acid, which robustly causes deterioration in the photosynthetic rate and the functioning of PS II (Sullivan et al. 2015). A vast quantity of GB is localized in the chloroplasts, where it shields the thylakoid membranes and modulates proper photosynthetic activities (Rathinasabapathi et al. 1994). It has also been confirmed that GB regulates cytoplasmic dehydration and maintains turgor in leaf cells exposed to varied abiotic stresses, thus sustaining proper photosynthetic action (Iqbal et al. 2008). Corroboratively, GB application also resulted in lowering the photosynthetic efficacy in Pb-exposed cotton plants (Bharwana et al. 2014). The protective role of GB is attributed to augmentation of stomatal movement and conductance, safeguarding RuBisCo enzyme activity, and conserving the ultrastructure of chloroplast in opposition to varied environmental cues (Nomura et al. 1998).

For estimating alterations in the photosynthetic activity in a plant, several parameters, such as carbon dioxide fixation, stomatal conductance, water use efficiency and transpiration rates in leaves, are generally analyzed. Ahmad et al (2020) observed that Cr toxicity in *Brassica oleracea* plants led to an extreme reduction in these traits with increasing levels of the HM in the soil. GB application drastically enhanced the efficacy of these traits as well as improving the content of chlorophyll pigments. HM-induced ROS generation restricts the fixation of CO_2 and lowers the content of 3-phosphoglyceric acid. This suppressed fixation of CO_2 consequently results in a decline in levels of $NADP^+$. Due to absence of this major electron acceptor in PS I, there is an augmentation in the intercellular levels of molecular oxygen, which is accompanied by over-production of hydrogen peroxide (Asada et al. 1999; Fariduddin et al. 2013).

The elevation in ROS also results in inhibition of protein synthesis and increased photoinhibition, thus hampering PS II repair work. It causes photodamage to PS II by injuring the reaction centers, i.e., D1 protein of PS II. The application of GB up-regulates the tolerance of photosynthetic machinery to photoinhibition by bringing about various alterations in biochemical processes. These alterations include: a) exclusion of photodamaged D1 protein, b) elevated transcription of *psbA* gene, which is one the most crucial precursors of D1 protein synthesis and translation, c) inclusion of precursors of polypeptide synthesis in the PS II complex and d) reconstitution of properly functioning PS II (Aro et al. 1993; Fariduddin et al. 2013). GB has also been increasingly confirmed to protect the CO_2 fixing enzymes RuBisCo and RuBisCo activase in abiotic stress-exposed plants, thus eventually sustaining CO_2 fixation and reducing the generation of ROS (Chen and Murata, 2011).

7.4.3 Glycine Betaine and ROS Homeostasis

Plants under HM stress generate superfluous and excessive ROS, such as hydrogen peroxide (H_2O_2) and superoxide anion (O^{2-}) (Farooq et al. 2016; Ali et al. 2020). In numerous plant species, HM ions bind to the protein groups and are also able to substitute certain cations at binding spots, which results in deactivation of antioxidant enzymes and an increase in ROS, eventually leading to oxidative stress. A sudden rise in intracellular ROS levels results from an imbalance between ROS scavenging and its production. For normal cell functioning, ROS must persist within compatible limits in plants. Antioxidants such as CAT, glutathione (GSH), peroxidase (POX) and SOD, present in plants, work as reducing agents, culminating in several oxidation reactions that disrupt cellular components by eradicating free radicals (Jabeen et al. 2015). Glycine betaine further enhances the expression of genes responsible for abiotic stress tolerance by a ROS scavenging mechanism (Annunziata et al. 2019). According to Fariduddin et al. (2013), expression of the ROS scavenging genes helps in synthesis of ROS scavenging enzymes, which in turn induce ROS homeostasis and alleviate oxidative stress by degradation and depression of ROS.

GB, as an osmolyte in plants, is reported to augment the antioxidant system, thus minimizing the deleterious consequences of HM toxicity (Ali et al. 2020). Tobacco cultivar Yunyan2, exposed to Cd stress, exhibited augmented ROS scavenging ability under the influence of exogenously supplied GB. The promotion of activity of CAT upon GB application suggested that improved tolerance of the cultivar is an outcome of its robust antioxidant enzyme system in comparison to the Guiyan1 cultivar, thus protecting plant cells and tissues from oxidative damage (He et al. 2019). A study conducted by Jabeen et al. (2015) revealed that when GB was applied to mung bean seedlings under Cr stress, a marked increase in the activity of SOD, CAT and POD was recorded in comparison to the control. This increase was an outcome of reduced electrolyte leakage and reduced HM (Cr) uptake by the seedlings. The ascorbate-glutathione cycle prevalent in plants also protects the plants from the negative effects of ROS. Successful GB treatment up-regulates the activity of enzymes involved in the ascorbate-glutathione cycle, i.e., ascorbate peroxidase (APX), monohydro ascorbate reductase (MDHAR) and dihydro ascorbate reductase (DHAR), in plants under HM stress (Ali et al. 2020).

7.4.4 Glycine Betaine and Protein Up-Regulation

GB plays an important role in plant metabolism, including various processes such as plant growth and development, signaling, protein synthesis, osmoprotection, redox balance and antioxidative defense mechanisms under abiotic stress without causing any harm to membranes and macromolecules (Ali et al. 2019; Annunziata et al. 2019). GB can control transcription and regulate gene expression under stress (Hasanuzzaman et al. 2019). GB has the capacity to reduce protein carbonylation, leading to detoxification of a toxic compound called methylglyoxal in cells (Hoque et al. 2008). Moreover, Li et al. (2011) explained that GB might enhance the synthesis of HSP70 protein, which works as a chaperone in protein metabolism and under abiotic stress. GB alleviates enzymatic activity in the plasma membrane for ion homeostasis, and interacts with protein complexes to maintain their structure and functions by remediating aggregation and/or mis-folding of proteins (Annunziata et al. 2019). They further suggested that GB stimulates the activities of various metabolic enzymes, including those responsible for the Krebs cycle.

In a study conducted by Carillo et al. (2008), content of GB was found to show a strong positive correlation with the enzyme phosphoenolpyruvate carboxylase, which helps in providing the carbon skeletons for nitrogen integration in amino acids. GB can regulate enzymes involved in signal transduction of various pathways by enhancing the expression of associated genes, e.g., putative receptor kinase, calmodulin, protein kinase, lipoxygenase, osmotin, monodehydroascorbate reductase, etc. (Ashraf and Foolad, 2007). GB transport inside the cell also facilitates the proline and γ-amino butyric acid (GABA) co-transport, since the glycine transporter is permeable to proline and GABA to some extent (Hasanuzzaman et al. 2019). GB can also up-regulate the biosynthesis and transport of hormones (Hasanuzzaman et al. 2019). In stressed plants, serine and glycine (intermediates in the photorespiratory cycle) are replaced by GB, suggesting that photorespiration helps in mobilization of carbon reserves *via* the glycolate oxidation pathway and produces compatible nitrogen solutes for acclimation of plants under stress (Di Martino et al. 2003). According to Ahanger et al. (2018), GB helps protect the carboxylase activity of RuBisCo, resulting in better carbon assimilation and photosynthetic efficiency (Ahanger et al. 2018).

GB also modulates the synthesis of GSH, a form of reduced sulfur, which protects cells from HM stress by increasing the sequestration of phytochelatin conjugates of HM inside the cell vacuole (Hossain et al. 2010; Gigolashvili and Kopriva 2014). Li et al. (2019) reported that GB accumulation in transgenic tomato plants under stressed conditions regulated the carbohydrate (sucrose) signaling and expression of genes responsive to low phosphate levels. In this study, *codA*-transgenic tomato plants having choline oxidase gene *codA* (administered from *Arthrobacter globiformis*) showed better GB and sucrose accumulation than wild type tomato plants (*Solanum lycopersicum* cv. 'Moneymaker'). Higher activities of sucrose phosphate synthase and sucrose synthase; and increased expression of the gene *SUC2* involved in sucrose transport in *codA*-transgenic plants, suggested that GB plays a role in sucrose metabolism in plants under stressed conditions. Gupta and Thind (2019) reported that exogenous application of GB increased the contents of total soluble sugars, which could be due to alterations in regulation of sugar metabolism, e.g., reduction in activities of sucrose phosphate synthase and sucrose synthase, in plants under abiotic stress.

Ali et al. 2015 observed reduced protein content in wheat plants subjected to Cr stress. Foliar application of GB under Cr stress considerably elevated the concentrations of protein in leaves and roots of the plant in comparison to Cr alone treatments. It was suggested that metal stress produced ROS, which led to protein modification/degradation. The decline in protein content was attributed to a negative outcome of Cr on photosynthetic pigments or because of elevated Cr uptake. GB under stress maintains the protein activity in chloroplasts, and the elevation in protein levels is a positive effect of GB on the chloroplast and overall plant growth. Exposure of cotton seedlings to Cd at toxic concentrations inhibited the soluble protein content in both leaf and root of the plants due to oxidative stress under these conditions (Farooq et al. 2016). Enhanced levels of GB were recorded in Cd-exposed cotton seedlings and were found to protect these proteins through a chaperone-like effect on protein folding and also to behave as a signaling molecule to initiate stress responses. Thus, improved protein content in cotton plants was observed when GB was supplied exogenously.

7.4.5 Glycine Betaine and Nutrient/Heavy Metal Uptake

Plants have established abundant mechanisms by which they can successfully maintain and control cell metal homeostasis by averting HM ion uptake and its accumulation at high concentrations. GB-mediated reduction in levels of HMs have been quantified in several plant species, including rice (Cao et al. 2013), wheat (Ali et al. 2015) and cotton (Farooq et al. 2016).

Exposure of cauliflower seedlings to high levels of Cr resulted in enhanced Cr uptake and its accumulation in the plant roots. GB application to the stressed seedlings reduced Cr uptake by providing protection to cell membranes from the damaging effects of the HM, which eventually resulted in reduced uptake of Cr into the cell cytoplasm (Ahmad et al. 2020). The authors also postulated that GB enhances nutrient uptake in plants under stress, which decreases Cr uptake, since Cr competes with several other essential nutrients. In an experiment conducted by Zouari et al. (2018) on olive trees, exogenous GB application reduced Pb uptake and transport to leaves from roots in comparison to Pb-treated trees. They confirmed that GB supplementation limits Pb content in the early stages of plant development, thus reducing the harmful effects of HM. Similar results were obtained by Bharwana et al. (2014) in cotton seedlings.

7.5 GENETIC TRANSFORMATION FOR GLYCINE BETAINE SYNTHESIS TO AMELIORATE HEAVY METAL STRESS

Genetic engineering of different plant species is one of the best techniques for increasing abiotic stress tolerance in the plants by raising the level of generation of different phytohormones, organic osmolytes and various antioxidant enzymes (Ali et al. 2020). The development of techniques in proteomics, genomics and genetic engineering has made it possible to produce transgenic plants containing various genes for better GB synthesis and increased abiotic stress tolerance (Khan et al. 2009). In transgenic plants, GB helps in abiotic stress tolerance by mechanisms such as protection of cellular macromolecules, controlling the expression of various endogenous genes and detoxification of ROS (Giri 2011). At present, genes or cDNAs of enzymes involved in GB biosynthetic pathway that are cloned for generating transgenic plants are from bacteria or plants. Genetic manipulation for introduction of the same has been attempted in higher plants. The strategy for generation of transgenic plants includes introduction of a relevant gene along with a promoter that guarantees high-level gene expression in the transgenic plants. For localization of enzymes of prokaryotes, the gene is modified such that the encrypted polypeptide after post-translational modification is transported into the chloroplast of the genetically engineered plant (Sakamoto and Murata 2000).

Major cereal crops such as wheat, barley and maize do not amass significant levels of GB under natural conditions, probably due to the fabrication of abnormal transcripts of BADH, the GB synthesizing enzyme. Rice cannot amass GB naturally, even though it has one CMO and two BADH encoding genes. The transcripts of BADH are translated in an unusual manner that leads to loss of functional domains, removal of the translational initiation codon and premature stop codons (Giri 2011). Several other crop plants are not capable of accumulating GB during stress. The identification and characterization of genes involved in the GB biosynthetic pathway has paved a way to engineer biosynthesis of GB in such non-accumulators, employing the transgenic approach to improve abiotic stress tolerance (Fariddudin et al. 2013).Cloning of genes such as *codA* and *BADH* encoding enzymes catalyzing GB biosynthesis has been reported, and many transgenic plants expressing the GB synthesis genes and exhibiting elevated abiotic stress tolerance have been generated (Kumar et al. 2019; Ali et al 2020). According to Khan et al. (2009), choline availability (exogenous and/or endogenous), type of transgene and promoter are the most important factors that influence the accumulation of GB in transgenic plants.

Wei et al. (2017) genetically engineered a tomato cultivar (*Solanum lycopersicum* cv. 'Moneymaker') using the choline oxidase gene *codA* isolated from *Arthrobacter globiformis*, causing increased accumulation of GB with a corresponding rise in the rate of photosynthesis and activities of antioxidant enzymes along with reduced production of ROS in comparison to the wild type tomato plants under salt stress. Fan et al. (2012) observed that in the transgenic plants of sweet potato (*Ipomoea batatas*) cultivar Sushu-2, overexpression of a chloroplastic gene *SoBADH* introduced from *Spinacia oleracea* (responsible for synthesis of the enzyme BADH during biosynthesis of GB) resulted in increased tolerance to various abiotic stresses, including salinity, low temperature and oxidative stress. They further noted that genetically engineered plants had a higher rate of photosynthesis, protected cell membranes, reduced generation of ROS and greater stimulated scavenging of ROS. These observations suggested that production of transgenic sweet potato plants with enhanced synthesis and accumulation of GB can provide tolerance to multiple types of abiotic stresses.

In a study conducted by Li et al. (2014), it was observed that tomato (*Lycopersicon esculentum*) cultivar 'Moneymaker' with *BADH* gene inserted from spinach under heat stress exhibited an increased rate of accumulation of GB, increased photosynthesis and greater tolerance to photoinhibition induced by excessive heat. It was found that alleviation of heat-related photoinhibition was associated with a higher level of D1 protein, leading to repair of PS II. Transgenic plants also showed lower levels of malondialdehyde, hydrogen peroxide, superoxide radicals, and relative electrolyte conductivity and higher levels of antioxidant enzyme activities in comparison to wild type plants under heat stress. Khan et al. (2009) reviewed the literature and found that various genes such as *codA*, *BADH*, *CMO*, *COX*, etc. were genetically engineered into different plant species for inducing tolerance to various abiotic stresses by GB accumulation.

Although substantial efforts have been made in transgenic plants to elevate the concentration of GB, the reported content of GB in these plants in comparison to natural GB accumulators is relatively low on exposure to abiotic stress. The factors that limit GB accumulation in the chloroplasts of the transgenic plants include availability of choline (endogenous) and its transport across the chloroplast membrane (Chen and Murata 2011). Studies regarding the approach of using genetic engineering to create transgenic plants with higher GB accumulation to combat abiotic stresses mainly focus on salinity, drought, heat, etc., but studies based on alleviation of HM stress are not available in the literature. However, this approach can be effectively used for mitigation of HM stress by targeting the genes that help increase GB biosynthesis in plants for the creation of more tolerant crop cultivars (Ali et al. 2020).

7.6 CONCLUSION AND FUTURE PROSPECTS

The occurrence of HMs in soil greatly affects crop yields; thus, understanding plant responses towards HM becomes imperative to avoid or reduce any negative consequences. Accumulation of organic osmoprotectants such as GB in different plants during environmental stresses helps them to recover quickly from stress. This chapter highlights the prime signaling events regulating plant responses under HM stress, leading to GB being amassed within the plant cells, which subsequently impacts the growth traits, antioxidant system, photosynthetic machinery and cellular metabolism, helping plants alleviate metal stress. Exogenous GB application diminishes HM toxicity by elevating the activity of important antioxidant enzymes, thereby controlling ROS bursts and reducing the oxidative stress. Improvement in growth, ROS scavenging capabilities, osmoregulation, buffered redox potential, membrane stabilization and regulation of stress responsive genes can be achieved by employing metabolic/genetic engineering approaches as well as exogenous applications aimed at inducing GB synthesis and accumulation in plant tissues. Conducting more experimental studies to understand the role of GB in improving nutrient uptake at the molecular level and its influence on physiochemical processes is a prerequisite. Gene manipulation and additional knowledge of GB-related genes and their by-products can help improve our understanding of the stress resistance and tolerance exhibited by plants under the influence of GB.

REFERENCES

Aamer M, Muhammad UH, Li Z, Abid A, Su Q, Liu Y, Adnan R, Muhammad AU, Tahir AK, Huang G. Foliar application of glycine betaine (Gb) alleviates the cadmium (Cd) toxicity in spinach through reducing Cd uptake and improving the activity of anti-oxidant system. *Appl Ecol Environ Res* 2018, 16: 7575–7583; doi 10.15666/aeer/1606_75757583

Ahanger MA, Alyemeni MN, Wijaya L, Alamri SA, Alam P, Ashraf M, Ahmad P. Potential of exogenously sourced kinetin in protecting *Solanum lycopersicum* from NaCl-induced oxidative stress through up-regulation of the antioxidant system, ascorbate-glutathione cycle and glyoxalase system. *PLoS One* 2018, 13(9): e0202175; https://doi.org/10.1371/journal.pone.0202175

Ahmad R, Ali S, Abid M, Rizwan M, Ali B, Tanveer A, Ahmad I, Azam M, Ghani MA. Glycine betaine alleviates the chromium toxicity in *Brassica oleracea* L. by suppressing oxidative stress and modulating the plant morphology and photosynthetic attributes. *Environ Sci Poll Res* 2020, 27(1): 1101–1111; https://doi.org/10.1007/s11356-019-06761-z

Ahmad R, Lim CJ, Kwon SY. Glycine betaine: a versatile compound with great potential for gene pyramiding to improve crop plant performance against environmental stresses. *Plant Biotechnol Rep* 2013, 7(1): 49–57; https://doi.org/10.1007/s11816-012-0266-8

Akram NA, Ashraf M, Al-Qurainy F. Amino levulinic acid-induced changes in some key physiological attributes and activities of antioxidant enzymes in sunflower (*Helianthus annuus* L.) plants under saline regimes. *Sci Horti* 2012, 142: 143–148; https://doi.org/10.1016/j.scienta.2012.05.007

Alasvandyari F, Mahdavi B, Hosseini SM. Glycine betaine affects the antioxidant system and ion accumulation and reduces salinity-induced damage in safflower seedlings. *Arch Biol Sci* 2017, 69(1): 139–147; https://doi.org/10.2298/ABS160216089A

Ali Q, Shahid S, Ali S, Javed MT, Iqbal N, Habib N, Hussain SM, Shahid SA, Noreen Z, Hussain AI, Haider MZ. Trehalose metabolism in plants under abiotic stresses. In *Approaches for Enhancing Abiotic Stress Tolerance in Plants* (2nd ed.) 2019. CRC Press. pp. 349–364; https://doi.org/10.1111/j.1744-7909.2008.00736.x

Ali S, Abbas Z, Seleiman MF, Rizwan M, YavaŞ İ, Alhammad BA, Shami A, Hasanuzzaman M, Kalderis D. Glycine betaine accumulation, significance and interests for heavy metal tolerance in plants. *Plants* 2020, 9: 896; https://doi.org/10.3390/plants9070896

Ali S, Chaudhary A, Rizwan M, Anwar HT, Adrees M, Farid M, Irshad MK, Hayat T, Anjum SA. Alleviation of chromium toxicity by glycinebetaine is related to elevated antioxidant enzymes and suppressed chromium uptake and oxidative stress in wheat (*Triticum aestivum* L.). *Environ Sci Poll Res* 2015, 22(14): 10669–10678; https://doi.org/10.1007/s11356-015-4193-4

Anjum NA, Singh HP, Khan MIR, Masood A, Per TS, Negi A, Batish DR, Khan NA, Duarte AC, Pereira E, Ahmad I. Too much is bad—an appraisal of phytotoxicity of elevated plant-beneficial heavy metal ions. *Env Sci Poll Res Int* 2014, 22(5): 3361–3382; doi:10.1007/s11356-014-3849-9

Annunziata MG, Ciarmiello LF, Woodrow P, Dell'Aversana E, Carillo, P. Spatial and temporal profile of glycine betaine accumulation in plants under abiotic stresses. *Front Plant Sci* 2019, 10: 230; https://doi.org/10.3389/fpls.2019.00230

Aro EM, McCaffery S, Anderson JM. Photoinhibition and D1 protein degradation in peas acclimated to different growth irradiances. *Plant Physiol* 1993, 103(3): 835–843; https://doi.org/10.1104/pp.103.3.835

Asada S, Fukuda K, Oh M, Hamanishi C, Tanaka S. Effect of hydrogen peroxide on the metabolism of articular chondrocytes. *Inflamm Res* 1999, 48(7): 399–403; https://doi.org/10.1007/s000110050478

Ashfaque F, Farooq S, Chopra P, Chhillar H, Khan MIR. Improving heavy metal tolerance through plant growth regulators and osmoprotectants in plants. In *Improving Abiotic Stress Tolerance in Plants*. Khan MIR, Singh A, Poor P (Eds.) 2020. CRC Press. p. 30.

Ashraf MFMR, Foolad MR. Roles of glycine betaine and proline in improving plant abiotic stress resistance. *Environ Exp Bot* 2007, 59(2): 206–216; https://doi.org/10.1016/j.envexpbot.2005.12.006

Banu MNA, Hoque MA, Watanabe-Sugimoto M, Matsuoka K, Nakamura Y, Shimoishi Y, Murata Y. Proline and glycinebetaine induce antioxidant defense gene expression and suppress cell death in cultured tobacco cells under salt stress. *J Plant Physiol* 2009, 166(2): 146–156; https://doi.org/10.1016/j.jplph.2008.03.002

Bharwana SA, Ali S, Farooq MA, Iqbal N, Hameed A, Abbas F, Ahmad MS. Glycine betaine-induced lead toxicity tolerance related to elevated photosynthesis, antioxidant enzymes suppressed lead uptake and oxidative stress in cotton. *Turk J Bot* 2014, 38(2): 281–292; https://doi.org/10.3906/bot-1304-65

Bhatti KH, Anwar S, Nawaz K, Hussain K, Siddiqi EH, Sharif RU, Talat A, Khalid A. Effect of exogenous application of glycinebetaine on wheat (*Triticum aestivum* L.) under heavy metal stress. *Middle-East J Sci Res* 2013, 14(1): 130–137; https://doi.org/10.5829/idosi.mejsr.2013.14.1.19550

Cao F, Liu L, Ibrahim W, Cai Y, Wu F. Alleviating effects of exogenous glutathione, glycinebetaine, brassinosteroids and salicylic acid on cadmium toxicity in rice seedlings (*Oryza sativa*). *Agrotechnol* 2013, 2(1): 107–112; https://doi.org/10.4172/2168-9881.1000107

Carillo P, Mastrolonardo G, Nacca F, Parisi D, Verlotta A, Fuggi A. Nitrogen metabolism in durum wheat under salinity: accumulation of proline and glycine betaine. *Funct Plant Biol* 2008, 35(5): 412–426; https://doi.org/10.1071/FP08108

Carrillo-Campos J, Riveros-Rosas H, Rodríguez-Sotres R, Munoz-Clares RA. Bona fide choline monoxygenases evolved in Amaranthaceae plants from oxygenases of unknown function: evidence from phylogenetics, homology modeling and docking studies. *PloS One* 2018, 13(9): e0204711; https://doi.org/10.1371/journal.pone.0204711

Chen T, Zhang Y, Wang H, Lu W, Zhou Z, Zhang Y, Ren L. Influence of pyrolysis temperature on characteristics and heavy metal adsorptive performance of biochar derived from municipal sewage sludge. *Biores Technol* 2014, 164: 47–54; https://doi.org/10.1016/j.biortech.2014.04.048

Chen TH, Murata N. Glycinebetaine protects plants against abiotic stress: mechanisms and biotechnological applications. *Plant Cell Environ* 2011, 34(1): 1–20; https://doi.org/10.1111/j.1365-3040.2010.02232.x

Cuypers A, Karen S, Jos R, Kelly O, Els K, Tony R, Nele H, Nathalie V, Yves G, Jan C, Jaco V. The cellular redox state as a modulator in cadmium and copper responses in *Arabidopsis thaliana* seedlings. *J Plant Physiol* 2011, 168(4): 309–316. https://doi.org/10.1016/j.jplph.2010.07.010

Di Martino C, Delfine S, Pizzuto R, Loreto F, Fuggi A. Free amino acids and glycine betaine in leaf osmoregulation of spinach responding to increasing salt stress. *New Phytol* 2003, 158(3): 455–463; https://doi.org/10.1046/j.1469-8137.2003.00770.x

Dikilitas M, Simsek E, Roychoudhury A. Role of proline and glycine betaine in overcoming abiotic stresses. *Protective Chemical Agents in the Amelioration of Plant Abiotic Stress: Biochemical and Molecular Perspectives* 2020, 1–23; https://doi.org/10.1002/9781119552154.ch1

Duman F, Aksoy A, Aydin Z, Temizgul R. Effects of exogenous glycinebetaine and trehalose on cadmium accumulation and biological responses of an aquatic plant (*Lemna gibba* L.). *Water Air Soil Pollut* 2011, 217(1): 545–556; https://doi.org/10.1007/s11270-010-0608-5

Fan W, Zhang M, Zhang H, Zhang P. Improved tolerance to various abiotic stresses in transgenic sweet potato (*Ipomoea batatas*) expressing spinach betaine aldehyde dehydrogenase. *PLoS One* 2012, 7(5): e37344; https://doi.org/10.1371/journal.pone.0037344

Fariduddin Q, Varshney P, Yusuf M, Ali A, Ahmad A. Dissecting the role of glycine betaine in plants under abiotic stress. *Plant Stress* 2013, 7(1): 8–18.

Farooq MA, Ali S, Hameed A, Bharwana SA, Rizwan M, Ishaque W, Farid M, Mahmood K, Iqbal Z. Cadmium stress in cotton seedlings: physiological, photosynthesis and oxidative damages alleviated by glycinebetaine. *S Afr J Bot* 2016, 104: 61–68; https://doi.org/10.1016/j.sajb.2015.11.006

Gigolashvili T, Kopriva S. Transporters in plant sulfur metabolism. *Front Plant Sci* 2014, 5: 442; https://doi.org/10.3389/fpls.2014.00442

Giri J. Glycinebetaine and abiotic stress tolerance in plants. *Plant Signal Behav* 2011, 6(11): 1746–1751; https://doi.org/10.4161/psb.6.11.17801

Gratão PL, Alves LR, Lima LW. Heavy metal toxicity and plant productivity: role of metal scavengers. In *Plant-Metal Interactions*, Springer, Cham, 2019, pp. 49–60; https://doi.org/10.1007/978-3-030-20732-8_3

Gupta N, Thind SK. Foliar application of glycine betaine alters sugar metabolism of wheat leaves under prolonged field drought stress. *Proc Natl Acad Sci India Sect B Biol Sci* 2019, 89(3): 877–884; https://doi.org/10.1007/s40011-018-1000-2

Hasanuzzaman M, Banerjee A, Bhuyan MB, Roychoudhury A, Al Mahmud J, Fujita M. Targeting glycinebetaine for abiotic stress tolerance in crop plants: physiological mechanism, molecular interaction and signaling. *Phyton* 2019, 88: 185–221; doi:10.32604/phyton.2019.07559

He X, Richmond ME, Williams DV, Zheng W, Wu F. Exogenous glycinebetaine reduces cadmium uptake and mitigates cadmium toxicity in two tobacco genotypes differing in cadmium tolerance. *Int J Mol Sci* 2019, 20(7): 1612; https://doi.org/10.3390/ijms20071612

Hoque MA, Banu MNA, Nakamura Y, Shimoishi Y, Murata Y. Proline and glycinebetaine enhance antioxidant defense and methylglyoxal detoxification systems and reduce NaCl-induced damage in cultured tobacco cells. *J Plant Physiol* 2008, 165(8): 813–824; https://doi.org/10.1016/j.jplph.2007.07.013

Hossain MA, Hasanuzzaman M, Fujita M. Up-regulation of antioxidant and glyoxalase systems by exogenous glycine betaine and proline in mung bean confer tolerance to cadmium stress. *Physiol Mol Biol Plant* 2010, 16(3): 259–272; https://doi.org/10.1007/s12298-010-0028-4

Hussain WS, Brajendra SN, Haribhushan A, Iqbal MJ. Compatible solute engineering in plants for abiotic stress tolerance-role of glycine betaine. *Curr Gen* 2013, 14(3): 157–165; https://doi.org/10.2174/1389202911314030001

Ibrahim M, Anjum A, Khaliq NA, Iqbal MU, Athar H. Four foliar applications of glycinebetaine did not alleviate adverse effects of salt stress on growth of sunflower. *Pak J Bot* 2006, 38(5):1561–1570.

Iqbal N, Ashraf M, Ashraf MY. Glycinebetaine, an osmolyte of interest to improve water stress tolerance in sunflower (*Helianthus annuus* L.): water relations and yield. *S Afr J Bot* 2008,74(2): 274–281; https://doi.org/10.1016/j.sajb.2007.11.016

Islam MM, Hoque MA, Okuma E, Banu MN, Shimoishi Y, Nakamura Y, Murata Y. Exogenous proline and glycine betaine increase antioxidant enzyme activities and confer tolerance to cadmium stress in cultured tobacco cells. *J Plant Physiol* 2009, 166(15): 1587–1597; https://doi.org/10.1016/j.jplph.2009.04.002

Jabeen N, Abbas Z, Iqbal M, Rizwan M, Jabbar A, Farid M, Ali S, Ibrahim M, Abbas F. Glycine betaine mediates chromium tolerance in mung bean through lowering of Cr uptake and improved antioxidant system. *Arch Agron Soil Sci* 2015, 62(5): 648–662; https://doi.org/10.1080/03650340.2015.1082032

Jalmi SK, Bhagat PK, Verma D, Noryang S, Tayyeba S, Singh K, Sharma D, Sinha AK. Traversing the links between heavy metal stress and plant signaling. *Front Plant Sci* 2018, 9: 12; https://doi.org/10.3389/fpls.2018.00012

Khan MS, Yu X, Kikuchi A, Asahina M, Watanabe KN. Genetic engineering of glycine betaine biosynthesis to enhance abiotic stress tolerance in plants. *Plant Biotechnol* 2009, 26(1): 25–134; https://doi.org/10.3389/fpls.2018.01995

Khan MIR, Chopra P, Chhillar H, Ahanger MA, Hussain SJ, Maheshwari C, 2021. Regulatory hubs and strategies for improving heavy metal tolerance in plants: chemical messengers, omics and genetic engineering. *Plant Physiol Biochem* 2021, 164: 260–278; https://doi.org/10.1016/j.plaphy.2021.05.006.

Khan MIR, Chopra P, Chhillar H, Ahanger MA, Hussain SJ, Maheshwari C, 2021a. Regulatory hubs and strategies for improving heavy metal tolerance in plants: chemical messengers, omics and genetic engineering. *Plant Physiol Biochem* 2021, 164: 260–278.

Kumar P, Tokas J, Singal HR. Amelioration of chromium VI toxicity in Sorghum (*Sorghum bicolor* L.) using glycine betaine. *Sci Rep* 2019, 9(1):1–5; https://doi.org/10.1038/s41598-019-52479-w

Kurepin LV, Ivanov AG, Zaman M, Pharis RP, Hurry V, Hüner NP. Interaction of glycine betaine and plant hormones: protection of the photosynthetic apparatus during abiotic stress. In *Photosynthesis: Structures, Mechanisms, and Applications*. 2017. Springer, Cham, pp. 185–202; https://doi.org/10.1007/978-3-319-48873-8_9

Ladyman JAR, Hitz WD, Hanson AD. Translocation and metabolism of glycine betaine by barley plants in relation to water stress. *Planta* 1980, 150: 191–196; https://doi.org/10.1007/BF00390825

Li D, Zhang T, Wang M, Liu Y, Brestic M, Chen TH, Yang X. Genetic engineering of the biosynthesis of glycine betaine modulates phosphate homeostasis by regulating phosphate acquisition in tomato. *Front Plant Sci* 2019, 9: 1995; https://doi.org/10.3389/fpls.2018.01995

Li M, Li Z, Li S, Guo S, Meng Q, Li G, Yang X. Genetic engineering of glycine betaine biosynthesis reduces heat-enhanced photoinhibition by enhancing antioxidative defense and alleviating lipid peroxidation in tomato. *Plant Mol Biol Rep* 2014, 32(1): 42–51; https://doi.org/10.1007/s11105-013-0594-z

Li S, Li F, Wang J, Zhang WEN, Meng Q, Chen TH, Murata N, Yang X. Glycinebetaine enhances the tolerance of tomato plants to high temperature during germination of seeds and growth of seedlings. *Plant Cell Environ* 2011, 34(11): 1931–1943; https://doi.org/10.1111/j.1365-3040.2011.02389.x

Lokhande VH, Suprasanna P. Prospects of halophytes in understanding and managing abiotic stress tolerance. In *Environmental Adaptations and Stress Tolerance of Plants in the Era of Climate Change* 2012. Springer, New York. pp. 29–56; https://doi.org/10.1007/978-1-4614-0815-4_2

Lou Y, Sun X, Chao YI, Han F, Sun MI, Wang T, Wang H, Song F, Zhuge YU. Glycine betaine application alleviates salinity damage to antioxidant enzyme activity in alfalfa. *Pak J Bot* 2019, 51(1): 19–25; https://doi.org/10.30848/PJB2019-1(3)

Lou Y, Yang Y, Hu L, Liu H, Xu Q. Exogenous glycinebetaine alleviates the detrimental effect of Cd stress on perennial ryegrass. *Ecotoxicology* 2015, 24(6): 1330–1340; https://doi.org/10.1007/s10646-015-1508-7

Mahdavian K, Ghaderian SM, Schat H. Pb accumulation, Pb tolerance, antioxidants, thiols, and organic acids in metallicolous and non-metallicolous *Peganum harmala* L. under Pb exposure. *Environ Exp Bot* 2016, 126: 21–31; https://doi.org/10.1016/j.envexpbot.2016.01.010

Malar S, Vikram SS, Favas PJ, Perumal V. Lead heavy metal toxicity induced changes on growth and antioxidative enzymes level in water hyacinths [*Eichhornia crassipes* (Mart.)]. *Bot Stud* 2016, 55(1): 54; https://doi.org/10.1186/s40529-014-0054-6

Malekzadeh P. Influence of exogenous application of glycinebetaine on antioxidative system and growth of salt-stressed soybean seedlings (*Glycine max* L.). Physiol Mol Biol Plant 2015, 21(2): 225–232; https://doi.org/10.1007/s12298-015-0292-4

Masood A, Per TS, Asgher M, Fatma M, Khan MIR, Rasheed F, Hussain SJ, Khan NA. Glycine betaine: role in shifting plants toward adaptation under extreme environments. In *Osmolytes and Plants Acclimation to Changing Environment: Emerging Omics Technologies* 2016. Springer, New Delhi. pp. 69–82; https://doi.org/10.1007/978-81-322-2616-1_5

Molla MR, Ali MR, Hasanuzzaman M, Al-Mamun MH, Ahmed A, Nazim-Ud-Dowla MA, Rohman MM. Exogenous proline and betaine-induced upregulation of glutathione transferase and glyoxalase I in lentil (*Lens culinaris*) under drought stress. *Not Bot Horti Agrobot Cluj-Napoca* 2014, 42(1): 73–80; https://doi.org/10.15835/nbha4219324

Nayyar H, Chander K, Kumar S, Bains T. Glycine betaine mitigates cold stress damage in chickpea. *Agron Sustain Dev* 2005, 25(3): 381–388; https://doi.org/10.1051/agro:2005033

Nomura M, Hibino T, Takabe T, Sugiyama T, Yokota A, Miyake H, Takabe T. Transgenically produced glycinebetaine protects ribulose 1, 5-bisphosphate carboxylase/oxygenase from inactivation in *Synechococcus* sp. PCC7942 under salt stress. *Plant Cell Physiol* 1998, 39(4): 425–432; https://doi.org/10.1093/oxfordjournals.pcp.a029386

Paradisone V, Barrameda-Medina Y, Montesinos-Pereira D, Romero L, Esposito S, Ruiz JM. Roles of some nitrogenous compounds protectors in the resistance to zinc toxicity in *Lactuca sativa* cv. *Phillipus* and *Brassica oleracea* cv. Bronco. *Acta Physiol Plant* 2015, 37(7): 1–8; https://doi.org/10.1007/s11738-015-1893-9

Rasheed R, Ashraf MA, Hussain I, Haider MZ, Kanwal U, Iqbal M. Exogenous proline and glycinebetaine mitigate cadmium stress in two genetically different spring wheat (*Triticum aestivum* L.) cultivars. *Braz J Bot* 2014, 37(4): 399–406; https://doi.org/10.1007/s40415-014-0089-7

Rathinasabapathi B, McCue KF, Gage DA, Hanson AD. Metabolic engineering of glycine betaine synthesis: plant betaine aldehyde dehydrogenases lacking typical transit peptides are targeted to tobacco chloroplasts where they confer betaine aldehyde resistance. *Planta* 1994, 193(2): 155–612; https://doi.org/10.1007/BF00192524

Raza MA, Saleem MF, Shah GM, Khan IH, Raza A. Exogenous application of glycinebetaine and potassium for improving water relations and grain yield of wheat under drought. *J Soil Sci Plant Nutr* 2014, 14(2): 348–364; http://dx.doi.org/10.4067/S0718-95162014005000028

Roychoudhury A, Banerjee A. Endogenous glycine betaine accumulation mediates abiotic stress tolerance in plants. *Trop Plant Res* 2016, 3(1): 105–111.

Rucińska-Sobkowiak R. Water relations in plants subjected to heavy metal stresses. *Acta Physiol Plant* 2016, 38(11):1–13; https://doi.org/10.1007/s11738-016-2277-5

Sakamoto A, Murata N. Genetic engineering of glycinebetaine synthesis in plants: current status and implications for enhancement of stress tolerance. *J Exp Bot* 2000, 51: 81–88; https://doi.org/10.1093/jexbot/51.342.81

Shahbaz M, Zia B. Does exogenous application of glycinebetaine through rooting medium alter rice (*Oryza sativa* L.) mineral nutrient status under saline conditions? *J Appl Bot Food Qual* 2011, 84(1): 54–60.

Sharma A, Shahzad B, Kumar V, Kohli SK, Sidhu GPS, Bali AS, Handa N, Kapoor D, Bhardwaj R, Zheng B. Phytohormones regulate accumulation of osmolytes under abiotic stress. *Biomolecules* 2019, 9(7): 285; https://doi.org/10.3390/biom9070285

Sheetal KR, Singh SD, Anand A, Prasad S. Heavy metal accumulation and effects on growth, biomass and physiological processes in mustard. *Indian J Plant Physiol* 2016, 21(2): 219–223; https://doi.org/10.1007/s40502-016-0221-8

Stepien A, Wojtkowiak K, Orzech K, Wiktorski A. Nutritional and technological characteristics of common and spelt wheats are affected by mineral fertilizer and organic stimulator nano-gro. *Acta Sci Pol Agricultura* 2016, 15(2): 49–63; https://doi.org/10.37660/aspagr

Sullivan MN, Gonzales AL, Pires PW, Bruhl A, Leo MD, Li W, Oulidi A, Boop FA, Feng Y, Jaggar JH, Welsh DG. Localized TRPA1 channel Ca^{2+} signals stimulated by reactive oxygen species promote cerebral artery dilation. *Sci Signal* 2015, 8(358):ra2; https://doi.org/10.1126/scisignal.2005659

Surabhi GK, Rout A. (2020). Glycine betaine and crop abiotic stress tolerance: an update. In Protective Chemical Agents in the Amelioration of Plant Abiotic Stress: Biochemical and Molecular Perspectives, John Wiley & Sons. pp. 24–52; https://doi.org/10.1002/9781119552154.ch2

Tian F, Wang W, Liang C, Wang X, Wang G, Wang W. Overaccumulation of glycine betaine makes the function of the thylakoid membrane better in wheat under salt stress. *Crop J* 2017, 5(1): 73–82; https://doi.org/10.1016/j.cj.2016.05.008

Türkan I, Demiral T. Recent developments in understanding salinity tolerance. *Environ Exp Bot* 2009, 67(1): 2–9; https://doi.org/10.1016/j.envexpbot.2009.05.008

Umezawa T, Fujita M, Fujita Y, Yamaguchi-Shinozaki K, Shinozaki K. Engineering drought tolerance in plants: discovering and tailoring genes to unlock the future. *Curr Opin Biotechnol* 2006, 17(2): 113–122; https://doi.org/10.1016/j.copbio.2006.02.002

Wang GP, Zhang XY, Li F, Luo Y, Wang W. Overaccumulation of glycine betaine enhances tolerance to drought and heat stress in wheat leaves in the protection of photosynthesis. *Photosynthetica* 2010, 48(1): 117–126; https://doi.org/10.1007/s11099-010-0016-5

Wei D, Zhang W, Wang C, Meng Q, Li G, Chen TH, Yang X. Genetic engineering of the biosynthesis of glycinebetaine leads to alleviate salt-induced potassium efflux and enhances salt tolerance in tomato plants. *Plant Sci* 2017, 257: 74–83; http://doi.org/10.1016/j.plantsci.2017.01.012

Yadu B, Chandrakar V, Meena RK, Keshavkant S. Glycinebetaine reduces oxidative injury and enhances fluoride stress tolerance via improving antioxidant enzymes, proline and genomic template stability in *Cajanus cajan* L. *S Afr J Bot* 2017, 111: 68–75; https://doi.org/10.1016/j.sajb.2017.03.023

Yancey PH. Organic osmolytes as compatible, metabolic and counteracting cytoprotectants in high osmolarity and other stresses. *J Exp Biol* 2005, 208(15): 2819–2830; https://doi.org/10.1242/jeb.01730

Zouari M, Elloumi N, Labrousse P, Rouina BB, Abdallah FB, Ahmed CB. Olive trees response to lead stress: exogenous proline provided better tolerance than glycine betaine. *S Afr J Bot* 2018, 118: 158–165; https://doi.org/10.1016/j.sajb.2018.07.008

8 Abiotic Stress and Its Role in Altering the Nutritional Landscape of Food Crops

Veda Krishnan, Muzaffar Hasan, Monika Awana, and Archana Singh*
Division of Biochemistry, ICAR-Indian Agricultural Research Institute (IARI), New Delhi, India
*Corresponding Author: E-mail: vedabiochem@gmail.com

CONTENTS

8.1 INTRODUCTION

Abiotic stresses are undoubtedly major threats not only to global food security but also to nutritional security, limiting the UN sustainable development goals (Ashmore et al., 2006; Pareek et al., 2020). Abiotic stresses due to heat, drought, salinity, reduced light level, and excess ultraviolet (UV) radiation are becoming even more evident in the twenty-second century, and globally, countries are striving towards developing tolerant or smart varieties (Lobell et al., 2008; Wassmann et al., 2009b; Khan et al., 2021a; Poor et al., 2021). Relevance on food crop yield has been well documented for many decades, as abiotic stress poses a greater challenge to food shortage (Lobell et al., 2008; Wassmann et al., 2009a). Even though reports suggested the possibility for changes in photosynthetic potential, gas exchange, water uptake, antioxidant activity, and nutrient translocation, less attention has been devoted to understand how such stress stimuli, as well as their persistence, impacts the nutritional spectrum of food crops (Blokhina et al., 2003; Morgan et al., 2004; Katerji et al., 2010). As plant-based foods are the backbone of any dietary culture, the change in crop fingerprints in terms of nutritional quality needs to be evaluated comprehensively. This chapter summarizes how abiotic stress types alter the macro/micronutrient composition in their matrix, which affects the ultimate nutritional and processing quality.

DOI: 10.1201/9781003159636-8

8.2 IMPACT OF ABIOTIC STRESS ON CARBOHYDRATES

Carbohydrates, especially the polymeric starch form, are known as a key determinant of plant fitness to survive under abiotic stress conditions. As starch is the major storage form, its remobilization to sugars has been observed as a common phenomenon to mitigate stress, as sugars also act as osmo-protectants (Thalmann & Santelia, 2017). In starchy staples, exposure to abiotic stresses in the grain filling stage has been most associated with reduction in starch accumulation: for example, terminal heat stress in wheat (Labuschagne et al., 2009), salinity stress in rice (Siscarlee et al., 1990), drought/heat in barley (Savin et al., 1996), freezing (Nagao et al., 2005), or ozone (O_3) stress (Fuhrer et al., 1992). In general, a major reduction in total starch content and yield recovery has been observed, correlating well with the differential expression of starch biosynthetic genes or altered enzyme activities (Prathap et al., 2019). Impeded conversion of sugars to starch under abiotic stresses has been reported in several crops (Khan & Abdullah, 2003; Keeling et al., 1993). On the gene expression level, decline in the transcription of starch synthase genes were reported in rice, wheat, and many other crops under heat and drought stress (Szucs et al., 2010; Yamakawa & Hakata, 2010). On the contrary, starchy tuber crops, halophytes, green algae, and *Arabidopsis* have been reported to have enhanced starch concentration when subjected to abiotic stress (Wang et al., 2013; Siaut et al., 2011; Kaplan & Guy, 2004), while potato tubers under salinity and heat stress responded differently (Silva et al., 2001; Debon et al., 1998). Surprisingly, cassava and sweet potato under drought and heat stress showed a decrease in starch content (Santisopasri et al., 2001; Kano & Mano, 2002) with a consistent increase in amylose fraction (Noda et al., 2001). Reports also endorse the fact that abiotic stresses alter the amylose–amylopectin ratio, which is a key determinant for the glycemic index (GI) of foods. The varying results reported are non-comparable due to differences in the experimental protocols (stress applied for varied durations), analytical methods, genetic origin, data presentation (wet or dry basis), as well as considering the fact that the enzymatic conversion rate of sugars to starch might be strongly influenced by different environmental factors (Silva et al., 2001; Kano & Mano, 2002). Among the starch biosynthetic enzymes, starch synthase or granule bound starch synthase (GBSS) activity has been found to be most affected by abiotic stresses like heat and drought (Noda et al., 2001; Debron et al., 1998). However, debranching enzymes like isoamylase, or degradative enzymes like β-amylase, which remobilize starch to maltose, has been found to be enhanced under such conditions (Streb et al., 2012; Kaplan & Guy, 2004; Kaplan et al., 2006; Neilson et al., 1997).

Among the non-structural carbohydrates, sugar concentration has shown the most inconsistent response to drought (Tsialtas et al., 2009) and salinity (Hajiboland et al., 2009). Drought and salinity stresses were reported to cause an increase in sugar profiles of fruits and vegetables like sugar beets, tomatoes, cucumbers, and grapes (Wu & Kubota, 2008; Sgherri et al., 2008). The altered starch degradation rate could be due to the changed photoperiods and differential carbon supply, while the observed sugar concentration has been correlated with its osmo-protective capability. Conversion of starch to sugar has also highlighted its role in regulating plant physiology, knowing its cross talk with abscisic acid (ABA)-dependent signaling, which increases guard cell turgor to promote stomatal opening, or generating organic acids to tolerate stress (Horrer et al., 2016). The role of the digestion-resistant fraction of starch, known as resistant starch (RS), as well as GI has been found to alter under abiotic stress like drought in rice (Prathap et al., 2019; Kumar et al., 2020). Impacts of O_3 on carbon metabolism, studied using tree species, revealed a significant increase in water-soluble carbohydrates and the lowest effect on polysaccharides in leaves (Chen et al., 2018). Furthermore, Yadav et al. (2019) reported the effect of elevated levels of CO_2 (EC) and elevated O_3 (EO) exposure on yield as well as quality of wheat cultivars; the starch content was found to be increased under EC and insignificant under EO conditions. Moreover, Zheng et al. (2014) underlined that prolonged exposure to O_3 decreased the photosynthate translocation and accumulation of starch in wheat.

8.3 IMPACT OF ABIOTIC STRESS ON PROTEIN CONTENT AND COMPOSITION

The nutritional significance of food crops, particularly legumes, is mainly determined by their total protein content; however, quality cereal proteins in the diet are also necessary to have a balanced amino acid intake. The sensitivity of the protein content of a food crop to abiotic stress depends on the genotype and the magnitude and duration of stress. Generally, most crop species have adapted to unfavorable environmental conditions by increasing their protein content. On the contrary, few reports have also shown a decline or a non-significant effect of stress types on protein profiles. Good et al. (1994) discovered an overall reduction in protein synthesis when drought increased, which improved after watering in *Brassica napus*. The soluble protein content was drastically reduced, whereas the concentration of free amino acids has been observed to increase in wheat under drought conditions (Abid et al., 2018). When plants encountered drought stress within the first 2 weeks of grain filling, the grain nitrogen content decreased, while drought stress had no such impact in the middle and final stages of grain filling. Panozzo et al. (2001) found an insignificant difference between irrigated and non-irrigated treatments of wheat in terms of the gliadin/glutenin ratio in some field trials performed in Victoria, Australia, while total protein fraction was higher in the non-irrigated condition. Mild drought boosted the protein content of mung bean and chickpea grain by 6–21% (Behboudian et al., 2001). However, extreme drought decreased the content of grain protein in lupins by 19–35% (Khalil & Ismael, 2010). Drought has also been reported to cause a decrease in minerals and total protein in common bean, while an increase has been observed in the case of fava beans (Ghanbari et al., 2013). Various grain proteins like albumins, globulins, and prolamins have been found to decline in 13 fava bean varieties evaluated under drought, while glutelin concentration and total protein content has been improved (Alghamdi, 2009). Other than the storage proteins, a substantial increase in dehydrin abundance proteins has also been observed under drought conditions (in greenhouse and field) in soybean grains (Fuganti-Pagliarini et al., 2017). In rice, drought stress reduced grain-soluble protein and insoluble protein by 30% and 35%, respectively, during the reproductive stage (Emam et al., 2014). Likewise, drought reduced protein content in maize kernels by 47% and 17% during the reproductive phase (Nawaz et al., 2016) and the vegetative phase (Nawaz et al., 2016; Ali & Ashraf, 2011; Ali et al., 2013), respectively. At the early grain filling stage, high temperatures enhanced the accumulation of all groups of storage proteins but lowered the accumulation of prolamins in rice at maturation (Lin et al., 2010). Similarly, heat stress in *Brassica napus* has been found to increase the levels of protein (Triboi & Triboi-Blondel, 2002). During grain filling, gluten protein synthesis, aggregation, and assembly are known to be influenced by heat stress. The accumulation of gluten was found to be initially accelerated but gradually decreased due to the inactivation of enzymes under heat (Altenbach et al., 2002). In peas, high temperature caused denaturation and aggregation of seed storage proteins (legumin, globulins, and vicilin) (Mession et al., 2013), which could be due to the loss of covalent and non-covalent interactions among the proteins (Sun & Arntfield, 2012). Similarly, heat stress has been reported to denature b-conglycinin in soybean (Iwabuchi & Yamauchi, 1984). It has also adversely affected the seed protein fractions, mainly globulins and albumins, in lentil seeds (Sita et al., 2018), thereby lowering the overall quality of seeds. Injury to protein metabolism and structure under heat stress has also suppressed the function of enzymes involved in protein synthesis, as observed in Andean lupin (Zou, 2009). Under salinity stress, high respiration rate results in a reduction in assimilate transport to grains, and as a consequence, a reduction in the total amount of stored carbohydrate and an increase in the protein content were observed (Poustini, 2002). But under water stress, increased protein content was observed in barley (Savin & Nicolas, 1996), maize (Oktem, 2008), peanuts (Dwivedi et al., 1996), and soybean. A reduction in total protein at high salinity (120 mM) was reported by Sonia & Mohammad (2013) in wheat, while Houshmand et al. (2014) revealed that salinity induces a substantial increase in grain protein content, wet and dry gluten content, and volume of sedimentation in wheat. A substantial increase in protein content in brown rice in *cv*Pokkali was observed in a study conducted by Ahmed et al. (1982) at high levels of salinity (6 and 8 dS/m), whereas the protein content remained unchanged at salt levels lower than this. It is considered that the higher the protein content, the lower the disruption of rice grain during the milling (Lee et al., 2009). In saline conditions, the grain protein content of chickpea and mung bean has been found to decline because of the disruption in nitrogen metabolism and/or decrease in NO_3 absorption from the soil solution. A decrease in grain protein content with increased salt stress (60 mM, 120 mM, and 240 mM) has been observed in fava beans (Qados, 2011). Similarly, a decrease in soluble protein content was also observed with increase in salt stress in cowpeas (Chen et al., 2007). Various studies showed that EC decreased the protein concentration in several crop species (Dong et al., 2018; Medek et al., 2017; Myers et al., 2014), which directly affects human nutrition (Toreti et al., 2020). Medek et al. (2017) reported that protein concentration has been found to be low in C_3 grains (barley, rice, and wheat: 14.1%, 7.6%, and 7.8%, respectively), vegetables (17.3%), potatoes (6.4%), and fruit (17.3%) when grown under EC conditions. In legume species, EC has also reported to be responsible for a minor reduction in protein content (3.5%), while no major effects have observed in oil crops and C_4 plants. In an experiment conducted in Southern Chile, Calderini et al. (2008) confirmed that a spike in UV radiation-B (UVR-B) did not affect wheat production and quality. No difference in grain protein and gluten content has been found between increased UV-B treated and control samples of wheat. On the contrary, a study carried out by Zu et al. (2004) on 10 wheat cultivars stated that UV-B greatly affects the grains' free amino acids, protein, and total sugar content. Hidema et al. (2005) found that an increase in UVR-B radiation changed the quality of grain protein in rice. Galantini et al. (2000) observed that fertilization (NPK) does not contribute to the dry matter but greatly affects the accumulation of nutrients in the wheat grain. Among the nutrients that influence grain quality, nitrogen is the most vital (Ehdaie & Waines, 2001; Shi et al., 2010). The application of high nitrogen results in increased yields, but there is still debate about the technical parameters of such yields. In addition to stimulating yields, by increasing the proportion of low molecular weight gluten, higher nitrogen levels degrade the quality of gluten in grains (Wooding et al., 2000). During the important phases of protein biosynthesis, potassium is an essential element, and its deficiency leads to a reduction in the protein content; this effect occurs despite an increased level of nitrogen and the accumulation of non-protein nitrogen (Rice, 2007). In sulfur-deficient soils, wheat grains usually accumulate arginine and asparagine but have lower amounts of sulfur-containing methionine and cysteine amino acids (Hesse et al., 2004). When sulfur is applied along with nitrogen, the gluten, methionine, and cysteine content was found to be substantially increased in wheat (Klikocka et al., 2016). In wheat, the protein content has also significantly improved under the increased O_3 condition (Broberg et al., 2015). Despite the 18% decrement in grain protein yield, a 6% increment in total grain protein content was observed in wheat (Feng et al., 2008).

8.4 IMPACT OF ABIOTIC STRESS ON LIPID CONTENT AND COMPOSITION

Other than carbohydrates and proteins, the nutritional and storage attributes depend mainly on the oil content and its fractions, like saturated or unsaturated fatty acids. Among the plant types, legumes have been most studied for understanding the alterations in oil content and composition under different climatic conditions. A high proportion of polyunsaturated fatty acids (PUFAs) have been reported to have beneficial health effects, like lowered risk of coronary heart disease and atherogenesis (Jackson et al., 1978). But PUFAs are also known to be prone to oxidation, resulting in an unpleasant flavor, considered as an unfavorable trait for processing applications (Branch et al., 1990). Most studies have reported the adverse effect of abiotic stresses on oil concentration and composition in oilseed crops with essential oils as an important constituent. In general, the saturation level of the oil fraction has been found to increase under abiotic stresses, with an average decrease in PUFAs.

Oil content has been found to decrease in most of the legumes under drought stress. But in olives, oil content has increased under drought stress, probably because olives are less susceptible to water deficit conditions (Wang & Frei, 2011). Drought stress has also been reported to affect the composition of the oil fractions, with an increase in oleic acid (C18:1) and a decrease in linoleic (C18:3) and linoleic (C18:2) acids. A substantial decrease in the proportion of PUFAs has also been reported in the oil fractions of groundnut (Dwivedi et al., 1996), sunflower (Flagella et al., 2002), and sage under drought stress (Bettaieb et al., 2009). Dornbos and Mullen (1992) studied the effect of drought and high air temperature on soybean seed protein and oil content and found that both stress types affected the proportion, quantity, and quality of the seed oil and protein. In soybean and peanut, oil content decreased along with an increased proportion of oleic acid (C18:1) under severe drought conditions, especially at the time of the grain filling (Dornbos & Mullen, 1992). This in turn compromised the oil stability. The grain oil content of lupin bean has also been found to decrease by 50–55% under drought stress (Carvalho et al., 2005). On the contrary, fava bean has not shown any change in fatty acid profile under drought stress. Tocopherols are the most prominent antioxidants, which are also found in edible oils, either inherently or added to hinder lipid oxidation.

In line with drought, salt stress has been reported to affect the oil content of food legumes by restricting the absorption of NO_3 and interfering with nitrogen metabolism (Farooq et al., 2018). In general, a negative correlation has been observed between grain protein and oil content when subjected to salinity stress. Ghassemi-Golezani et al. (2010)found a linear relation between oil content and salt stress. A significant decrease in the proportion of PUFAs has been seen under salt stress in the oil fractions of olive (Chartzoulakis, 2005), sunflower (Di Caterina et al., 2007), cotton (Ahmad et al., 2007), sage (Taarit et al., 2010), and coriander (Neffati & Marzouk, 2008).

Among the abiotic stress types, heat stress is known to have a favorable impact on the total oil content of food crops, with a few exceptions. For instance, oil content significantly increased in soybeans and peanuts by 37% and 20%, respectively, under heat stress, whereas a 23% reduction in oil content has been reported in the case of kidney beans (Thomas et al., 2009; Wolf et al., 1982). Prasad et al. (2000)reported that high daytime soil and air temperature significantly increased the oil content compared with the optimum daytime soil and air temperature. Heat stress not only influences the oil content of food legumes but also effectively changes the ratio of fatty acids by inducing structural alterations in the fatty acids. Under elevated temperature, oleic acid (C18:1) concentration was found to increase, while the concentration of linoleic acid (C18:2) decreased. High air temperature during the soybean seed filling stage decreased the proportion of linoleic acid (C18:2) and linolenic acid (C18:3) by 11.1%, while the oleic acid (C18:1) content increased up to 10.0% (Dornbos & Mullen, 1992). A similar relation has also been reported in linseed between high air temperature and fatty acid composition, where the proportion of linoleic and linolenic acids decreased under high air temperature due to the inactivation of oleic acid (C18:1) desaturase enzyme, which converts oleic acid into linoleic and linolenic acids (Wilson et al., 1980; Dornbos & Mullen, 1992). The reduction of PUFAs like linoleic acid (C18:2) and linolenic acids (C18:3) with a concomitant rise of monounsaturated fatty acids like oleic acid (C18:1) illustrates the restriction of the dehydrogenation reaction at elevated temperature. It is speculated that oxygen is the rate limiting factor for the dehydrogenation reaction employed by the desaturase enzyme, and increased temperature leads to reduction in solubility of O_2 in the cytoplasm. The temperature-dependent decline in the enzymatic activity could be explained by two factors: the enzyme's heat stability and the effects of temperature on the seed's internal oxygen concentration, which is a central regulator of oleate desaturase enzyme activity. Transport of fatty acids from the plastids to the cytosol has also been proposed as a process that could be influenced by these environmental stresses in addition to enzymatic desaturation of fatty acids (Flagella et al., 2002).

Reckless use of industrial wastewater as a source of irrigation for agricultural land has led to the accumulation of heavy metals, i.e., cadmium (Cd), mercury (Hg), chromium (Cr), lead (Pb), zinc (Zn), and nickel (Ni) (Pokhrel et al., 2009). Crops cultivated in soils polluted with heavy metal ions have been reported to accumulate heavy metals in their grains, which may reach the food chain, resulting in severe health hazards. In soybean, the oil content was reduced by exposure to a high concentration of heavy metals (Cd, Hg). However, the reduction of oil content was higher with an individual application of Cd or Hg rather than a combined application, revealing the antagonistic effect of Cd and Hg on the oil content of soybean (Khan et al., 2013). Heavy metal contamination of the soil enhanced the content of saturated fatty acids like palmitic acid (C16:0), stearic acid (C18:0), and reduced PUFAs, which altered the oil profile of soybean, resulting in a low-grade oil (Dornbos & Mullen, 1992; Khan et al., 2013). On the other hand, EC(above 250 ppm ambient CO_2) has not been found to have an effect on altering the content of oil in food crops, while in the case of mung beans, a significant increase in ω-3-fatty acids with a concomitant decrease in the ratio ofω-3 to ω-6 fatty acids has been observed (Wang et al., 2003). The significant rise in the ratio of ω-3 to ω-6 fatty acids suggested the potential of mung bean to provide ω-3fatty acids in response to EC level.

The substantial changes observed in oil content and fatty acid profiles during various abiotic stresses has been attributed to increasing fitness and survival mechanisms. In most cases, oil content decreased under environmental stress, especially PUFAs, as mentioned in Table 8.1, and it must be remembered

TABLE 8.1
Impact of Abiotic Stresses on Nutritional Spectrum of Food Crops

Plant Species	Stress Type	Change in Nutrient	References
		Carbohydrates Including Starch and Sugars	
Oryza sativa Japonica and Indica	Drought	Soluble sugar ↑sucrose ↑starch ↓	Yang et al.(2001)
Triticale	Drought	Starch ↓	He et al.(2012)
Glycine max	Drought	Hexose ↑ sucrose ↓ starch ↓	Liu et al.(2004)
Phaseolus vulgaris	Drought	Starch ↓	Gonzalez-Cruz & Pastenes (2012)
Hordeum vulgare	Drought	Glucose ↑ fructose ↑ sucrose ↓ starch ↓	Villadsen et al.(2005)
Arabidopsis thaliana	Drought	Starch ↓	Harb et al.(2010)
Triticum aestivum	Elevated CO_2 (EC)	Starch ↓ (EC)	Yadav et al.(2019)
Arabidopsis thaliana	Freezing(−3 °C to −15 °C)	Glucose ↑ fructose ↑ maltose ↑ sucrose ↑ starch ↓	Pressel et al.(2006)
Arabidopsis thaliana	Heat	Soluble sugars ↓ starch ↓	Vasseur et al. (2011)
Solanum tuberosum	Low temperature	Reducing sugars ↑	Sitnicka & Orzechowski (2014)
Solanum tuberosum	Osmotic stress (100, 200, 300, 500 mM mannitol)	Sucrose ↑ starch ↓	Geigenberger et al. (1997)
Arabidopsis thaliana	Osmotic stress (300 mM mannitol) ABA treatment (100 mM)	Glucose ↑ fructose ↑ maltose ↑ sucrose ↑ starch ↓	Thalmann et al. (2016)
Arabidopsis thaliana	Salt stress (150 mM) ABA treatment (25 mM)	Sucrose ↑ maltose ↑ starch ↓	Kempa et al. (2008)
		Protein (Content and Composition)	
Vigna radiata	Salinity stress (1500 mg/l, 3000 mg/l, 4500 mg/ l, 6000 mg/l)	Total protein content ↓	Qados (2010)

TABLE 8.1 (Continued)
Impact of Abiotic Stresses on Nutritional Spectrum of Food Crops

Plant Species	Stress Type	Change in Nutrient	References
Vigna radiata	Drought	Total protein content ↑	Sadeghipour(2009)
Vigna radiata	Elevated carbon dioxide (700 ppm)	Total protein content ↓	Mishra & Aggarwal (2014)
Pisum sativum	Heat stress	Legumin ↓ globulins ↓ vicilin ↓	Mession et al. (2013)
Glycine max	Salinity stress (3, 6 and 9 ds/m NaCl)	Total protein content↓	Ghassemi-Golezani et al. (2010)
Glycine max	Drought	Dehydrin abundance proteins ↑	Fuganti-Pagliarini et al. (2017)
Glycine max	Heat stress	b-conglycinin ↓	Iwabuchi & Yamauchi (1984)
Triticum aestivum	Salinity stress (60 mM, 120 mM NaCl) 10 dS/m NaCl	Total protein content ↑	Sonia and Mohammad (2013) Houshmand et al. (2014)
Triticum aestivum	Drought	Total protein content ↑	Ozturk & Aydin (2004) Flagella et al. (2010)
Triticum aestivum	Heat stress	Total protein content ↑	Asseng & Milroy (2006)
Triticum aestivum	Elevated carbon dioxide	Total protein content ↓	Medek et al. (2017)
Triticum aestivum	Ultraviolet radiation	Total protein content ↑	Zu et al. (2004) Yao et al. (2014)
Triticum aestivum	Nutrient deficiency (nitrogen, sulfur)	Total protein content ↓	Klikocka et al. (2016)
Triticum aestivum	Ozone	Total protein content ↑	Broberg et al. (2015)
Oryza sativa	Salinity stress (12 ds/ m)	Total protein content ↑	Ahmed et al. (1982)
Oryza sativa	Heat stress (35/30 °C day/night)	Storage proteins ↑ prolamins ↓	Lin et al. (2010)
Oryza sativa	Ultraviolet radiation	Total protein content ↑	Hidema et al. (2005)
Oryza sativa	Drought	Soluble protein ↓ insoluble protein ↓	Emam et al. (2014)
Hordeum vulgare	Drought	Total protein content ↑	Savin & Nicolas (1996)
Hordeum vulgare	Heat stress (20/32 °C)	Total protein content ↑	Pettersson & Eckersten (2007)
Zea mays	Drought	Total protein content ↑	Oktem (2008)
Cicer arietinum	Drought	Total protein content ↓	Ashrafi et al. (2015) Khamssi (2011)
Cicer arietinum	Heat stress	Total protein content ↑	Blum (1988)
Vigna unguiculata	Salinity stress (75 mM NaCl)	Total protein content ↓	Chen et al. (2007)
Vicia faba	Drought	Albumins ↓ globulins ↓ prolamins ↓ glutelin ↑	Alghamdi (2009)
		Lipids (Oil Content and Composition)	
Glycine max	Salinity (NaCl; 3 dS/ m)	Oil content ↓	Ghassemi-Golezani et al. (2010)
Glycine max	Salinity (NaCl; 6 dS/ m)	Oil content ↓	Ghassemi-Golezani et al. (2010)
Glycine max	Salinity (NaCl; 9 dS/ m)	Oil content ↓	Ghassemi-Golezani et al. (2010)

(continued)

TABLE 8.1 (Continued)
Impact of Abiotic Stresses on Nutritional Spectrum of Food Crops

Plant Species	Stress Type	Change in Nutrient	References
Glycine max	Salinity (NaCl; 9 dS/m)	Oil content ↓	Ghassemi-Golezani et al. (2009)
Vicia faba	Water stress	Fat content ↑	Al-Suhaibani (2009)
Glycine max	Drought	Oil content ↓	Dornbos & Mullen (1992)
Arachis hypogaea	Heat stress (26/20 °C)	Oil content ↑ oleic acid ↑	Dornbos & Mullen (1992)
Glycine max	Heat stress (40/30 °C)	Oleic acid ↑ linolenic acid ↓	Golombek et al. (1995)
Glycine max	Heat stress (33/28 °C)	Oil content ↑ oleic acid ↑	Wolf et al. (1982)
Phaseolus vulgaris	Heat stress (34/24 °C)	Oil content ↓	Thomas et al. (2009)
Glycine max	Heat stress (35 °C)	Oil content ↑	Dornbos & Mullen (1992)
Glycine max	Heavy metal (Cd 0.1 mM)	Oil content ↓	Khan et al. (2013)
Glycine max	Heavy metal (Cd 0.5 mM)	Oil content ↓	Khan et al. (2013)
Glycine max	Heavy metal (Cd 1.0 mM)	Oil content ↓	Khan et al. (2013)
Glycine max	Heavy metal (Hg 0.1 mM)	Oil content ↓	Khan et al. (2013)
Glycine max	Heavy metal (Hg 0.5 mM)	Oil content ↓	Khan et al. (2013)
Glycine max	Heavy metal (Hg 1.0 mM)	Oil content ↓	Khan et al. (2013)
Vigna radiata	Elevated CO_2 (667 µmol/mol)	Palmitic acid ↓	Ziska et al. (2007)
Vigna radiata	Elevated CO_2 (667 µmol/mol)	ω-6-fatty acid ↓	Ziska et al. (2007)
Vigna radiata	Elevated CO_2 (667 µmol/mol)	ω-3-fatty acid ↑	Ziska et al. (2007)
		Minerals and Types	
Triticum aestivum	Elevated ozone (O_3)	Mineral accumulation ↑	Pleijel et al. (2006)
Triticum aestivum	Elevated ozone (O_3) and CO_2	Zn content ↑	Pleijel & Danielsson (2009)
Zea mays	Elevated ozone (O_3)	Zn, Fe, Cu ↑	Garcia et al. (1983)
Solanum tuberosum	Elevated ozone (O_3)	K, Mg ↑	Piikki et al. (2007)
Brassica	Elevated ozone (O_3)	K, Mg, Ca, Zn ↓	Singh et al. (2009)
Brassica napus	Drought	Ca, Zn, Cu, Mg ↑ P, K ↓	Mamnabi et al. (2020)
Triticum aestivum	Salt stress	K:Na ↑	Singh et al. (2015)
		Vitamins and Types	
Triticum aestivum	Salinity stress	Total vitamin content ↓	Ashraf (2014)
Lactuca sativa	Drought	Vit. E ↑	Oh et al. (2010)
Oryza sativa	Elevated carbon dioxide	Total vitamin content ↑	Zhu et al. (2018)
Vigna radiata	Osmotic stress (50 mM, 100 mM, 150 mM NaCl)	Vit. C ↑	Televiciute et al. (2020
Vigna radiata	Heat stress (5 °C, 40 °C, 45 °C)	Vit. C ↑	Televiciute et al. (2020)

TABLE 8.1 (Continued)
Impact of Abiotic Stresses on Nutritional Spectrum of Food Crops

Plant Species	Stress Type	Change in Nutrient	References
Lens culinaris	Osmotic stress (50 mM, 100 mM, 150 mM NaCl)	Vit. C ↑	Televiciute et al. (2020)
Solanum lycopersicum	Drought	Vit. C ↑	Favati et al. (2009)
Daucus carota	Drought	Vit. A ↑ Vit. C ↑	Sorenson et al. (1997)
Capsicum frutescens	Drought	Vit. C ↓	Ahmed et al. (2014)
Vigna unguiculata	Drought	Vit. C ↓	Nair et al. (2008)
Triticum aestivum	Drought	Vit. A ↓	Abid et al. (2018)
		Phytochemicals (Total and Types)	
Phaseolus vulgaris	Drought	Epicatechin ↑, naringenin ↑, cinnamic acid ↑, benzoic acid ↑, β-sitosterylglucoside ↓	Herrera et al. (2019)
Phaseolus vulgaris	Nutrient deficiency (nitrogen)	Total phenolic content ↓	Sanchez et al. (2000)
Triticum aestivum	Drought	α-tocopherol ↑	Herbinger et al. (2002)
Glycine max	Drought	α-tocopherol ↑	Britz & Kremer (2002)
Hordeum vulgare	Salinity	Total phenolic content ↑	Ahmed et al. (2013)
Fagopyrum esculentum	Salinity stress (10, 50, 100, and 200 mM NaCl)	Isoorientin ↑ rutin ↑ vitexin ↑	Lim et al. (2012)
Oryza sativa indica, japonica, glaberrima	Elevated temperature	α-tocopherol ↑ γ-tocopherol ↑	Britz et al. (2007)
Oryza sativa	Elevated carbon dioxide (375 and 550 μmol/mol)	Sinapic acid ↓ apigenin ↓ tocopherols ↓ tocotrienols ↓	Goufo et al. (2014)
Lactuca sativa	Drought	Caftaric acid ↑ caffeic acid ↓ chlorogenic acid ↓ chichoricacid ↓	Galieni et al. (2015)
Lactuca sativa	Drought	Chicoric acid ↑	Oh et al. (2010)
Lactuca sativa	Heat stress (40 °C)	Caffeic acid ↑	Oh et al. (2009)
Lactuca sativa	Chilling stress (4 °C)	Chicoric acid ↑	Oh et al. (2009)
Nicotiana tabacum	Drought	α-tocopherol ↑	Liu et al. (2008)
		Dietary Fiber (Total and Types)	
Hordeum vulgare	Drought	β-glucan ↓	Guler (2003)
Hordeum vulgare	Nutrient deficiency (nitrogen)	β-glucan ↓	Guler (2003)
Hordeum vulgare	Heat	No change in β-glucan	MacNicol et al. (1993)
Triticum aestivum	Drought	β-glucan ↓ arabinoxylan ↑	Rakszegi et al. (2014)
Triticum aestivum	Elevated temperature	Arabinoxylan ↑	Laurentin & Douglas (2003)
Triticum aestivum	Nutrient deficiency (nitrogen)	β-glucan ↓	Rakszegi et al. (2014)
Triticum aestivum	Heat	Arabinoxylan ↑	Hong et al. (1989)
Zea mays	Drought	Total fiber content ↑	Ali & Ashraf (2011)

that these are just general patterns with frequent exceptions. Exploiting the positive effects of stresses on crop quality without compromising yield could thus be an efficient strategy for the future.

8.5 IMPACT OF ABIOTIC STRESS ON MINERAL PROFILES

Micronutrients, being major cofactors of metabolic enzymes, have a substantial role in metabolism, and their deficiency has been known as hidden hunger. Mineral profiles of crops have been studied to investigate the effect of mineral (heavy metals, soil pollutants, etc.) and non-mineral stresses (salinity, O_3), but reports do not give a consistent trend. The abiotic factors could either directly alter mineral homeostasis or affect absorption inhibitors like phytic acid, which limits mineral availability in the matrix. Among the stress types, EO has been observed to result in higher accumulation of minerals in wheat grains (Pleijel et al., 2006). Pleijel and Danielsson (2009) reported the presence of enhanced Zn content in positive correlation with protein concentration under EC/EO. Unlike O_3, CO_2 was found to stimulate yields and lead to dilution in protein and Zn concentrations. O_3 stress in maize has been found to have minor effects on macronutrient (K, P, Mg) concentrations but significant increase in Zn, Fe, and Cu (Garcia et al., 1983). In line with this, Piikki et al. (2007) found that O_3 exposure increased K and Mg in potato tubers, while Ca was found to be unaffected. The possible role of O_3-induced alteration in the mineral landscape of crops could be that the total biomass accumulation itself has been reduced due to impeded carbon assimilation and thus, mineral uptake (Wang & Frei, 2011). Singh et al. (2009) reported decreased concentrations of K, Zn, Ca, and Mg in mustard seeds on exposure to O_3stress. During drought, a significant change in mineral mobilization has also been observed due to restricted water uptake along with mineral elements (Oktem, 2008). Ti et al. (2010) reported the role of water deficit stress in increasing Ca (28%), Zn (33%), Cu (18%), and Mg (11%) concentrations in maize. In contrast, a significant decrease in P (17%) and K (17%) content was found, which could be due to lowered soil moisture content. Salt stress has been reported to increase Na and Cl ions in many plant species (De Pascale et al., 2005; Keutgen & Pawelzik, 2009), which could be balanced by the presence of K, implying that excessive Na absorption has resulted in reduced K uptake. Thus, saline-tolerant smart crops have been visualized that would have the ability to maintain high K:Na ratios under salt stress (Munns et al., 2006; Singh et al., 2014) and to regulate osmotic functions. The observed differences in mineral accumulation under various stresses have been attributed to the altered expression of zinc-regulated transporter (ZIP), natural resistance-associated macrophage protein (NRAMP), copper transporter (COPT), yellow stripe like (YSL), heavy metal ATPase (HMA), metal tolerance protein (MTP), and other vascular transporters. Further understanding stress physiology by knowing the metabolic key players will assist in developing tolerant crops in the future.

8.6 IMPACT OF ABIOTIC STRESS ON VITAMINS

Vitamins are also considered a vital component of grains for the proper retention of their nutritional values, in addition to sugars, proteins, and minerals (Chen et al., 2012; Heinemann et al., 2005). Although there is limited knowledge on the impact of abiotic stress on vitamins in grains, evidence in wheat has shown that environmental factors have significantly impacted the content of vitamins B1, B3, and B6 (Shewry et al., 2011). Salinity has significantly reduced the vitamin content in wheat (Ashraf, 2014), while no such variation has been observed in brown rice after salt stress imposition (Nam et al., 2018). Vitamin E has been found to be increased in lettuce under water deficit conditions (Oh et al., 2010). Studies have also shown a strong link between an increased level of CO_2 and vitamin content (Zhu et al., 2018). Osmotic stress induction has been reported to increase the vitamin C content in lentil seeds and mung bean sprouts. The vitamin C content of alfalfa seeds was predominantly affected by 100 mM NaCl solution, resulting in a 2.3% increase in vitamin C content. When seeds were subjected to temperature stress at various temperatures, vitamin C was found to increase in mung bean and alfalfa seeds. The vitamin C content was observed to be 6.9% and

1.2% higher at 40 and 45 °C, respectively, as compared with control samples. In contrast, temperature stress in germinated lentil seeds did not significantly affect the vitamin C content (Televiciute et al., 2020). Several studies carried out on tomato under drought revealed an increase in vitamin C content, correlated as an antioxidant to quench the stress signaling molecules (Veit-Kohler et al., 1999; Favati et al., 2009). In agreement with this, an investigation carried out by Sorenson et al. (1997) showed that vitamin A (carotenes) and vitamin C (ascorbate) content was found to increase in carrots when subjected to extreme drought stress. On the other hand, the vitamin C content was decreased in *Ocimum gratissimum* L. (Edeoga et al., 2010), fruits (Ahmed et al., 2014), and *Vigna unguiculata* L. seeds (Nair et al., 2008). Furthermore, it was also speculated that the reduction in vitamin C content under water stress might be due to destruction caused by oxygen and its derived species (Fang & Xiong, 2015). Drought imposed at a particular growth stage showed no variation in vitamin C, while another study carried out by Abid et al. (2018) showed that when wheat was subjected to extreme and moderate drought stress, vitamin A content was reduced, and the reduction was very prominent in susceptible cultivars. Therefore, with regard to nutritional attributes, a systemic approach can be applied to determine the variations in vitamin content under various environmental stresses. Fingerprinting for other vitamins other than B, C, and A must be carried out to elucidate the correlation that exists between diet and health.

8.7 IMPACT OF ABIOTIC STRESS ON PHYTOCHEMICAL POOL

Protective phytochemicals play an important role in abiotic stress resilience, yet how their composition alters in response to environmental conditions has not been widely investigated. Among the abiotic stresses, water stress has been found to have an adverse effect on the growth of plants and on crop productivity. Apart from the negative effect, a few positive effects on crop quality, such as higher accumulation of phytochemicals and in turn, enhanced antioxidant activity, may be expected, but contrary effects have also been reported (Ripoll et al., 2014). During common bean cultivation, the application of an extreme drought enhanced the content of 3 types of flavanones, 2 flavanols, 5 flavones, 15 flavonols, and 6 isoflavones. The flavonoid profile of the common bean revealed that luteolin was increased 5.1-fold, followed by kaempferol 3-O-(6-malonyl-glucoside) (2.8-fold), daidzein (2.5-fold), and quercetin 3-O-glucoside-xyloside (2.5-fold), as compared with the full irrigation treatment. The content of 5-hydroxycinnamic acids and 6-hydroxybenzoic acids has also been found to increase under extreme drought conditions. A profound effect was found on p-coumaric acid (2.5-fold), followed by ferulic acid (2.2-fold), as well as in the case of ferulic acid 4-O-glucoside (2.1-fold) (Herrera et al., 2019). Caftaric acid has been found to be increased, while other phenolics, like caffeic acid, chlorogenic acid, chicoric acid, and rutin, have been found to decrease in lettuce under water stress (Galieni et al., 2015; Oh et al., 2010). Oh et al. (2009) reported similar results, where heat and chilling stress increased the caffeic acid and chicoric acid content in lettuce. An analysis by Herbinger et al. (2002) stated that α-tocopherol and glutathione concentrations increased in wheat cultivars under drought stress. A general increase in α-tocopherol levels has also been reported in peanuts, tobacco (Liu et al., 2008), and soybeans (Britz & Kremer, 2002) under drought conditions. Being major antioxidant moieties, the increased expression of tocopherols has been reported as a stress-countering mechanism in many crops. In the case of tomato, a reduction in phenolic content was observed under drought stress (Coyago-Cruz et al., 2017). *Cucumis sativus* L. plants, when exposed to different water regimes, showed an increase in total anthocyanin content (Sonnenberg et al., 2013) due to anthocyanin hydroxylation stimulation by increasing the gene expression levels (Castellarin et al., 2007). Ahmed et al. (2013) reported an increase of phenolics and total anthocyanin content with increasing salt concentrations in barley. Lim et al. (2012) stated that buckwheat treated after 7 days of cultivation with 10, 50, and 100 mM salt concentration had 57%, 121%, and 153%, respectively, higher total phenolic content as compared with control. In addition, when purslane was subjected to different salt concentrations, total polyphenol content and total flavonoid content increased by 8–35% and 35%, respectively (Alam et al., 2015). During cultivation, elevated temperatures have been shown to alter phytochemical profiles, such as increasing α-tocopherol and α-tocotrienol while reducing γ-tocopherol and γ-tocotrienolol in brown rice (Britz et al., 2007). Flavonoid concentration and composition are sensitive to temperature stress (Arcas et al., 2000). Mori and his team have shown that high temperatures not only downregulate the anthocyanin biosynthetic genes but also degrade them, which leads to the inhibition of mRNA transcription. By increasing phenylalanine ammonia lyase (PAL) activity (Teklemariam & Blake, 2004), UV-B has been shown to promote the aggregation of a wide variety of phytochemicals (Burritt & MacKenzie, 2003). The availability of nutrients has also been implicated in plant phytochemical responses. Increase in flavonoids, flavonols, and ascorbic acid were observed under nitrogen deficiency in *Arabidopsis* and tomatoes (Kandlbinder et al., 2004; Bongue-Bartelsman & Phillips, 1995). Low nitrogen levels have been found to inhibit the activity of PAL, an important enzyme involved in phenolic compound biosynthesis, which results in reduced foliar accumulation in French bean plants (Sanchez et al., 2000). Phosphorus deficiency in plants has also been known to cause similar phytochemical responses (Kandlbinder et al., 2004; Stewart et al., 2001). EC reduces the quality of rice grains as measured by the phytochemical content, including total phenolics, flavonoids, tocopherol, and tocotrienols (Goufo et al., 2014). High CO_2 levels were also found to increase several phenolic compounds and flavonoid glycosides in lettuce (Becker & Klaring, 2016). Earlier studies reported that a decline in light intensity enhances the production of total flavonoids and phenols in herbs (Graham, 1998), while a decrease in total phenolic compounds was observed in *Beta vulgaris* (Stagnari et al., 2014). Phytochemicals play a significant role in regulating inflammatory cascades in humans; thus, they have potential health benefits, and therefore, fingerprinting these compounds under stress is of utmost importance.

8.8 IMPACT OF ABIOTIC STRESS ON DIETARY FIBER CONTENT

Dietary fiber, although not extensively studied, has great relevance in the age of lifestyle disorders. Other than fresh fruits and vegetables, whole grain cereals are the major sources of dietary fiber (Steer et al., 2008). In cereals, the composition and quantity of dietary fiber are known to be affected by both environmental and genetic factors, with the latter being the most prominent (Gebruers et al., 2010; Henry, 1989; Shewry et al., 2010). Depending on the stage of crop development, drought stress has a variable impact on dietary fiber content, especially arabinoxylans. Among cereals, barley grain is one of the rich sources of dietary fiber, but the predominant component is β-glucan, accounting for about70%of the total dietary fiber. In three winter wheat varieties, Rakszegi et al. (2014) reported the impact of both drought and heat stress on the composition and content of dietary fibers (arabinoxylan and β-glucan). Both the stresses were found to decrease the grain β-glucan content but increase the arabinoxylan content. But surprisingly, decreased arabinoxylan content was observed in the drought-tolerant variety Plainsman V when subjected to drought stress. Fiber levels have been found to rise by 19% in maize under field conditions when drought stress was introduced in the vegetative stage (Ali & Ashraf, 2011; Ali et al., 2013), whereas drought stress (60%) lowered the fiber content by 16% during the reproductive stage (Nawaz et al., 2016). Early studies have found that heat and drought stress increased the arabinoxylan concentration in wheat during flowering (Hong et al., 1989), whereas after flowering, the observed effect was reversed under high-temperature–drought conditions (Laurentin & Douglas, 2003). Coles et al. (1997) reported that the concentration of arabinoxylan in wheat under mild drought increased after flowering, but under extreme drought conditions it decreased, and the degree of stress also appeared to be significant in determining the concentration of arabinoxylan under drought conditions. According to Zhang et al. (2010), the concentration of arabinoxylan in the wheat grain, especially the water extractable arabinoxylan fraction, increased during high flowering temperatures. The grain has also been reported to have higher levels of dietary fiber components when grown under hot and dry (Hungary) as compared with cool and wet (UK) conditions when water and temperature stress have likely occurred at the same point, as is common under natural conditions in Europe (Gebruers et al., 2010). Guler (2003) showed reduced β-glucan content in barley grains when irrigation was used. Although it has been proved that under hot and dry conditions, the content of total β-glucan (Morgan & Riggs, 1981; Swanston et al., 1997) and water extractable β-glucan (Nilssen et al., 2008) is higher, a short and very extreme period of high-temperature stress can reduce the content of β-glucan (Savin et al., 1997).In a study carried out in Turkey, β-glucan content was shown to be altered by nitrogen application and availability of water in barley (Guler, 2003). In contrast, Rakszegi et al. (2014) reported that high levels of nitrogen increased the β-glucan content in wheat grain, even though a starch fraction, RS, also known as a prebiotic fiber, has been found to increase under abiotic stress conditions.

8.9 ALTERED NUTRITIONAL PROFILE AND ITS EFFECT ON PROCESSING QUALITY

The implications of abiotic stress in food and processing quality have been neglected, and most of the altered biochemical components have been addressed only as food contaminants. But, the effect of stress conditions, which result in altered profiles in the nutritional matrix of crops in terms of lipid oxidation products and hydrogenation of polyunsaturated fatty acids as well as trans fatty acids, have huge impacts on the processing quality of food products (Halford et al., 2015). The concentration of metabolites like free amino acids and sugars also has profound effects on the processing properties of the grain. The Maillard reaction, an umbrella term for a series of non-enzymatic reactions that takes place only at high temperatures in cereals using such free amino acids and sugars as substrates, has been reported (Curtis et al., 2014). The reaction products are mainly known as processing contaminants. They include melanoidin pigments and a complex mixture of compounds such as pyrazines, pyrroles, furan, oxazoles, thiazoles, and thiophenes, which impart flavor and aroma (Mottram, 2007). Acrylamide, being a group 2A type probable carcinogen (International Agency for Research on Cancer, 1994), even though not present in high enough amounts to cause threats, is still a matter of concern. Probable reasons behind acrylamide build-up have been correlated with sulfur availability, which affected the concentration of free asparagine in wheat and barley grains. Atmospheric EC (700 ppm compared with 350 ppm) has been shown to reduce protein content in wheat, which affected dough quality (poorer dough with decreased loaf volume) (Halford et al., 2015). Changes in protein profile, especially glutens, play a role, as dietary intolerance to this protein subtype is an important problem for celiac patients (Shewry, 2009). Reports also endorse the fact that abiotic stresses alter the amylose–amylopectin ratio in cereals, which is a key determinant for the GI of foods (Prathap et al., 2019; Kumar et al., 2020). Considering the existence of chronic hyperglycemic conditions and associated sequalae, the index of starch bioavailability altered by abiotic stress is also a matter of concern. Along with these, the formation of furan, another possible carcinogen, from Maillard reactions and levels of associated chemicals like hydroxymethylfurfuryl have also been found to be increasing, and the abiotic stress mechanism plays a major role in this. However, the actual risk posed by the presence of very low levels of furan in food is still not well known. Specific abiotic stresses are being identified that affect grain composition in ways that have implications for food quality and safety, and hence, it is vital to understand how these stresses interact with genetic as well as metabolic factors. Integrating such knowledge gained from basic understanding will assist researchers (agronomists, breeders, biochemists, food technologists, nutritionists, and health physiologists) to work hand in hand to address this global issue.

8.10 ADVANCEMENT IN OMICS: UNRAVELING THE STRESS-INDUCED ALTERATIONS IN THE NUTRITIONAL SPECTRUM OF FOOD CROPS

Recent improvements in plant metabolic research using *omic* platforms provide a comprehensive idea about how plant metabolism and the underlying complex "metabolic regulation" respond to various environmental stresses (Obata & Fernie, 2012).Such nutritional fingerprints have been comprehensively collected in metabolome databases, which have been used for mapping populations to classify QTLs controlling the amount of a specific metabolite and to check its coincidence with QTLs for yield and/or genes involved in metabolic pathways (Gupta & Varshney, 2005). Integrated study of transcriptomic and metabolic profiles of rice reproductive organs under drought and heat stress has shown that sugar metabolism is a key determinant of tolerance/susceptibility of floral organs to stresses. In a study, Moroberekan (heat-sensitive) anthers were found to have lower amounts of myo-inositol and sucrose, which were associated with the upregulation of the carbon starved anther (CSA) gene and down-regulation of the monosaccharide transporter (MST8) and cell wall invertase (INV4) genes. Heat-tolerant rice (Nagina 22) anthers showed elevated amounts of glucose, sucrose, maltose, myo-inositol, glucose-6-P, and fructose-6-P under combined stress, with considerable upregulation of the MST8 and INV4 genes and down-regulation of the CSA genes (Li et al., 2015).

A recent study carried out using combined QTL and metabolite QTL (mQTL) analysis of the rice recombinant inbred lines "Lemont" × "Teqing" revealed two QTL hotspots (C74A on chromosome 3 and CDO497 on chromosome 7), which had reverse effects on nitrogen (amino acid) and carbon (sugar) metabolites (Li et al., 2016). In addition, the CDO497 locus co-located with sugar signaling in barley 2 (SUSIBA2), a novel WRKY78 (WRKY transcriptional factor) that regulates the rate of downstream sugar-inducible genes such as sucrose transporter 5 (SUT5), sucrose synthase 1 (SUS1), and UTP glucose-1-phosphate uridylyltransferase (UGP1), resulting in the increment of starch content and sink strength in rice grains, which is essential for high performance under several environmental stresses (Li et al., 2016; Su et al., 2015). Overexpression of HVA1 barley gene, which encodes for late embryogenesis of abundant proteins in wheat, has been shown to improve the efficiency of water use and accumulation of biomass under drought and finally resulted in increased drought tolerance in transgenic wheat lines (Sivamani et al., 2000). Proteomic studies in Chinese Spring wheat with *Aegilops longissima* chromosome substitution line CS-1Sl (1B), illustrated the role of albumins and globulins under drought conditions, and further highlighted the use of 1Sl chromosome as a promising gene resource to improve drought tolerance in wheat (Zhou et al., 2016). In particular, the drought-inducible expression of essential biosynthetic genes (LKR/SDH, P5CS1, BCAT2, and ADC2) is strongly associated with the drought-inducible accumulation of saccharopine, proline, branch-chain amino acids (BCAAs), and agmatine, respectively, which are governed by endogenous ABA levels (Jain et al., 2019). Babu et al. (2014) carried out association mapping for opaque 2 modifiers, which affects the tryptophan content; this led to the identification of two QTLs, one for tryptophan and other for protein content.

Goufo et al. (2017) observed that metabolites such as galactinol, quercetin, and proline play key roles in plant adjustment in dry cowpea under stress. Similarly, differential variation in different metabolites such as sucrose, aspartate, myo-inositol, allantoin (in leaves), pinitol, and 2-oxoglutaric acid (in root nodule) in soybean (Silvente et al., 2012) and gamma-aminobutyric acid, choline, allantoin, phenylalanine, and proline in chickpea (Khan et al., 2018; Khan et al., 2021a) provides the basis for plant responses under drought conditions. When *Spinacia oleracea* betaine aldehyde dehydrogenase (SoBADH) gene was transferred into the sweet potato (*Ipomoea batatas*) cultivar "Sushu-2", increased accumulation of glycine betaine and osmotic adjustment efficiency were reported in the transgenic lines, resulting in increased maintenance of cell membrane integrity, lowered reactive oxygen species (ROS) production, and enhanced photosynthetic activity under low temperature, salinity, and oxidative stress as compared with wild-type plants (Fan et al., 2012). In order to determine genes associated with anthocyanin biosynthesis in *Arabidopsis*, an integrated approach including metabolome and transcriptome analysis has been carried out to investigate the activation-tagged mutant and overexpressors of the MYB TF, *PAP1* gene (Tohge et al., 2005). The co-expression information of the *Arabidopsis* transcriptome available in the ATTED-II database has been used to determine novel genes associated with lipid metabolism, resulting in the identification of the novel gene UDP-glucose pyrophosphorylase 3 (UGP3), which is required in the first step of sulfo-lipid biosynthesis (Okazaki et al., 2009).All genes directly linked to flavonoid biosynthesis have been identified by co-expression study (Yonekura-Sakakibara et al., 2008). In *Scutellaria baicalensis*, SbMYB8 has been proven to enhance drought tolerance by controlling flavonoid biosynthesis (Yuan et al., 2015). Usually, MYB-type transcription factors (TFs) interact with other TFs like bHLH (Gonzalez et al., 2008) and WD40 proteins to control the synthesis of anthocyanins (Ferreyra et al., 2012; Walker et al., 1999). Metabolomic analysis in soybean showed overexpression of GmCHS5 and GmMYB173S59D, which enhances the salt tolerance and aggregation of Cyanid-3-arabinoid chloride, dihydroxy B-ring flavonoid (Pi et al., 2018). As the food matrix alters during abiotic stresses, more research is needed using food omics in an integrated way in the future.

8.11 CONCLUSION

The effect of abiotic stress on grain quality in terms of composition and its implications for food safety are addressed in this chapter. Altered food matrixes in terms of major (carbohydrate, protein, and lipids) and minor (phytochemicals, vitamins, fiber, free amino acids, sugars, and fatty acids) components were observed with positive and negative trends. We comprehensively elaborated and conceptually explained the effect of different abiotic stresses on altering the nutritional profiles. But, as the outcomes depend on a multitude of genetic, physiological, environmental, and experimental factors, reports sometimes seemed to be contradictory and non-comparable. The observed common threads have been compiled as a figure depicting the impact of five different stress types on altering the nutritional matrix of crops along with their processing quality (Figure 8.1). More than altering the nutritional quality, the modified profile also has known to affect the processing

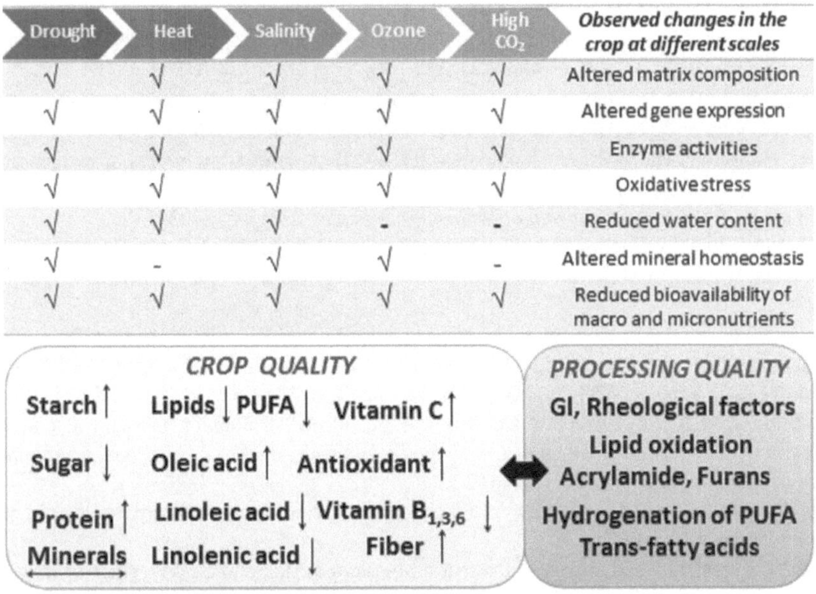

FIGURE 8.1 Comprehensive illustration depicts the prevalent impact of different abiotic stress factors like drought, heat, salinity, ozone (O_3), and high CO_2 on nutritional and processing quality parameters. Here, "prevalent impact" is considered based on the majority reports on a certain parameter, while exceptions have been discussed in the text and in Table 8.1.

Note: Upward arrows: increase; downward arrow: decrease; horizontal two way: non-significant change.

quality and the potential for the formation of processing contaminants, such as acrylamide, furan, hydroxymethylfurfuryl, and trans fatty acids are also discussed. Integrated omics approaches unraveled mQTLs controlling specific primary and secondary metabolites that can assist in developing tolerant cultivars in near future. Identification of specific environmental stresses that affect grain composition in ways having implications for food quality and safety, it's vital to understand how the stress factors interact with genetic factors for shaping the overall grain composition in major food crops.

REFERENCES

Abid, M., Ali, S., Qi, L. K., Zahoor, R., Tian, Z., Jiang, D., Snider, J. L., & Dai, T. (2018). Physiological and biochemical changes during drought and recovery periods at tillering and jointing stages in wheat (*Triticum aestivum* L.). *Scientific Reports*, 8(1), 1–15.

Ahmad, P., Jamsheed, S., Hameed, A., Rasool, S., Sharma, I., Azooz, M. M., & Hasanuzzaman, M. (2014). Drought stress induced oxidative damage and antioxidants in plants. In: *Oxidative Damage to Plants: Antioxidant Networks and Signaling*, Ahmad, P. (Ed.), Academic Press Waltham, MA, USA, 345–367.

Ahmad, S., Anwar, F., Hussain, A. I., Ashraf, M., & Awan, A. R. (2007). Does soil salinity affect yield and composition of cottonseed oil? *Journal of the American Oil Chemists Society*, 84(9), 845–851.

Ahmed, I. M., Cao, F., Han, Y., Nadira, U. A., Zhang, G., & Wu, F. (2013). Differential changes in grain ultrastructure, amylase, protein and amino acid profiles between Tibetan wild and cultivated barleys under drought and salinity alone and combined stress. *Food Chemistry*, 141(3), 2743–2750.

Ahmed, I. U., Bhuiya, M. R., & Wadud, A. (1982). A study of the effect of different levels of salinity on the quality on IR 8. *Journal of the Asiatic Society of Bangladesh (Bangladesh)*, 8, 37–44.

Alam, M. A., Juraimi, A. S., Rafii, M. Y., Hamid, A. A., Aslani, F., & Alam, M. Z. (2015). Effects of salinity and salinity-induced augmented bioactive compounds in purslane (*Portulaca oleracea* L.) for possible economical use. *Food Chemistry*, 169, 439–447.

Alghamdi, S. S. (2009). Chemical composition of faba bean (*Vicia faba* L.) genotypes under various water regimes. *Pakistan Journal of Nutrition*, 8(4), 477–482.

Ali, Q., & Ashraf, M. (2011). Exogenously applied glycine betaine enhances seed and seed oil quality of maize (*Zea mays* L.) under water deficit conditions. *Environmental and Experimental Botany*, 71(2), 249–259.

Ali, Q., Anwar, F., Ashraf, M., Saari, N., & Perveen, R. (2013). Ameliorating effects of exogenously applied proline on seed composition, seed oil quality and oil antioxidant activity of maize (*Zea mays* L.) under drought stress. *International Journal of Molecular Sciences*, 14(1), 818–835.

Al-Suhaibani, N. A. (2009). Influence of early water deficit on seed yield and quality of faba bean under arid environment of Saudi Arabia. *Am-Eurasian Journal of Agricultural and Environmental Sciences*, 5(5), 649–654.

Altenbach, S. B., DuPont, F. M., Kothari, K. M., Chan, R., Johnson, E. L., & Lieu, D. (2002). Temperature, water and fertilizer influence the timing of key events during grain development in a US spring wheat. *Journal of Cereal Science*, 37(1), 9–20.

Arcas, M. C., Botia, J. M., Ortuno, A. M., & Del Rio, J. A. (2000). UV irradiation alters the levels of flavonoids involved in the defence mechanism of *Citrus aurantium* fruits against *Penicillium digitatum*. *European Journal of Plant Pathology*, 106(7), 617–622.

Ashmore, M., Toet, S., & Emberson, L. (2006). Ozone–a significant threat to future world food production? *New Phytologist*, 170, 201–204.

Ashraf, M. (2014). Stress-induced changes in wheat grain composition and quality. *Critical Reviews in Food Science and Nutrition*, 54(12), 1576–1583.

Ashrafi, V., Pourbozorg, H., Kor, N. M., Ajirloo, A. R., Shamsizadeh, M., & Shaaban, M. (2015). Study on seed protein and protein profile pattern of chickpea (*Cicer arietinum* L.) by SDS-PAGE under drought stress and fertilization. *International Journal of Life Sciences*, 9(5), 87–90.

Asseng, S., & Milroy, S. P. (2006). Simulation of environmental and genetic effects on grain protein concentration in wheat. *European Journal of Agronomy*, 25(2), 119–128.

Babu, B. K., Agrawal, P. K., Pandey, D., & Kumar, A. (2014). Comparative genomics and association mapping approaches for opaque2 modifier genes in finger millet accessions using genic, genomic and candidate gene-based simple sequence repeat markers. *Molecular Breeding*, 34, 1261–1279.

Becker, C., & Klaring, H. P. (2016). CO_2 enrichment can produce high red leaf lettuce yield while increasing most flavonoid glycoside and some caffeic acid derivative concentrations. *Food Chemistry*, 199, 736–745.

Behboudian, M. H., Ma, Q., Turner, N. C., & Palta, J. A. (2001). Reactions of chickpea to water stress: yield and seed composition. *Journal of the Science of Food and Agriculture*, 81(13), 1288–1291.

Bettaieb, I., Zakhama, N., Wannes, W. A., Kchouk, M. E., & Marzouk, B. (2009). Water deficit effects on *Salvia officinalis* fatty acids and essential oils composition. *Scientia Horticulturae*, 120(2), 271–275.

Blokhina, O., Virolainen, E., & Fagerstedt, K. V. (2003). Antioxidants, oxidative damage and oxygen deprivation stress: a review. *Annals of Botany*, 91(2), 179–194.

Blum, A. (1988). *Plant breeding for stress environments*. CRC Press, Boca Raton, Florida, 223.

Bongue-Bartelsman, M., & Phillips, D. A. (1995). Nitrogen stress regulates gene expression of enzymes in the flavonoid biosynthetic pathway of tomato. *Plant Physiology and Biochemistry (Paris)*, 33(5), 539–546.

Branch, W. D., Nakayama, T., & Chinnan, M. S. (1990). Fatty acid variation among US runner-type peanut cultivars. *Journal of the American Oil Chemists Society*, 67(9), 591–593.

Britz, S. J., & Kremer, D. F. (2002). Warm temperatures or drought during seed maturation increase free α-tocopherol in seeds of soybean (*Glycine max* [L.] Merr.). *Journal of Agricultural and Food Chemistry*, 50(21), 6058–6063.

Britz, S. J., Prasad, P. V. V., Moreau, R. A., Allen, L. H., Kremer, D. F., & Boote, K. J. (2007). Influence of growth temperature on the amounts of tocopherols, tocotrienols, and γ-oryzanol in brown rice. *Journal of Agricultural and Food Chemistry*, 55(18), 7559–7565.

Broberg, M. C., Feng, Z., Xin, Y., & Pleijel, H. (2015). Ozone effects on wheat grain quality–A summary. *Environmental Pollution*, 197, 203–213.

Burritt, D. J., & Mackenzie, S. (2003). Antioxidant metabolism during acclimation of *Begonia×erythrophylla* to high light levels. *Annals of Botany*, 91(7), 783–794.

Calderini, D. F., Lizana, X. C., Hess, S., Jobet, C. R., & Zuniga, J. A. (2008). Grain yield and quality of wheat under increased ultraviolet radiation (UV-B) at later stages of the crop cycle. *The Journal of Agricultural Science*, 146(1), 57–64.

Carvalho, I. S. D., Chaves, M., & Pinto Ricardo, C. (2005). Influence of water stress on the chemical composition of seeds of two lupins (*Lupinus albus* and *Lupinus mutabilis*). *Journal of Agronomy and Crop Science*, 191(2), 95–98.

Castellarin, S. D., Pfeiffer, A., Sivilotti, P., Degan, M., Peterlunger, E., & Di Gaspero, G. (2007). Transcriptional regulation of anthocyanin biosynthesis in ripening fruits of grapevine under seasonal water deficit. *Plant Cell and Environment*, 30(11), 1381–1399.

Chartzoulakis, K. S. (2005). Salinity and olive: growth, salt tolerance, photosynthesis and yield. *Agricultural Water Management*, 78(1–2), 108–121.

Chen, C., Tao, C., Peng, H., & Ding, Y. (2007). Genetic analysis of salt stress responses in asparagus bean (*Vigna unguiculata* (L.) ssp. sesquipedalis Verdc.). *Journal of Heredity*, 98(7), 655–665.

Chen, Y., Wang, M., & Ouwerkerk, P. B. (2012). Molecular and environmental factors determining grain quality in rice. *Food and Energy Security*, 1(2), 111–132.

Chen, Z., Cao, J., Yu, H., & Shang, H. (2018). Effects of elevated ozone levels on photosynthesis, biomass and non-structural carbohydrates of *Phoebe bournei* and *Phoebe zhennan* in subtropical China. *Frontiers in Plant Science*, 9, 1764.

Coles, G. D., Hartunian-Sowa, S. M., Jamieson, P. D., Hay, A. J., Atwell, W. A., & Fulcher, R. G. (1997). Environmentally-induced variation in starch and non-starch polysaccharide content in wheat. *Journal of Cereal Science*, 26(1), 47–54.

Coyago-Cruz, E., Corell, M., Stinco, C. M., Hernanz, D., Moriana, A., & Melendez-Martinez, A. J. (2017). Effect of regulated deficit irrigation on quality parameters, carotenoids and phenolics of diverse tomato varieties (*Solanum lycopersicum* L.). *Food Research International*, 96, 72–83.

Curtis, T. Y., & Halford, N. G.(2014). Food security: the challenge of increasing wheat yield and the importance of not compromising food safety. *Annals of Applied Biology*, 164, 354–372.

De Pascale, S., Maggio, A., & Barbieri, G. (2005). Soil salinization affects growth, yield and mineral composition of cauliflower and broccoli. *European Journal of Agronomy*, 23(3), 254–264.

Debon, S. J. J., Tester, R. F., Millam, S., & Davies, H. V. (1998). Effect of temperature on the synthesis, composition and physical properties of potato microtuber starch. *Journal of the Science of Food and Agriculture*, 76, 599–607.

Di Caterina, R., Giuliani, M. M., Rotunno, T., De Caro, A., & Flagella, Z. (2007). Influence of salt stress on seed yield and oil quality of two sunflower hybrids. *Annals of Applied Biology*, 151(2), 145–154.

Dong, J., Gruda, N., Lam, S. K., Li, X., & Duan, Z. (2018). Effects of elevated CO_2 on nutritional quality of vegetables: a review. *Frontiers in Plant Science*, 9, 924.

Dornbos, D. L., & Mullen, R. E. (1992). Soybean seed protein and oil contents and fatty acid composition adjustments by drought and temperature. *Journal of the American Oil Chemists Society*, 69(3), 228–231.

Dwivedi, S. L., Nigam, S. N., Rao, R. N., Singh, U., & Rao, K. V. S. (1996). Effect of drought on oil, fatty acids and protein contents of groundnut (*Arachis hypogaea* L.) seeds. *Field Crops Research*, 48(2–3), 125–133.

Edeoga, H. O., Osuagwu, G. G. E., & Osuagwu, A. N. (2010). The influence of water stress (drought) on the mineral and vitamin potential of the leaves of *Ocimum gratissimum* L. *Recent Research in Science and Technology*, 2(2), 27–33.

Ehdaie, B., & Waines, J. G. (2001). Sowing date and nitrogen rate effects on dry matter and nitrogen partitioning in bread and durum wheat. *Field Crops Research*, 73(1), 47–61.

Emam, M. M., Khattab, H. E., Helal, N. M., & Deraz, A. E. (2014). Effect of selenium and silicon on yield quality of rice plant grown under drought stress. *Australian Journal of Crop Science*, 8(4), 596–605.

Fan, W., Zhang, M., Zhang, H., & Zhang, P. (2012). Improved tolerance to various abiotic stresses in transgenic sweet potato (*Ipomoea batatas*) expressing spinach betaine aldehyde dehydrogenase. *PLoS One*, 7(5), e37344.

Fang, Y., & Xiong, L.(2015). General mechanisms of drought response and their application in drought resistance improvement in plants. *Cellular and Molecular Life Sciences*, 72(4), 673–689.

Farooq, M., Hussain, M., Usman, M., Farooq, S., Alghamdi, S. S., & Siddique, K. H. (2018). Impact of abiotic stresses on grain composition and quality in food legumes. *Journal of Agricultural and Food Chemistry*, 66(34), 8887–8897.

Favati, F., Lovelli, S., Galgano, F., Miccolis, V., Di Tommaso, T., & Candido, V. (2009). Processing tomato quality as affected by irrigation scheduling. *Scientia Horticulturae*, 122(4), 562–571.

Feng, Z., Kobayashi, K., & Ainsworth, E. A. (2008). Impact of elevated ozone concentration on growth, physiology, and yield of wheat (*Triticum aestivum* L.): a meta-analysis. *Global Change Biology*, 14(11), 2696–2708.

Ferreyra, F. M. L., Rius, S., & Casati, P. (2012). Flavonoids: biosynthesis, biological functions, and biotechnological applications. *Frontiers in Plant Science*, 3, 222.

Flagella, Z., Rotunno, T., Tarantino, E., Di Caterina, R., & De Caro, A. (2002). Changes in seed yield and oil fatty acid composition of high oleic sunflower (*Helianthus annuus* L.) hybrids in relation to the sowing date and the water regime. European *Journal* of *Agronomy*, 17(3), 221–230.

Flagella, Z., Giuliani, M. M., Giuzio, L., Volpi, C., & Masci, S. (2010). Influence of water deficit on durum wheat storage protein composition and technological quality. *European Journal of Agronomy*, 33(3), 197–207.

Fuganti-Pagliarini, R., Ferreira, L. C., Rodrigues, F. A., Molinari, H. B., Marin, S. R., Molinari, M. D., Marcolino-Gomes, J., Mertz-Henning, L. M., Farias, J. R., de Oliveira, M. C., & Nepomuceno, A. L. (2017). Characterization of soybean genetically modified for drought tolerance in field conditions. *Frontiers in Plant Science*, 8, 448.

Fuhrer, J., Grimm, A. G., Tschannen, W., & Shariat-Madari, H. (1992). The response of spring wheat (*Triticum aestivum* L.) to ozone at higher elevations: II. Changes in yield, yield components and grain quality in response to ozone flux. *New Phytologist*, 121(2), 211–219.

Galantini, J. A., Landriscini, M. R., Iglesias, J. O., Miglierina, A. M., & Rosell, R. A. (2000). The effects of crop rotation and fertilization on wheat productivity in the Pampean semiarid region of Argentina: 2. Nutrient balance, yield and grain quality. *Soil and Tillage Research*, 53(2), 137–144.

Galieni, A., Di Mattia, C., De Gregorio, M., Speca, S., Mastrocola, D., Pisante, M., & Stagnari, F. (2015). Effects of nutrient deficiency and abiotic environmental stresses on yield, phenolic compounds and antiradical activity in lettuce (*Lactuca sativa* L.). *Scientia Horticulturae*, 187, 93–101.

Garcia, W. J., Cavins, J. F., Inglett, G. E., Heagle, A. S., & Kwolek, W. F. (1983). Quality of corn grain from plants exposed to chronic levels of ozone. *Cereal Chemistry*, 60(5), 388–391.

Gebruers, K., Dornez, E., Bedo, Z., Rakszegi, M., Fras, A., Boros, D., Courtin, C. M., & Delcour, J. A. (2010). Environment and genotype effects on the content of dietary fiber and its components in wheat in the healthgrain diversity screen. *Journal of Agricultural and Food Chemistry*, 58(17), 9353–9361.

Geigenberger, P., Reimholz, R., Geiger, M., Merlo, L., Canale, V., & Stitt, M. (1997). Regulation of sucrose and starch metabolism in potato tubers in response to short term water deficit. *Planta*, 201, 502–518.

Ghanbari, A. A., Mousavi, S. H., Gorji, A. M., & Idupulapati, R. A. O. (2013). Effects of water stress on leaves and seeds of bean (*Phaseolus vulgaris* L.). *Turkish Journal of Field Crops*, 18(1), 73–77.

Ghassemi-Golezani, K., Ghanehpoor, S., & DabbaghMohammadi-Nasab, A. (2009). Effects of water limitation on growth and grain filling of faba bean cultivars. *Journal of Food Agriculture and Environment*, 7(3), 442–447.

Ghassemi-Golezani, K., Taifeh-Noori, M., Oustan, S., Moghaddam, M., & Seyyed-Rahmani, S. (2010). Oil and protein accumulation in soybean grains under salinity stress. *Notulae Scientia Biologicae*, 2(2), 64–67.

Golombek, S. D., Sridhar, R., & Singh, U. (1995). Effect of soil temperature on the seed composition of three Spanish cultivars of groundnut (*Arachis hypogaea* L.). *Journal of Agricultural and Food Chemistry*, 43(8), 2067–2070.

Gonzalez, A., Zhao, M., Leavitt, J. M., & Lloyd, A. M. (2008). Regulation of the anthocyanin biosynthetic pathway by the TTG1/bHLH/Myb transcriptional complex in *Arabidopsis* seedlings. *The Plant Journal*, 53(5), 814–827.

Gonzalez-Cruz, J., & Pastenes, C. (2012). Water-stress-induced thermotolerance of photosynthesis in bean (*Phaseolus vulgaris* L.) plants: the possible involvement of lipid composition and xanthophyll cycle pigments. *Environmental and Experimental Botany*, 77, 127–140.

Good, A. G., & Zaplachinski, S. T. (1994). The effects of drought stress on free amino acid accumulation and protein synthesis in *Brassica napus*. *Physiologia Plantarum*, 90(1), 9–14.

Goufo, P., Moutinho-Pereira, J. M., Jorge, T. F., Correia, C. M., Oliveira, M. R., Rosa, E. A., António, C., & Trindade, H. (2017). Cowpea (*Vigna unguiculata* L. Walp.) metabolomics: osmoprotection as a physiological strategy for drought stress resistance and improved yield. *Frontiers in Plant Science*, 8, 586.

Goufo, P., Pereira, J., Figueiredo, N., Oliveira, M. B. P., Carranca, C., Rosa, E. A., & Trindade, H. (2014). Effect of elevated carbon dioxide (CO_2) on phenolic acids, flavonoids, tocopherols, tocotrienols, γ-oryzanol and antioxidant capacities of rice (*Oryza sativa* L.). *Journal of Cereal Science*, 59(1), 15–24.

Graham, T. L. (1998). Flavonoid and flavonol glycoside metabolism in *Arabidopsis*. *Plant Physiology and Biochemistry*, 36(1–2), 135–144.

Guler, M. (2003). Barley grain β-glucan content as affected by nitrogen and irrigation. *Field Crops Research*, 84(3), 335–340.

Gupta, P. K., & Varshney, R. K. (2005). Cereal genomics. In: *Cereal genomics*, https://doi.org/10.1007/1-4020-2359-6

Hajiboland, R., Joudmand, A., & Fotouhi, K. (2009). Mild salinity improves sugar beet (*Beta vulgaris* L.) quality. *Acta Agriculturae Scandinavica Section B–Soil and Plant Science*, 59, 295–305.

Halford, N. G., Curtis, T. Y., Chen, Z., & Huang, J. (2015). Effects of abiotic stress and crop management on cereal grain composition: implications for food quality and safety. *Journal of Experimental Botany*, 66(5), 1145–1156.

Harb, A., Krishnan, A., Ambavaram, M. M. R., & Pereira, A. (2010). Molecular and physiological analysis of drought stress in Arabidopsis reveals early responses leading to acclimation in plant growth. *Plant Physiology*, 154, 1254–1271.

He, J. F., Goyal, R., Laroche, A., Zhao, M. L., & Lu, Z. X. (2012). Water stress during grain development affects starch synthesis, composition and physicochemical properties in triticale. *Journal of Cereal Science*, 56, 552–560.

Heinemann, R. J. B., Fagundes, P. L., Pinto, E. A., Penteado, M. V. C., & Lanfer-Marquez, U. M. (2005). Comparative study of nutrient composition of commercial brown, parboiled and milled rice from Brazil. *Journal of Food Composition and Analysis*, 18(4), 287–296.

Henry, R. J. (1989). Factors influencing the rate of modification of barleys during malting. *Journal of Cereal Science*, 10(1), 51–59.

Herbinger, K., Tausz, M., Wonisch, A., Soja, G., Sorger, A., & Grill, D. (2002). Complex interactive effects of drought and ozone stress on the antioxidant defence systems of two wheat cultivars. *Plant Physiology and Biochemistry*, 40(6–8), 691–696.

Herrera, M. D., Acosta-Gallegos, J. A., Reynoso-Camacho, R., & Perez-Ramirez, I. F. (2019). Common bean seeds from plants subjected to severe drought, restricted-and full-irrigation regimes show differential phytochemical fingerprint. *Food Chemistry*, 294, 368–377.

Hesse, H., Nikiforova, V., Gakiere, B., & Hoefgen, R. (2004). Molecular analysis and control of cysteine biosynthesis: integration of nitrogen and sulphur metabolism. *Journal of Experimental Botany*, 55(401), 1283–1292.

Hidema, J., Zhang, W., Yamamoto, M., Sato, T., & Kumagai, T. (2005). Changes in grain size and grain storage protein of rice (*Oryza sativa* L.) in response to elevated UV-B radiation under outdoor conditions. *Journal of Radiation Research*, 46(2), 143–149.

Hong, B. H., Rubenthaler, G. L., & Allan, R. E. (1989). Wheat pentosans. I. Cultivar variation and relationship to kernel hardness. *Cereal Chemistry*, 66(5), 369–373.

Horrer, D., Flutsch, S., Pazmino, D., Matthews, J. S. A., Thalmann, M., Nigro, A., Leonhardt, N., Lawson, T., & Santelia, D. (2016). Blue light induces a distinct starch degradation pathway in guard cells for stomatal opening. *Current Biology*, 26, 362–370.

Houshmand, S., Arzani, A., & Mirmohammadi-Maibody, S. A. M. (2014). Effects of salinity and drought stress on grain quality of durum wheat. *Communications in Soil Science and Plant Analysis*, 45(3), 297–308.

International Agency for Research on Cancer (1994). *IARC Monographs on the Evaluation of Carcinogenic Risks to Humans. Volume 60. Some industrial chemicals*. International Agency for Research on Cancer (IARC), Lyon.

Iwabuchi, S., & Yamauchi, F. (1984). Effects of heat and ionic strength upon dissociation-association of soybean protein fractions. *Journal of Food Science*, 49(5), 1289–1294.

Jackson, R. L., Taunton, O. D., Morrisett, J. D., & Gotto Jr, A. M. (1978). The role of dietary polyunsaturated fat in lowering blood cholesterol in man. *Circulation Research*, 42(4), 447–453.

Jain, D., Ashraf, N., Khurana, J. P., & Kameshwari, M. S. (2019). The "Omics" approach for crop improvement against drought stress. In: *Genetic enhancement of crops for tolerance to abiotic stress: Mechanisms and approaches, Vol. I*. Springer, Cham, 183–204.

Kandlbinder, A., Finkemeier, I., Wormuth, D., Hanitzsch, M., & Dietz, K. J. (2004). The antioxidant status of photosynthesizing leaves under nutrient deficiency: redox regulation, gene expression and antioxidant activity in *Arabidopsis thaliana*. *Physiologia Plantarum*, 120(1), 63–73.

Kano, Y., & Mano, K. (2002). The effects of night soil-temperatures on diurnal changes in carbohydrate contents in roots and stems of sweet potatoes (*Ipomoea batatas Poir.*). *Journal of the Japanese Society Horticultural Science*, 71, 747–751.

Kaplan, F., & Guy, C.L. (2004). Beta-amylase induction and the protective role of maltose during temperature shock. *Plant Physiology*, 135, 1674–1684.

Kaplan, F., Sung, D. Y., & Guy, C. L. (2006). Roles of beta-amylase and starch breakdown during temperatures stress. *Physiologia Plantarum*, 126, 120–128.

Katerji, N., Rana, G., & Mastrorilli, M. (2010). Modelling of actual evapotranspiration in open top chambers (OTC) at daily and seasonal scale: multi-annual validation on soybean in contrasted conditions of water stress and air ozone concentration. *European Journal of Agronomy*, 33, 218–230.

Keeling, P. L., Bacon, P. J., & Holt, D. C. (1993). Elevated temperature reduces starch deposition in wheat endosperm by reducing the activity of soluble starch synthase. *Planta*, 191(3), 342–348.

Kempa, S., Krasensky, J., Dal Santo, S., Kopka, J., & Jonak, C. (2008). A central role of abscisic acid in stress-regulated carbohydrate metabolism. *PLoS ONE*, 3, e3935.

Keutgen, A. J., & Pawelzik, E. (2009). Impacts of NaCl stress on plant growth and mineral nutrient assimilation in two cultivars of strawberry. *Environmental and Experimental Botany*, 65(2–3), 170–176.

Khalil, S. E., & Ismael, E. G. (2010). Growth, yield and seed quality of *Lupinus termis* as affected by different soil moisture levels and different ways of yeast application. *Journal of American Science*, 6(8), 141–153.

Khamssi, N. N. (2011). Grain yield and protein of chickpea (*Cicer arietinum* L.) cultivars under gradual water deficit conditions. *Research Journal of Environmental Sciences*, 5(6), 611–616.

Khan, M. I. R., Palakolanu, S. R., Chopra, P., Rajurkar, A. B., Gupta, R., Iqbal, N., & Maheshwari, C. (2021a).Improving drought tolerance in rice: ensuring food security through multi-dimensional approaches. *Physiologia Plantarum*, 172(2), 645–668. doi: 10.1111/ppl.13223. Epub 2020 Oct 27. PMID: 33006143.

Khan, M. I. R., Jalil, S. U., Chopra, P., Chhillar, H., Ferrante, A., Khan, N. A., & Ansari, M. I. (2021b). Role of GABA in plant growth, development and senescence. *Plant Gene*, 26, 100283.

Khan, M. A., & Abdullah, Z. (2003). Salinity–sodicity induced changes in reproductive physiology of rice (*Oryza sativa*) under dense soil conditions. *Environmental and Experimental Botany*, 49(2), 145–157.

Khan, N., Bano, A., Rathinasabapathi, B., & Babar, M. A. (2018). UPLCHRMS-based untargeted metabolic profiling reveals changes in chickpea (*Cicer arietinum*) metabolome following long-term drought stress. *Plant Cell and Environment*, 42(1), 115–132.

Khan, R., Srivastava, R., Abdin, M. Z., & Manzoor, N. (2013). Effect of soil contamination with heavy metals on soybean seed oil quality. *European Food Research and Technology*, 236(4), 707–714.

Klikocka, H., Cybulska, M., Barczak, B., Narolski, B., Szostak, B., Kobialka, A., Nowak, A., & Wojcik, E. (2016). The effect of sulphur and nitrogen fertilization on grain yield and technological quality of spring wheat. *Plant Soil and Environment*, 62(5), 230–236.

Kumar, A., Dash, G. K., Barik, M., Panda, P. A., Lal, M. K., Baig, M. J., & Swain, P. (2020). Effect of drought stress on resistant starch content and glycemic index of rice (*Oryza sativa* L.). *Starch-Starke*, 72(11–12), 1900229.

Labuschagne, M. T., Elago, O., & Koen, E. (2009). The influence of temperature extremes on some quality and starch characteristics in bread, biscuit and durum wheat. *Journal of Cereal Science*, 49(2), 184–189.

Laurentin, A. M., & Douglas, E. (2003). Dietary fibre in health and disease. *Nutrition Bulletin*, 28, 69–73.

Lee, I. S., Lee, J. O., & Ge, L. (2009). Comparison of terrestrial laser scanner with digital aerial photogrammetry for extracting ridges in the rice paddies. *Survey Review*, 41(313), 253–267.

Li, B., Zhang, Y., Mohammadi, S. A., Huai, D., Zhou, Y., & Kliebenstein, D. J. (2016). An integrative genetic study of rice metabolism, growth and stochastic variation reveals potential C/N partitioning loci. *Scientific Reports*, 6(1), 1–13.

Li, X., Lawas, L. M., Malo, R., Glaubitz, U., Erban, A., Mauleon, R., Heuer, S., Zuther, E., Kopka, J., Hincha, D. K., & Jagadish, K. S. (2015). Metabolic and transcriptomic signatures of rice floral organs reveal sugar starvation as a factor in reproductive failure under heat and drought stress. *Plant, Cell and Environment*, 38(10), 2171–2192.

Lim, J. H., Park, K. J., Kim, B. K., Jeong, J. W., & Kim, H. J. (2012). Effect of salinity stress on phenolic compounds and carotenoids in buckwheat (*Fagopyrum esculentum* M.) sprout. *Food Chemistry*, 135(3), 1065–1070.

Lin, C. J., Li, C. Y., Lin, S. K., Yang, F. H., Huang, J. J., Liu, Y. H., & Lur, H. S. (2010). Influence of high temperature during grain filling on the accumulation of storage proteins and grain quality in rice (*Oryza sativa* L.). *Journal of Agricultural and Food Chemistry*, 58(19), 10545–10552.

Liu, F., Jensen, C. R., & Andersen, M. N. (2004). Drought stress effect on carbohydrate concentration in soybean leaves and pods during early reproductive development: its implication in altering pod set. *Field Crops Research*, 86, 1–13.

Liu, X., Hua, X., Guo, J., Qi, D., Wang, L., Liu, Z., Jin, Z., Chen, S., & Liu, G. (2008). Enhanced tolerance to drought stress in transgenic tobacco plants overexpressing VTE1 for increased tocopherol production from *Arabidopsis thaliana*. *Biotechnology Letters*, 30(7), 1275–1280.

Lobell, D. B., Burke, M. B., Tebaldi, C., Mastrandrea, M. D., Falcon, W. P., & Naylor, R. L. (2008). Prioritizing climate change adaptation needs for food security in 2030. *Science*, 319, 607–610.

Macnicol, P. K., Jacobsen, J. V., Keys, M. M., & Stuart, I. M. (1993). Effects of heat and water stress on malt quality and grain parameters of Schooner barley grown in cabinets. *Journal of Cereal Science*, 18(1), 61–68.

Mamnabi, S., Nasrollahzadeh, S., Ghassemi-Golezani, K., & Raei, Y. (2020). Improving yield-related physiological characteristics of spring rapeseed by integrated fertilizer management under water deficit conditions. *Saudi Journal of Biological Sciences*, 27(3), 797–804.

Medek, D. E., Schwartz, J., & Myers, S. S. (2017). Estimated effects of future atmospheric CO_2 concentrations on protein intake and the risk of protein deficiency by country and region. *Environmental Health Perspectives*, 125(8), 087002.

Mession, J. L., Sok, N., Assifaoui, A., & Saurel, R. (2013). Thermal denaturation of pea globulins (*Pisum sativum* L.) molecular interactions leading to heat-induced protein aggregation. *Journal of Agricultural and Food Chemistry*, 61(6), 1196–1204.

Mishra, A. K., & Agrawal, S. B. (2014). Cultivar specific response of CO_2 fertilization on two tropical mung bean (*Vigna radiata* L.) cultivars: ROS generation, antioxidant status, physiology, growth, yield and seed quality. *Journal of Agronomy and Crop Science*, 200(4), 273–289.

Morgan, A. G., & Riggs, T. J. (1981). Effects of drought on yield and on grain and malt characters in spring barley. *Journal of the Science of Food and Agriculture*, 32(4), 339–346.

Morgan, P. B, Bernacchi, C. J., Ort, D. R., & Long, S. P. (2004). An in vivo analysis of the effect of season-long open-air elevation of ozone to anticipated 2050 levels on photosynthesis in soybean. *Plant Physiology*, 135, 2348–2357.

Mottram, D. S. (2007). The Maillard reaction: source of flavour in thermally processed foods. In: *Flavours and fragrances: chemistry, bioprocessing and sustainability*, Berger, RG (Ed.), Springer-Verlag, Berlin, 269–284.

Munns, R., James, R. A., & Läuchli, A. (2006). Approaches to increasing the salt tolerance of wheat and other cereals. Journal of E*xperi*mental B*otany*, 57(5), 1025–1043.

Myers, S. S., Zanobetti, A., Kloog, I., Huybers, P., Leakey, A. D., Bloom, A. J., Carlisle, E., Dietterich, L. H., Fitzgerald, G., Hasegawa, T., & Usui, Y. (2014). Increasing CO_2 threatens human nutrition. *Nature*, 510(7503), 139–142.

Nagao, M., Minami, A., Arakawa, K., Fujikawa, S., & Takezawa, D. (2005). Rapid degradation of starch in chloroplasts and concomitant accumulation of soluble sugars associated with ABA-induced freezing tolerance in the moss *Physcomitrella patens*. *Journal of Plant Physiology*, 162, 169–180.

Nair, A. S., Abraham, T. K., & Jaya, D. S. (2008). Studies on the changes in lipid peroxidation and antioxidants in drought stress induced cowpea (*Vigna unguiculata* L.) varieties. *Journal of Environmental Biology*, 29(5), 689–691.

Nam, K. H., Kim, D. Y., Shin, H. J., Pack, I. S., & Kim, C. G. (2018). Changes in the metabolic profile and nutritional composition of rice in response to NaCl stress. *Korean Journal of Agricultural Science*, 45(2), 154–168.

Nawaz, F., Naeem, M., Ashraf, M. Y., Tahir, M. N., Zulfiqar, B., Salahuddin, M., Shabbir, R. N., & Aslam, M. (2016). Selenium supplementation affects physiological and biochemical processes to improve fodder yield and quality of maize (*Zea mays* L.) under water deficit conditions. *Frontiers in Plant Science*, 7, 1438.

Neffati, M., & Marzouk, B. (2008). Changes in essential oil and fatty acid composition in coriander (*Coriandrum sativum* L.) leaves under saline conditions. *Industrial Crops and Products*, 28(2), 137–142.

Nielsen, T. H., Deiting, U., & Stitt, M. (1997). A beta-amylase in potato tubers is induced by storage at low temperature. *Plant Physiology*, 113, 503–510.

Nilssen, A. K., Sahlstrom, S., Knutsen, S. H., Holtekjolen, A. K., & Uhlen, A. K. (2008). Influence of growth temperature on content, viscosity and relative molecular weight of water-soluble β-glucans in barley (*Hordeum vulgare* L.). *Journal of Cereal Science*, 48(3), 670–677.

Noda, T., Kobayashi, T., & Suda, I. (2001). Effect of soil temperature on starch properties of sweet potatoes. *Carbohydrate Polymers*, 44, 239–246.

Obata, T., & Fernie, A. R. (2012). The use of metabolomics to dissect plant responses to abiotic stresses. *Cellular and Molecular Life Science*, 69(19), 3225–3243.

Oh, M. M., Carey, E. E., & Rajashekar, C. B. (2009). Environmental stresses induce health promoting phytochemicals in lettuce. *Plant Physiology and Biochemistry*, 47(7), 578–583.

Oh, M. M., Carey, E. E., & Rajashekar, C. B. (2010). Regulated water deficits improve phytochemical concentration in lettuce. *Journal of the American Society for Horticultural Science*, 135(3), 223–229.

Okazaki, Y., Shimojima, M., Sawada, Y., Toyooka, K., Narisawa, T., Mochida, K.,Tanaka, H., Matsuda, F., Hirai, A., Hirai, M. Y., Ohta, H., & Saito, K. (2009). A chloroplastic UDP-glucose pyrophosphorylase from *Arabidopsis* is the committed enzyme for the first step of sulfolipid biosynthesis. *The Plant Cell*, 21(3), 892–909.

Oktem, A. (2008). Effect of water shortage on yield, and protein and mineral compositions of drip irrigated sweet corn in sustainable agricultural systems. *Agricultural Water Management*, 95(9), 1003–1010.

Ozturk, A., & Aydin, F. (2004). Effect of water stress at various growth stages on some quality characteristics of winter wheat. *Journal of Agronomy and Crop Science*, 190(2), 93–99.

Panozzo, J. F., Eagles, H. A., & Wootton, M. (2001). Changes in protein composition during grain development in wheat. *Australian Journal of Agricultural Research*, 52(4), 485–493.

Pareek, A., Joshi, R., Gupta, K. J., Singla-Pareek, S. L., & Foyer, C. (2020). Sensing and signalling in plant stress responses: ensuring sustainable food security in an era of climate change. *New Phytologist*, 228(3), 823–827.

Pettersson, C. G., & Eckersten, H. (2007). Prediction of grain protein in spring malting barley grown in northern Europe. *European Journal of Agronomy*, 27(2–4), 205–214.

Pi, E., Zhu, C., Fan, W., Huang, Y., Qu, L., Li, Y., Zhao, Q., Ding, F., Qiu, L., Wang, H., Poovaiah, B. W., & Du, L. (2018). Quantitative phosphoproteomic and metabolomic analyses reveal GmMYB173 optimizes flavonoid metabolism in soybean under salt stress. *Molecular and Cellular Proteomics*, 17(6), 1209–1224.

Piikki, K., Vorne, V., Ojanperä, K., & Pleijel, H. (2007). Impact of elevated O_3 and CO_2 exposure on potato (*Solanum tuberosum* L. cv. Bintje) tuber macronutrients (N, P, K, Mg, Ca). Agriculture, Ecosystems and *Environment*, 118(1–4), 55–64.

Pleijel, H., & Danielsson, H. (2009). Yield dilution of grain Zn in wheat grown in open-top chamber experiments with elevated CO_2 and O_3 exposure. *Journal of Cereal Science*, 50(2), 278–282.

Pleijel, H., Eriksen, A. B., Danielsson, H., Bondesson, N., & Sellden, G. (2006). Differential ozone sensitivity in an old and a modern Swedish wheat cultivar – grain yield and quality, leaf chlorophyll and stomatal conductance. *Environmental and Experimental Botany*, 56(1), 63–71.

Pokhrel, D., Bhandari, B. S., & Viraraghavan, T. (2009). Arsenic contamination of groundwater in the Terai region of Nepal: an overview of health concerns and treatment options. *Environment International*, 35(1), 157–161.

Poór, P., Nawaz, K., Gupta, R., Ashfaque, F., & Khan, M. I. R.(2021). Ethylene involvement in the regulation of heat stress tolerance in plants. *Plant Cell Reports*, 1–24. doi: 10.1007/s00299-021-02675-8

Poustini, K. (2002). Evaluation of the response of 30 wheat cultivars to salinity. *Iranian Journal of Agriculture Science*, 3, 57–64.

Prasad, P. V., Craufurd, P. Q., & Summerfield, R. J. (2000). Effect of high air and soil temperature on dry matter production, pod yield and yield components of groundnut. *Plant and Soil*, 222(1), 231–239.

Prathap, V., Ali, K., Singh, A., Vishwakarma, C., Krishnan, V., Chinnusamy, V., & Tyagi, A. (2019). Starch accumulation in rice grains subjected to drought during grain filling stage. *Plant Physiology and Biochemistry*, 142, 440–451.

Pressel, S., Ligrone, R., & Duckett, J. G. (2006). Effects of de- and rehydration on food conducting cells in the moss *Polytrichum formosum*: a cytological study. *Annals of Botany*, 98, 67–76.

Qados, A. M. A. (2011). Effect of salt stress on plant growth and metabolism of bean plant *Vicia faba* (L.). *Journal of the Saudi Society of Agricultural Sciences*, 10(1), 7–15.

Qados, A. M. S. A. (2010). Effect of arginine on growth, nutrient composition, yield and nutritional value of mung bean plants grown under salinity stress. *Nature*, 8, 30–42.

Rakszegi, M., Lovegrove, A., Balla, K., Lang, L., Bedo, Z., Veisz, O., & Shewry, P. R. (2014). Effect of heat and drought stress on the structure and composition of arabinoxylan and β-glucan in wheat grain. *Carbohydrate Polymers*, 102, 557–565.

Rice, R. W. (2007). *The physiological role of minerals in the plant. Mineral nutrition and plant disease*. APS Press, St. Paul, MI, 9–29.

Ripoll, J., Urban, L., Staudt, M., Lopez-Lauri, F., Bidel, L. P., & Bertin, N. (2014). Water shortage and quality of fleshy fruits—making the most of the unavoidable. *Journal of Experimental Botany*, 65(15), 4097–4117.

Sadeghipour, O. (2009). The influence of water stress on biomass and harvest index in three mung bean (*Vigna radiata* (L.) R. Wilczek) cultivars. *Asian Journal of Plant Sciences*, 8(3), 245–249.

Sanchez, E., Soto, J. M., Garcia, P. C., Lopez-Lefebre, L. R., Rivero, R. M., Ruiz, J. M., & Romero, L. (2000). Phenolic and oxidative metabolism as bioindicators of nitrogen deficiency in French bean plants (*Phaseolus vulgaris* L. cv. Strike). *Plant Biology*, 2(3), 272–277.

Santisopasri, V., Kurotjanawong, K., Chotineeranat, S., Piyachomkwan, K., Sriroth, K., & Oates, C. G. (2001). Impact of water stress on yield and quality of cassava starch. *Industrial Crops and Products*, 13, 115–129.

Savin, R., & Nicolas, M. E. (1996). Effects of short periods of drought and high temperature on grain growth and starch accumulation of two malting barley cultivars. *Functional Plant Biology*, 23(2), 201–210.

Savin, R., Stone, P. J., Nicolas, M. E., & Wardlaw, I. F. (1997). Effects of heat stress and moderately high temperature on grain growth and malting quality of barley. *Australian Journal of Agricultural Research*, 48, 615–624.

Sgherri, C., Kadlecova, Z., Pardossi, A., Navari-Izzo, F., & Izzo, R. (2008). Irrigation with diluted seawater improves the nutritional value of cherry tomatoes. *Journal of Agricultural and Food Chemistry*, 56, 3391–3397.

Shewry, P. R. (2009). Wheat. *Journal of Experimental Botany*, 60, 1537–1553.

Shewry, P. R., Freeman, J., Wilkinson, M., Pellny, T., & Mitchell, R. A. (2010). Challenges and opportunities for using wheat for biofuel production. *Energy Crops*, 13–26.

Shewry, P. R., Van Schaik, F., Ravel, C., Charmet, G., Rakszegi, M., Bedo, Z., & Ward, J. L. (2011). Genotype and environment effects on the contents of vitamins B1, B2, B3, and B6 in wheat grain. *Journal of Agricultural and Food Chemistry*, 59(19), 10564–10571.

Shi, R., Zhang, Y., Chen, X., Sun, Q., Zhang, F., Romheld, V., & Zou, C. (2010). Influence of long term nitrogen fertilization on micronutrient density in grain of winter wheat (*Triticum aestivum* L.). *Journal of Cereal Science*, 51(1), 165–170.

Siaut, M., Cuine, S., Cagnon, C., Fessler, B, Nguyen, M., Carrier, P., Beyly, A., Beisson, F., Triantaphylides, C., Li-Beisson, Y., & Peltier, G. (2011). Oil accumulation in the model green alga *Chlamydomonas reinhardtii*: characterization, variability between common laboratory strains and relationship with starch reserves. *BMC Biotechnology*, 11(1), 1–15.

Silva, J. A. B., Otoni, W. C., Martinez, C. A., Dias, L. M., & Silva, M. A. P. (2001). Microtuberization of Andean potato species (*Solanum* spp.) as affected by salinity. *Scientia Horticulturae-Amsterdam*, 89, 91–101.

Silvente, S., Sobolev, A. P., & Lara, M. (2012). Metabolite adjustments in drought tolerant and sensitive soybean genotypes in response to water stress. *PLoS ONE*, 7(6), e38554.

Singh, A., Bhushan, B., Gaikwad, K., Yadav, O. P., Kumar, S., & Rai, R. D. (2015). Induced defence responses of contrasting bread wheat genotypes under differential salt stress imposition. *Indian Journal of Biochemistry and Biophysics*, 52, 75–85.

Singh, M., Kumar, J., Singh, V. P., & Prasad, S. M. (2014). Plant tolerance mechanism against salt stress: the nutrient management approach. *Biochemical Pharmacol*ogy, 3, e165.

Singh, P., Agrawal, M., & Agrawal, S. B. (2009). Evaluation of physiological, growth and yield responses of a tropical oil crop (*Brassica campestris* L. var. Kranti) under ambient ozone pollution at varying NPK levels. *Environmental Pollution*, 157(3), 871–880.

Siscar-Lee, J. J. H., Juliano, B. O., Qureshi, R. H., & Akbar, M. (1990). Effect of saline soil on grain quality of rices differing in salinity tolerance. *Plant Foods for Human Nutrition*, 40, 31–36.

Sita, K., Sehgal, A., Bhandari, K., Kumar, J., Kumar, S., Singh, S., Siddique, K. H., & Nayyar, H. (2018). Impact of heat stress during seed filling on seed quality and seed yield in lentil (*Lens culinaris Medikus*) genotypes. *Journal of the Science of Food and Agriculture*, 98(13), 5134–5141.

Sitnicka, D., & Orzechowski, S. (2014). Cold-induced starch degradation in potato leaves—intercultivar differences in the gene expression and activity of key enzymes. *Biologia Plantarum*, 58, 659–666.

Sivamani, E., Bahieldin, A., Wraith, J. M., Al-Niemi, T., Dyer, W. E., Ho, T. H. D., & Qu, R. (2000). Improved biomass productivity and water use efficiency under water deficit conditions in transgenic wheat constitutively expressing the barley HVA1 gene. *Plant Science*, 155, 1–9.

Sonia, K., & Mohammad, S. (2013). Effect of salt stress on grain reserve composition in ten durum wheat cultivars. *Journal of Stress Physiology and Biochemistry*, 9(3), 113–121.

Sonnenberg, D., Ndakidemi, P. A., & Laubscher, C. (2013). The effects of various drip fertigated water quantities on flavonoid and anthocyanin content on hydroponically cultivated *Cucumis sativa* L. *International Journal of Physical Sciences*, 8(19), 1012–1016.

Sorensen, J. N., Jorgensen, U., & Kuhn, B. F. (1997). Drought effects on the marketable and nutritional quality of carrots. *Journal of the Science of Food and Agriculture*, 74(3), 379–391.

Stagnari, F., Galieni, A., Cafiero, G., & Pisante, M. (2014). Application of photo-selective films to manipulate wavelength of transmitted radiation and photosynthate composition in red beet (*Beta vulgaris* var. conditiva Alef.). *Journal of the Science of Food and Agriculture*, 94(4), 713–720.

Steer, T., Thane, C., Stephen, A., & Jebb, S. (2008). Bread in the diet: consumption and contribution to nutrient intakes of British adults. *Proceedings of the Nutrition Society*, 67(OCE8).

Stewart, A. J., Chapman, W., Jenkins, G. I., Graham, I., Martin, T., & Crozier, A. (2001). The effect of nitrogen and phosphorus deficiency on flavonol accumulation in plant tissues. *Plant Cell and Environment*, 24(11), 1189–1197.

Streb, S., Eicke, S., & Zeeman, S. C. (2012). The simultaneous abolition of three starch hydrolases blocks transient starch breakdown in *Arabidopsis*. *The Journal of Biological Chemistry*, 287, 41745–41756.

Su, J., Hu, C., Yan, X., Jin, Y., Chen, Z., Guan, Q., Wang, Y., Zhong, D., Jansson, C., Wang, F., Schnurer, A., & Sun, C. (2015). Expression of barley SUSIBA2 transcription factor yields high-starch low-methane rice. *Nature*, 523(7562), 602–606.

Sun, X. D., & Arntfield, S. D. (2012). Molecular forces involved in heat-induced pea protein gelation: effects of various reagents on the rheological properties of salt-extracted pea protein gels. *Food Hydrocolloids*, 28(2), 325–332.

Swanston, J. S., Ellis, R. P., Perez-Vendrell, A., Voltas, J., & Molina-Cano, J. L. (1997). Patterns of barley grain development in Spain and Scotland and their implications for malting quality. *Cereal Chemistry*, 74(4), 456–461.

Szucs, A., Jager, K., Jurca, M. E., Fabian, A., Bottka, S., Zvara, A., Barnabas, B., & Feher, A. (2010). Histological and microarray analysis of the direct effect of water shortage alone or combined with heat on early grain development in wheat (*Triticum aestivum*). *Physiologia Plantarum*, 140(2), 174–188.

Taarit, M. B., Msaada, K., Hosni, K., & Marzouk, B. (2010). Changes in fatty acid and essential oil composition of sage (*Salvia officinalis* L.) leaves under NaCl stress. *Food Chemistry*, 119(3), 951–956.

Teklemariam, T. A., & Blake, T. J. (2004). Phenylalanine ammonia-lyase-induced freezing tolerance in jack pine (*Pinus banksiana*) seedlings treated with low, ambient levels of ultraviolet-B radiation. *Physiologia Plantarum*, 122(2), 244–253.

Televiciute, D., Taraseviciene, Z., Danilcenko, H., Barcauskaite, K., Kandaraite, M., & Paulauskiene, A. (2020). Changes in chemical composition of germinated leguminous under abiotic stress conditions. *Food Science and Technology*, 40, 415–421.

Thalmann, M., & Santelia, D. (2017). Starch as a determinant of plant fitness under abiotic stress. *New Phytologist*, 214(3), 943–951.

Thalmann, M., Pazmino, D., Seung, D., Horrer, D., Nigro, A., Meier, T., Kolling, K., Pfeifhofer, W. H., Zeeman, S. C., & Santelia, D. (2016). Regulation of leaf starch degradation by abscisic acid is important for osmotic stress tolerance in plants. *The Plant Cell*, 28, 1860–1878.

Thomas, J. M. G., Prasad, P. V. V., Boote, K. J., & Allen Jr, L. H. (2009). Seed composition, seedling emergence and early seedling vigour of red kidney bean seed produced at elevated temperature and carbon dioxide. *Journal of Agronomy and Crop Science*, 195(2), 148–156.

Ti, D. G., Sui, F. G., Nie, S. A., Sun, N. B., Xiao, H. A., & Tong, C. L. (2010). Differential responses of yield and selected nutritional compositions to drought stress in summer maize grains. *Journal of Plant Nutrition*, 33, 1811–1818.

Tohge, T., Nishiyama, Y., Hirai, M. Y., Yano, M., Nakajima, J. I., Awazuhara, M., Inoue, E., Takahashi, H., Goodenowe, D.B., & Kitayama, M. (2005). Functional genomics by integrated analysis of metabolome and transcriptome of *Arabidopsis* plants over-expressing an MYB transcription factor. *The Plant Journal*, 42(2), 218–235.

Toreti, A., Deryng, D., Tubiello, F. N., Muller, C., Kimball, B. A., Moser, G., Boote, K., Asseng, S., Pugh, T. A., Vanuytrecht, E., & Rosenzweig, C. (2020). Narrowing uncertainties in the effects of elevated CO_2 on crops. *Nature Food*, 1(12), 775–782.

Triboi, E., & Triboi-Blondel, A. M. (2002). Productivity and grain or seed composition: a new approach to an old problem. *European Journal of Agronomy*, 16(3), 163–186.

Tsialtas, J. T., Soulioti, E., Maslaris, N., & Papakosta, D. (2009). Genotypic response to re-growth of defoliated sugar beets after re-watering in a water-limited environment: effects on yield and quality. *International Journal of Plant Production*, 3, 1–18.

Vasseur, F., Pantin, F., & Vile, D. (2011). Changes in light intensity reveal a major role for carbon balance in *Arabidopsis* responses to high temperature. *Plant Cell and Environment*, 34, 1563–1576.

Veit-Kohler, U., Krumbein, A., & Kosegarten, H. (1999). Effect of different water supply on plant growth and fruit quality of *Lycopersicon esculentum*. *Journal of Plant Nutrition and Soil Science*, 162(6), 583–588.

Villadsen, D., Rung, J. H., & Nielsen, T. H. (2005). Osmotic stress changes carbohydrate partitioning and fructose-2,6-bisphosphate metabolism in barley leaves. *Functional Plant Biology*, 32, 1033–1043.

Walker, A. R., Davison, P. A., Bolognesi-Winfield, A. C., James, C. M., Srinivasan, N., Blundell, T. L., Esch, J.J., Marks, M.D., & Gray, J. C. (1999). The TRANSPARENT TESTA GLABRA1 locus, which regulates trichome differentiation and anthocyanin biosynthesis in *Arabidopsis*, encodes a WD40 repeat protein. *The Plant Cell*, 11(7), 1337–1349.

Wang, T. L., Domoney, C., Hedley, C. L., Casey, R., & Grusak, M. A. (2003). Can we improve the nutritional quality of legume seeds? *Plant Physiology*, 131(3), 886–891.

Wang, X., Chang, L., Wang, B., Wang, D., Li, P., Wang, L., Yi, X., Huang, Q., Peng, M., & Guo, A. (2013). Comparative proteomics of *Thellungiella halophila* leaves from plants subjected to salinity reveals the importance of chloroplastic starch and soluble sugars in halophyte salt tolerance. *Molecular and Cellular Proteomics*, 12, 2174–2195.

Wang, Y., & Frei, M. (2011). Stressed food–The impact of abiotic environmental stresses on crop quality. *Agriculture Ecosystems and Environment*, 141(3–4), 271–286.

Wassmann, R., Jagadish, S. V. K., Sumfleth, K., Pathak, H., Howell, G., Ismail, A., Serraj, R., Redona, E., Singh, R. K., Heuer, S., & Donald, L. S. (2009b). Regional vulnerability of climate change impacts on Asian rice production and scope for adaptation. In: *Advances in agronomy*, Academic Press, 91–133.

Wassmann, R., Jagadish, S. V. K., Heuer, S., Ismail, A., Redona, E., Serraj, R., Singh, R. K., Howell, G., Pathak, H., Sumfleth, K., & Donald, L. S. (2009a). Climate change affecting rice production: the physiological and agronomic basis for possible adaptation strategies. *Advances in Agronomy*, 101, 59–122.

Wilson, R. F., Weissinger, H. H., Buck, J. A., & Faulkner, G. D. (1980). Involvement of phospholipids in polyunsaturated fatty acid synthesis in developing soybean cotyledons. Plant Physiology, 66(4), 545–549.

Wolf, R. B., Cavins, J. F., Kleiman, R., & Black, L. T. (1982). Effect of temperature on soybean seed constituents: oil, protein, moisture, fatty acids, amino acids and sugars. *Journal of the American Oil Chemists Society*, 59(5), 230–232.

Wooding, A. R., Kavale, S., MacRitchie, F., Stoddard, F. L., & Wallace, A. (2000). Effects of nitrogen and sulfur fertilizer on protein composition, mixing requirements, and dough strength of four wheat cultivars. *Cereal Chemistry*, 77(6), 798–807.

Wu, M., & Kubota, C. (2008). Effects of high electrical conductivity of nutrient solution and its application timing on lycopene, chlorophyll and sugar concentrations of hydroponic tomatoes during ripening. *Scientia Horticulturae-Amsterdam* 116, 122–129.

Yadav, A., Bhatia, A., Yadav, S., Kumar, V., & Singh, B. (2019). The effects of elevated CO_2 and elevated O_3 exposure on plant growth, yield and quality of grains of two wheat cultivars grown in north India. *Heliyon*, 5(8), e02317.

Yamakawa, H., & Hakata, M. (2010). Atlas of rice grain filling-related metabolism under high temperature: joint analysis of metabolome and transcriptome demonstrated inhibition of starch accumulation and induction of amino acid accumulation. *Plant and Cell Physiology*, 51, 795–809.

Yang, J., Zhang, J., Wang, Z., & Zhu, Q. (2001). Activities of starch hydrolytic enzymes and sucrose-phosphate synthase in the stems of rice subjected to water stress during grain filling. *Journal of Experimental Botany*, 52, 2169–2179.

Yao, X., Chu, J., He, X., & Si, C. (2014). Grain yield, starch, protein, and nutritional element concentrations of winter wheat exposed to enhanced UV-B during different growth stages. *Journal of Cereal Science*, 60(1), 31–36.

Yonekura-Sakakibara, K., Tohge, T., Matsuda, F., Nakabayashi, R., Takayama, H., Niida, R., Watanabe-Takahashi, A., Inoue, E., & Saito, K. (2008). Comprehensive flavonol profiling and transcriptome coexpression analysis leading to decoding gene–metabolite correlations in *Arabidopsis. The Plant Cell*, 20(8), 2160–2176.

Yuan, Y., Qi, L., Yang, J., Wu, C., Liu, Y., & Huang, L. (2015). A *Scutellaria baicalensis* R2R3-MYB gene, SbMYB8, regulates flavonoid biosynthesis and improves drought stress tolerance in transgenic tobacco. *Plant Cell, Tissue and Organ Culture (PCTOC)*, 120(3), 961–972.

Zhang, B., Liu, W., Chang, S. X., & Anyia, A. O. (2010). Water-deficit and high temperature affected water use efficiency and arabinoxylan concentration in spring wheat. *Journal of Cereal Science*, 52(2), 263–269.

Zheng, Y., Cheng, D., & Simmons, M. (2014). Ozone pollution effects on gas exchange, growth and biomass yield of salinity-treated winter wheat cultivars. *Science of the Total Environment*, 499, 18–26.

Zhou, J., Ma, C., Zhen, S., Cao, M., Zeller, F. J., Hsam, S. L., & Yan, Y. (2016). Identification of drought stress related proteins from 1S l (1B) chromosome substitution line of wheat variety Chinese Spring. *Botanical Studies*, 57(1), 1–10.

Zhu, C., Kobayashi, K., Loladze, I., Zhu, J., Jiang, Q., Xu, X., Liu, G., Seneweera, S., Ebi, K. L., Drewnowski, A., & Ziska, L. H. (2018). Carbon dioxide (CO_2) levels this century will alter the protein, micronutrients, and vitamin content of rice grains with potential health consequences for the poorest rice-dependent countries. *Science Advances*, 4(5), eaaq1012.

Ziska, L. H., Palowsky, R., & Reed, D. R. (2007). A quantitative and qualitative assessment of mung bean (*Vigna mungo* (L.) Wilczek) seed in response to elevated atmospheric carbon dioxide: potential changes in fatty acid composition. *Journal of the Science of Food and Agriculture*, 87(5), 920–923.

Zou, L. (2009). Effects of gradual and sudden heat stress on seed quality of Andean lupin, Lupinus mutabilis. Doctoral dissertation, The University of Helsinki, Helsinki.

Zu, Y., Li, Y., Chen, J., & Chen, H. (2004). Intraspecific responses in grain quality of 10 wheat cultivars to enhanced UV-B radiation under field conditions. *Journal of Photochemistry and Photobiology B: Biology*, 74(2–3), 95–100.

9 Plant Transcription Factors from Halophytes and Their Role in Salinity and Drought Stress Tolerance

Priyanka S. Joshi[1,2], *Ankita Dave*[1,2], *Parinita Agarwal*[1]*, *and Pradeep K. Agarwal*[1,2]
[1]Division of Plant Omics, CSIR-Central Salt and Marine Chemicals Research Institute (CSIR-CSMCRI), Council of Scientific & Industrial Research (CSIR), Gijubhai Badheka Marg, Bhavnagar, (Gujarat), India
[2]Academy of Scientific and Innovative Research, CSIR-CSMCRI, Gijubhai Badheka Marg, Bhavnagar, (Gujarat), India
*Corresponding author: E-mail: parinita.a@rediffmail.com

CONTENTS

ABBREVIATIONS

ABA abscisic acid
ABRE ABA-responsive element
bZIP basic leucine zipper
BHLH basic helix–loop–helix
CBF CRT binding factor
CK2 casein kinase 2
CRT C-RepeaT
DRE dehydration-responsive element
DREB dehydration-responsive element-binding factor
ERD1 early responsive to dehydration stress
HTH helix–turn–helix
MAPK mitogen-activated protein kinase

DOI: 10.1201/9781003159636-9

MYB	myeloblastosis oncogene
MYC	myelocytomatosis oncogene
NACRS	NAC recognition site
NLS	nuclear localizing sequence
PKC	protein kinase C
ROS	reactive oxygen species
SA	salicylic acid
TF	transcription factor

9.1 INTRODUCTION

Crop production is greatly affected by limited freshwater resources and increased soil salinization. Salinity and drought are the most significant factors decreasing crop production worldwide (Ma *et al.*, 2020). Around 800 million hectares (ha) of land is affected by salt, which contributes more than 6% of the total land area of the world. An area of nearly 45 million ha (20%), out of 230 million ha of irrigated land, has been damaged by high salinity (FAO, 2016). By the end of this century, global drought incidences are also expected to increase by 20%, especially in the Central and South American corn belt as well as in central and western Europe (Edmeades, 2013; Singh *et al.*, 2015). Moreover, Mediterranean regions and the Middle East have also been shown to be among the 'hot-spots' of severe drought in regional climate change models (Saadi *et al.*, 2015; Barlow *et al.*, 2016). About 26% of the world's population is suffering from malnutrition and food insecurity; moreover, it is also anticipated that by 2050, the world population will reach nine billion (FAO *et al.*, 2019). Looking at the current climate change scenario, the occurrence of drought events and the degree of global land salinization are anticipated to increase; hence, the development of salt and drought-tolerant crop varieties by breeding and genetic modification is absolutely crucial for future food security.

On the basis of adaptation to salinity, plants are categorized into glycophytes and halophytes. Plants that can complete their life cycle under saline conditions are classified as halophytes, while glycophytes cannot grow in more than 4 dcm^2 (40 mM NaCl) soil salinity. Many classifications of halophytes have been done based on general ecological behaviour and distribution, plant growth response to salinity, the capacity to take in salt, etc. (Waisel, 1972). Halophytes are classified as obligate or facultative on the basis of their requirement of salt for growth and development. For optimum growth, obligate halophytes require salt, whereas facultative halophytes can grow in the presence or absence of salt. Members of the Chenopodiaceae family are obligate halophytes, while Gramineae, Cypraceae and Juncaceae family members, along with many dicotyledonous species, are facultative halophytes (Cushman, 2001; Sabovljevic and Sabovljevic, 2007). On the basis of the resistance mechanism, halophytes have been categorized into euhalophytes (salt-accumulators or halosucculants), recretohalophytes or salt-includers (exorecretohalophytes and endorecretohalophytes), and pseudohalophytes (salt-excluders; Breckle, 1995; Figure 9.1).

Euhalophytes can adapt to high salinity, as they can uptake and transport salt and water to leaves and/or stems. The increased cell size forms succulence, which helps them to dilute excess salts in the plant body (Munns *et al.*, 1983). *Plantago*, *Salsola*, *Suaeda*, etc. are examples of leaf succulence with fleshy and globular leaves, enlarged mesophyll cells, smaller intercellular spaces and increase in the amount of leaf sap (Shabala and Mackay, 2011), while *Salicornia*, *Arthrocnemum*, etc. are stem succulent and are mostly leafless or have reduced leaves. These halophytes have storage tissue in their stem, where they have large vacuoles for the storage of water and Na^+. Recretohalophytes can accumulate or excrete excessive salts from metabolically active tissues through specialized salt-secreting structures like salt glands and salt bladders, which helps in ion homeostasis and regulation of osmotic pressure (Zhou *et al.*, 2001; Zhang *et al.*, 2003). On the basis of excretion or accumulation of extra salt, recretohalophytes are further classified into exorecretohalophytes and endorecretohalophytes. The exorecretohalophytes have epidermal salt glands present on leaves and stems, which help them secrete excessive salts from the plant body. *Crassa*, *Tamarix*, *Spartina*, etc.

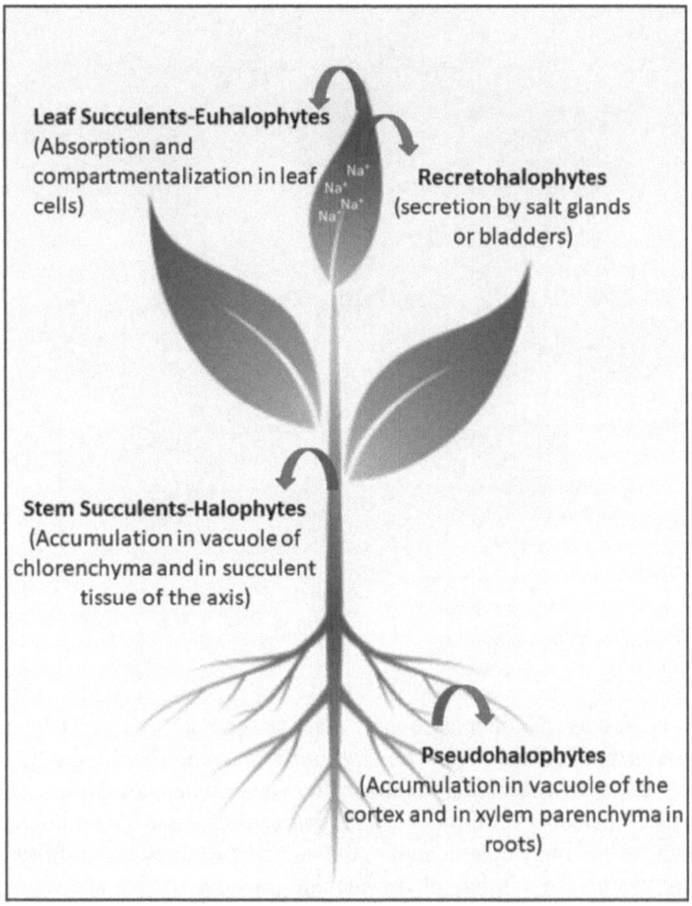

FIGURE 9.1 Different salinity tolerance mechanisms in halophytes.

are some examples of exorecretohalophytes. On the other hand, endorecretohalophytes sequester salts in their enlarged vacuole present in the salt bladders on the aerial parts of the plant and include *Atriplex*, *Mesembryanthemum*, *Chenopodium*, etc. species. Pseudohalophytes show optimum growth in non-saline conditions but are also tolerant to high salt concentration. They do not have specialized morphological and anatomical features and accumulate Na^+ in the roots and xylem parenchyma (Howard and Mendelssohn, 1999). *Phragmites*, *Artemisia* and *Juncus* species are examples of pseudohalophytes.

Halophytes have been studied extensively to understand the anatomical, physiological and biochemical mechanisms behind their salt tolerance as well as for the development of genetically improved crops. Even though halophytes account for only 1% of world flora, they show high diversity in their distribution among taxa, response to stress and habitat (Flowers *et al.*, 2010). The anatomical, morphological and physiological adaptations help halophytes to grow and complete their life cycle under salinity stress. Halophytes are widely used as crops, for desalination of degraded soils by cultivation, for retrieval of value-added products, and to isolate potential genes and improve abiotic stress tolerance by overexpression in glycophytes (Nikalje *et al.*, 2017). The salt accumulation potential of halophytes is also used for the remediation of toxic heavy metals from the soil. Moreover, saline soils are utilized for cultivation of halophytes that have potential as biofuel sources, like *Atriplex*, *Batis*, *Distchlis*, *Euphorbia tirucalli*, *Salicornia* and *Suaeda* (Sharma *et al.*, 2016; Rozema and Flowers, 2008; Hastilestari *et al.*, 2013; Akinshina *et al.*, 2016). Use of halophytes

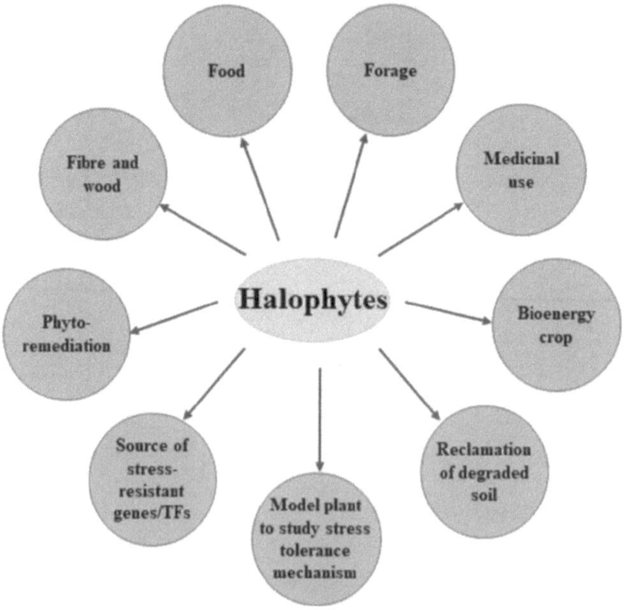

FIGURE 9.2 Application of halophytes.

as animal forage and fodder has increased over time (Figure 9.2). Several halophytes, such as *Panicum*, *Atriplex*, *Haloxylon*, *Kochia*, *Spartina alterniflora* and *Salicornia bigelovii*, are cultivated for forage in countries with low rainfall and low-quality soil (Ahmad and Malik, 2002; Koyro *et al.*, 2013). Halophytes accumulate several biochemical compounds, such as amino acids, carotenoids, flavonoids, phenolics, etc., which are essential to the human diet (Fiedor and Burda, 2014).

The salt adaptive mechanisms in halophytes include selective uptake and transport of ions, ion compartmentalization, development of succulence, osmotic adjustment, regulation of redox and energetic status, and antioxidant response. In addition to this, halophytes have exceptional molecular responses, which make them more tolerant to abiotic stress than glycophytes. The efficient perception of stress signals and rapid activation of downstream processes are crucial for effective salt adaptation mechanisms, which are further controlled by a cascade of molecular networks. Secondary messengers like Ca^{2+}, inositol phosphate and reactive oxygen species as well as phytohormones such as ABA, ethylene and salicylic acid play a crucial role in signal transduction mechanisms (Nikalje *et al.*, 2017; Tuteja, 2007). Many transcription factors (TFs), like CBF1, DREB1A, ATHB7, NAC, WRKY, MYB, etc., are upregulated during stress and are known to regulate plant processes like cell division and cell expansion, which subsequently regulate growth under stress conditions (Zhu, 2001). A complicated signalling network is activated via synergistic action of various TFs in both temporal and spatial ways for the expression of an array of stress-responsive genes to alleviate stress after its perception by plants (Golldack *et al.*, 2011). TFs play a critical role in the complex network of signalling pathways by manoeuvring a large array of downstream genes associated with both abiotic and biotic stress, growth, development, metabolism, and senescence of plants (Agarwal *et al.*, 2013; Nakashima *et al.*, 2014). The use of genetic engineering by overexpressing TFs is one of the robust techniques to produce stress-tolerant crop varieties, which can improve agricultural yield in adverse conditions as compared with conventional breeding techniques.

9.2 TRANSCRIPTOME STUDY OF POTENTIAL HALOPHYTE PLANTS

Transcriptomics has rapidly grown as a study due to the development of next generation sequencing (NGS). Earlier transcriptome analysis was based on microarray, quantitative polymerase chain reaction (qPCR) or real-time PCR techniques, which were costlier and more time consuming than NGS. NGS technology can be used for gene expression profiling, genome annotation and discovery of non-coding RNAs (Mutz *et al.*, 2013). Soil salinity is a global challenge for food security and sustainable agriculture. The study of salt tolerance mechanisms can give us an insight that may help to engineer better crops. As the NGS technologies have developed over time, RNA sequencing has become the most established method for gene expression profiling (McGettigan, 2013). This method can identify transcripts and their expression at different time intervals and in different conditions. Whole transcriptome sequencing of many halophyte plants under salinity stress have been carried out to uncover molecular mechanisms involved in salinity stress tolerance.

Aeluropus littoralis is a halophyte plant with small genome size (342 Mb) and high salt tolerance ability, which makes it a suitable candidate for salinity studies. The study of *A. littoralis* using a *de novo* RNA sequencing method indicated a salt concentration-dependent response (Younesi-Melerdi *et al.*, 2020). Well-known salt stress dependent pathways were induced at 400 mM salt, while lesser contributing pathways, like cell cycle and DNA replication, were induced at 200 mM salt. Homology search for differentially expressed transcripts suggested that 67 TFs were identified, of which ERF and WRKY were most abundant (Younesi-Melerdi *et al.*, 2020; Table 9.1). In *Chenopodium glaucum*, bHLH TF was found to be upregulated during salt and drought stress in coordination with a calcium-dependent protein kinase (CDPK) (Wang *et al.*, 2017c). As many as 11,242 TFs were found

TABLE 9.1
Identification of Transcription Factors from Halophytes in Response to Salt Stress through Transcriptome Analysis

Halophyte Name	No. of Transcription Factors	Major TFs Differentially Expressed During Salt Stress	Regulation During Salt Stress	Reference
Aeluropus littoralis	Total identified TFs – 67	ERF, WRKY, C2H2, NAC, bZIP, MYB-related and MYB	ERF (16) upregulated and downregulated (7) during 400 mM salt stress. ERF (4) downregulated under 200 mM salt stress	Younesi-Melerdi *et al.* (2020)
Chenopodium glaucum	–	bHLH, GATA	Upregulated under salt and drought treatment	Wang *et al.* (2017c)
Chrysanthemum lavandulifolium	Total identified TFs – 185, WRKY(13), NAC (17)	ERF, bHLH, MYB4, bZIP, C3HC4	Upregulated during salt stress	Huang *et al.* (2019)
Oryza coarctata	Total identified TFs – 122 MYB (19), bZIP (17), HB (16), HSF (11), and MYB-related (9) TFs	MYB, bZIP, HB, HSF, and MYB-related TFs	bHLH and MYB-related TF upregulated under salinity stress	Bansal *et al.* (2020)
Phragmites karka	Total identified TFs – 11,242 MYB (1086), WRKY (1057), Nin-like (917), bZIP (829), C3H (737) and C2H2 (663)	ERF, MYB, C2H2, Zn finger, FAR1, NAC, bHLH, Trihelix TFs and AP2/ERF	ARF, C3H, MYB, C2H2, and FAR1 upregulated in leaves, NAC and Trihelix TF families upregulated in roots, and bHLH TFs downregulated in roots under salt stress	Nayak *et al.* (2020)

(*continued*)

TABLE 9.1 (Continued)
Identification of Transcription Factors from Halophytes in Response to Salt Stress through Transcriptome Analysis

Halophyte Name	No. of Transcription Factors	Major TFs Differentially Expressed During Salt Stress	Regulation During Salt Stress	Reference
Porteresia coarctata	Total identified TFs – 2749, the MYB domain (197), bHLH (150)- and NAC (147)-domain TFs	MYB, bHLH, AP2-EREBP, WRKY, bZIP and NAC	NAC, MYB/MYB-related and WRKY upregulated during salt stress	Garg *et al.* (2014)
Pugionium cornutum	–	WRKYs, MYBs, ZFs, or bZIPs/HD-ZIPs, NAC62, MYB59	Upregulated during salt stress	Cui *et al.* (2019)
Salicornia persica	Total identified TFs – 41 HD-ZIP (5), MBF (4)	HD-ZIP, MBF	Upregulated during salt stress	Aliakbari *et al.* (2020)
Sedum lineare	HD-ZIP (2), NAC (4), MYB (2), WRKY (8), DREB (2), ERF (2)	ERFs, NAC, WRKYs, MYB, DREB, and Homeodomain-leucine Zipper (HD-Zip)	WRKY40 and WRKY41 were downregulated while other TFs were upregulated	Song *et al.* (2019)
Spartina alterniflora	–	SaAF2 (NAC)	Upregulation under salt treatment	Ye *et al.* (2020)
Spartina alterniflora	Total identified TFs – 4462	ARF, AUX/IAA, CCAAT, E2F-DP, HSF, BBR/BPC and BSD, GeBP, SBP	HSF, BBR/BPC and BSD upregulated in leaves, ARF, AUX/IAA, CCAAT, E2F-DP were upregulated in roots	Bedre *et al.* (2016)
Sporobolus virginicus	Six bZIP genes	bZIP, C3H	Three bZIP upregulated and other three downregulated under salt stress	Yamamoto *et al.* (2015)
Suaeda fruticosa	Total identified TFs – 3110. MYB (134), ERF (80), bZIP (71), NAC (70), CAMTA (44)	MYB, CAMTA, MADS-box and bZIP	Upregulated in salt, cold and drought stress	Diray-Arce *et al.* (2019)
Sueda rigida	Three ERF, two WRKY and four bHLH TF	ERFs, WRKY, bHLH	ERF, WRKY and bHLH genes were upregulated during salt stress	Han *et al.* (2020)

Note: DREB: dehydration-responsive element-binding, ERF: ethylene-responsive factor, bHLH: basic helix–loop-helix TF, SBP: squamosa-promoter binding protein, GeBP: GL1 enhancer binding protein, HSF: heat specific factor, BBR/BPC: barley B recombinant/Basic Pentacysteine protein, HD-Zip: homeodomain-leucine zipper during transcriptome analysis of *Phragmites karka*, which were differentially expressed during salt stress, like C2H2, ARF, MYB, NAC, C3H, bHLH, etc. MYB, C3H and C2H2 were upregulated in leaves, while NAC and trihelix TF were upregulated in roots during salt stress (Nayak *et al.*, 2020). Functional annotation in *Porteresia coarctata* revealed 2749 TFs, including MYB, bHLH, AP2-EREBP, WRKY, bZIP and NAC, commonly involved in stress response (Garg *et al.*, 2014). Similar studies in *Salicornia persica* revealed that 41 TFs were differentially expressed, among which MBF and HD-ZIP were upregulated during salt stress (Aliakbari *et al.*, 2020). In *Sporobolus verginicus*, 4462 TFs belonging to 54 families were identified. Interestingly, a comparative study of rice and *Arabidopsis* transcriptome revealed six genes specific to *S. verginicus*. All of them showed root-specific expression, and five of them encoded bZIP type TF (Yamamoto *et al.*, 2015). In another study in *Sueda fruticose,* 3110 TFs were differentially expressed, like MYB, ERF, CAMTA, NAC, etc., which were upregulated during salt, cold and drought stress (Diray-Arce *et al.*, 2019).

9.3 THE POTENTIAL OF TRANSCRIPTION FACTORS FROM HALOPHYTES FOR ABIOTIC STRESS TOLERANCE

The response of the plant to stress is a three-stage process: perception of stress, signal transduction from membrane receptors to the nucleus via mediators, and regulation of gene expression (Hoang *et al.*, 2014). The TFs are the key component participating in plant responses, and they regulate an array of downstream genes to enhance stress tolerance. TFs have a DNA-binding domain and a transcriptional regulatory domain via which they bind to several *cis*-elements in the promoters of their target genes and regulate expression of an array of genes; thus, they are key regulators of gene expression. The whole-genome sequencing of *Arabidopsis* has identified more than 1600 TFs, which contribute approximately 6% of the total number of genes (Gong *et al.*, 2004). Several TFs, like bZIP, bHLH, DREB, MYC, MYB, NAC, etc., are characterized into different families and superfamilies based on the presence of DNA-binding domains. Moreover, phytohormones, especially ABA, also play an essential role in the regulation of several processes, such as germination, growth and development, as well as several stress-responsive genes involved in biotic and abiotic responses, such as phosphatases, kinases, LEA (late embryogenesis abundant) proteins, water transporters like aquaporins, and antioxidative enzymes like catalase, superoxide dismutase and peroxidase, as well as TFs like APETALA2/ethylene-responsive element-binding factor (AP2/ERF), basic leucine zipper (bZIP), NAC and MYB TF families (Fujita *et al.*, 2011; Hoang *et al.*, 2014; Nakashima *et al.*, 2012; Vishwakarma *et al.*, 2017). After the perception of stress signals by plants, ABA-dependent and ABA-independent signalling pathways are activated (Shinozaki and Yamaguchi- Shinozaki, 1996). Some TFs regulate stress via either ABA-dependent or independent pathways, while some show cross-talk between both pathways (Figure 9.3). Several TFs have been isolated from halophytes and overexpressed *in planta* to develop stress-tolerant plants and study the mechanism behind stress tolerance (Table 9.2).

9.4 DREBs

Dehydration-responsive element-binding (DREBs) are plant-specific ABA-independent TFs belonging to the AP2/ERF (APETALA2/ ethylene-responsive element-binding) TFs family (Liu *et al.*, 2007). DREBs bind to the DRE/CRT *cis*-element and hence, are also called CRT-binding factors (CBFs; Yamaguchi-Shinozaki and Shinozaki, 1994) as well as to the GCC box, present in the promoters of many PR (pathogens related) genes (Gu *et al.*, 2000). They comprise two subclasses: DREB1/CBF, which is induced by cold stress and involved in low-temperature stress signalling, and DREB2, which is induced by drought stress and involved in drought and salt stress signalling; however, cross-talk between both pathways exists (Liu *et al.*, 1998). The DREB TFs have a single conserved ERF/AP2 DNA-binding domain consisting of ~60 amino acids, wherein valine

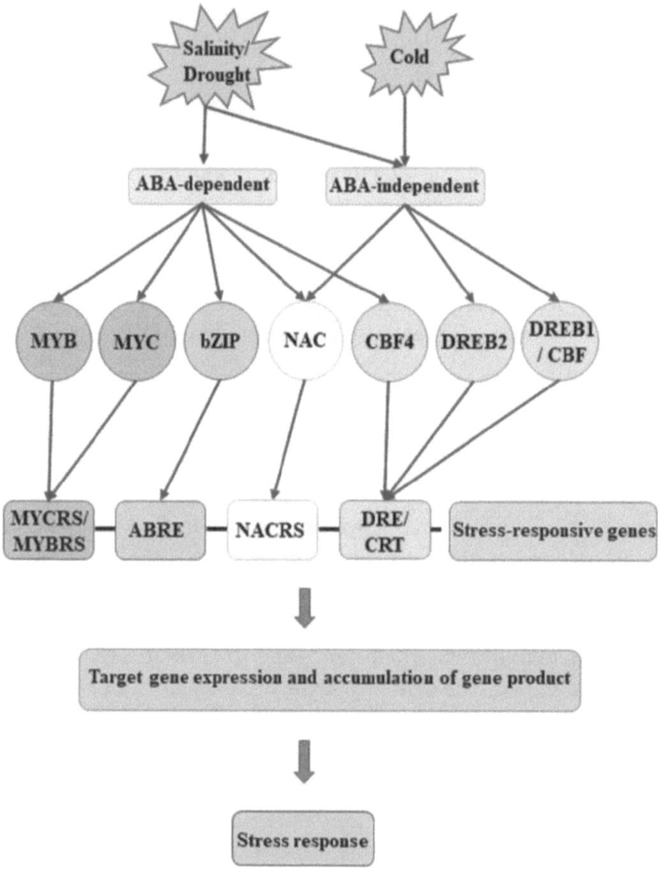

FIGURE 9.3 Schematic representation of transcription factor and transcriptional regulatory network involved in signal transduction of abiotic stress. MYC, MYB and bZIP act via an ABA-dependent pathway only. NAC regulate expression of downstream genes via both ABA-dependent and ABA-independent pathways. DREB2 TF regulate salinity and drought tolerance via an ABA-independent pathway. DREB1/CBF also regulate cold tolerance via a different ABA-independent pathway. CBF4 is the only DREB TF to regulate drought and salinity tolerance via an ABA-dependent pathway.

MYCRS, MYBRS and NACRS: MYC, MYB and NAC Recognition Sequence. ABRE: ABA Recognition Element. DRE/CRT: Dehydration responsive Element/C-RepeaT

TABLE 9.2
Genetic Engineering of Transcription Factors from Halophytes for Abiotic Stress Tolerance

Candidate TF	Source of TF	Transgenics	Phenotype of Transgenics	Reference
		DREB/CRF TF		
AhDREB1	*Atriplex hortensis*	Tobacco	Salt and drought stress tolerance	Shen *et al.* (2003)
AhDREB1	*A. hortensis*	*Populus tomentosa*	Salt stress tolerance	Du *et al.* (2012)
BdDREB2	*Buchloe dactyloides*	Tobacco	Drought stress tolerance	Zhang *et al.* (2014)
LbDREB	*Limonium bicolor*	*Populus ussuriensis*	Salt stress tolerance	Zhao *et al.* (2018)

TABLE 9.2 (Continued)
Genetic Engineering of Transcription Factors from Halophytes for Abiotic Stress Tolerance

Candidate TF	Source of TF	Transgenics	Phenotype of Transgenics	Reference
LcDREB2	*Leymus chinensis*	*A. thaliana*	Salt stress tolerance	Peng et al. (2013)
LcDREB3a	*L. chinensis*	*A. thaliana*	Salt and drought stress tolerance	Xianjun *et al.* (2011)
SbDREB2A	*Salicornia brachiata*	Tobacco	Salt and drought stress tolerance	Gupta *et al.* (2014)
SsDREB	*Suaeda salsa*	Tobacco	Salt and drought stress tolerance	Zhang *et al.* (2015)
ThCRF1	*Tamarix hispida*	*T. hispida*, *A. thaliana*	Salt stress tolerance	Qin *et al.* (2017)
ThDREB	*T. hispida*	Tobacco, *T. hispida*	Salt and drought stress tolerance	Yang *et al.* (2017)
TsCBF1	*Thellungiella halophila*	*Zea mays*	Drought stress tolerance	Zhang *et al.* (2010)
		MYB TF		
AmMYB1	*Avicennia marina*	Tobacco	Salt stress tolerance	Ganesan *et al.* (2012)
ChiMYB	*Chrysanthemum indicum*	*A. thaliana*	Salt and drought stress tolerance	He *et al.* (2016)
CmMYB2	*Chrysanthemum morifolium*	*A. thaliana*	Salt and drought stress tolerance	Shan *et al.* (2012)
LbTRY	*L. bicolor*	*A. thaliana*	Salt sensitivity	Leng *et al.* (2020)
LcMYB1	*L. chinensis*	*A. thaliana*	Salt stress tolerance	Cheng *et al.* (2013)
SbMYB15	*S. brachiata*	Tobacco	Salt and drought stress tolerance	Shukla *et al.* (2015)
ThMYB8	*T. hispida*	*T. hispida*, *A. thaliana*	Salt stress tolerance	Liu *et al.* (2021)
ThMYB13	*T. hispida*	*T. hispida*	Salt stress tolerance	Zhang *et al.* (2018)
NAC TF				
AlNAC1	*Aeluropus lagopoides*	Tobacco	Drought stress tolerance	Joshi *et al.* (2021)
AmNAC1	*A. marina*	Tobacco	Salt, drought and cold stress tolerance	Murugesan *et al.* (2020)
BoNAC019	*Brassica oleracea*	*A. thaliana*	Drought sensitivity	Wang *et al.* (2018)
DgNAC1	*Chrysanthemum*	Tobacco	Salt stress tolerance	Liu *et al.* (2011)
DgNAC1	*Chrysanthemum*	*Chrysanthemum*	Salt stress tolerance	Wang *et al.* (2017a)
SlNAC8	*Suaeda liaotungensis*	*A. thaliana*	Salt and drought stress tolerance	Wu *et al.* (2018)
ThNAC13	*T. hispida*	*T. hispida*, *A. thaliana*	Salt and drought stress tolerance	Wang *et al.* (2017b)
TsNAC1	*T. halophila*	*T. halophile*, *A. thaliana*	Salt, drought, cold and oxidative stress tolerance	Liu *et al.* (2018)
		Other TFs (GRAS, WRKY, bZIP)		
HcSCL13	*Halostachys caspica*	*A. thaliana*	Salt stress tolerance	Zhang *et al.* (2020)
RtWRKY1	*Reaumuria trigyna*	*A. thaliana*	Salt stress tolerance	Du *et al.* (2017)
ThbZIP1	*T. hispida*	Tobacco	Salt stress tolerance	Wang *et al.* (2010)
ThbZIP1	*T. hispida*	*Arabidopsis thaliana*	Salt stress tolerance	Ji *et al.* (2013), Ji *et al.* (2015)

and glutamic acid, at the 14th and 19th position, respectively, play a determining role in DNA-binding specificity. The AP2 domain is considered to be plant specific; however, its homologs have been reported in some cyanobacteria, viruses and bacteriophages, suggesting functional conservation with plant AP2/ERF proteins. Although DREB has been characterized in many plants (mostly glycophytes), data on the structure of this gene in halophytes is not sufficient to perform a sequence comparison between both groups of plants. Characterization of the DREB gene from the halophyte *Sueda salsa* revealed that SsDREB proteins contain valine residues at the 14th position and leucine residues (not glutamic acid residues) at the 19th position in the AP2/ERF domains (Sun et al., 2015). Similarly, *Atriplex hortensis* and *A. halimus* also had leucine in the 19th position instead of glutamic acid. The replacement has been also been found in the other known DREB A-6 group proteins like *Asparagus officinalis*, *Aloe vera* and *Zea mays* (Huang *et al.* 2012; Wang *et al.* 2010; Khedr *et al.* 2011). A valine residue replacement at the 19th position in the AP2/ERF domains is also found in DREB1-type factors from the monocotyledons rice, wheat and barley (Dubouzet *et al.* 2003; Qin *et al.* 2004). Therefore, it can be speculated that the 14th residue is more important and more conserved than the 19th residue in the regulation of the DRE-binding activity of DREB.

In addition, other amino acids, arginine, arginine, tryptophan, glutamic acid, arginine and tryptophan, at the 6th, 8th, 10th, 16th, 25th and 27th positions, respectively, facilitate DNA-binding activity by making direct contact with DNA and are known to be conserved in DREB1-type TFs (Akhtar *et al.*, 2012). A Ser/Thr-rich conserved region adjacent to the ERF/AP2 domain is accountable for phosphorylation of DREB proteins. In *Sueda salsa* it was observed that the serine-rich segments in the C-terminal regions of the two SsDREB proteins might function as phosphorylation sites that regulate transcription activity of SsDREBs. The serine-rich segment in the N-terminal region of SsDREBb may be responsible for controlling its DRE-binding ability by phosphorylation. Sequence analysis of SsDREBb identified a glutamine-rich region next to the serine-rich segment in the C-terminal. In general, glutamine-rich regions located in the C-terminal region can act as essential structural motifs required for transcriptional activation activity (Bruhn *et al.* 1992; Dai *et al.* 2003). The PKRPAGRTKFRETRHP DREB1/CBF1-type consensus NLS sequence distinguishes DREBs from other ERF/AP2 proteins. The LWSY and DSAW motif present at the end of the C-terminal region and the ERF/AP2 domain, respectively, are conserved in most DREB1 TFs (Agarwal *et al.*, 2006). In addition to this, the PEST sequence (RSDASEVTSTSSQSEVCTVETPGCV), which acts as a signal peptide for protein degradation along with target sites for protein kinases such as PKC and CK2, was found to be present in the negative regulatory domain of *Arabidopsis* DREB2A genes, which could not enhance the tolerance of transgenic plants. However, deletion of this region changed DREB2A to a constitutive active form (Rogers *et al.*, 1986; Sakuma *et al.*, 2006), while this negative regulatory domain was found to be absent in several DREBs, like PgDREB2A, SbDREB2A and ZmDREB2A, which can enhance the tolerance of transgenics plants without deletion (Agarwal *et al.*, 2006).

The expression of the DREB1 group is regulated at the transcriptional level, while the DREB2 group is regulated at transcriptional, post-transcriptional and post-translational levels. Most DREB1s/CBFs are induced by cold, as the DRE/CRT *cis*-element is present in their promoter region, which further activates several cold-responsive genes (Seki *et al.*, 2001). Thus, the expression of DREB1/CBF is controlled and co-ordinated inside the sub-family. In addition to DRE/CRT elements, several other important *cis*-elements, like ICE1 (inducer of CBF expression), MYC recognition elements, HOS1 (high expression of somatically responsive genes), conserved motif 2 (CM2) elements and G-box, present in the promoter region of DREB1/CBF TFs, regulate their expression at the transcriptional level (Agarwal *et al.*, 2017). The presence of several essential *cis*-elements in the promoter of the DREB2A TFs group, like coupling element 3 (CE3), heat shock element (HSE) and ABA-responsive element (ABRE), regulates their expression at the transcriptional level (Shen *et al.*, 1996; Kim *et al.*, 2011). All DREB TFs are ABA-independent, while only CBF4 is ABA-responsive, recognizes CRT/DRE elements and works in the ABA-dependent pathway (Agarwal *et al.*,

2006). Alternate splicing regulates DREB2-type TFs at the post-transcriptional level, while phosphorylation and ubiquitination regulate at the post-translational level (Agarwal *et al.*, 2017). Several protein–protein interaction studies have shown that DREB TFs interact with TFs, such as DREB2C with ABF2, ABF3 and ABF4, DERB2A with ABF2 and ABF4, HaDREB2 with HaHSP17.6G1 and with other proteins like Arabidopsis DREB2A, DREB2B and DREB2C with RCD1 (RADICAL-INDUCED CELL DEATH1), and DREB2A with SRO1 (SIMILAR TO RCD) and SRO5 (Lee *et al.*, 2010; Almoguera *et al.*, 2009; Sato *et al.*, 2014; Jaspers *et al.*, 2009).

The DREB genes from glycophytes such as *Arabidopsis* and rice have been well characterized. However, there are few reports on the cloning and characterization of DREB genes from halophytes, which may enhance our knowledge of the functions of DREB. *AhDREB1* from *Atriplex hortensis* halophyte increased salt and drought tolerance in transgenic tobacco and salt tolerance in transgenic *Populus tomentosa* plants (Shen *et al.*, 2003; Du *et al.*, 2012). *BdDREB2* from *Buchloe dactyloides* increased the drought tolerance of transgenic tobacco, while *LbDREB* from *Limonium bicolor* increased the salt tolerance of transgenic *Populus ussuriensis* plants (Zhang *et al.*, 2014; Zhao *et al.*, 2018). Transgenic *Arabidopsis* overexpressing *LcDREB2* from *Leymus chinensis* showed enhanced salt tolerance and upregulation of *LcSAMDC* (S-adenosyl-methionine decarboxylase). Moreover, *LcDREB2* also showed binding to the promoter of *LcSAMDC*, suggesting that *LcSAMDC2* is the downstream gene of *LcDREB2* (Peng *et al.*, 2013), while overexpression of *LcDREB3a* increased the salt and drought tolerance of *Arabidopsis* transgenics in an ABA-dependent manner, as transcript expression of *LcDREB3a* increased in response to ABA, suggesting that *LcDREB3a* is involved in both ABA-dependent and ABA-independent stress signal transduction pathways of *L. chinensis* (Xianjun *et al.*, 2011). Overexpression of *SbDREB2A* from the succulent halophyte *Salicornia brachiata* increased the salt and drought tolerance of tobacco transgenics by upregulating stress-responsive TFs like *HSF2* and *ZFP*, dehydrins like *ERD10B*, *ERD10D* and *LEA5*, heat shock genes like *Hsp18*, *Hsp26* and *Hsp70*, as well as signalling components like *PLC3* and *Ca^{2+}/calmodulin* (Gupta *et al.*, 2014). Overexpression of *SsDREB* from the succulent halophyte *Suaeda salsa* enhanced the salt and drought stress tolerance of transgenic tobacco (Zhang *et al.*, 2015). *ThCRF1* from the woody halophyte *Tamarix hispida* increased the salt tolerance of *T. hispida* and *Arabidopsis* transgenic plants (Qin *et al.*, 2017), while tobacco transgenics overexpressing *ThDREB* also showed enhanced salt and drought stress tolerance (Yang *et al.*, 2017). *TsCBF1* from *Thellungiella halophila* improved the tolerance as well as the grain yield of transgenic maize during drought stress (Zhang *et al.*, 2010).

9.5 MYBs

The MYB TF is one of the largest TF families in plants and is also present in all eukaryotes, including animals, fungi and slime moulds (Dubos *et al.*, 2010). It is characterized by the presence of a highly conserved MYB domain, which is involved in DNA binding and protein–protein interaction, and a variable C-terminal region involved in regulatory activity of the protein. The MYB domain comprises up to four repeats (R) of a 52-amino-acid sequence, each forming three α-helices. A helix–turn–helix (HTH) structure is formed by the second and third helices of each repeat, with three regularly spaced tryptophan (or hydrophobic) residues, forming a hydrophobic core in the three-dimensional HTH structure (Ogata *et al.*, 1996).

On the basis of the number of repeats, MYB proteins are classified into four groups: 4R-MYB, R1R2R3-type MYB (3R-MYB), R2R3-type MYB (2R-MYB) and 1R MYB. 4R-MYB proteins consist of either four R1 or R2 repeats and are the smallest group of the MYB family in plants, while the second class, 3R-MYB, contain three repeat R1R2R3 and are also called R1R2R3-type MYB. The 2R-MYB contains R2R3 repeats and is also known as R2R3-type MYB. The structure of the R2R3-MYB protein consists of a N-terminal DNA-binding domain and a C-terminal conserved amino acid sequence (Stracke *et al.*, 2001). On the basis of the C-terminal domain, R2R3-MYB is characterized into 22 subgroups (Kranz *et al.*, 1998). 1R-MYB is the second largest group of the

MYB family, containing an intact or partial repeat, and is also known as MYB-related protein. It is further divided into subgroups: CCA1-like, I-box-binding-like, CPC-like, TBP- and TRF-like, and other MYB-related proteins (Yanhui *et al.*, 2006). A large number of studies have demonstrated the important regulatory role of MYB TFs in various processes, such as regulation of cell fate and identity, seed and floral development, primary and secondary metabolism, as well as defence and stress (Dubos *et al.*, 2010; Li *et al.*, 2015).

The activity of MYB TFs is regulated at the post-transcriptional level by alternate splicing and microRNAs. MYB TFs are shown to be targeted by miR159 and are involved in anther/pollen development, germination and drought stress, while the targets of TAS4-siR81 are involved in the regulation of anthocyanin biosynthesis (Dubos *et al.*, 2010). QsMYB1 from *Quercus suber* was regulated by alternate splicing, and the two splice variants were differently regulated by abiotic stress conditions (Almeida *et al.*, 2013). Moreover, post-translational modifications such as phosphorylation regulate MYB activity at the post-translational level. Phosphorylation by either cyclin-dependent kinase complex or mitogen-activated protein kinase (MAPK) has been shown to regulate the activity of several MYB TFs, like NtMYB2, PtMYB4 and AtMYB46. In addition to this, ubiquitination also targets MYB TFs for proteasomal degradation, for example, AtMYB18 is ubiquitinated by COP1 (Seo *et al.*, 2003). Moreover, MYB TFs are also known to form homo- and hetero-dimers and have been shown to interact with other TFs/genes.

AmMYB1 from the salt-tolerant mangrove tree *Avicennia marina* improved the salt tolerance of tobacco transgenics in an ABA-dependent manner, as transcript expression of *AmMYB1* increased in response to ABA (Ganesan *et al.*, 2012). The deduced amino acid sequence of the AmMYB1 protein revealed the presence of a single MYB repeat. AmMYB1 protein contained a putative nuclear localization signal and highly conserved tryptophan residues separated by 18–19 amino acids. It also exhibits transactivation activity, and its activation domain is located in the C-terminal region rich in both proline and acidic amino acids (Asp and Glu) (Ganesan *et al.*, 2012). Overexpression of *ChiMYB* from *Chrysanthemum indicum* increased the salt tolerance of *Arabidopsis* transgenics by improving germination, growth and photosynthesis. Moreover, several stress-responsive genes, like *COR15*, *RAB18* and *RD29A*, as well as ABA signalling genes like *ABI1* and *ABA1* were upregulated in transgenics, suggesting regulation of salt stress in the ABA-dependent pathway (He *et al.*, 2016). *Arabidopsis* transgenic plants overexpressing *CmMYB2* from *Chrysanthemum morifolium* showed enhanced tolerance to salt and drought stress, increased sensitivity to ABA and delayed flowering, suggesting that *CmMYB2* regulates abiotic stress in the ABA-dependent signal transduction pathway (Shan *et al.*, 2012). Overexpression of *LbTRY*, a MYB TF from *Limonium bicolor*, increased the sensitivity to salt stress of *Arabidopsis* transgenics by inhibiting germination and root growth, decreasing osmolyte accumulation and increasing ion accumulation (Leng *et al.*, 2020). *LcMYB1* from sheepgrass, *Leymus chinensis*, was induced in response to ABA, drought and salt and slightly by cold, and increased the salt tolerance of *Arabidopsis* transgenics (Cheng *et al.*, 2013). Overexpression of *SbMYB15* from the extreme halophyte *Salicornia brachiata* increased the salt and drought tolerance of tobacco transgenics in an ABA-independent manner, as the transcript of *SbMYB15* was unresponsive to exogenous ABA (Shukla *et al.*, 2015). The *in silico* analysis revealed that the SbMYB44 protein contains the conserved R2R3 imperfect repeats, two SANT domains and post-translational modification sites (Shukla *et al.*, 2015). The 14 ThMYB protein sequences have a conserved MYB domain, and the conserved domains have high similarity. The N-terminus of ThMYB 2, 3, 4, 8, 9 and 13 protein sequences contain two conserved MYB domains. The conserved domain of ThMYB12 exists at the end of the peptide chain. The conserved domains of ThMYB 5, 6, 7, 10 and 11 were located in the middle of the peptide chain. ThMYB1 and ThMYB14 contain a conserved domain of MYB-type in the N-terminus. ThMYB8 and ThMYB13 from *Tamarix hispida* increased the salt tolerance of transgenic plants in an ABA-dependent manner (Liu *et al.*, 2021; Zhang *et al.*, 2018).

9.6 NACs

NAC TFs are unique and one of the largest plant-specific TF families, named after the first three proteins characterized (NAM [no apical meristem] from petunia, ATAF1/2 [Arabidopsis transcription activation factor 1/2] and CUC2 [cup-shaped cotyledon]), to contain a common conserved N-terminal domain. A typical NAC protein contains a highly conserved 150–160-amino-acid N-terminal NAC domain, further divided into five sub-domains A–E, with DNA-binding ability, nuclear localization, and homo- or hetero-dimerization characteristics, and the C-terminal region, a variable transcriptional regulatory region, acting as transcription activator or repressor (Ooka *et al.*, 2003). The NAC TFs can be divided into two classes on the basis of structural features: typical NAC TFs and atypical NAC TFs, wherein the typical NAC TFs contain a conserved N-terminal NAC domain and a divergent C-terminal region, while the atypical NAC TFs contain additional conserved domains/motifs at the C-terminal region or a complete absence of the C-terminal region (Puranik *et al.*, 2012). The membrane-associated NAC TFs, designated as NTLs (NAC with transmembrane motif1-like), contain additional transmembrane motifs in the C-terminal region and are identified as atypical NAC TFs (Seo *et al.*, 2008). Tran *et al.* (2004) recognized CATGT as NACRS with CACG as core-binding sequence and isolated three NACs, i.e., ANAC015, ANAC055 and ANAC072, using erd1 promoter NACRS as bait in Y1H (yeast 1 hybrid). Since then, many independent studies have shown binding of different NAC TFs to NACRS, indicating that the binding sequence for stress-inducible NAC TFs might be conserved among plant species. In addition to NACRS, NAC TFs also showed binding to other *cis*-elements like CBNACBS (calmodulin-binding NAC binding sequence), iron deficiency responsive IDE2 sequence and SNBE (secondary wall NAC binding element), indicating recognition and binding to other *cis*-elements for the regulation of multiple downstream genes (Shao *et al.*, 2015). NAC TFs regulate gene expression during stress via both ABA-dependent and ABA-independent pathways. Moreover, the NAC TFs themselves are tightly regulated at the transcriptional, post-transcriptional and post-translational levels. Numerous studies have shown that NAC TFs play key roles in diverse biological processes, like cell division, formation of secondary walls, shoot apical meristem, flower development, and leaf senescence as well as in biotic and abiotic stress responses (Shao *et al.*, 2015). Thus, the NAC TF family is a potential candidate for genetic engineering.

AlNAC1 TF from a recretohalophyte, *Aeluropus lagopoides*, contained highly conserved NAC domains with well-identified A–E sub-domains and a highly variable transcription regulatory domain at the C-terminal region. The AlNAC1 protein contained a bi-partite nuclear localization sequence (NLS) in the C and D sub-domains. The crude protein containing recombinant AlNAC1 protein showed binding to the NACRS *cis*-element of the *erd1* promoter in an EMSA (electrophoretic mobility shift assay) study. Overexpression of *AlNAC1* increased the drought tolerance of transgenic tobacco plants in an ABA-dependent manner. The transgenics showed enhanced growth, membrane stability, increased accumulation of osmolytes and antioxidative enzyme activity, and decreased accumulation of ROS. Moreover, the endogenous ABA level increased in transgenics during drought stress, which further increased stomatal closure, thus reducing water loss and transpiration rate and maintaining the electron transport rate, which further reduced ROS accumulation, thus protecting photosynthetic pigments and machinery (Joshi *et al.*, 2021). *AmNAC1* from *Avicennia marina* also contained a highly conserved NAC domain harbouring bipartite NLS in the C and D sub-domains and a variable C-terminal region. The transcript expression of *AmNAC1* increased in response to salinity, drought, cold and exogenous ABA treatments (Ganesan *et al.*, 2008; Kumar *et al.*, 2018). Tobacco transgenics overexpressing *AmNAC1* showed better germination, increased root length, and better survival rate during salt, drought and cold stresses (Murugesan *et al.*, 2020). *Brassica oleracea BoNAC019*, a homolog of *AtNAC019*, was shown to be localized in the nucleus, functions as a transcription activator and was upregulated by salinity, drought, ABA and oxidative stresses. *BoNAC019* negatively regulates drought stress by upregulating ABA catabolism-related genes and downregulating ABA signalling genes, anthocyanin biosynthetic genes, antioxidant enzymatic genes and several stress-responsive genes. Thus, the *Arabidopsis* transgenics showed reduced endogenous ABA level and decreased sensitivity to ABA as well as decreased accumulation of anthocyanin, which further decreased drought tolerance (Wang *et al.*, 2018). *DgNAC1*, from chrysanthemum, was localized in the nucleus and upregulated by salinity, drought, cold and ABA treatments. The tobacco plants overexpressing *DgNAC1* showed higher germination and survival rates in the presence of salinity stress (Liu *et al.*, 2011). Overexpression of *DgNAC1* in chrysanthemum also enhanced the salt tolerance of transgenic plants by increasing survival rate, membrane stability, antioxidative enzyme activities and proline accumulation. Several stress-responsive genes, like *DgCu/ZnSOD*, *DgCAT*, *DgP5CS* and *DgEF1a*, were also upregulated in transgenic plants (Wang *et al.*, 2017a). *SlNAC8* from *Suaeda liaotungensis* halophyte contained a highly conserved NAC domain in the N-terminal region and an activation domain in the C-terminal region. *SlNAC8* was localized in the nucleus and upregulated by salt and drought stress. Overexpression of *SlNAC8* in *Arabidopsis* increased drought and salt stress tolerance by increasing the germination rate, growth and survival rate of transgenic plants. Transgenics also showed increased antioxidative enzyme activities, proline accumulation and membrane stability as well as increased expression of several stress-responsive genes, like *COR47*, *GSTF6*, *NYC1*, *RD20*, *RD29A* and *RD29B* (Wu *et al.*, 2018). *ThNAC13*, from *Tamarix hispida*, is a nuclear protein containing an N-terminal conserved NAC domain and a C-terminal transactivation domain. ThNAC13 showed binding to both NACRS and CBNAC (calmodulin-binding NAC) *cis*-elements. Overexpression of *ThNAC13* increased the salt and drought tolerance of transgenic *Tamarix* and *Arabidopsis* plants in an ABA-dependent manner, as transgenics also showed enhanced germination and growth in the presence of ABA (Wang *et al.*, 2017b). Overexpression of *TsNAC1* from the extreme halophyte *Thellungiella halophila* increased tolerance to salinity, drought and low temperature as well as oxidative stress by upregulating stress-responsive genes and TFs like MYB HYPOCOTYL ELONGATION-RELATED and HOMEOBOX12 as well as ion transporters like *T. halophila* H^+-PPASE1. Interestingly, *TsNAC1* reduced growth of transgenics by downregulating growth and cell expansion-related genes like LIGHT-DEPENDENT SHORT HYPOCOTYLS1 and UDP-XYLOSYLTRANSFERASE2 (Liu *et al.*, 2018).

9.7 OTHER TRANSCRIPTION FACTORS

GRAS TFs are a large family of plant-specific TFs and were named after the first three identified proteins, GAI (GIBBERELLIN-ACID INSENSITIVE), RGA (REPRESSOR of GA1) and SCR (SCARECROW; Pysh *et al.*, 1999). The whole-genome analysis of *Arabidopsis* has identified at least 33 GRAS TFs. They can be divided into several subfamilies, like DELLA, LISCL, PAT, SCR, SCL3 and SHR, on the basis of sequence alignment and phylogenetic analysis (Tian *et al.*, 2004). The GRAS proteins contain two leucine-rich areas flanking a VHIID motif, named from the most prominent amino acid, which is present in all members; however, only histidine and aspartic acid are completely conserved. The expression of GRAS TF is regulated by microRNA. For example, co-transformation of *Nicotiana benthamiana* leaves with miR171/miRNA39 and AtSCL6 or AtSCL22 demonstrated miRNA-dependent cleavage of the mRNAs (Llave *et al.*, 2002). The GRAS TFs are involved in diverse functions, like root development, axillary meristem development, shoot meristem maintenance, phytochrome signalling and gibberellin signal transduction (Bolle, 2004). Overexpression of *HcSCL13*, a GRAS transcription factor, from *Halostachys caspica* increased the growth and salt stress resistance of *Arabidopsis* transgenic plants. The transgenics showed enhanced fresh weight and root elongation as well as chlorophyll content in the presence of salt stress. Moreover, the transcriptomic analysis revealed that genes related to plant growth, development, phytohormone signalling and abiotic/light stimulus/organic substance/phytohormones were altered in transgenics under normal and salt stress conditions (Zhang *et al.*, 2020).

WRKY TF is one of the largest plant-specific TF families, which contains extremely conserved WRKYGQK heptapeptide at the N-terminus and is hence named WRKY, as well as a novel metal chelating zinc finger at the C-terminus. On the basis of the number of WRKY motifs present and the constitution of their zinc-finger motif, WRKY TFs are classified into three groups: Group I, Group II and Group III. Group I contains two WRKY domains and a C2H2-type zinc-finger motif (C-X4–5-C-X22–23-H-XH); group II contains a single WRKY domain and a C2H2-type zinc-finger motif similar to group I,; while group III possesses a single WRKY domain like group II and a C2HC zinc-finger motif (C-X7-C-X23-H-X-C). Furthermore, on the basis of extra amino acid motifs present outside the WRKY domain, Group II is further divided into a–e subgroups. Moreover, several additional motifs, like Gln- and Pro-rich stretches, Ser-/Thr-rich stretches, nuclear localization signal (NLS), kinase domains, leucine zippers and pathogenesis-related TIR-NBS-LRR domains have also been reported (Jiang and Deyholos, 2006). WRKY proteins bind specifically to the W-box *cis*-element via the WRKY domain; however, the binding preference depends on the flanking sequence present outside the W-box motif (Agarwal *et al.*, 2011). Since the W-box motif is present in the promoter region of WRKY itself, WRKY TFs show autoregulation by binding to self or cross-regulation by binding of other WRKY TFs. Moreover, WRKY interacts with other WRKY through homo- or hetero-dimerization as well as with other TFs/proteins. WRKY has been shown to interact with other proteins, like calmodulin, 14–3-3, VQ proteins, E3 ubiquitin ligases, chromatin remodelling proteins such as Arabidopsis histone deacetylase 19, and the chloroplast/plastid-localized ABA receptor, ABAR (Tripathi *et al.*, 2014). Overexpression of *RtWRKY1* in Arabidopsis increased the salt tolerance of transgenics by enhancing growth, maintaining ion and osmotic homeostasis, and improving the antioxidant system as well as regulating several downstream genes involved in ion transport, like *AtAPX1* and *AtSOS1*, antioxidant-related genes, like *AtCAT1* and *AtSOD1*, and genes involved in proline metabolism, like *AtP5CS1*, *AtP5CS2*, *AtPRODH1* and *AtPRODH2* (Du *et al.*, 2017).

The bZIPs (basic leucine zippers) are typical ABRE-binding TFs extensively distributed in eukaryotes, which bind to ABRE elements present in the promoters of downstream genes and regulate gene expression (Yamaguchi-Shinozaki and Shinozaki, 2005). They contain an extremely conserved basic DNA-binding domain, which comprises 16 basic amino acid residues, and a leucine zipper motif involved in dimerization. They are the largest known ABA-inducible DNA-binding protein family in plants and are involved in several plant processes, like light response cell elongation, primary root growth, seed maturation, tissue and organ differentiation, flower development and plant senescence as well as regulation of abiotic stress (Khan *et al.*, 2018). After perception of an ABA signal by specialized receptors, several groups of Ser/Thr kinases transduce the signal and phosphorylate bZIP TFs, which are then activated. The activated bZIP TFs then bind to ABA-responsive elements (ABREs) or GC-rich coupling elements (CEs) and regulate expression of their target downstream genes. The bZIP protein comprises a leucine zipper domain involved in dimerization and a positively charged DNA-binding domain (Roychoudhury and Paul, 2012). Post-translational modifications like phosphorylation, sumoylation and ubiquitination regulate the activity of bZIP TFs. Moreover, bZIP also interacts with other TFs and proteins like NAC TFs, NFY, CDPK, Ash2 and WDR5a via protein–protein interaction (Banerjee and Roychoudhury, 2017). Overexpression of *ThbZIP1* from *Tamarix hispida* increased salt tolerance in tobacco transgenics and the salt and drought tolerance of Arabidopsis transgenics (Wang *et al.*, 2010; Ji *et al.*, 2013). *Arabidopsis* transgenics showed altered expression of genes involved in processes like response to stimulus, catalytic activity, binding functions and metabolic processes when exposed to exogenous ABA. Also, GO analysis showed that ABA increased several processes, like structural molecule activity, membrane-enclosed lumen, organelle parts, and reproductive processes, while salt stress inhibited them. On the contrary, salt stress enhanced immune system and multi-organism processes, while ABA inhibited them. Moreover, *ThABF1* was shown to interact with and regulate the expression of *ThbZIP1* (Ji *et al.*, 2015).

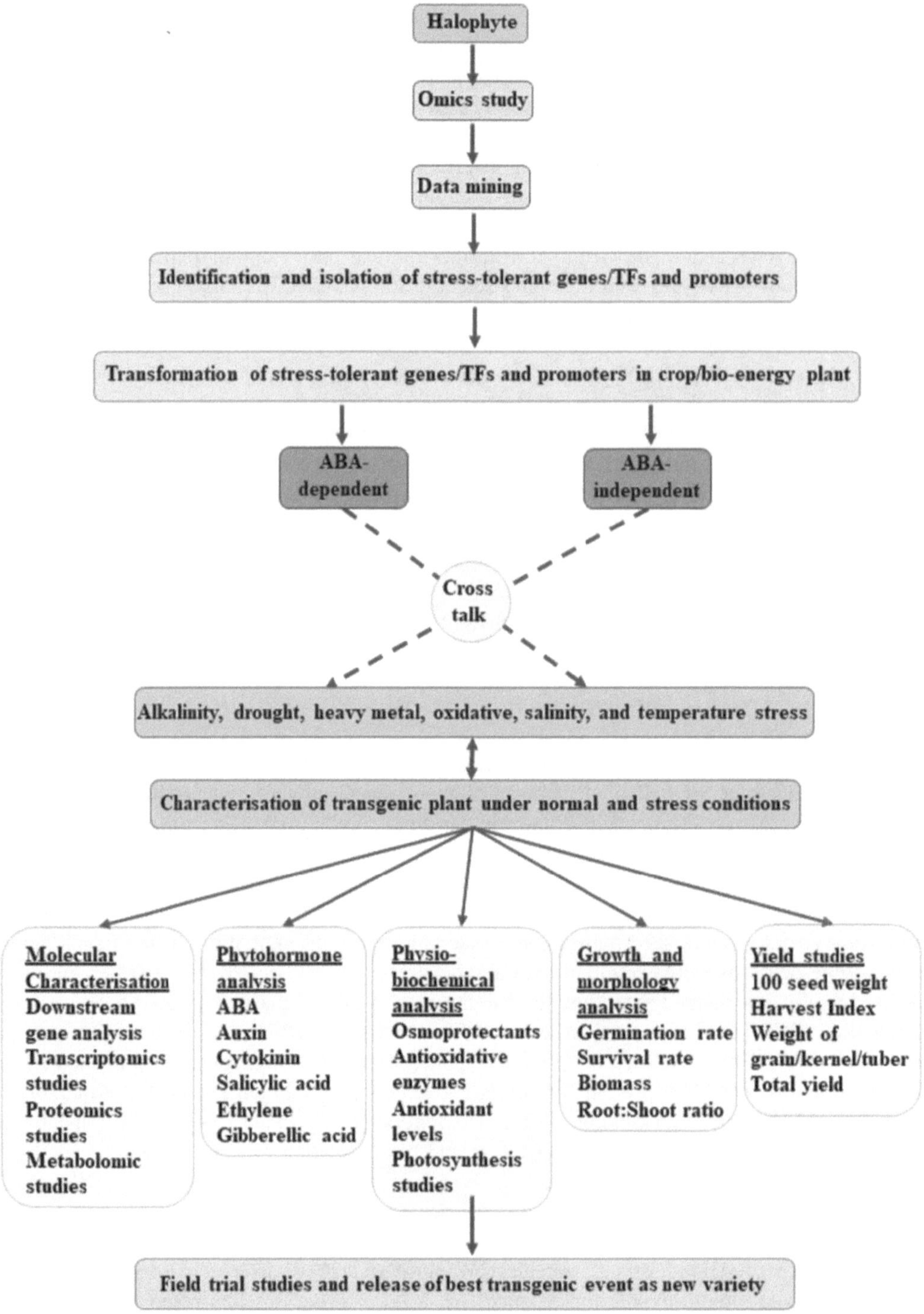

FIGURE 9.4 Schematic representation of approach to develop stress-tolerant transgenic crops using genes/TFs from halophyte plants.

9.8 CONCLUSION AND FUTURE PERSPECTIVE

Climate change and the associated yield losses are the greatest challenge to humankind for food security. The development of new stress-tolerant varieties that can support sustainable agriculture should be the first priority of agricultural research. For this, a better and deeper understanding of the physiological and molecular mechanisms behind the response and adaptation of plants to environmental stress is necessary. Halophytes are the model plants to understand the genetic networks and are also the ideal source of gene/TFs and promoters to generate genetically engineered stress-tolerant crop varieties through a transgenic approach (Figure 9.4). However, very few TFs from halophytes have been studied. To exploit the natural genetic variability of halophytes for stress tolerance, more studies of TFs from halophytes are needed, which can further be transformed to crop plants, and large-scale field evaluations should be carried out to develop and utilize stress-tolerant crop varieties for human consumption.

ACKNOWLEDGEMENTS

CSIR-CSMCRI Communication No 42/2021. P.A. acknowledges the financial support from DSTWOS-A scheme. All the authors acknowledge CSIR, New Delhi for financial support.

REFERENCES

Agarwal P, Reddy MP, Chikara J (2011) WRKY: its structure, evolutionary relationship, DNA-binding selectivity, role in stress tolerance and development of plants. *Molecular Biology Reports* 38(6):3883–3896

Agarwal PK, Agarwal P, Reddy MK, Sopory SK (2006) Role of DREB transcription factors in abiotic and biotic stress tolerance in plants. *Plant Cell Reports* 25(12):1263–1274

Agarwal PK, Gupta K, Lopato S, Agarwal P (2017) Drought responsive element binding transcription factors and their applications for the engineering of stress tolerance. *Journal of Experimental Botany* 68:2135–2148

Agarwal PK, Shukla PS, Gupta K, Jha B (2013) Bioengineering for salinity tolerance in plants: state of the art. *Molecular Biotechnology* 54(1):102–123

Akhtar M, Jaiswal A, Taj G, Jaiswal JP, Qureshi MI, Singh NK (2012) DREB1/CBF transcription factors: their structure, function and role in abiotic stress tolerance in plants. *Journal of Genetics* 91(3):385–395

Akinshina N, Azizov A, Karasyova T, Klose E (2016) On the issue of halophytes as energy plants in saline environment. *Biomass and Bioenergy* 91:306–311

Aliakbari M, Razi H, Alemzadeh A, Tavakol E (2020) RNA-seq transcriptome profiling of the halophyte *Salicornia persica* in response to salinity. *Journal of Plant Growth Regulation* 40: 707–721

Almeida T, Pinto G, Correia B, Santos C, Gonçalves S (2013) QsMYB1 expression is modulated in response to heat and drought stresses and during plant recovery in *Quercus suber*. *Plant Physiology and Biochemistry* 73:274–281

Almoguera C, Prieto-Dapena P, Díaz-Martín J, Espinosa JM, Carranco R, Jordano J (2009) The HaDREB2 transcription factor enhances basal thermotolerance and longevity of seeds through functional interaction with HaHSFA9. *BMC Plant Biology* 9(1):1–2

Banerjee A, Roychoudhury A (2017) Abscisic-acid-dependent basic leucine zipper (bZIP) transcription factors in plant abiotic stress. *Protoplasma* 254(1):3–16

Bansal J, Gupta K, Rajkumar MS, Garg R, Jain M (2020) Draft genome and transcriptome analyses of halophyte rice Oryza coarctata provide resources for salinity and submergence stress response factors. *Physiologia Plantarum*. https://doi.org/10.1111/ppl.13284

Barlow M, Zaitchik B, Paz S, Black E, Evans J, Hoell A (2016) A review of drought in the Middle East and southwest Asia. *Journal of Climate* 23:8547–8574

Bedre R, Mangu VR, Srivastava S, Sanchez LE, Baisakh NJBg (2016) Transcriptome analysis of smooth cordgrass (*Spartina alterniflora* Loisel), a monocot halophyte, reveals candidate genes involved in its adaptation to salinity. *BMC Genomics* 17(1):657

Bolle C (2004) The role of GRAS proteins in plant signal transduction and development. *Planta* 218(5):683–692
Breckle SW (1995) How do halophytes overcome salinity. *Biology of Salt Tolerant Plants* 23:199–203
Bruhn L, Hwang-Shum JJ, Sprague GF Jr (1992) The N-terminal 96 residues of MCM1, a regulator of cell type-specific genes in *Saccharomyces cerevisiae*, are sufficient for DNA binding, transcription activation, and interaction with alpha 1. *Molecular and Cellular Biology* 12:3563–3572
Cheng L, Li X, Huang X, Ma T, Liang Y, Ma X, Peng X, Jia J, Chen S, Chen Y, Deng B (2013) Overexpression of sheepgrass R1-MYB transcription factor LcMYB1 confers salt tolerance in transgenic Arabidopsis. *Plant Physiology and Biochemistry* 70:252–260
Cui Y-N, Wang F-Z, Yang C-H, Yuan J-Z, Guo H, Zhang J-L, Wang S-M, Ma QJG (2019) Transcriptomic profiling identifies candidate genes involved in the salt tolerance of the xerophyte *Pugionium cornutum. Genes* 10(12):1039
Cushman JC (2001) Osmoregulation in plants: implications for agriculture. *American Zoologist* 41(4):758–769
Dai S, Petruccelli S, Ordiz MI, Zhang Z, Chen S, Beachy RN (2003) Functional analysis of RF2a, a rice transcription factor. *Journal of Biological Chemistry* 278:36396–36402
Diray-Arce J, Knowles A, Suvorov A, O'Brien J, Hansen C, Bybee SM, Gul B, Khan MA, Nielsen BL (2019) Identification and evolutionary characterization of salt-responsive transcription factors in the succulent halophyte *Suaeda fruticosa. PLOS ONE* 14(9):e0222940.
Du C, Zhao P, Zhang H, Li N, Zheng L, Wang Y (2017) The *Reaumuria trigyna* transcription factor RtWRKY1 confers tolerance to salt stress in transgenic Arabidopsis. *Journal of Plant Physiology* 215:48–58
Du N, Liu X, Li Y, Chen S, Zhang J, Ha D, Deng W, Sun C, Zhang Y, Pijut PM (2012) Genetic transformation of *Populus tomentosa* to improve salt tolerance. *Plant Cell, Tissue and Organ Culture (PCTOC)* 108(2):181–189
Dubos C, Stracke R, Grotewold E, Weisshaar B, Martin C, Lepiniec L (2010) MYB transcription factors in Arabidopsis. *Trends in Plant Science* 15(10):573–581
Dubouzet JG, Sakuma Y, Ito Y, Kasuga M, Dubouzet EG, Miura S, Seki M, Shinozaki K, Yamaguchi-Shinozaki K (2003) OsDREB genes in rice, *Oryza sativa* L., encode transcription activators that function in drought-, high-salt- and cold-responsive gene expression. *Plant Journal* 33:751–763
Edmeades GO (2013) *Progress in Achieving and Delivering Drought Tolerance in Maize – an Update* (p. 130). ISAAA, Ithaca, NY
FAO, IFAD, WHO WFP, UNICEF, *et al.* (2019) *The State of Food Security and Nutrition in the World 2019: Safeguarding against Economic Slowdowns and Downturns*
FAO (2016) FAO Soils Portal. Available at: www.fao.org/soils-portal/soil-management/management-of-some-problem-soils/salt-affected-soils/more-information-on-salt-affected-soils/en/
Fiedor J, Burda K (2014) Potential role of carotenoids as antioxidants in human health and disease. *Nutrients* 6(2):466–488
Flowers TJ, Galal HK, Bromham L (2010) Evolution of halophytes: multiple origins of salt tolerance. *Functional Plant Biology* 37:604–612
Fujita Y, Fujita M, Shinozaki K, Yamaguchi-Shinozaki K (2011) ABA-mediated transcriptional regulation in response to osmotic stress in plants. *Journal of Plant Research* 124(4):509–525
Ganesan G, Sankararamasubramanian HM, Harikrishnan M, Ashwin G, Parida A (2012) A MYB transcription factor from the grey mangrove is induced by stress and confers NaCl tolerance in tobacco. *Journal of Experimental Botany* 63(12):4549–4561
Ganesan G, Sankararamasubramanian HM, Narayanan JM, Sivaprakash KR, Parida A (2008) Transcript level characterization of a cDNA encoding stress regulated NAC transcription factor in the mangrove plant *Avicennia marina. Plant Physiology and Biochemistry* 46(10):928–934
Garg R, Verma M, Agrawal S, Shankar R, Majee M, Jain MJDr (2014) Deep transcriptome sequencing of wild halophyte rice, *Porteresia coarctata*, provides novel insights into the salinity and submergence tolerance factors. *DNA Research* 21(1):69–84
Golldack D, Lüking I, Yang O (2011) Plant tolerance to drought and salinity: stress regulating transcription factors and their functional significance in the cellular transcriptional network. *Plant Cell Reports* 30(8):1383–1391
Gong W, Shen YP, Ma LG, Pan Y, Du YL, Wang DH, Yang JY, Hu LD, Liu XF, Dong CX, Ma L (2004) Genome-wide ORFeome cloning and analysis of Arabidopsis transcription factor genes. *Plant Physiology* 135(2):773–782

Gu YQ, Yang C, Thara VK, Zhou J, Martin GB (2000) Pti4 is induced by ethylene and salicylic acid, and its product is phosphorylated by the Pto kinase. *The Plant Cell* 12(5):771–785

Gupta K, Jha B, Agarwal PK (2014) A dehydration-responsive element binding (DREB) transcription factor from the succulent halophyte *Salicornia brachiata* enhances abiotic stress tolerance in transgenic tobacco. *Marine Biotechnology* 16(6):657–673

Han Z-J, Sun Y, Zhang M, Zhai J-T (2020) Transcriptomic profile analysis of the halophyte *Suaeda rigida* response and tolerance under NaCl stress. *Scientific Reports* 10(1):15148.

Hastilestari BR, Mudersbach M, Tomala F, Vogt H, Biskupek-Korell B, Van Damme P, Guretzki S, Papenbrock J (2013) *Euphorbia tirucalli* L. – Comprehensive characterization of a drought tolerant plant with a potential as biofuel source. *PLOS ONE* 8(5):e63501

He M, Wang H, Liu YZ, Gao WJ, Gao YH, Wang F, Zhou YW (2016) Cloning and characterization of ChiMYB in *Chrysanthemum indicum* with an emphasis on salinity stress tolerance. *Genetics and Molecular Research* 15:1–7

Hoang XL, Thu NB, Thao NP, Tran LS (2014) Transcription factors in abiotic stress responses: Their potentials in crop improvement. In *Improvement of Crops in the Era of Climatic Changes* (pp. 337–366). Springer, New York

Howard RJ, Mendelssohn IA (1999) Salinity as a constraint on growth of oligohaline marsh macrophytes. I. Species variation in stress tolerance. *American Journal of Botany* 86(6):785–794

Huang GT, Ma SL, Bai LP, Zhang L, Ma H, Jia P, Liu J, Zhong M, Guo ZF (2012) Signal transduction during cold, salt and drought stresses in plants. *Molecular Biology Reports* 39:969–987

Huang H, Liu Y, Pu Y, Zhang M, Dai S (2019) Transcriptome analysis of *Chrysanthemum lavandulifolium* response to salt stress and overexpression a K+ transport ClAKT gene-enhanced salt tolerance in transgenic Arabidopsis. *Journal of the American Society for Horticultural Science* 144(4):219.

Jaspers P, Blomster T, Brosché M, Salojärvi J, Ahlfors R, Vainonen JP, Reddy RA, Immink R, Angenent G, Turck F, Overmyer K (2009) Unequally redundant RCD1 and SRO1 mediate stress and developmental responses and interact with transcription factors. *The Plant Journal* 60(2):268–279

Ji X, Liu G, Liu Y, Nie X, Zheng L, Wang Y (2015) The regulatory network of ThbZIP1 in response to abscisic acid treatment. *Frontiers in Plant Science* 6:25

Ji X, Liu G, Liu Y, Zheng L, Nie X, Wang Y (2013) The bZIP protein from *Tamarix hispida*, ThbZIP1, is ACGT elements binding factor that enhances abiotic stress signalling in transgenic Arabidopsis. *BMC Plant Biology* 13(1):1–3

Jiang Y, Deyholos MK (2006) Comprehensive transcriptional profiling of NaCl-stressed Arabidopsis roots reveals novel classes of responsive genes. *BMC Plant Biology* 6:25

Joshi PS, Agarwal P, Agarwal PK (2021) Overexpression of AlNAC1 from recretohalophyte *Aeluropus lagopoides* alleviates drought stress in transgenic tobacco. *Environmental and Experimental Botany* 181:104277

Khan SA, Li MZ, Wang SM, Yin HJ (2018) Revisiting the role of plant transcription factors in the battle against abiotic stress. *International Journal of Molecular Sciences* 19(6):1634

Khedr AHA, Serag MS, Nemat-Alla MM, Abo-Elnaga AZ, Nada RM, Quick WP, Abogadallah GM (2011) A DREB gene from the xero-halophyte *Atriplex halimus* is induced by osmotic but not ionic stress and shows distinct differences from glycophytic homologues. *Plant Cell, Tissue and Organ Culture* 106:191–206

Kim JS, Mizoi J, Yoshida T, Fujita Y, Nakajima J, Ohori T, Todaka D, Nakashima K, Hirayama T, Shinozaki K, Yamaguchi-Shinozaki K (2011) An ABRE promoter sequence is involved in osmotic stress-responsive expression of the DREB2A gene, which encodes a transcription factor regulating drought-inducible genes in Arabidopsis. *Plant and Cell Physiology* 52(12):2136–2146

Koyro HW, Hussain T, Huchzermeyer B, Khan MA (2013) Photosynthetic and growth responses of a perennial halophytic grass *Panicum turgidum* to increasing NaCl concentrations. *Environmental and Experimental Botany* 91:22–29

Kranz HD, Denekamp M, Greco R, Jin H, Leyva A, Meissner RC, Petroni K, Urzainqui A, Bevan M, Martin C, Smeekens S (1998) Towards functional characterisation of the members of the R2R3-MYB gene family from *Arabidopsis thaliana*. *The Plant Journal* 16(2):263–276

Kumar MA, Ganesan G, Rama SS, Ajay P (2018) NAC transcription factor differentially regulated by abiotic stresses and salicylic acid in the mangrove plant *Avicennia marina* (Forsk.) Vierh. *Research Journal of Biotechnology* 13:65–71

Lee SJ, Kang JY, Park HJ, Kim MD, Bae MS, Choi HI, Kim SY (2010) DREB2C interacts with ABF2, a bZIP protein regulating abscisic acid-responsive gene expression, and its overexpression affects abscisic acid sensitivity. *Plant Physiology* 153(2):716–727

Leng B, Wang X, Yuan F, Zhang H, Lu C, Chen M, Wang B (2020) Heterologous expression of the *Limonium bicolor* MYB transcription factor LbTRY in *Arabidopsis thaliana* increases salt sensitivity by modifying root hair development and osmotic homeostasis. *Plant Science* 302:110704

Li C, Ng CK, Fan LM (2015) MYB transcription factors, active players in abiotic stress signaling. *Environmental and Experimental Botany* 114:80–91

Liu C, Wang B, Li Z, Peng Z, Zhang J (2018) TsNAC1 is a key transcription factor in abiotic stress resistance and growth. *Plant Physiology* 176(1):742–756

Liu N, Zhong NQ, Wang GL, Li LJ, Liu XL, He YK (2007) Cloning and functional characterization of PpDBF1 gene encoding a DRE-binding transcription factor from *Physcomitrella patens*. *Planta* 226(4):827–838

Liu Q, Kasuga M, Sakuma Y, Abe H, Miura S, Yamaguchi-Shinozaki K, Shinozaki K (1998) Two transcription factors DREB1 and DREB2 with an EREBP/AP2 DNA binding domain separate two cellular signal transduction pathways in drought- and low temperature-responsive gene expression, respectively, in Arabidopsis. *The Plant Cell* 10(8):1391–1406

Liu QL, Xu KD, Zhao LJ, Pan YZ, Jiang BB, Zhang HQ, Liu GL (2011) Overexpression of a novel chrysanthemum NAC transcription factor gene enhances salt tolerance in tobacco. *Biotechnology Letters* 33(10):2073

Liu ZY, Li XP, Zhang TQ, Wang YY, Wang C, Gao CQ (2021) Overexpression of ThMYB8 mediates salt stress tolerance by directly activating stress-responsive gene expression. *Plant Science* 302:110668

Llave C, Xie Z, Kasschau KD, Carrington JC (2002) Cleavage of Scarecrow-like mRNA targets directed by a class of Arabidopsis miRNA. *Science* 297:2053–2056

Ma Y, Dias MC, Freitas H (2020) Drought and salinity stress responses and microbe-induced tolerance in plants. *Frontiers in Plant Science* 11:59191

Malik KA (2002) Prospects for saline agriculture in Pakistan: Today and tomorrow. In Ahmad R, Malik KA (eds) *Prospects for Saline Agriculture*. Tasks for Vegetation Science, vol 37. Springer, Dordrecht. https://doi.org/10.1007/978-94-017-0067-2_1

McGettigan PA (2013) Transcriptomics in the RNA-seq era. *Current Opinion in Chemical Biology* 17(1):4–11.

Munns R, Greenway H, Kirst GO (1983) Halotolerant eukaryotes. In *Physiological Plant Ecology III* (pp. 59–135). Springer, Berlin, Heidelberg

Murugesan AK, Somasundaram S, Mohan H, Parida AK, Alphonse V, Govindan G (2020) Ectopic expression of AmNAC1 from *Avicennia marina* (Forsk.) Vierh. confers multiple abiotic stress tolerance in yeast and tobacco. *Plant Cell, Tissue and Organ Culture* 142(1):51–68

Mutz K-O, Heilkenbrinker A, Lönne M, Walter J-G, Stahl F (2013) Transcriptome analysis using next-generation sequencing. *Current Opinion in Biotechnology* 24 (1):22–30

Nakashima K, Takasaki H, Mizoi J, Shinozaki K, Yamaguchi-Shinozaki K (2012) NAC transcription factors in plant abiotic stress responses. *Biochimica et Biophysica Acta (BBA)-Gene Regulatory Mechanisms* 1819(2):97–103

Nakashima K, Yamaguchi-Shinozaki K, Shinozaki K (2014) The transcriptional regulatory network in the drought response and its crosstalk in abiotic stress responses including drought, cold, and heat. *Frontiers in Plant Science* 5:170

Nayak SS, Pradhan S, Sahoo D, Parida A (2020) De novo transcriptome assembly and analysis of *Phragmites karka*, an invasive halophyte, to study the mechanism of salinity stress tolerance. *Scientific Reports* 10(1):5192

Nikalje G, Nikam T, Suprasanna P (2017) Looking at halophytic adaptation to high salinity through genomics landscape. *Current Genomics* 18(6):542–552

Ogata K, Kanei-Ishii C, Sasaki M, Hatanaka H, Nagadoi A, Enari M, Nakamura H, Nishimura Y, Ishii S, Sarai A (1996) The cavity in the hydrophobic core of Myb DNA-binding domain is reserved for DNA recognition and trans-activation. *Nature Structural Biology* 3(2):178–187

Ooka H, Satoh K, Doi K, Nagata T, Otomo Y, Murakami K, Matsubara K, Osato N, Kawai J, Carninci P, Hayashizaki Y (2003) Comprehensive analysis of NAC family genes in *Oryza sativa* and *Arabidopsis thaliana*. *DNA Research* 10(6):239–247

Peng X, Zhang L, Zhang L, Liu Z, Cheng L, Yang Y, Shen S, Chen S, Liu G (2013) The transcriptional factor LcDREB2 cooperates with LcSAMDC2 to contribute to salt tolerance in *Leymus chinensis*. *Plant Cell, Tissue and Organ Culture* 113(2):245–256

Puranik S, Sahu PP, Srivastava PS, Prasad M (2012) NAC proteins: regulation and role in stress tolerance. *Trends in Plant Science* 17(6):369–381

Pysh LD, Wysocka-Diller JW, Camilleri C, Bouchez D, Benfey PN (1999) The GRAS gene family in Arabidopsis: sequence characterization and basic expression analysis of the SCARECROW-LIKE genes. *Plant Journal* 18:111–119

Qin F, Sakuma Y, Li J, Liu Q, Li YQ, Shinozaki K, Yamaguchi-Shinozaki K (2004) Cloning and functional analysis of novel DREB1/CBF transcription factor involved in cold-responsive gene expression in *Zea mays* L. *Plant Cell Physiology* 45:1042–1052

Qin L, Wang L, Guo Y, Li Y, Ümüt H, Wang Y (2017) An ERF transcription factor from *Tamarix hispida*, ThCRF1, can adjust osmotic potential and reactive oxygen species scavenging capability to improve salt tolerance. *Plant Science* 265:154–166

Rogers S, Wells R, Rechsteiner M (1986) Amino acid sequences common to rapidly degraded proteins: the PEST hypothesis. *Science* 234(4774):364–368

Roychoudhury A, Paul A (2012) Abscisic acid-inducible genes during salinity and drought stress. *Advances in Medicine and Biology* 51:1–78

Rozema J, Flowers T (2008) Crops for a salinized world. *Science* 322:1478–1480

Saadi S, Todorovic M, Tanasijevic L, Pereira LS, Pizzigalli C, Lionello P (2015) Climate change and Mediterranean agriculture: impacts on winter wheat and tomato crop evapotranspiration, irrigation requirements and yield. *Agricultural Water Management* 147:103–115

Sabovljevic M, Sabovljevic A (2007) Contribution to the coastal bryophytes of the Northern Mediterranean: are there halophytes among bryophytes. *Phytologia Balcanica* 2:131–135

Sakuma Y, Maruyama K, Osakabe Y, Qin F, Seki M, Shinozaki K, Yamaguchi-Shinozaki K (2006) Functional analysis of an Arabidopsis transcription factor DREB2A involved in drought responsive gene expression. *Plant Cell* 18(5):1292–1309

Sato H, Mizoi J, Tanaka H, Maruyama K, Qin F, Osakabe Y, Morimoto K, Ohori T, Kusakabe K, Nagata M, Shinozaki K (2014) Arabidopsis DPB3-1, a DREB2A interactor, specifically enhances heat stress-induced gene expression by forming a heat stress-specific transcriptional complex with NF-Y subunits. *The Plant Cell* 26(12):4954–4973

Seki M, Narusaka M, Abe H, Kasuga M, Yamaguchi-Shinozaki K, Carninci P, Hayashizaki Y, Shinozaki K (2001) Monitoring the expression pattern of 1300 Arabidopsis genes under drought and cold stresses by using a full-length cDNA microarray. *The Plant Cell* 13(1):61–72

Seo HS, Yang JY, Ishikawa M, Bolle C, Ballesteros ML, Chua NH (2003) LAF1 ubiquitination by COP1 controls photomorphogenesis and is stimulated by SPA1. *Nature* 423(6943):995–999

Seo PJ, Kim SG, Park CM (2008) Membrane-bound transcription factors in plants. *Trends in Plant Science* 13(10):550–556

Shabala S, Mackay A (2011) Ion transport in halophytes. *Advances in Botanical Research* 57:151–199

Shan H, Chen S, Jiang J, Chen F, Chen Y, Gu C, Li P, Song A, Zhu X, Gao H, Zhou G (2012) Heterologous expression of the chrysanthemum R2R3-MYB transcription factor CmMYB2 enhances drought and salinity tolerance, increases hypersensitivity to ABA and delays flowering in *Arabidopsis thaliana*. *Molecular Biotechnology* 51(2):160–173

Shao H, Wang H, Tang X (2015) NAC transcription factors in plant multiple abiotic stress responses: progress and prospects. *Frontiers in Plant Science* 6:902

Sharma R, Wungrampha S, Singh V, Pareek A, Sharma MK (2016) Halophytes as bioenergy crops. *Frontiers in Plant Science* 7:1372

Shen Q, Zhang P, Ho TH (1996) Modular nature of abscisic acid (ABA) response complexes: composite promoter units that are necessary and sufficient for ABA induction of gene expression in barley. *The Plant Cell* 8(7):1107–1119

Shen YG, Zhang WK, Yan DQ, Du BX, Zhang JS, Liu Q, Chen SY (2003) Characterization of a DRE-binding transcription factor from a halophyte *Atriplex hortensis*. *Theoretical and Applied Genetics* 107(1):155–161

Shinozaki K, Yamaguchi-Shinozaki K (1996) Molecular responses to drought and cold stress. *Current Opinion in Biotechnology* 7(2):161–167

Shukla PS, Gupta K, Agarwal P, Jha B, Agarwal PK (2015) Overexpression of a novel SbMYB15 from *Salicornia brachiata* confers salinity and dehydration tolerance by reduced oxidative damage and improved photosynthesis in transgenic tobacco. *Planta* 242(6):1291–1308

Singh B, Bohra A, Mishra S, Joshi R, Pandey S (2015) Embracing new-generation 'omics' tools to improve drought tolerance in cereal and food-legume crops. *Biologia Plantarum* 59(3):413–428

Song Y, Yang X, Yang S, Wang JJMG (2019) Transcriptome sequencing and functional analysis of *Sedum lineare* Thunb. upon salt stress. *Genomics* 294(6):1441–1453

Stracke R, Werber M, Weisshaar B (2001) The R2R3-MYB gene family in *Arabidopsis thaliana. Current Opinion in Plant Biology* 4(5):447–456

Sun XB, Ma HX, Jia XP, Chen Y, Ye XQ (2015) Molecular cloning and characterization of two novel DREB genes encoding dehydration-responsive element binding proteins in halophyte *Suaeda salsa. Genes & Genomics* 37(2):199–212

Tian C, Wan P, Sun S, Li J, Chen M (2004) Genome-wide analysis of the GRAS gene family in rice and Arabidopsis. *Plant Molecular Biology* 54(4):519–532

Tran LS, Nakashima K, Sakuma Y, Simpson SD, Fujita Y, Maruyama K, Fujita M, Seki M, Shinozaki K, Yamaguchi-Shinozaki K (2004) Isolation and functional analysis of Arabidopsis stress-inducible NAC transcription factors that bind to a drought-responsive cis-element in the early responsive to dehydration stress 1 promoter. *The Plant Cell* 16(9):2481–2498

Tripathi P, Rabara RC, Rushton PJ (2014) A systems biology perspective on the role of WRKY transcription factors in drought responses in plants. *Planta* 239(2):255–266

Tuteja N (2007) Abscisic acid and abiotic stress signalling. *Plant Signalling & Behaviour* 2(3):135–138

Vishwakarma K, Upadhyay N, Kumar N, Yadav G, Singh J, Mishra RK, Kumar V, Verma R, Upadhyay RG, Pandey M, Sharma S (2017) Abscisic acid signalling and abiotic stress tolerance in plants: a review on current knowledge and future prospects. *Frontiers in Plant Science* 8:161

Waisel Y (1972) *Biology of Halophytes*. Academic Press, New York and London

Wang J, Lian W, Cao Y, Wang X, Wang G, Qi C, Liu L, Qin S, Yuan X, Li X, Ren S (2018) Overexpression of BoNAC019, a NAC transcription factor from *Brassica oleracea*, negatively regulates the dehydration response and anthocyanin biosynthesis in Arabidopsis. *Scientific Reports* 8(1):1–5

Wang K, Zhong M, Wu YH, Bai ZY, Liang QY, Liu QL, Pan YZ, Zhang L, Jiang BB, Jia Y, Liu GL (2017a) Overexpression of a chrysanthemum transcription factor gene DgNAC1 improves the salinity tolerance in chrysanthemum. *Plant Cell Reports* 36(4):571–581

Wang L, Li Z, Lu M, Wang Y (2017b) ThNAC13, a NAC transcription factor from *Tamarix hispida*, confers salt and osmotic stress tolerance to transgenic Tamarix and Arabidopsis. *Frontiers in Plant Science* 8:635

Wang Y, Gao C, Liang Y, Wang C, Yang C, Liu G (2010) A novel bZIP gene from *Tamarix hispida* mediates physiological responses to salt stress in tobacco plants. *Journal of Plant Physiology* 167(3):222–230

Wang J, Cheng G, Wang C, He Z, Lan X, Zhang S, Lan H (2017c) The bHLH transcription factor CgbHLH001 is a potential interaction partner of CDPK in halophyte *Chenopodium glaucum. Scientific Reports* 7(1):8441.

Wu D, Sun Y, Wang H, Shi H, Su M, Shan H, Li T, Li Q (2018) The SlNAC8 gene of the halophyte *Suaeda liaotungensis* enhances drought and salt stress tolerance in transgenic *Arabidopsis thaliana. Gene* 662:10–20

Xianjun P, Xingyong M, Weihong F, Man S, Liqin C, Alam I, Lee BH, Dongmei Q, Shihua S, Gongshe L (2011) Improved drought and salt tolerance of *Arabidopsis thaliana* by transgenic expression of a novel DREB gene from *Leymus chinensis. Plant Cell Reports* 30(8):1493–1502

Yamaguchi-Shinozaki K, Shinozaki K (1994) A novel cis-acting element in an Arabidopsis gene is involved in responsiveness to drought, low-temperature, or high-salt stress. *The Plant Cell* 6(2):251–264

Yamaguchi-Shinozaki K, Shinozaki K (2005) Organization of cis-acting regulatory elements in osmotic-and cold-stress-responsive promoters. *Trends in Plant Science* 10(2):88–94

Yamamoto N, Takano T, Tanaka K, Ishige T, Terashima S, Endo C, Kurusu T, Yajima S, Yano K, Tada YJ (2015) Comprehensive analysis of transcriptome response to salinity stress in the halophytic turf grass *Sporobolus virginicus. Frontiers in Plant Science* 6:241

Yang G, Yu L, Zhang K, Zhao Y, Guo Y, Gao C (2017) A ThDREB gene from *Tamarix hispida* improved the salt and drought tolerance of transgenic tobacco and *T. hispida. Plant Physiology and Biochemistry* 113:187–197

Yanhui C, Xiaoyuan Y, Kun H, Meihua L, Jigang L, Zhaofeng G, Zhiqiang L, Yunfei Z, Xiaoxiao W, Xiaoming Q, Yunping S (2006) The MYB transcription factor superfamily of Arabidopsis: expression analysis and phylogenetic comparison with the rice MYB family. *Plant Molecular Biology* 60(1):107–124

Ye W, Wang T, Wei W, Lou S, Lan F, Zhu S, Li Q, Ji G, Lin C, Wu XJP (2020) The full-length transcriptome of Spartina alterniflora reveals the complexity of high salt tolerance in monocotyledonous halophyte. *Plant Cell Physiology* 61(5):882–896

Younesi-Melerdi E, Nematzadeh G-A, Pakdin-Parizi A, Bakhtiarizadeh MR, Motahari SA (2020) De novo RNA sequencing analysis of *Aeluropus littoralis* halophyte plant under salinity stress. *Scientific Reports* 10(1):9148.

Zhang DY, Yin LK, Pan BR (2003) A review on the study of salt glands of Tamarix. *Acta Botanica Boreali-Occidentalia Sinica* 23:190–194

Zhang P, Yang P, Zhang Z, Han B, Wang W, Wang Y, Cao Y, Hu T (2014) Isolation and characterization of a buffalograss (*Buchloe dactyloides*) dehydration responsive element binding transcription factor, BdDREB2. *Gene* 536(1):123–128

Zhang S, Li N, Gao F, Yang A, Zhang J (2010) Over-expression of TsCBF1 gene confers improved drought tolerance in transgenic maize. *Molecular Breeding* 26(3):455–465

Zhang S, Li X, Fan S, Zhou L, Wang Y (2020) Overexpression of HcSCL13, a *Halostachys caspica* GRAS transcription factor, enhances plant growth and salt stress tolerance in transgenic Arabidopsis. *Plant Physiology and Biochemistry* 151:243–254

Zhang T, Zhao Y, Wang Y, Liu Z, Gao C (2018) Comprehensive analysis of MYB gene family and their expressions under abiotic stresses and hormone treatments in *Tamarix hispida. Frontiers in Plant Science* 9:1303

Zhang X, Liu X, Wu L, Yu G, Wang X, Ma H (2015) The SsDREB transcription factor from the succulent halophyte *Suaeda salsa* enhances abiotic stress tolerance in transgenic tobacco. *International Journal of Genomics.* doi: 10.1155/2015/875497

Zhao H, Zhao X, Li M, Jiang Y, Xu J, Jin J, Li K (2018) Ectopic expression of *Limonium bicolor* (Bag.) Kuntze DREB (LbDREB) results in enhanced salt stress tolerance of transgenic *Populus ussuriensis* Kom. *Plant Cell, Tissue and Organ Culture* 132(1):123–136

Zhou S, Han JL, Zhao KF (2001) Advance of study on recretohalophytes. *Chinese Journal of Applied and Environmental Biology* 7(5):496–501

Zhu JK (2001) Cell signalling under salt, water and cold stresses. *Current Opinion in Plant Biology* 4(5):401–406

10 Plant Abiotic Stress Tolerance on the Transcriptomics Atlas

*Geetha Govind[1], Jayant Kulkarni[2], Harshraj Shinde[3], Ambika Dudhate[4], Ashish Srivastava[2,5], and P. Suprasanna[5]**

[1]Dept. of Crop Physiology, College of Agriculture, Hassan, UAS-B, Karnataka, India

[2]Nuclear Agriculture & Biotechnology Division, Bhabha Atomic Research Centre, Mumbai, India

[3]Environmental Epigenetics and Genetics Group (EEGG), Department of Horticulture, College of Agriculture Food and Environment, University of Kentucky, Lexington, USA

[4]College of Pharmacy, University of Kentucky, Lexington, USA

[5*]Homi Bhabha National Institute, Anushaktinagar, Mumbai, India

*Corresponding author: email: penna888@yahoo.com

CONTENTS

10.1 INTRODUCTION

Climate change is now recognized worldwide, and the Intergovernmental Panel on Climate Change (IPCC) has presented considerable evidence that the past three decades are successively warmer (surface temperature) than the previous decades (IPCC, 2014). In addition, climate projections are predicting increase in CO_2 and water vapor along with changes in temperature and rainfall pattern. The most evident climatic change is increase in the surface temperature because of increase in greenhouse gases like CO_2, methane, nitrous oxide, hydrofluorocarbons, perfluorocarbons and sulfur hexafluoride (IPCC, 2014). By the end of this century, the global mean temperature could rise by 1.8 to 4 °C. Climate change has also resulted in a high rate of land degradation, desertification,

DOI: 10.1201/9781003159636-10

and infertile, saline, dry and unhealthy soils. Countries may lose their arable land due to climate change; e.g., South America may lose 1–21%, Africa 1–18%, Europe 11–17% and India 20–40% (Zhang and Cai, 2011). Around 15 billion tons of fertile soil is lost every year due to human interference and climate change, thereby influencing 1.5 billion lives. According to the United Nations Environment Programme (UNEP, 2017), around 500 million hectares of agricultural land is lost due to only drought and desertification. Water will be scarce due to increased use by crops and prevalance of drought. Competition for land will increase, as marginal and other lands will be climatically unsuitable for food production. The surface temperatures are projected to further increase over the 21st century under all emission scenarios. Heat waves are predicted to occur more often, with greater magnitude, and last longer. Similarly, precipitation will be extreme, more intense and frequent (Field et al., 2012).

Over the past several decades, the average surface temperature has risen by 0.74 °C from the pre-industrial period to the period of 2006–2015 (Allen et al., 2018). It is predicted that the CO_2 concentration will increase two-fold (to 800 ppm) by the end of this century (Vaughan et al., 2018; Raza et al., 2019). In the short term, elevated CO_2 could be beneficial in increasing yield by affecting photosynthesis. The C3 plants (wheat, rice, soybean, cotton, potato) are more responsive to higher CO_2 compared with C4 plants (corn, sugarcane, sorghum, pearl millet). Contrastingly, increased CO_2 is also shown to negatively affect protein content, minerals and nutritional quality (Fernando et al., 2012).

10.1.1 Food Security

Food security depends on not just sufficient production of food but also access to food. According to the Food and Agriculture Organization (FAO, 1996), food security is achieved when "all people at all times have physical and economic access to sufficient, safe and nutritious food to meet their dietary needs and food preferences for an active and healthy life". With the population predicted to increase to 9 billion by 2050, the food requirement is expected to increase by 85% (FAO, 2017). Discovering how the changes in the abiotic and biotic factors affect yield and identifying the crop stage most vulnerable to these changes in various crops is crucial to combat food security. For example, 2010 summer heat waves in Russia and severe drought in Australia (2007–2008), southeast China (2013) and neighboring countries resulted in increased yield loss, led to an increase in commodity prices and pushed food security to its limits. Together with changes in rainfall pattern, increased frequency and magnitude of drought or flood will result in agricultural loss. Warmer climates are predicted to increase incidences of pest and diseases and their shift in geographical distribution. It is estimated that high temperature will result in yield reductions of 6%, 3.2%, 3.1% and 7.4% for wheat, rice, soybean and maize, respectively (Zhao et al., 2017). For each degree rise in temperature, wheat yield is predicted to reduce by 6–10%, maize yield by 8.3%, rice yield by 2.6% and sorghum yield by 7.8% (Kjellstrom et al., 2018; Raza et al., 2019).

10.1.2 Environmental Stresses and Crop Yield

In nature, plants are exposed to various abiotic stresses like drought, waterlogging, high or low temperature, salinity, ultraviolet (UV)-B, high or low light, etc. (Suzuki et al., 2014). The physiology of plants is greatly affected by fluctuations in the environmental conditions outside their physiological range of adaptation. There is a negative impact on yield for all major crops with an increase in 3 °C above ambient temperature. It is projected that each decade of climate change is expected to reduce mean yields by roughly 1%. The demand can only be met by increasing productivity by 14% per decade. Analyzing data on grain productivity at the global scale, it is observed that an increase in CO_2 will increase global yields by 1.8% per decade, while an increase in temperature will reduce global yields by 1.5% per decade (Lobell and Gourdji, 2012). Crop yield is more affected in tropical areas than in high-latitude areas and the severity increases with increase in degree of climate change. Climate change predictions indicate that yields of major crops will suffer by 2050 (Nelson et al., 2010). In Africa, the yield of wheat, maize, sorghum and millet is predicted to reduce by 17%, 5%, 15% and 10%, respectively. Considering a scenario of an increase in global mean temperature of 4 °C or more, it is predicted that the yield of maize and beans will decrease by 19% and 68%, respectively, in Sub-Saharan Africa. The trends in yield reduction are similar even for perennial trees like apples, cherries, etc. (Stocke et al., 2010; Lobell and Field, 2011).

Climate change significantly affects crop production by altering plant physiology, reducing yield and reducing product quality (Rosenzweig et al., 2014; Collier et al., 2014). The major abiotic factor affecting yield is drought. The extent of impact on yield depends on the stage of plants exposed to drought, and the magnitude and severity of the drought spell. For example, in wheat, yield was reduced by 1–30% under mild drought during the post-anthesis stage and by up to 92% under prolonged drought during the flowering and grain filling stage. In mash bean, drought reduced the yield by 26–57%, and in soybean it resulted in 42% reduction (De Oliveira et al., 2013; Raza et al., 2019).

Climate change will greatly increase abiotic stress, i.e., high temperature, drought and salt stress, affecting crop yield or production and quality. Understanding the molecular response of plants to various abiotic stresses individually and in combination is essential to manipulate plants' responses to changing environmental conditions (Dhankher and Foyer, 2018). Physiological, biochemical, cellular and molecular studies have shed light on how plants respond to abiotic stress factors like drought stress, salinity, extreme temperature and heavy metal toxicity. The molecular responses have been elucidated at structural and regulatory levels, including stress sensors, signaling, ionic homeostasis, osmotic adjustment, hormone signaling and reactive oxygen species (ROS) in plant systems (Figure 10.1).

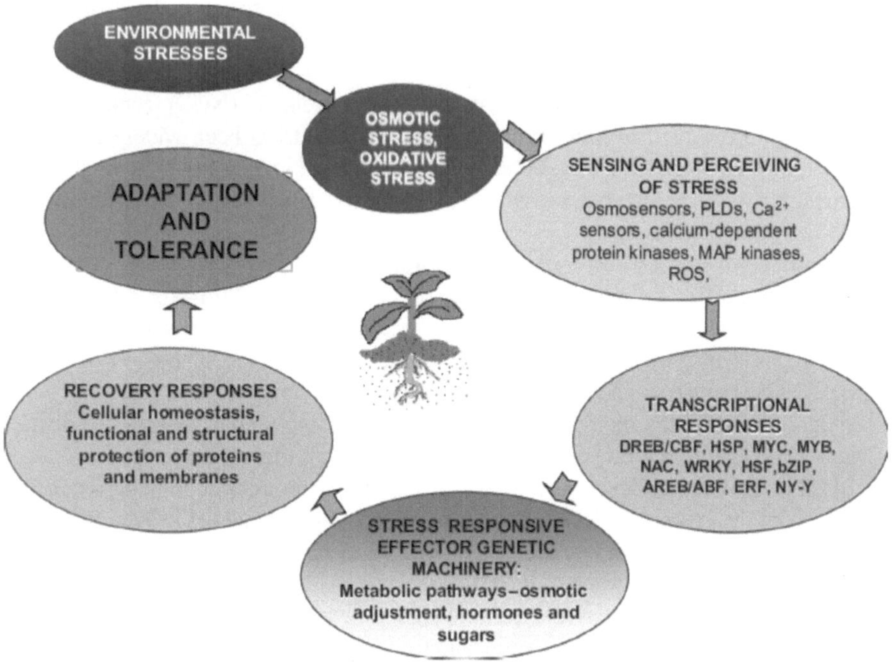

FIGURE 10.1 Diverse and coordinated plant responses to environmental stresses.

(From Suprasanna 2020.)

Understanding the molecular response of plants to various abiotic stresses individually and in combination is essential to manipulate plants' response to changing environmental conditions. In recent years, omics technology has been promising in discovering genes governing specific traits. Transcriptomics has gained a lot of attention (Nejat et al., 2018) and various platforms are available, e.g., microarray, RNA-seq, etc., to study the whole-genome-level changes in expression profile and identify differentially regulated genes in response to stresses. Here we present an overview of the transcriptomics of abiotic stress responses and discuss recent transcriptomic studies on crop responses to salt, drought and heat stresses, which are significant for developing tolerance mechanisms and designing climate-smart crops.

10.2 TRANSCRIPTOMICS

"Transcriptome" indicates a collection of mRNAs in a specific organ, tissue or cell of a specific organism under certain physiological conditions at a particular developmental or growth stage (Dong and Chen, 2013). Therefore, a transcriptomic study reveals the gene expression pattern at the whole-genome level under various stress conditions, which is useful in understanding the complex regulatory network involved in plant adaptation or tolerance to specific stress. Under stresses like salinity, drought, heavy metals and extreme temperatures, plants modify their transcriptome to trigger stress tolerance machinery (Wang et al., 2020). Transcriptomic studies can predict factors responsible for stress tolerance and disclose complex relationships among signal transduction, metabolic pathways and other defense responses, which is important for understanding stress tolerance mechanisms and improving stress tolerance in other plants (Wang et al., 2020).

In the last couple of decades, transcriptomic approaches have developed right from the initial northern blotting to RNA sequencing (RNA-seq) through several other techniques, like expressed sequence tags (EST), semiquantitative reverse transcription polymerase chain reaction (RT-PCR), quantitative RT-PCR, differential display RT-PCR, serial analysis of gene expression (SAGE),massively parallel signature sequencing (MPPS) and microarrays, etc. (Figure 10.2; Agarwal et al., 2014; Lowe et al., 2017).

After the invention of high-throughput next-generation sequencing (NGS) technology, there has been tremendous improvement in transcriptome research, and it has been widely used to explore the transcriptomic pools of various model plants, crop plants and halophytes (Yao et al., 2020). The single-cell transcriptome has also been established to analyze plant stress responses (Rich-Griffin et al., 2020; Shaw et al., 2021).

Northern blotting was the earliest technique for transcript analysis, based on the hybridization of labeled probes to localize a single RNA sample on agarose gel and transfer it to the membrane (Agarwal et al., 2014). After the invention of PCR, it became possible to perform transcript quantification of a few genes; semiquantitative PCR and quantitative real-time PCR are used (Becker-Andre and Hahlbrock, 1989). To overcome the limitations of gaining data on a few transcripts, ESTs came into the limelight, because this method can quantify transcripts without sequencing of the whole genome (Marra et al., 1998) but it lacks the accuracy of transcript-level detection. To overcome the limitations of ESTs, SAGE was developed to manage enhanced throughput of the tags generated and allow quantitation of the expression level of rare transcripts (Velculescu et al., 1995). There are some advanced variants of SAGE, which include LongSAGE, SuperSAGE, DeepSAGE and cap analysis of gene expression (CAGE) (Agarwal et al., 2014).

Another tag-based method is massively parallel signature sequencing (MPSS), which allows the identification of millions of transcripts at a time, and this technique is more sensitive in terms of low-abundance transcript detection (Brenner et al., 2000). The limitation of this technique is difficulty in the cloning of RNA on microbeads. After the invention of this technology, it was not freely available to researchers until NGS was developed (Agarwal et al., 2014).

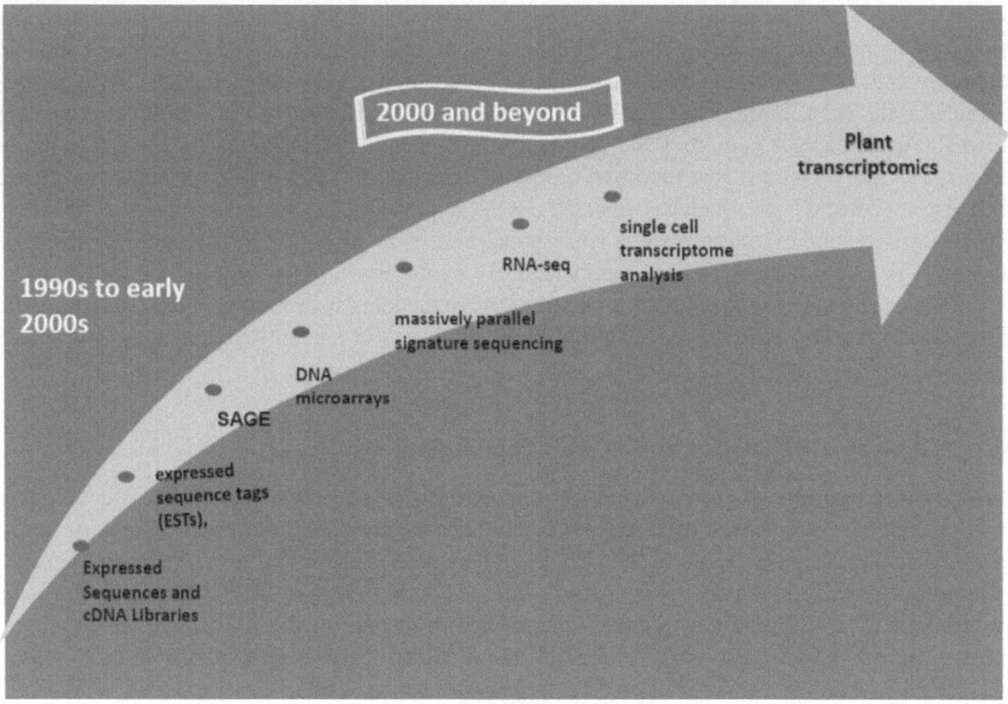

FIGURE 10.2 An overview of the progressive advancement in transcriptomic technologies in plants.

Microarray allows simultaneous analysis of thousands of transcripts at greatly reduced cost. With this technique, the capability of transcript detection was significantly enhanced from a few genes to a whole transcriptome, and the method is based on the quantification of the abundance of a particular set of RNAs through their hybridization to complementary probes (Pozhitkov et al., 2007). The first plant genome on GeneChip® was reported in Arabidopsis (Zhu and Wang, 2000), followed by studies in other plants like rice, loblolly pine and strawberry (Agarwal et al., 2014). Microarray enhances the application of transcriptomic studies to the next level by analyzing regulatory networks and pathways (Sasidharan et al., 2013).

NGS has been the major landmark in the field of transcriptomics, as RNA-seq by NGS allows analysis of transcripts even in the absence of a reference genome. It also provides other advantages, like reduced cost, detection of splice variants and novel transcripts, analysis of low-abundance transcripts, detections of noncoding RNA, detection of single nucleotide polymorphisms (SNPs), etc. RNA-seq technology advancements also allow analysis of a single-cell transcriptome, in which the transcriptomes of single cells are directly investigated in fixed tissues (Lee et al., 2014).

In 2006, the first RNA-seq data was published, comprising 105 transcripts sequenced by 454 technology (Bainbridge et al., 2006). In 2007, the first report on Arabidopsis transcriptomics by NGS was released (Weber et al., 2007). But its applicability has been increased after the introduction of Solexa/Illumina technologies in 2008, because this allowed the estimation of 109 transcript sequences (Mortazavi et al., 2008; Wilhelm et al., 2008; Ozsolak and Milos, 2011; Lappalainen et al., 2013). Different platforms used for sequencing include 454 (Roche), Illumina (Illumina), SOLiD (Thermo Fisher Scientific), Ion Torrent (Thermo Fisher Scientific) and PacBio (Pacbio) (Quail et al., 2012). A huge amount of data is generated after the sequencing; therefore, it is necessary to process it to gain useful information using a combination of different bioinformatics software. Processes usually fall into four major steps: 1) quality control, 2)alignment, 3) quantification and 4) differential expression (Van Verk et al., 2013).

A more recent advancement in the field of RNA-seq is direct RNA sequencing by nanopore technologies (Oikonomopoulos et al., 2020). This technology was first introduced by Oxford Nanopore Technologies to analyze mRNA by using adapters developed to capture polyadenylated RNA strands (Garalde et al., 2018; Smith et al., 2020). A device commercially known as MinION, sequences native RNA and enables genuine RNA-seq data to be generated (Garalde et al., 2018).The major advantage of this method is that modified bases are detected at the time of sequencing, which are otherwise not detected by sequencing of cDNA. RNA sequencing has many advantages: (1) it is not restricted to detection of transcripts that correspond to the existing genomic sequence, (2) it reveals the precise position of transcription boundaries, (3) it has very low background signal and (4) it does not have any upper limit of quantification. RNA sequencing is the first method that allows the entire transcriptome to be surveyed with high throughput and in a quantitative manner (Mortazavi et al., 2008).

10.3 TRANSCRIPTOMIC CHANGES UNDER ABIOTIC STRESS CONDITIONS

The study of transcriptomes has a significant impact on all the facets of biological sciences, as it provides the ability to analyze differences in gene expression of multiple mRNAs both quantitatively and qualitatively (Tan et al., 2009). The key objectives of transcriptomics are to catalogue all the transcripts (mRNAs, noncoding RNAs and small RNAs), determine their transcriptional status, determine the 5′ and 3′ end sites of the genome, post-transcriptional modifications and splicing patterns. Transcriptomics also aims to quantify the modulations in gene expression levels during different stress conditions and developmental stages (Wang et al., 2009; Reddy et al., 2016, Li et al., 2018). This technology has enabled linking of different metabolic pathways, signal transduction and defense response, all of which are crucial in understanding the mechanism of plant stress tolerance. Plants respond to stress through a complex array of cellular, molecular and physiological processes, and their metabolic networks are intricate.

10.3.1 Salt Stress

Transcriptomics data from salt stress-treated plants can provide guidance on which candidate genes and pathways to manipulate for developing salt-tolerant varieties. Such studies on evaluating the effects of salt stress on growth and development at different developmental stages and tissues provide valuable information about the biological processes and how they are coordinated during stress. A meta-analysis of 96 public microarray datasets of different rice genotypes at the seedling stage was carried out to identify salt-responsive genes and understand the molecular mechanism of salt tolerance in rice (Kong et al., 2019). Around 5559 DEGs were regulated under salt stress and 3210 DEGs were regulated during recovery process. Signaling networks consisting of calcium signal transduction pathway, MAPK cascade, multiple hormone signals, transcription factors and regulators, protein kinases and other functional proteins were identified to play essential role in salt stress response in rice. In addition, TNG67 meets its energy demand by shifting to the fermentation pathway under salt stress, thereby increasing the Calvin cycle to repair the damaged photosystem during recovery (Kong et al., 2019). A study with reciprocally crossed plants of salt-tolerant Horkuch and high-yielding salt-sensitive IR29 was carried out using RNA-seq to identify the genes associated with salt tolerance. The number of differentially expressed genes (DEGs) was higher in leaves compared to root tissues. Gene expression was reduced in leaves of the sensitive genotype while it increased in the tolerant genotype. The reciprocal crosses revealed a strong maternal cytoplasmic effect on gene expression, which was most evident in roots (Razzaque et al., 2019).

Genes functioning in phosphorylation, electron carriers, transporters and cation transmembrane activities were upregulated. In the sensitive plants, upregulation of genes coding for redox proteins and apoptosis-related proteins was observed, whereas tolerant plants showed upregulated genes coding for membrane sensor proteins, enzymes of the signaling pathway, trehalose production, G-protein-coupled receptor, photosynthesis-related enzymes, Golgi body recycling, and prolamin precursor proteins involved in protein refolding (Razzaque et al., 2019). Similar transcriptome analysis of roots and stems of rice under salt stress identified carbohydrate and amino acid metabolism and induced secondary-metabolite-related genes along with phenolic and flavonoid accumulation in roots (Chandran et al., 2019). Genome-wide meta-analysis, using microarray and RNA-seq data (46 meta-QTL (quantitative trait loci) positions), identified 3449 DEGs, of which 1286, 86, 1729 and 348 DEGs were regulated in root, shoot, seedling and leaf tissue, respectively. Of these, 23 candidates were detected in Saltol QTL and hotspot-regions overlying QTLs for yield and ion homeostasis. Candidate genes detected were pectinesterase, peroxidase, transcription regulator, high-affinity potassium transporter, cell wall organization, protein serine/threonine phosphatase, and CBS domain containing protein, suggesting that salt-tolerant genotypes use sensing and signaling, transcriptional regulation, ion homeostasis and ROS scavenging under salt stress (Mansuri et al., 2020). The circadian clock has also been shown to regulate plant abiotic stress tolerance. One of the rice clock components, OsPRR73, was shown to fine-tune salt tolerance by recruiting HDAC10 to transcriptionally repress OsHKT2:1, thereby regulating sodium accumulation (Wei et al., 2021).

Gene expression dynamics in the salt-tolerant *Arabidopsis pumila* was studied out using second- and third-generation (single-molecule real-time [SMRT]) sequencing methods. Of the 8075 DEGs identified as being regulated under stress (as compared with control), 483 were transcription factors and 1157 were transport proteins. In addition, genes for oxidation-reduction and transmembrane transport were also upregulated. Orphan, MYB, HB, bHLH, C3H, PHD, bZIP, ARF and NAC transcription factors were among the highly enriched DEGs. In addition, ABCB1, CLC-A, CPK30, KEA2, KUP9, NHX1, SOS1, VHA-A and VP1 transport proteins were highly regulated under salt stress. On the other hand, the MAPKs (MAPKKK18) were continuously upregulated (Yang et al., 2018).

Studies in the salt-tolerant plants have shown the regulation of antioxidants, transcription factors, signaling, ion and metabolic homeostasis, and transporters (Das et al., 2015). Transcriptome analysis of salt-tolerant and susceptible genotypes of rice under salt stress provided insight into the regulation differences at the transcript level. Some of the important genes regulated in rice under salt stress were *OsSOS1*, *OsNHX1*, *OsHKT2:1*, *OsCAX1*, *OsAKT1*, *OsKCO1*, *OsTPC1*, *OsCLC1*, *OsNRT1:2*, *OSCDPK7*, *OsMAPK4*, *CaMBP*, *GST*, *LEA*, *V-ATPase*, *OSAP1* and *HBP1B* (Kumar et al., 2013). Around 507 transcripts were exclusively differentially expressed in salt-tolerant genotype Pokkali compared with salt-susceptible genotypes IR64. Members of the bHLH, MYB and C2H2 family of transcription factors were enriched in Pokkali and exhibited differential expression. In addition, genes involved in wax and terpenoid metabolism were upregulated in Pokkali. The authors also observed transcript isoforms (alternatively spliced products) that are genotype-specific and modulated by the stress response. Two transcript isoforms coding for glutathione S-transferase displayed enhanced expression in Pokkali under salt stress and several novel transcripts specific to Pokkali (Shankar et al., 2016). Transcripts induced under salt stress have been studied earlier using microarray. The Illumina HiSeq 2000 platform was used to profile the transcriptome of leaves and roots of rice seedlings under salt stress in Dongxiang wild rice (*Oryza rufipongon* Griff). Nearly 6867 DEGs (2216 upregulated and 4651 downregulated) were identified in leaves and 4988 DEGs (3105 upregulated and 1883 downregulated) in roots (Zhou et al., 2016).

Cellular, physiological and transcriptome analysis of salt-tolerant and sensitive rice cultivars revealed differential responses of expressed genes. Salt-tolerant cultivars showed activation of adaptive mechanisms that limited sodium distribution to root and old leaves, and activation of the regulatory mechanism of photosynthesis in young leaves. In contrast, the sensitive cultivar initiated genes of senescence and cell death (Formentin et al., 2018). An understanding of molecular mechanisms responsible for salt tolerance has also been obtained through a comparative transcriptomics using leaf samples of rice collected at three different time points (Wang et al., 2018). A total of 5273 DEGs were identified between salt-tolerant and sensitive genotypes of rice. This identified 1375 novel genes, of which 286 DEGs were exclusively found in the tolerant cultivar. Of these two genes, the disease resistance response 206 and TIFY10A were validated for further analysis (Wang et al., 2018). *OsSPL10*, a positive regulator of trichome formation, negatively regulates salt tolerance and is preferentially expressed in young panicle and stems (Lan et al., 2019).

Comparative transcriptome analysis in maize using salt-tolerant and sensitive inbred lines (L87) identified upregulation of 1856 unique DEGs that included HSP70s and aquaporins in the 70 kDa family. Genes involved in hormonal (abscisic acid (ABA), ethylene, jasmonic acid (JA) and salicylic acid (SA)) mediated signal transduction were differentially regulated between tolerant and sensitive varieties. Genes coding for ROS scavenging, WRKY transcription factor, and SnRK2, a component of ABA signaling, are among the upregulated DEGs in a salt-tolerant maize variety (Wang et al., 2019; Wang et al., 2020).

Wheat arrays and GeneChip have identified many abiotic stress-responsive genes (Kawaura et al., 2008; Schreiber et al., 2009). Transcriptome analysis of wheat introgression lines (Shanrong No.3-SR3-salt tolerant derivative of asymmetric somatic hybrid of Jinan 177-JN177 and *Thinopyrum ponticum*) using microarray revealed the upregulation of genes associated with stress response, unsaturated fatty acids, flavonoid synthesis, and pentose phosphate metabolism in salt-tolerant lines (Liu et al., 2012). RNA-seq analysis of the transcriptome of wheat cultivars exposed to salt stress has identified genes involved in providing energy to drive sodium into vacuoles (V-ATPase), energy metabolism (ATP citrate synthase), ROS scavengers and signaling genes (Cbl-interacting protein kinase) that play a role in salt tolerance of Kharchia, a local variety (Goyal et al., 2016). Introduction of genes coding for *AtNAC*, *AtMYB*, a histone-lysine N-methyltransferase (*AtSDG16*) and gene for root hair development (*TaRSL4*) enhanced salt tolerance of bread wheat (Zhang et al., 2016). Comparison of the transcriptome of the salt-sensitive wild type with salt-tolerant mutant bread wheat identified the oxidation-reduction process to be the most important for salt tolerance. Genes like arginine decarboxylase, polyamine oxidase and hormone-related genes were enriched in the salt-tolerant mutant. In addition, a new pathway for salt response – butanoate metabolism – has also been identified (Xiong et al., 2017). RNA-seq analysis was carried out to understand the phenotypic differences observed between two wheat varieties (salt-tolerantXY60 – greener leaves and longer roots than ZM175) under salt stress. The DEGs of salt-tolerant varieties were mainly associated with polyunsaturated fatty acid (PUFA) metabolism in old and new leaves, while in the sensitive variety (ZM175), they were associated with photosynthesis and energy metabolism. However, the glucosinolate biosynthesis was regulated in the roots of both varieties. The JA levels were higher in the salt-tolerant variety. It is proposed that salt tolerance of the tolerant wheat cultivar is coordinated by PUFAs, which probably enhance the photosynthetic system and JA-related pathways (Luo et al., 2019).

Similarly, the root transcriptome of an Iranian salt-tolerant wheat cultivar, Arg, exposed to salt stress was studied to understand the molecular basis of salt tolerance in wheat (Amirbakhtiar et al., 2019). Nearly 26,171 novel transcripts and 5128 DEGS were identified by RNA-seq under salt stress. The DEGs were classified into transporters, phenylpropanoid biosynthesis, transcription factors, glycosyltransferase, glutathione metabolism, and hormone signal transduction. Transcription factors belonging to MYB, NAC, bHLH, WRKY, bZIPs and AP2/ethylene-responsive transcription factor (ERF), which control genes regulating osmotic, ionic and oxidative stress occurring due to salinity, were among the DEGs. Genes coding for aquaporins, late embryogenesis abundant (LEA) proteins, dehydrins, P5CS, SOS1, K^+ transporters, ATP-binding cassette (ABC) transporters, catalase, glutaredoxin and GST, which deal with osmotic, ionic and oxidative stress, displayed enhanced expression, while proline oxidase was downregulated (Amirbakhtiar et al., 2019). Dynamic changes in the transcriptome of QTL regions under salt stress were analyzed to identify candidate genes regulating salt stress tolerance in bread wheat using massive 3′-ends sequencing.

The salt-tolerant genotype upregulated calcium-binding and cell wall synthesis genes earlier than the susceptible genotype, along with upregulation of some ABC transporters and sodium–calcium exchanges. Photosynthesis-related and calcium-binding genes were downregulated, mainly because of increased oxidative stress in the susceptible genotype. Salt-responsive genes within the QTL intervals based on linkage disequilibrium were identified as a potential component of salt tolerance (Duarte-Delgado et al., 2020).

RNA-seq analysis of leaf transcriptome changes under long-term salt stress identified 4506 salt-responsive unigenes in the diploid progenitor of bread wheat D-genome, *Aegilops tauschii* (Mansouri et al., 2019). The DEGs were shown to be involved in regulating ion homeostasis, the signaling process, carbohydrate metabolism and post-translational modifications. In addition, the content of starch and sucrose was regulated under salt stress. Most of the DEGs (82%) were mapped to known salt tolerance QTLs (Mansouri et al., 2019).

Using barley gene chips, many abiotic stress-responsive genes have been identified (Walia et al., 2007a). Differential gene expression screening of barley (*Hordeum vulgare*) leaf transcriptome using mRNA-seq identified 48 and 62 significantly up- and downregulated transcripts, respectively, under acute salt stress. Early responsive to dehydration, also known as LEA, chloroplast localized lipoxygenase 2.1, plasma membrane bound ATPase, an ODORANT1 homologue and protein phosphatase 2C were highly upregulated under salt stress. An aquaporin was among the top highly downregulated gene under acute salt stress, indicating reduced water transport to leaf under stress (Ziemann et al., 2013). RNA-seq analysis has also been carried out in wild salt-tolerant barley (*Hordeum spontaneum*). The transcriptome profiling of leaves from salt-stressed plants identified response to external stimuli and the electron carrier activity group as being most differentially expressed. In addition, electron transport and exchange, flavonoid biosynthesis, ROS scavenging, ethylene production, signaling network and protein folding were highly expressed under salt stress (Bahieldin et al., 2015). RNA-seq analysis of hexaploid hulless oats identified 65,000 DEGs under salt stress at different stages. ABC transporters, plant hormone signal transduction, plant–pathogen interactions, starch and sucrose metabolism, arginine and proline metabolism, and other secondary metabolite pathway genes were enriched in response to salt stress. Transcription factors belonging to MYB, MYB-related, bHLH, WRKY and NAC families were among the DEGs. Around 8% of the DEGs encoded membrane transporter-related proteins, which included ABC transporters, ATPase, ion transporters and aquaporin, indicating oats' ability to maintain and enhance transport in response to salt stress (Wu et al., 2017).

Transcriptomic analysis using RNA-seq of foxtail millet (*Setaria italica* L.) cultivars contrasting for their salt tolerance (Yugu2 and An04 are salt-tolerant and susceptible, respectively) identified 8887 and 12,249 DEGS in response to salt stress, of which 3149 were common to both cultivars. Ion transport, redox homeostasis, phytohormone metabolism, signaling and secondary metabolism were enriched in the salt-tolerant cultivar. Phenylpropanoid, flavonoid and lignin biosynthesis pathways and lysophospholipids, in addition to an efficient ion channel and antioxidant system, are important in determining the tolerance phenotype of salt-tolerant cultivars (Pan et al., 2020). Similar RNA-seq analysis of salt-tolerant (ICMB 01222) and susceptible (ICMB 081) pearl millet (*Pennisetum glaucum*) varieties identified a total of 11,627 DEGs in both lines. Ubiquitin-mediated proteolysis and phenylpropanoid biosynthesis pathways were upregulated in the tolerant line, and glycolysis/gluconeogenesis and genes for ribosomes were downregulated in the susceptible line. The DEGs involved in pathways such as hormone and MAPK signaling, cysteine and methionine metabolism were upregulated. Plant-specific transcription factor family genes, SBPs (squamosa promoter binding proteins) were differentially expressed only in the tolerant lines (Shinde et al., 2018).

Global transcriptome analysis of chickpea in response to salt stress in salt-tolerant (JG 62) and susceptible (ICCV 2) genotypes indicated that a total of 5545 genes were differentially regulated in the salt-tolerant genotype (Garg et al., 2015). Key enzymes of metabolic pathways (carbohydrate and lipid metabolism, photosynthesis, redox homeostasis, cell wall component biogenesis, protein modification, regulation of transcription, nucleic acid and nitrogen compounds, metabolic processes and generation of precursor metabolites/energy) were highly regulated by salt stress. In addition, during the vegetative stage, genes associated with lipid transport, protein targeting, DNA conformation change and glucan metabolic processes were significantly enriched in the salt-tolerant cultivar. Nearly 47% of transcription factor coding genes were regulated, and a large number of the WRKY and NAC family of transcription factors were differentially expressed under salt stress. Transcription factors regulating ABA, auxin, gibberellic acid (GA) and cytokinin were also regulated under salt stress (Garg et al., 2015).

Whole-genome analysis of genes regulated under salt stress in alfalfa (*Medicago sativa*) and its close relative (*Medicago truncatula*) has been carried out (Zahaf et al., 2012; Postnikova et al., 2013). RNA-seq analysis of root samples from salt-tolerant (AZ-GERM SALT-II) and susceptible (AZ-88NDC) genotypes of alfalfa identified 1165 genes, of which 86 transcription factors were regulated in response to salt stress. The gene ontology (GO) terms or molecular functions like transport, reproduction, nuclease activity, DNA binding and transcription factor activity were regulated only in the root of the susceptible alfalfa genotype under salt stress. Genes involved in the extracellular region (apoplast) were highly upregulated in the tolerant genotype, while genes in response to stress and kinase activity were repressed in the susceptible genotype. Around 604 DEGs were specific to the salt-tolerant line of which 12 coded for ABC transporters, which play a role in the transport of stress-related secondary metabolites. More than 10 genes regulating the phenylpropanoid pathway were enriched in tolerant lines (Postnikova et al., 2013).

RNA-seq analysis of the model halophytic crop quinoa (*Chenopodium quinoa* Willd.) was used to study the genome-wide transcriptional changes induced under salt stress. A total of 2416 DEGs were identified whose expression was higher 5 days after exposure to salt stress. Stress- and ABA-responsive pathways, including others for oxidation-reduction, protein phosphorylation, protein kinase activity, cofactor binding and cell wall organization processes, were highly regulated under salt stress (Ruiz et al., 2019).

The molecular mechanism of salt tolerance in potato was investigated by RNA-seq analysis of the salt-tolerant tetraploid potato Longshu No. 5. A total of 5508 DEGs were regulated in response to salt stress, of which 274 encoded transcription factors. The transcription factors belonged to 13 families: zinc finger proteins, AP2/ERF, MYB, bHLH, ZIP, WRKY, NAC, TCP, heat stress transcription factor (HSF), HB, NFY, ARF, and MADS. DEGs involved in nucleic acid binding, transporter activity, ion or molecule transport, ion binding, kinase activity and oxidative phosphorylation were significantly enriched. Nearly 158 genes regulating osmoregulation, carbohydrate metabolism and redox regulation were differentially expressed under salt stress. Protein kinases, i.e., RLKs, MAPKs, CDPKs, and CIPKs, were highly regulated in response to stress (Li et al., 2020). Comparative analysis of sweet potato transcriptome using salt-tolerant (Xushu 22) and susceptible (Xushu 32) cultivars was carried out. A total of 16,396 DEGs were identified, of which 1764 DEGs were specifically expressed in the tolerant genotype. Regulation of processes such as ion accumulation, stress signaling, transcriptional regulation, redox reaction, hormone signaling and transduction, and secondary metabolite accumulation plays a role in salt tolerance (Meng et al., 2020).

In Salix transcriptomic analysis, ethylene-regulated transcription factors were detected in response to salt stress that were associated with upregulation of levels of amino acids, sucrose, inositol, stress proteins and ROS scavenging enzymes. Genes involved in plant hormone pathways and beta-alanine, galactose and betalain metabolism were also upregulated (Zhou et al., 2020). Transcriptome analysis of eggplant revealed that the response to salt stress is genotype and organ specific. Upregulation of *C_2C_2-CO-like*, *WRKY*, *MYB*, *NAC* family transcription factor, *AKT1*, *KAT1* and *SOS1* was observed in the salt-tolerant genotype. The overexpression of *SmAKT1* improved tolerance of yeast and Arabidopsis akt1 mutant plants to salt stress (Li et al., 2019).

10.3.2 Drought Stress

Drought is one of the major abiotic stresses affecting plant growth, development and yield. The outcome of drought stress depends on the severity and duration of stress and the age and developmental stage of the crop (Bartels and Sunkar, 2005). Most studies on transcriptomic analysis have tried to capture the gene regulation under drought stress in various crops. Nevertheless, there are many areas that still need to be explored, such as the stage of plant growth exposed and the duration, comparison among genotypes, etc.

In *Arabidopsis* more than 800 data sets using Affymetrix microarray have been examined to study gene expression in response to various environmental stress conditions since 2003. In addition to using meta-analysis, meta-regression on transcriptomic data was carried out using these data sets. Meta-analysis of 10 data sets of gene expression in response to water stress identified 4015 DEGs under water stress. Meta-analysis identified changes in signaling pathway genes, with a large effect size for *HAI2/AIP1*, which was among the top 100DEGs in the individual studies. This protein (HAI2/AIP1) is a negative regulator of osmoregulatory solute accumulation during drought stress and a positive regulator of ABA signaling. In addition, RCAR3 (binding partner of HAI2/AIP1) was significantly downregulated and SnRK2-6 (calcium independent ABA activated protein kinase) was upregulated under drought stress in a meta-analysis (Rest et al., 2016).

In rice, alternate wetting and drying is found to be highly efficient in rice production compared with continuous flooding. Transcriptome analysis revealed the regulation of genes involved in lignin biosynthesis and phytohormone signal transduction pathway under alternate wetting and drying (Song et al., 2020). Transcription factors belonging to bHLH, bZIP, NAC, WRKY, and HSF family were among the regulated genes (Song et al., 2020). Transcriptomic and proteomic analysis of rice genotypes contrasting in their response to drought stress identified differential expression of *Med37c* (mediator, ATP dependent molecular chaperone-assist folding of unfolded proteins), *RSOsPR10* (accumulate in cortex cells and protect inner vascular system of roots during drought) genes and chitinase to play a role in drought tolerance (Anupama et al. 2019). Transcriptomic analysis of 12 rice genotypes contrasting in their drought tolerance revealed differences in the temporal expression pattern of DEGs. Tolerant genotypes differentially expressed fewer genes with a high proportion of them being upregulated. Biological processes like raffinose, fucose and trehalose metabolism were regulated specifically, and processes like protein and histone deacetylation, protein peptidyl–propyl isomerization, transcriptional attenuation, ferric iron transport, etc., were regulated in the early time point in the tolerant genotypes, which contributed to their drought tolerance. The temporal differences in transcriptomic features between susceptible and tolerant genotypes contribute to their response to drought stress (Xia et al., 2020). Similarly, RNA-seq analysis of rice genotypes (leaves) contrasting in their drought tolerance nature (IR64 drought susceptible and Apo drought tolerant) identified fewer regulated genes in the drought-tolerant genotype compared with the drought-susceptible genotype. IR64 extensively regulated genes with functions associated with signal transduction, protein binding and receptor activity, while Apo highly expressed genes associated with oxygen binding function and the peroxisome pathway (Ereful et al., 2020). Microarray analysis of florets of different sizes from rice plants subjected to water stress during the reproductive stage identified more than 1000 drought-responsive genes. Genes that function in tapetum, microspore development, cell wall formation or expansion, starch synthesis, GA signaling and ABA catabolism were highly regulated under drought stress, suggesting a strong interaction between reproductive development, phytohormone signaling and carbohydrate metabolism. Failure of male gametophyte development is mainly due to limitation in sugar utilization (Jin et al., 2013).

Gene expression studies in rice using whole-transcriptome strand-specific RNA sequencing (ssRNA-seq) detected 6885 transcripts, and of these, 238 lncRNAs (long non-coding RNA) were found to be involved in drought memory response (Li et al., 2019). Arabidopsis and maize plants exposed to drought stress, showed upregulation of stress memory genes that function in ABA-and

JA-regulated pathways (Ding et al., 2014; Virlouvet and Fromm, 2015). Rice genotypes, Heena and Kiran, contrasting for drought tolerance were identified by screening of 106 rice varieties, and these were used to study transcriptional alterations under drought stress. Although both varieties accumulated DEGs, 1033 and 936 were uniquely regulated in Heena and Kiran, respectively. The genes involved in phytohormone signaling, photosynthesis, antioxidative mechanisms and stress genes like *LEA*, *DREB*, and transcription factors like *AP2/ERF*, *MYB*, *WRKY*, *bHLH* were significantly regulated (Tiwari et al., 2021). These transcription factors, along with *bZIP* and *NAC*, are also found to be regulated in microarray analysis of drought-stressed rice plants (Zhu et al., 2010).

Whole-genome-level regulation of gene expression in response to drought stress at various developmental stages (seedling, booting, anthesis, grain filling) has identified many drought-responsive genes in wheat using microarray and RNA-seq (Aprile et al., 2009, 2013; Li et al., 2012; Liu et al., 2015). Recently, regulation of gene expression in wheat under drought stress in field conditions during the reproductive stage was analyzed using RNA-seq (Ma et al., 2017). This study has revealed that drought stress during early reproductive stages rather than drought during flowering had a greater impact on yield. Around 309 DEGs that play a critical role in floral development, photosynthetic activity and stomatal movement were identified in response to drought at various developmental stages. Galactose, sucrose and starch metabolism, flavonoid biosynthesis and circadian rhythms were regulated at the pistil and stamen differentiation stage. At the anther differentiation stage, carbon fixation, carbon, glyoxylate and dicarboxylate metabolism were regulated. At the tetrad stage, the porphyrin and chlorophyll metabolism was regulated, while starch and sucrose metabolism was regulated at the grain formation stage. Most DEGs were identified at the pistil and stamen differentiation stage (Ma et al., 2017). Genomic and RNA-seq data analysis of drought-tolerant (NI5439) and drought-susceptible (WL711) wheat genotype identified 45,139 DEGs, 1380 transcription factors, 288 miRNAs and 640 pathways as being regulated under drought (Iquebal et al., 2019). These genes control sensing drought, root growth, and regulation of uptake, purine and thiamine metabolism, stomatal closure and senescence. Some of the known drought-responsive genes, like *ARF*, *WRKY*, *HSF*, *MPK3/MPK6* signaling, *AP2/ERF*, *MYB*, *NAC* transcription factors and genes encoding RNA-binding proteins, peroxidase, lipoxygenase and LRR receptor like serine/threonine proteins, are among the DEGs (Iquebal et al., 2019).

Comparative transcriptome analysis of a drought-tolerant pearl millet genotype (PRLT2/89-33) exposed to drought at both vegetative and flowering stages has been carried out (Shivhare et al., 2020). Around 510 and 495 DEGs were found to be unique to the vegetative and the flowering stage, respectively, under drought stress. The study identified upregulation of flavonoid, lignin biosynthesis, phenylpropanoid, pigment biosynthesis and secondary metabolite pathways during the flowering stage under both control and stress conditions. However, the stress-response-related biological processes that were upregulated included the oxidation-reduction process, signal transduction and response to extracellular stimulus in both stages of plant growth. In addition, 69 and 49 transcription factors were differentially expressed in the vegetative and reproductive stage, respectively. In the vegetative stage, transcription factors belonging to ARF, ARR-B, NAC, WRKY and HSF family were highly regulated, while in the reproductive stage, the AP2, ERF, bHLH, bZIP, NAC, and WRKY family of transcription factors were highly regulated. Nearly 466 genes were commonly regulated at both vegetative and reproductive stages under drought. These included genes associated with the hormones auxin, cytokinin, ethylene, ABA, and gibberellin. Among them, genes related to terpene biosynthesis, flavonoids and anthocyanin pigments were highly expressed under the reproductive stage compared with the vegetative stage (Shivhare et al., 2020). Transcriptome sequencing has also been carried out for pearl millet exposed to drought stress at the germination or vegetative stage (Choudhary et al., 2015; Dudhate et al., 2018; Shinde et al., 2018; Jaiswal et al., 2018).

In maize, transcriptomic analysis using microarray and RNA-seq under drought during the seedling stage and in different tissues has identified regulation of many genes, which belonged to ABA-dependent and other signaling pathways. A study by Wang et al. (2019) has shown the coordination of organ-specific and whole-plant drought response in a transcriptomic analysis of three different organs (ear leaf, ear and kernels) from plants subjected to drought stress. Compared with control, 2357 and 1402 genes in ear leaf, 1136 and 689 genes in ear, and 2627 and 3565 genes in the kernel, respectively, were found to be upregulated and downregulated. Mainly, metabolic and osmotic adjustment, protein-folding acclimation, and ABA-dependent and NAC-mediated stress response pathways were regulated in all organs under stress. In the ear leaf, photosynthetic proteins and net photosynthesis-regulating genes (genes involved in PSII, PSI, cytochrome b6/f complex, electron transport, ATPase complex, antenna proteins) were drastically reduced under stress. Cell division, cell growth and primordium development under drought were delayed in ear, while endosperm differentiation under drought was delayed in kernels. Transcriptome analysis identified that auxin played a major role in the regulation of stress response. In addition, enzymes involved in sucrose, trehalose, raffinose and proline biosynthesis and genes encoding peroxidase, thioredoxin and oxidative stress tolerance were highly induced in all the three organs under stress. Small HSP, HSP-101, HSP-70, LEA and other drought-induced proteins were also induced under stress (Wang et al., 2019). Similarly, transcript profiling of leaves, ears and tassels exposed to drought stress identified the regulation of metabolism and signaling pathways in response to drought (Miao et al., 2017; Danilevskaya et al., 2019).

Transcript profiling (RNA-seq) of leaves and roots from sorghum plants subjected to drought stress has identified 510–559 and 3368–5093 DEGs in leaves and roots, respectively, under mild and severe drought stress. Some of the transcription factors regulated by drought stress were HSF, ERF, NAM, ATAF 1/2, CUC2, NAC, WRKY, HD-ZIP, bHLH, and MYB. In addition, HSPs, LEAs, chaperones, aquaporins, and expansins, along with genes involved in ABA signaling and ROS scavenging, were regulated (Zhang et al., 2019). Similarly, transcript analysis using RNA-seq in sorghum genotypes contrasting for water use efficiency (WUE) and drought tolerance identified a greater number of drought-related genes were regulated in the sensitive genotype. The main genes upregulated belong to the oxidation-reduction reaction, antioxidants, secondary metabolism, the photosynthesis and carbon fixation process, lipids and carbon metabolism in sensitive genotypes (Fracasso et al., 2016). RNA-seq analysis of drought-tolerant (SC56) and sensitive (Tx7000) varieties identified overexpression of genes that play a role in defense against oxidative stress, like SOD1, SOD2, VTC1, MDAR1, MSRB2, and ABC1K1, in drought-tolerant genotypes even under control conditions. In addition, transmembrane transport and growth-promoting genes were heightened in the drought-tolerant genotype. Under drought stress, SOD1, RCI3, VTE1, UCP1, FD1 and FD2, CIPK1, CRK7, and SAUL1,which play a role in enhancing antioxidant capacity and regulatory factors and repress premature senescence, were upregulated in the drought-tolerant genotype (Azzouz-Olden et al., 2020).

Genes involved in protein synthesis, energy and defense were highly regulated in osmo-primed seeds of *Brassica oleracea* (Soeda et al., 2005). Seed priming is known to alter the seed metabolic profile and gene regulation and improve stress tolerance (Varier et al., 2010; Hussain et al., 2016; Srivastava et al. 2021). However, the molecular mechanism of priming is still unclear (Kaczmarek et al., 2017). Available microarray data was analyzed in barley varieties exposed to drought stress to identify genes involved in seed priming-induced drought tolerance. The ERF/AP2, C2C2-Dof, and bHLH transcription factor families were involved in priming-induced drought tolerance.

Nearly 116 and 173 genes were up- and downregulated, respectively, in microarray analysis in soybean under drought stress. LEA proteins, dehydrins, transcription factors and HSPs were among those upregulated (Luo et al., 2005). Using the 454 pyrosequencing platform, around 6313 and 5858 unigenes were identified from leaf and root tissue, respectively, from 1-month-old cotton plants subjected to drought stress. Regulator genes like WRKY, ERF, AP2, EREBP, MYB, and LEA were highly expressed in leaves. Genes involved in root development, like RHD3, LBD and WRKY75, were highly expressed in root tissue. In addition, chlorophyll a/b binding protein and photosystem-related proteins were highly expressed in roots compared with leaves (Ranjan and Sawant, 2015;

Ranjan et al., 2012a, b). Similarly, under drought stress, several genes and major biochemical pathways were upregulated in root tissues of *Gossypium raimondii*. The pathways regulated under water stress were starch and sucrose metabolism, glycolysis–gluconeogenesis, amino sugar and nucleotide sugar metabolism, galactose metabolism, flavonoid biosynthesis, carotenoid biosynthesis and oxidative phosphorylation (Bowman et al., 2013). Using high-throughput NGS of three cotton species (*G. hirsutum*, *G. rboretum*, *G. barbadense* L.) under drought stress identified 6968 DEGs. Transcription factors associated with ethylene-responsive genes (*ICE1*, *MYB44*, *FAMA*, etc.) along with genes linked to ABA response (*NCED*, *PYL*, *PP2C*, and *SRK2*), ROS, *HSP*, *YSL3* and *Odorant1* gene were regulated under drought (Hasan et al., 2019). Genes encoding enzymes involved in pectin modification and cytoskeleton proteins, transcription factors (AP2-EREBP, WRKY, NAC, and C2H2), osmoprotectants, ion transporters, HSPs, hormone biosynthesis and signal transduction (ABA, ethylene, JA) were upregulated, while phenylpropanoid and flavonoid biosynthesis, pentose and glucuronate interconversion, and starch and sucrose metabolism were downregulated during fiber elongation in cotton under drought stress, as observed by comparative transcriptomic analysis (Padmalatha et al., 2012). In important oilseed crop sesame, You et al. (2019) identified DEGs involved in protein processing in endoplasmic reticulum, plant hormone signaling, photosynthesis, lipid metabolism, and amino acid metabolism. More genes were differentially regulated in the drought-sensitive genotype compared with the drought-tolerant genotype. Genes enriched in the drought-tolerant genotype are associated with stress response, amino acid metabolism and ROS scavenging (You et al., 2019).

Similar studies on the identification of DEGs using transcriptomic platforms are also carried out in non-crop species. For example, in Masson pine, large-scale transcriptome sequencing using Illumina technology identified 1695 and 1550 DEGs in response to moderate and severe drought stress. Genes related to oxidation-reduction, plant hormone signal transduction, metabolism, and metabolic pathways were highly regulated. Transcription factors associated with circadian rhythm (HY5, LHY), signal transduction (ERF) and defense response (WRKY) were upregulated under stress. The functional genes involved in osmotic adjustment (P5CS), ABA biosynthesis and signal transduction (NCED, PYL, PP2C, SnRK), and ROS scavenging (GPX, GST, GSR) were also among the upregulated genes (Du et al., 2018). Similar transcriptomic analysis using RNA-seq in tea plant identified 5955 DEGs. Genes involved in ABA, ethylene, and JA biosynthesis and signaling were upregulated in leaves under drought stress. A large number of protein kinases and transcription factors, along with genes for starch decomposition, mannitol, trehalose and sucrose synthesis, were upregulated, while genes related to starch synthesis were downregulated under drought stress in tea (Liu et al., 2016). Transcriptomic changes in drought-tolerant (Saba) and drought-sensitive (Grand Naine) banana were detected by mRNA-seq analysis. Around 16% and 9.5% of DEGs belong to transcription factors (46 families) and protein kinase (82 families), respectively. In addition, DEGs involved in carbohydrate degradation, protein modification, lipid metabolism, alkaloid biosynthesis, glycan metabolism, amino acid biosynthesis, cofactors, nucleotide sugars, hormones, terpenoids, and other secondary metabolites were regulated under drought stress (Muthuswamy et al., 2016).

10.3.3 Heat Stress

Plant responses to high-temperature stress are manifested at physiological, biochemical and molecular levels (Reddy et al., 2016; Saini et al. 2021). Plant response patterns in the vegetative and reproductive tissue vary based on exposure to heat stress. Heat stress affects reproductive development, early phase anther and ovule development, and meiosis. There has been a great deal of information generated for heat-stress-induced transcriptomic changes in model plants like *Arabidopsis*, *Boechera* and *Physcomitrella*. Targeting shoots or whole plants to heat stress (40 °C) resulted in a transient decrease in ABA with a slight increase in cytokinin and cytokinin signaling pathway genes. Shoot apices responded by the stimulation of photosynthesis and carbohydrate metabolism genes. Heat stress to roots induced the expression of GRF9, associated with plastid biosynthesis of isopentenyl diphosphate (Dobrá et al., 2015). Transcriptional and physiological responses to heat are sensitive to the time of day. Nearly 75% of heat-responsive transcripts display time of day-dependent response. Novel night-responsive genes and *cis*-regulatory elements responsible for heat stress responses have been identified in Arabidopsis. The heat response network is different between heat shock during the day and at night. For example, HSFA1b is induced only in the night but not in the day. Among the top 100 targets, 16 were molecular chaperones, which were more highly expressed during the day, and four among them were specific to the day. Similarly, HSF7A and AT2G44130 (Kelch repeat F-box protein) displayed higher expression during the night. Transcripts like DREB2A, HSP70, HSP90.1, HSP90.2, HsfA7B, HsfA2, MBF1c, etc. are induced by heat shock during both the day and the night (Grinevich et al., 2019). Genes participating in the unfolded protein response (UPR-bZIP28 and 60) were found to be highly enriched in the heat stressed reproductive tissue. Loss of function of these genes (bZIP28/60) displayed their protective role in maintaining fertility and silique length. In addition, 455 transcription factors belonging to 54 different families are significantly regulated by heat stress (Zhang et al., 2017).

The male gametophyte is the most heat sensitive compared with the whole plant, as evidenced by the RNA-seq analysis of pollens from different species subjected to heat stress (Fragkostefanakis et al., 2016; Muller and Rieu, 2016; Begcy et al., 2019). In response to heat stress, pollen significantly regulated 2102 transcripts, of which 89 coded for transcription factors and 27 were specific to pollens. Pollen of heat-sensitive Arabidopsis carrying a knockout of cyclic nucleotide gated cation channel 16 displayed a ninefold higher number of transcripts than wild-type pollens under heat stress. High expression of genes related to cell wall or membrane dynamics was observed in knockout lines (Ishka et al., 2018). Nearly 66 novel and 246 annotated intergenic expressed loci (XLOCs) of unknown function were identified by RNA-seq analysis of developing and germinating Arabidopsis pollens exposed to heat stress. Comparison with heat-stress leaf RNA-seq identified 74% of the 312 XLOCs to be specific to pollens and 42% to be responsive to heat stress (Rutley et al., 2021).

Adaptive response of Arabidopsis to prolonged heat stress was analyzed by ribosome profiling and RNA-seq. Expression of many genes was regulated at both transcriptional and translational level. Genes that displayed disproportionate ratios of mRNA to ribosome protected fragments (RPF) are likely candidates, whose expression is controlled at either the transcriptional or the translational level. Nearly 723 and 1150 genes were regulated only transcriptionally and translationally, respectively (Lukoszek et al., 2016). Chloroplast RNA-seq (ChloroSeq), a bioinformatics pipeline analysis of the organellar transcriptome, revealed a global reduction in splicing and editing efficiency leading to increased abundance of chloroplast transcripts, including genic, intergenic, and antisense transcripts, in response to heat stress (Castandet et al., 2016).

RNA-seq analysis of *Boechera*, a highly thermotolerant plant compared with Arabidopsis, was carried out to identify genes responsible for heat stress tolerance. There was a significant difference in the response of Arabidopsis and Boechera to identical heat stress conditions. Heat shock genes were highly upregulated in Arabidopsis, while novel genes coding for uncharacterized function, proteins associated with unfolded proteins, endoplasmic reticulum (ER) stress response, genes that lack orthologs in other genomes, and genes that are protective in photosynthetic capacity were differentially regulated in Boechera (Ian et al., 2020).

The moss *Physcomitrella patens* is used as a model organism to study the adaptation of plants to various abiotic stresses. Plants exposed to mild stress of 32 °C were more tolerant to subsequent stress of 35 °C compared with non-acclimated plants. The heat shock response was shown to be occurring by the regulation of specific calcium-permeable channels in plasma membrane (Saidi et al., 2009). Transcriptome-level dynamic heat shock response was studied by exposing plants to a continuous temperature rise (mild – 30 °C and high – 37 °C) without acclimation. Most of the upregulated genes during the initial 1 h encode heat shock proteins, while prolonged exposure for

24 h encodes for the thiamine biosynthesis pathway with repression of photosynthesis and carbon portioning genes. Most of the genes contained the HsfA1E binding motif within their promoter (Elzanati et al., 2020).

In crop plants like rice, co-expression network analysis is used to identify the functional association of co-expressed genes in response to heat stress (Sarkar et al., 2014). Protein kinase activity or protein serine-threonine kinase activity, and ROS were upregulated 10 min after heat stress. Transcription factor binding sites for HSFs, bZIPs and DREBs were found in genes regulated by heat stress (Sarkar et al., 2014). RNA-seq analysis of heat-stressed and control plants at anthesis stage from heat-tolerant (N22) and sensitive cultivars identified a number of transcription factor families, signal transduction and metabolic pathway genes as being regulated (downregulated). Around 630 DEGs were identified in response to heat stress. Expression of genes coding for HSFs and HSPs was highly upregulated. Regulated expression of protective chaperones in anthers is required during anthesis to overcome stress damage and ensure fertilization (González-Schain et al., 2016). Transcriptome analysis using RNA-seq of heat-tolerant (HT54) and heat-sensitive (HT13) rice lines identified upregulation of 1438 and 1599 genes and downregulation of 1237 and 1754 genes in heat-sensitive and tolerant lines, respectively, in response to heat stress. Nearly 3059 genes were specifically regulated in the heat-tolerant line, of which 314displayed significant regulation. Of these, 91 genes were not functionally annotated, while the other 223 genes belonged to oxidation, transportation, metabolism, defense response and transcription pathways (Fang et al., 2018).

Physiological and transcriptomic analysis of heat-tolerant (SDWG005) and heat-sensitive rice (MH101) lines that differ in pollen viability, pollination characteristics and antioxidant enzyme activity in leaves and spikelets in response to heat stress was carried out. Regulation of genes coding for cellular activities such as response to abiotic stress and metabolic reorganization was observed in the heat-tolerant line. However, genes coding for DNA replication and DNA repair proofreading were downregulated in heat-sensitive lines. Nearly 77 and 11 DEGs involved in lignin and flavonoid biosynthesis pathways, respectively, were upregulated in heat-tolerant compared with heat-sensitive lines, indicating that lignin and flavonoid biosynthetic pathways play important roles in heat resistance in rice during meiosis (Cai et al., 2020). Stable anther structure may contribute to the heat-tolerant nature of tolerant lines. Transcriptomic analysis revealed 3559 DEGs in heat-tolerant (SDWG005) anthers at anthesis during heat stress. Genes coding for transcription factors, nucleic acid synthesis and metabolism, protein synthesis and modification, hormone signal transduction, ROS elimination and photosynthesis were upregulated (Liu et al., 2020). Transcriptomic and proteomic analysis was carried out in II YOU838 (II8) hybrid rice and its parents Fu Hui 838 (F8) and II-32A (II3) on exposure to heat stress of 42 °C to understand heat stress tolerance and heterosis. The time-dependent expression pattern and response of the hybrid were different from its parents under similar heat stress conditions. Genes encoding response to stimuli, cell communication, and metabolic and transcription factor activity were upregulated, while genes encoding photosynthesis and signal transduction were downregulated. Nearly 35 unique differentially abundant proteins, including bHLH transcription factor, calmodulin binding transcription activator, HSP and chaperonin 60, were identified by proteomic analysis of hybrid rice. The HSF–HSP network was identified as contributing to the heat tolerance of hybrid rice (Wang et al., 2020).

In wheat, transcriptomic analysis of flag leaf and grains identified 1705 and 17 common DEGs, respectively, in heat-stressed plants. The transcripts regulated by heat stress are different between flag leaf and grains, with regulatory mechanisms in flag leaf being more complex and tightly regulated. Zeatin, brassinosteroid, and flavonoid biosynthesis pathways seem to play important roles in regulating heat tolerance in wheat. In wheat grains of heat-stressed plants, genes involved in alpha-linolenic acid and glycerophospholipid metabolism and brassinosteroid, zeatin and unsaturated fatty acid biosynthesis were upregulated, while genes involved in nitrogen, amino sugar and nucleotide sugar, starch and sucrose metabolism, phenylpropanoid biosynthesis and carbon fixation were downregulated. In flag leaf of heat-stressed plants, genes involved in glutathione, glycerolipid, starch and sucrose metabolism, phenylpropanoid and flavonoid biosynthesis were upregulated, while genes coding for apoptosis were downregulated (Su et al., 2019). Similarly, comparative transcriptome profiling was carried out in developing grains of three wheat genotypes contrasted for heat stress tolerance to identify genes regulating heat stress tolerance. Most key genes were downregulated in the heat-sensitive genotypes. Genes coding for 6-phosphogluconate dehydrogenase, S6 RPS6-2 ribosomal protein, peptidylprolylisomerase, plasma membrane proton ATPase, heat shock cognate-70, FtsH protease, Rubisco activase B, methionine synthase, cytochrome C (class I) and HMW-glutenin were upregulated in heat-tolerant genotypes. In heat-sensitive genotypes, genes coding for β-glucanase (1,3 and 1,4), triose phosphate isomerase and calnexin were upregulated, while genes coding for Na^{+}/H^{+} antiporter, glucose-1-phosphate adenylyltransferase, ips1 riboregulator and AraC family transcriptional regulator were downregulated, suggesting differential expression of different sets of genes between heat-tolerant and sensitive genotypes (Rangan et al., 2020). High-throughput RNA-seq data obtained for wheat genotypes contrasted for heat tolerance (heat tolerant: HD2967 and heat sensitive: BT-Schomburgk) was subjected to a systems biology approach, i.e., weighted gene co-expression network analysis, to identify important key genes regulating heat tolerance. Around 522 and 668 DEGs were identified in the heat-sensitive and tolerant genotype, respectively. Hub genes of highly expressed module belonged to cysteine rice receptor kinase 26. Similarly, Remorin family protein was identified in the heat-tolerant variety (Mishra et al., 2020). Genome-wide analysis of alternative splicing responses to heat stress identified 3576 genes exhibiting change in the alternative splicing pattern. In addition, differential alternative splicing was observed among wheat homologous genes in response to heat stress, with more alternative splicing occurring in B genome compared with A and D genome. Around 40% of the genes induced by heat stress undergoing alternative splicing were subjected to transcriptional regulation (Liu et al., 2018).

In maize, RNA-seq-based transcriptomic studies were conducted to analyze the heat stress response of plants. Nearly 1857 DEGs were regulated in response to heat stress, with 1029 and 828 genes being upregulated and downregulated, respectively. Pathway analysis of DEGs revealed that protein processing in the endoplasmic reticulum pathway, including HSPs (Hsp40, Hsp70, Hsp90, Hsp100) and small HSPs, plays a central role in the heat stress response of maize plants. In addition, 167 transcription factors that belong to various families, i.e., MYB, AP2-EREBP, bZIP, bHLH, NAC and WRKY, were regulated in response to heat stress (Qian et al., 2019).

In potato, microarray hybridization studies with leaves and stolons from heat stressed plants (only shoot or root zone) have identified a total of 5093 DEGs. Among the DEGs, 1963 genes were regulated in the leaves of treated plants in a tissue-specific manner, with a higher number of HSPs upregulated in the leaf tissue. Around 281 DEGs coded for signaling proteins, with 24 of them involved in light signaling interconnected with temperature. The genes regulated in response to heat stress included those for altered source capacity and sink strength. Similarly, 1467 transcripts in potato tubers were regulated in response to heat stress. Genes coding for energy, nucleotide, lipid, DNA and hormone metabolism were downregulated in response to heat stress (Hastilestari et al., 2018). Integration of transcriptomic and metabolomics data of potato plants in response to heat stress identified 448 upregulated and 918 downregulated genes along with 325 and 219 metabolites in positive and negative ionization modes, respectively. The DEGs induced in response to stress were genes regulating photosynthesis, cell wall degradation, heat response, RNA processing and protein degradation. In addition, both transcript and metabolite profiling identified flavone and flavonol biosynthesis as being upregulated in response to heat stress (Liu et al., 2021).

10.4 COMBINED ABIOTIC STRESSES

In natural field conditions, plants are exposed to various abiotic and biotic stresses simultaneously. More than 300 stress genes are conserved across organisms, and these play a vital role in defense and repair against environmental stress (Kultz, 2005). The evolution of stress genes is rapid, to help the organism adapt to a changing environment. There are also reports of genes that are specific to organisms or species. For example, half of the osmo-responsive genes in Arabidopsis are plant specific, while antifreeze proteins have evolved in different phyla separately, etc. (Rabbani et al., 2003; Cheng, 1998). Plants respond to various stresses in a coordinated manner. Most of the time, the signals are interconnected and complex, involving regulation of many metabolic pathways. The transcriptome changes in response to combined abiotic stress can be unrelated, agonistic or antagonistic compared with the plant's response to individual stresses. A study of transcriptome changes in 10 Arabidopsis ecotypes exposed to cold, heat, highlight, salt and flagellin individually or as a combination of two stresses was carried out. Nearly 61% of the changes in the transcriptome under combined stress (double stress) were unpredictable compared with the response to individual stress. In addition, only 5–10% of the transcriptome response was antagonistic between combined and individual stress (Rasmussen et al., 2013).

Most of the studies have focused on studying the effect of an individual or single stress on crop and model plants; and there have been few studies on the effect of combinations of various stresses (Mittler, 2006; Atkinson and Urwin, 2012). For example, a couple of studies have shown the effect of combined heat and drought stress (Rizhsky et al., 2004). Meta-analysis of data from plants' response to individual stress and combined stress would help decipher and understand the complete components of stress response. This is also vital to understand the signaling cross-talk that occurs in response to various stresses, leading to a molecular response to alter the physiological and biochemical response, and ultimately altering the stress response of the individual (Rizhsky et al., 2004; Koussevitzky et al., 2008; Atkinson et al., 2013; Ramegowda and Senthil-Kumar, 2015). Molecular response to individual stresses and combinations of stresses could be entirely different. Hence, it is important to understand the regulation under combined stress (Mittler, 2006; Atkinson and Urwin, 2012).

The transcriptome of maize seedling (leaf) exposed to drought, salt, heat or cold stress in comparison with non-stressed seedling was profiled. Around 167 genes, of which 57 and 110 were up- and downregulated, respectively, compared with control, were commonly regulated in all the four abiotic stresses. Among them, 10 transcription factors belonging to ERF, NAC, ARF, MYB and HD-ZIP family were upregulated, and two TFs belonging to bZIP and MYB-related, were downregulated in all four abiotic stresses, suggesting that a common mechanism operates in response to various abiotic stresses. In addition, genes like Vp14, F-box protein and dehydrin, and genes related to hormones ABA and JA and fatty acid biosynthesis, were commonly regulated in all four abiotic stresses. Pairwise comparison identified 355, 203 and 197 DEGs in salt-drought, drought-heat and heat-cold, respectively. Few genes were common between drought-cold, salt-heat and salt-cold (Li et al., 2017). The majority of photosynthetic genes were downregulated, while cell rescue and transcription factors were upregulated, in response to both cold and salt stress in potato (Evers et al., 2012). In Arabidopsis, several genes regulated by phosphorus were also regulated by sucrose (Muller et al., 2007).

Combined stress imposition of high light and high temperature was found to regulate the induction of nearly 76stress-inducible genes in Arabidopsis, of which HsfA2 was the key candidate gene, which increased tolerance in HsfA2-overexpressing plants (Nishizawa et al., 2006). In sunflower, nearly 129 genes were differentially expressed in response to both high-light and high-temperature stress, of which 65 were specific to combined stress. Only 33 and 24 genes were also regulated by high light and high temperature, respectively. These genes mainly function in the regulation of cell structure and growth, translation, signal transduction (PP2C, calmodulin, 14-3-3 protein, kinase interacting protein family, HD-ZIP and Zn finger protein), transport activity (aquaporin like), protein folding and degradation in both leaves and immature seeds, indicating that high light and high temperature regulated protein synthesis and energy metabolism in sunflower. Upregulation of a gene involved in sterol biosynthesis, which is crucial to embryo growth and development, was observed in embryonic tissue under high-light and high-temperature stress (Hewezi et al., 2008).

Comparative transcriptomic analysis is a feasible option to identify genes that are commonly regulated or unique to combinations or specific stresses, respectively. Analysis of 390 microarray samples in Arabidopsis identified 28% of the differentially expressed genes as being commonly regulated by drought and cold stress. These genes regulate biological processes of photosynthesis, respiration, hormonal response, signal transduction, water deprivation, and metabolic processes. Forty-three transcription factors belonging to families like WRKY, NAC, MYB, AP2/ERF, bHLH and bZIP, which regulated 56% of the DEGs, were upregulated in response to a combination of drought and cold stress. These genes are involved in ABA signaling and stomatal closure, thereby minimizing water loss. The stress is also shown to reduce photosynthetic activity and shifts the metabolic pathway to meet the energy demand (Sharma et al., 2018).

Meta-analysis of microarray data (88 samples) of drought, cold, salinity, alkalinity, chilling, and waterlogging stress in cotton was carried out to identify genes commonly regulated by various abiotic stresses. Around 663 and 832 genes were significantly upregulated and downregulated, respectively, in comparison with control across datasets. Alcohol dehydrogenase 1 and uvrABC system A were highly upregulated, while HSP and chlorophyll a-b binding protein were highly downregulated. Broadly, the ubiquitin-dependent protein catabolic process, response to salt stress, ethylene-activated signaling pathway, pathways involved in biosynthesis of secondary metabolites, ribosomes, amino acid biosynthesis, hormone biosynthesis and signal transduction were highly regulated. Transcription factors belonging to 25 families, for example, ERF, NAC, bZIP, G2-like, WRKY, MYB,C2H2 and bHLH, were highly upregulated, and S1Fa-like, ARF and GRAS family were downregulated. Similarly, 26 protein kinase genes belonging to AGC, CMGC, RLK-Pelle, STE and TKL family were identified among the DEGs. Most of the photosynthesis, chloroplast stroma-related DEGs were downregulated (Tahmasebi et al., 2019).

Meta-analysis of eight publicly available rice abiotic stress transcriptomes led to the identification of genes commonly expressed under drought, salinity, submergence, metal, and other abiotic stresses. Around 1175 genes were commonly expressed, while 12,821 and 42,877 genes displayed expression only under individual stresses or were neutral in expression, respectively. Nearly 30 genes were expressed across all abiotic stresses. Of these, 28 displayed a tissue-specific expression pattern, of which nine (OS05G0350900, OS02G0612700, OS05G0104200, OS03G0596200, OS12G0225900, OS07G0152000, OS08G0119500, OS06G0594700, and Os01g0393100) were expressed in response to combined abiotic stress and high tissue-specific expression in seed, flower, seedling, germination, leaf, and root. These 30 novel abiotic stress-responsive genes regulate various stress pathways under combined abiotic stress. Some of them regulate accumulation of osmolytes (proline, lysine, glycine, betaine, sucrose, fructose, myo-inositol, and mannitol) that are known to play a major role in abiotic stress tolerance and avoidance. Combined abiotic stress also regulated the genes in both an ABA-dependent and an independent manner to modulate the expression of those involved in ROS homeostasis (Muthuramalingam et al., 2017). Genes like MATH (TRAF homology)-BTA and NAC domain are identified as the key players in response to combined abiotic stresses. MATH-BTB plays a significant role in activating ubiquitination during the regulation of various developmental, physiological processes, including resistance to biotic and abiotic stresses (Zapata et al., 2007; Kushwaha et al., 2016). NAC transcription factors are commonly induced by combined abiotic stresses that play a role in abiotic stress tolerance (Wu et al., 2009; Shao et al., 2015).

In another study involving rice plants exposed to cold, iron toxicity and salt stress, comparative analysis of RNA-seq data of rice leaves revealed 420 DEGs common to all three stress levels. The common genes were found to mainly affect the process of photosynthesis. Around 990, 753, and 572 were common to cold-salt, iron-salt and cold-iron stress, respectively. The expression pattern of genes was mostly similar between iron and salt stress but contrasted with that under cold stress: for example, cytochrome p450 like, ATP synthase subunit, SOD, stress-response protein nst1, chaperone protein htpg family, and drought induced protein rdi. However, the expression pattern of chloroplast-related genes was similar under all three stresses, indicating that they could be primary targets of stress (Amaral et al., 2016).

Meta-analysis of microarray data of sunflower plants exposed to drought, heat, salt, oxidative, cold stress and oomycete (*Plasmopara halstedii*) pathogen was carried out to identify common stress-responsive genes. Although 526 upregulated and 4440 downregulated genes were identified, it was interesting to see that none of the genes were commonly expressed under cold and oxidative stress. The greatest number of genes regulated in common were found between oxidative stress and pathogen infection. Most genes functioning in protein, ATP, DNA and RNA, Zn ion-binding, and as oxidoreductases, hydrolases, ligases, kinases, transcription factors and membrane transporters were upregulated. In addition, 29 genes involved in ROS-regulated and oxidative stress tolerance were commonly regulated by all the stresses. Transcription factors like C2H-ZF, MYB, MYC2, ERF12, and ERD6 were upregulated in a sunflower genotype tolerant to abiotic stresses (Ramu et al., 2016).

Combined drought and salt stress has also resulted in differential expression of 827 mRNAs, 12 miRNAs, and 364 proteins in cucumber seeds post germination. Of the 827 genes, 810 displayed similar expression patterns, while 17 displayed opposite expression patterns under drought and salt stress, respectively. In addition, 130 of the 827 genes had corresponding proteins in proteomic data. These common genes function in signal transduction of hormones, photosynthesis, and proline and arginine metabolism. The pathways enriched under both drought and salt stress belonged to porphyrin and chlorophyll metabolism, degradation of limonene and pinene, glyoxylate and dicarboxylate metabolism, biosynthesis and degradation of fatty acids, carbon metabolism, carbon fixation, biosynthesis of secondary metabolites and alpha-linolenic acid metabolism. The proteome data indicated that starch and sucrose metabolism, glycolysis or gluconeogenesis, carotenoid biosynthesis, steroid hormone biosynthesis, and arginine and proline metabolism are the common stress responses (Du et al., 2021).

10.5 TRANSCRIPTOMICS APPLICATIONS

In plants, exploration of candidate genes influencing any important trait can be approached through transcriptome analysis. Characterization of differential gene expression from plant tissues subjected to stress has gained much attention in transcriptome analyses (Weber, 2015). The identification of genes involved in a particular trait enhances the possibility of promoting crop improvement through genetic engineering (Sedeek et al., 2019). In rice, molecular mechanisms responsible for aerobic adaptation were elucidated by RNA sequencing. In this study, a number of transcripts responsible for aerobic adaptation were found to be higher in roots than in leaves. Candidate genes,i.e.,*MADS4*, *MADS5*, *MADS6*, *MADS7* and *MADS15*, sugar transporters, were highly and uniquely expressed in the aerobic-adapted cultivar of rice under aerobic conditions, indicating their role in adaptation (Phule et al., 2019). In pearl millet, root transcriptome analysis was performed to study the response of pearl millet roots to drought stress. This study revealed metabolic pathways such as photosynthesis, plant hormone signal transduction and mitogen-activated protein kinase signaling as a root-specific drought-responsive pathway (Dudhate et al., 2018). Another study, by Shinde et al. (2018), found that the pearl millet salinity stress-tolerant genotype accumulates higher sugar in leaves during salinity stress than the susceptible genotype, and in transcriptome analysis, enrichment of sugar-metabolism-related pathways was noted only in the salinity-tolerant genotype. This study suggested sugar as a key molecule for stress tolerance in pearl millet (Shinde et al., 2018). This study in pearl millet also discovered differentially expressed salinity stress-related candidate genes, and one of these, the NAC transcription gene PgNAC21, was overexpressed in *Arabidopsis thaliana*. Overexpression lines exhibited a higher degree of salinity stress tolerance than the wild type (Shinde et al., 2019). In sweet potato, RNA-seq was performed to identify several candidate gene families, such as basic helix–loop–helix (bHLH), basic leucine zipper (bZIP), cystein2/histidine2 (C2H2), C3H, ERF, homo domain-leucine zipper (HD-ZIP), MYB, NAC (NAM, ATAF1/2 and CUC2) and WRKY in response to drought stress (Arisha et al., 2020). Solanaceae plants produce steroidal glycoalkaloids (SGAs) as a host defense against stress. SGAs are a class of potentially toxic compounds but are beneficial for host resistance. Transcriptome analysis of potato revealed that the genes responsible for SGA production are induced by abiotic stress (Zhang et al., 2019). In apple, the plant hormone signal transduction pathway was identified as the largest functional pathway induced by abiotic stress (Li et al., 2019). Waterlogging conditions in maize activated metabolic pathways like energy production, programmed cell death, and ethylene responsiveness (Arora et al., 2017). In chickpea, several enzymes involved in carbohydrate metabolism, photosynthesis, and redox homeostasis were differentially expressed by drought and/or salinity stresses (Garg et al. 2016).

Microarray-based transcriptome analysis of bread wheat under drought stress identified four highly homologous wheat NAC genes, among which the TaNAC69gene showed high expression in root during stress. The functional analyses of *TaNAC69* in transgenic wheat showed more shoot biomass under drought stress conditions than in wild type. The study suggested that *TaNAC69* is involved in wheat adaptation to drought stress (Xue et al., 2011). High temperature or heat stress greatly affects crop production and productivity (Hatfield and Prueger, 2015). In *Brachypodium distachyon*, RNA-seq was performed to reveal the molecular mechanism in response to heat stress. The study identified DEGs in leaves of seedlings, leaves and inflorescences. Gene ontology analysis showed that the DEGs were responsible for heat stress response and protein folding (Chen and Li, 2017). Transcriptomic analysis of banana showed that six Rho-like GTPase (ROP) genes (*MaROP-3b*, *-5a*, *-5c*, *-5f*, *-5g* and *-6*) were highly expressed in response to cold, salt and drought stress. Among these six genes, *MaROP5g* was the most highly expressed in response to salt stress. Functional characterization of *MaROP5* using the model plant *Arabidopsis thaliana* showed that *Arabidopsis* plants overexpressing *MaROP5g* had longer primary roots and increased survival rates than wild type under salt stress. The enhanced expression of salt overly athwayy genes and calcium-signaling pathway genes in *MaROP5g*-overexpressing *Arabidopsis thaliana* reflected the enhanced tolerance to salt stress (Miao et al., 2018).

Heavy metal pollution has become a global problem (Tchounwou et al., 2012). In cotton, genes responsible for cadmium stress were elucidated by RNA-seq. Most of the genes differentially regulated during cadmium stress were associated with catalytic activity and binding action, especially metal iron binding (Chen et al. 2019). In barley, transcriptional-level RNA-seq analysis was performed to investigate the response of barley to copper and cobalt stress. Metabolic pathways such as anthocyanin biosynthesis, MAPK signaling, glutathione biosynthesis, phenylalanine metabolism, photosynthesis, arginine biosynthesis, fatty acid elongation, and plant hormone signal transduction biosynthesis were induced in response to metal stress (Lwalaba et al., 2021).

Besides RNA sequencing, small RNA has also gained a lot of attention. Small RNAs, especially microRNAs (miRNAs), regulate the expression of messenger RNA (mRNA) via mRNA degradation or transcription repression (Samad et al., 2017; Voinnet, 2009). A study in pearl millet identified that miRNAs and their target genes regulate the auxin response pathway under salinity stress in pearl millet (Shinde et al., 2020). In several plants, microRNA 397 has been reported to affect growth and yield. In banana, Musa-miR397 has been found to be associated with biomass and stress tolerance (Patel et al., 2019). Similar findings were noted in soybean, where miRNAs and their targets regulate the auxin signaling pathway during salinity stress (Sun et al., 2016). All these studies show that transcriptome analysis has potential in the identification of genes or small RNAs responsible for abiotic stress tolerance (Table 10.1). In future, these abiotic stress-responsive candidate genes or small RNAs can be successfully used to impart abiotic stress tolerance in plants through transgenic approaches.

TABLE 10.1
Selected Studies of Transcriptome Analysis of Plants in Response to Different Abiotic Stresses

Plant	Stress Condition	Molecular Response	Reference
Rice	Aerobic adaptation	Upregulation of MADS family transcription factors and transporters	Phule et al. (2019)
Pearl millet	Salinity	Key regulation of sugar metabolism-related pathways	Shinde et al. (2018)
Pearl millet	Drought	Upregulation of photosynthesis and plant hormone signaling pathways	Dudhate et al. (2018)
Pearl millet	Salinity	MicroRNA and their targets activate auxin signaling pathway	Shinde et al. (2020)
Maize	Waterlogging	Pathways like energy production and ethylene responsiveness are uniquely expressed under waterlogged stress.	Arora et al. (2017)
Apple	Drought, cold and high salinity	Plant hormonal, signal transduction pathways	Li et al. (2019)
Soyabean	Salinity	MiRNAs along with auxin signaling play a key role in regulating root development	Sun et al. (2016)
Chickpea	Drought and/or salinity	Differential expression of carbohydrate metabolism, photosynthesis, redox homeostasis	Garg et al. (2016)
Potato	Heat	Upregulation of heat shock proteins	Tang et al. (2020)
Sweet potato	Drought	Regulation of candidate gene families such as bHLH, bZIP, C2H2	Arisha et al. (2020)
Banana	Salinity	Higher expression of six ROP genes Longer primary roots and increased survival rates in *MaROP5g*-overexpressing Arabidopsis plants	Miao et al. (2018)
Bread wheat	Drought	Identification of four wheat NAC genes. Transgenic wheat expressing *TaNAC69* showed more shoot biomass	Xue et al. (2011)
Cotton	Cadmium	Differential regulation of Cd-responsive genes associated with catalytic activity and binding action	Chen et al. (2019)
Barley	Copper and cobalt	Upregulation of metabolic pathways of anthocyanin biosynthesis, MAPK signaling, photosynthesis	Lwalaba et al. (2021)
Brachypodium distachyon	Heat	Differential regulation of genes responsible for heat stress response and protein folding	Chen and Li (2017)

10.6 TRANSCRIPTOMICS IN HALOPHYTES

Salt stress is a major abiotic stress, and it drastically reduces crop yield around the world by limiting seed germination, plant growth and crop productivity across the globe (Yang and Guo, 2018). Most plants are glycophytes, which are adversely affected by high salt stress (Mantri et al., 2012). In contrast to glycophytes, halophytes can survive under extreme conditions like high salinity, drought and heavy metal stress (Nikalje and Suprasanna, 2018). During evolution, they have adapted different morphological, anatomical, and physiological strategies to grow under high salinity (Grigore et al.,

2014). Therefore, understanding the mechanism of salinity tolerance in halophytes can be useful for improving salinity tolerance in crop plants. Various halophytes have been explored to understand salt stress responses at the physiological, molecular, biochemical, metabolic, proteomic, and ionomic levels (Kumari et al., 2015; Nikalje et al., 2018). However, plant tolerance to salt stress is a multigenic trait, and therefore it is difficult to understand the mechanism of salinity tolerance (Nongpiur et al., 2016). In the last two decades, advanced genomic tools like NGS technology have been developed to aid the understanding of gene expression patterns in plants under various stresses through transcriptomic studies (Wang et al., 2020) (Table 10.2). Transcriptome studies have

TABLE 10.2
Key Findings from Transcriptomic Studies in Different Halophytes

Plant	Key finding	References
Achnatherum splendens	Genes related to transcription factors, ion transporters, cellular communication and metabolism were differentially expressed under salt stress.	Liu et al. (2016)
Aeluropus littoralis	Genes related to transcription factors (ERF, WRKY, C2H2, NAC, bZIP, MYB), MAPK and phytohormones signaling, synthesis of osmolytes, and ionic homeostasis were differentially expressed.	Younesi-Melerdi et al. (2020)
Atriplex centralasiatica	Genes related to various transcription factors, ion transporters, ROS scavenging and ABA-dependent signaling pathway were differentially expressed under salt stress.	Yao et al. (2020)
Avicennia officinalis	Genes related to ethylene and auxin signaling were upregulated and those involved in ABA signaling were downregulated.	Krishnamurthy et al. (2017)
Beta vulgaris ssp. *maritima*	Genes related to osmoprotection, membrane transport, molecular chaperoning, redox metabolism or protein synthesis were differentially expressed under salt stress.	Skorupa et al. (2016)
Caragana microphylla	Genes related to photosynthesis and chlorophyll were upregulated.	Kim et al. (2021)
Elaeagnus angustifolia	Downregulation of genes involved in photosynthetic pathways.	Lin et al. (2018)
Glehnia littoralis	Genes involved in secondary metabolite biosynthetic pathways and plant signal transduction pathways were altered. Various transcription factors were also differentially expressed.	Li et al. (2018)
Halogeton glomeratus	Vacuolar H^+ ATPase, ATP citrate synthase α chain protein 3, vegetative cell wall protein, peroxidase, serine hydroxyl methyl transferase, abscisic stress ripening protein (ASR) and betaine aldehyde dehydrogenase (BADH) were induced under salt stress. Other genes involved in ion transport, ROS scavenging, energy metabolism and hormone response pathways were also altered.	Wang et al. (2015)
Halogeton glomeratus	Genes related to signal transduction and various transporters were differentially expressed.	Yao et al. (2018)
Halogeton glomeratus	Genes related to transporters (NHD1), carbohydrate metabolism, and energy production and conversion were differentially expressed.	Wang et al. (2018)

(*continued*)

TABLE 10.2 (Continued)
Key Findings from Transcriptomic Studies in Different Halophytes

Plant	Key finding	References
Helianthus tuberosus	Genes related to carbohydrate metabolism, ribosome activation and translation, oxidation-reduction and ion binding were differentially expressed.	Zhang et al. (2018)
Hordeum marinum	Genes related to ion transporters (SOS1, HKT1;1, HKT1;5 and HKT2;2) and ROS detoxification genes were differentially expressed.	Huang et al. (2018)
Ipomoea imperati	Genes related to ABA signaling pathway, various transporters, transcription factors, antioxidant enzymes, and enzymes associated with metabolism of synthesis and catalysis were differentially expressed.	Luo et al. (2017)
Iris halophila Pall.	Hormone-dependent signaling pathways, sodium/potassium ion transporters, flavonoid biosynthesis pathway and some transcription factors were upregulated under salt stress.	Liu et al. (2018)
Iris lactea var. *chinensis*	Genes related to transcription factors (WRKY, AP2/ERF, Zinc finger, bHLH, MYB, NAC, etc.), transporters (ABC, amino acids, nitrate, potassium, proton, sugar, etc.) and ROS scavenging (GPX, POD, GLP, GST, SOD, CAT, etc.) were differentially expressed.	Gu et al. (2018)
Karelinia caspica	Genes related to ABA metabolism, transport, sensing and signaling were differentially expressed.	Zhang et al. (2014)
Kochia sieversiana	Genes related to aquaporins, antioxidant defense (Cu/Zn-SOD, CAT, Mn-SOD and GST1) and K transporters were differentially expressed.	Zhao et al. (2017)
Limonium bicolor	Genes related to ion transport, vesicles, reactive oxygen species scavenging, the abscisic acid-dependent signaling pathway and transcription factors were differentially expressed.	Yuan et al. (2016)
Mesembryanthemum crystallinum	Genes related to ion transport (NHD1, CHX19 upregulated and HKT1 and KT1 downregulated), proline biosynthesis, glycolysis, myo-inositol metabolism, and ABA synthesis were differentially expressed under salt stress in epidermal bladder cells (EBC).	Oh et al. (2015)
Millettia pinnata	Genes related to oxidation-reduction and cellular amino acid metabolism were differentially expressed under salt stress.	Huang et al. (2012)
Nitraria sibirica	Genes related to amino sugar and nucleotide sugar metabolism, plant hormone signal transduction, alanine, aspartate and glutamate metabolism, and alpha-linolenic acid metabolism were upregulated, and those related to phenylpropanoid and starch and sucrose metabolism were downregulated.	Li et al. (2017)
Phragmites karka	In shoot, transcription factors (ethylene-responsive transcription factor, MYB, C2CH Zn finger, FAR1 and AP2/ERF), heat shock proteins, chaperones, glutathione S-transferase and genes for the 26 S proteasome pathway, and in roots, transcription factors (ERF, NAC, WRKY, CCCH and MYB), ribosomal proteins, kinases, transporters (polyol and ion) and antioxidants were differentially expressed.	Nayak et al. (2020)

TABLE 10.2 (Continued)
Key Findings from Transcriptomic Studies in Different Halophytes

Plant	Key finding	References
Porteresia coarctata	Transcription factor-encoding genes were differentially expressed, including MYB, bHLH, AP2-EREBP, WRKY, bZIP and NAC. Genes related to amino acid biosynthesis, hormone biosynthesis, secondary metabolite biosynthesis, carbohydrate metabolism and cell wall structures were also altered.	Garg et al. (2014)
Reaumuria trigyna	Genes related to ion transport (HKT1, KUP, AKT, HAK, SOS, NHX, V-H^+-ATPase, CNGC, CHX, KEA, KOE, PM-H^+-ATPase and V-H^+-PPase) and ROS scavenging system (GLR, APX, MDAR, POD and SOD) were differentially expressed under salt stress.	Dang et al. (2013)
Salicornia europaea	Genes related to cell wall metabolism and lignin biosynthetic pathways were altered to promote the development of the xylem. Salt treatment activated genes related to electron transfer encoded by the chloroplast chromosome.	Fan et al. (2013)
Salicornia europaea L	Genes that are involved in osmotic adjustment and ion homeostasis were significantly altered under salt stress, including cation transporters and proteins for the synthesis of low-molecular-weight compounds.	Ma et al. (2013)
Salicornia persica	Genes related to transcription factors, protein kinases, ABA signaling and transporters were differentially expressed.	Aliakbari et al. (2020)
Spartina alterniflora	Genes related to transcription factors were differentially expressed under salt stress.	Ye et al. (2020)
Spartina alterniflora Loisel	Genes related to CDPK/SnRK family of protein kinases, casein kinase family, cyclin-dependent kinase family, serine-threonine protein kinase family, mitogen-activated protein kinases family and protein phosphatase family were upregulated. Potassium, ABC and hydrogen transporters and Na^+/H^+ antiporter were highly upregulated.	Bedre et al. (2016)
Sporobolus virginicus	Dehydrin and aquaporin, transporters such as cation, amino acid, and citrate transporters, and H^+-ATPase were upregulated in both shoots and roots. A gene involved in proline biosynthesis was upregulated in both root and shoot. Five bZIP-type transcription factor-related genes were upregulated in root.	Yamamoto et al. (2015)
Suaeda fruticosa	Genes related to transporters, kinases and phosphatases, transcription factors, hormones, photosynthetic genes, detoxifiers and osmolytes were differentially expressed.	Diray-Arce et al. (2015)
Suaeda glauca	Genes involved in signal transduction (protein phosphatase 2C 24 and 2C 8 and ABA signal transduction), transporters (oligopeptide and ABC transporters), cell wall and growth, ROS scavenging (APX, POX, SOD, CAT and GR) and transcription factors (WRKY and bHLH) were differentially expressed.	Jin et al. (2016)

(*continued*)

TABLE 10.2 (Continued)
Key Findings from Transcriptomic Studies in Different Halophytes

Plant	Key finding	References
Suaeda maritima	Genes related to transcription factors, cell wall metabolism, carbohydrate metabolism, ion relation, redox responses and G-protein, phosphoinositide and hormone signaling were differentially expressed.	Gharat et al. (2016)
Suaeda rigida	Genes related to nitrogen compound transport, organic substance transport and intracellular protein transport were downregulated, while genes related to cell wall biogenesis, cell wall assembly, extracellular matrix organization and plant-type cell wall organization were upregulated.	Han et al. (2020)
Suaeda salsa	Genes related to Na^+/H^+ exchanger, V-H^+ ATPase, choline monooxygenase, potassium and chloride channels, antioxidant system (Fe-SOD, glutathione, L-ascorbate and flavonoids), and plant hormone signal transduction pathways (auxin, ethylene and jasmonic acid) were upregulated under salt stress.	Guo et al. (2019)
Zoysia japonica Steud.	Genes belonging to Ca^{2+} sensor family, ROS scavenging and various transcription factors (ERF, bZIP, NAC, WRKY, MYB and bHLH) were differentially expressed under salt stress.	Xie et al. (2015)
Zoysia macrostachya	Genes related to plant hormone signal transduction, formation and development of salt glands, and ROS - responsive proteins were differentially expressed.	Wang et al. (2020)

been used as an important strategy to understand gene expression patterns under salt stress (Ma et al., 2013; Li et al., 2018). Earlier attempts were made to understand salt-induced transcriptomic responses in model glycophytic plants and crop plants like Arabidopsis (Jiang and Deyholos, 2006), maize (Qing et al., 2009), and rice (Walia et al., 2007b). From these studies, various salt-responsive genes and signal transduction pathways were identified and studied for their role in salt tolerance.

Various reports suggested that both halophytes and glycophytes have similar salt tolerance regulatory mechanisms and that there are quantitative rather than qualitative differences (Anjum et al., 2012; Rai et al., 2012). It may be because of the higher expression of key genes involved in salt tolerance mechanisms in halophytes. It means that promoters also play a crucial role in salt tolerance in halophytes (Mishra and Tanna, 2017). However, a few reports have suggested that amino acid substitution in halophytic transporters is responsible for salt tolerance by improving cation selectivity (Ali et al., 2016, 2018). Therefore, the focus has shifted to studying transcriptomic responses in halophytes to understand the mechanism of salinity tolerance. Various transcriptomic studies in halophytes have identified and isolated diverse salt-responsive genes, and are functionally validated using transgenic approach (Krishnamurthy et al., 2017, 2020).

Regulation of abiotic stress-induced ROS is essential to adapt under oxidative stress induced by extreme conditions. Transcriptomic studies in halophytes like *Halogeton glomeratus*, *Atriplex centralasiatica* and *Hordeum marinum* revealed induction of genes involved in the ROS scavenging system. Glutathione S-transferases (GSTs), peroxidases (PODs) and serine hydroxymethyltransferase (SHMT) genes were upregulated in *Halogeton glomeratus* (Wang et al., 2015). A study in *Iris lactea* found that genes involved in the glutathione (GSH)-ascorbate cycle, the catalase (CAT) pathway, the glutathione peroxidase (GPX) pathway and the peroxiredoxin/thioredoxin (PrxR/Trx) pathway were differentially expressed (Gu et al., 2018). Enzymes involved in antioxidant defense, like ascorbate peroxidase (APX), glutamate receptor (GLR), GST and GPX were upregulated under salt stress in *Atriplex centralasiatica* (Yao et al., 2020). Luo et al. (2017) reported, in *Ipomoea imperati*, that most of the ROS scavenging enzymes were upregulated in roots but not in leaf. Transcriptomic studies of *Hordeum marinum* (halophyte) and *H. vulgare* (glycophyte) revealed that the halophyte *H. marinum* showed higher expression of genes related to ROS detoxification than the glycophyte *H. vulgare* (Huang et al., 2018). In other halophytes, like *Atriplex centralasiatica* and *Hordeum marinum*, genes involved in the ROS scavenging system were significantly altered (Yao et al., 2020; Huang et al., 2018).

For maintaining ionic homeostasis under salt stress, plants exclude or compartmentalize excess Na^+ in the vacuole and regulate K^+ uptake (Munns and Tester, 2008). Transcriptomic studies in halophytes confirm this efficient mechanism to maintain ionic homeostasis, as the expression of various transporters, like high-affinity potassium transporters (HKTs), ABC transporters, K^+ channel (KAT), high-affinity K^+ transporter (HAK), Arabidopsis K^+ transporter (AKT), and tonoplast Na^+/H^+ antiporter (NHX),was differentially expressed under salt stress (Huang et al., 2018; Zhao et al., 2017; Diray-Arce et al., 2015; Liu et al., 2018;Gu et al., 2018). Members of the HKT family of transporters were upregulated in the root of *Iris lactea* var. *chinensis*, shoots of *Iris halophila* Pall, and whole seedling of *Atriplex centralasiatica* (Gu et al., 2018; Liu et al., 2018; Yao et al., 2020). However, they were downregulated in the leaf of *Reaumuria trigyna* (Dang et al., 2013). Enhanced expression of antiporter NHX1 and SOS1, which leads to better regulation of Na^+ under salt stress, was also reported in a number of halophytes (Ma et al., 2013; Yao et al.,2018; Yuan et al., 2016; Dang et al., 2013; Liu et al., 2016). Antiporters need an electrochemical H^+ gradient, which is generated by the activity of V-ATPase. Transcriptomics in halophytes like *Halogeton glomeratus*, *Festuca rubra* ssp. *Litoralis* and *Reaumuria trigyna* indicated higher expression of V-ATPase under salt stress (Diédhiou et al., 2009; Dang et al., 2013; Wang et al., 2015). Jin et al. (2016) reported that in transcriptomic data of *Suaeda glauca*, genes belonging to HKT and NHX family were not differentially expressed under salt stress, but their basal level of expression was enough to cope with salinity stress. In the halophytic turfgrass *Sporobolus virginicus*, amino acid transmembrane transporters, including proline-specific permease, were differentially expressed under salt stress (Yamamoto et al., 2015). Other solute transporters, like sugar and peptide transporters, were upregulated in *Ipomoea imperati* under salt stress (Luo et al., 2017).

The ABC transporters, one of the largest gene families which transport a diverse range of molecules across the plasma membrane, were also differentially expressed under salt stress in various halophytes (Jin et al., 2016). Genes from the ABC transporter family were downregulated in *Iris lacteal* (Gu et al., 2018); however, they were upregulated in *Atriplex centralasiatica* (Yao et al., 2020) and the leaf of *Spartina alterniflora* (Bedre et al., 2016). In addition to these transporters, some K^+ transporters were also differentially expressed under salt stress in halophytes. Expression of potassium channel KAT was induced under salt stress in *Iris halophila* (Liu et al., 2018) and *Atriplex centralasiatica* (Yao et al., 2020). K^+ uptake capacity in halophytes was also enhanced by the upregulation of K^+ transporters like HAK and AKT (Dang et al., 2013). Transcriptome studies in halophytes also revealed regulation, uptake and accumulation of K^+, Na^+, and Cl^+ by altering the expression of salt-responsive transporters (Bedre et al., 2016).

Another key finding from transcriptome studies is the regulation of various transcription factors. Transcription factors are essential for response to environmental stimuli and regulate the expression of genes by binding to promoters. Therefore, under salt stress, alterations in the expression of various transcription factors in halophytes indicate their functional relevance in salinity stress (Yuan et al., 2016; Li et al., 2018). Transcriptomic studies in *Glehnia littoralis* identified 1661 unigenes of transcription factors, and of these 151 were DEGs, including MYB, NAC, ERF, bHLH, and bZIP (Li et al., 2018). Similarly, studies in *Iris lactea* reported 247 DEGs encoding transcription factors. These transcription factors belong to 19 different subfamilies, like bHLH, APETALA2/

ethylene-responsive transcription factor (AP2/ERF), HSF, WRKY dehydration-responsive element-binding protein (DREB), bZIPs, etc. (Gu et al., 2018). Other important transcription factors, like bZIP, bHLH, HSF, WRKY, C2H2, ARF, bHLH, HSF, MADS, MBF, OFP, ZFP,AP2/EREBP, HD-ZIP, HB12, and MBF1C, were found to be differentially expressed under salt stress in halophytes (Garg et al., 2014; Yuan et al., 2016; Liu et al., 2018; Luo et al., 2017; Aliakbari et al., 2021). In a diverse range of halophytes, WRKY transcription factors were differentially expressed under salt stress. WRKY TFs play an important role in various signaling processes and interact with different proteins for transcriptional reprogramming (Diray-Arce et al., 2015; Jin et al., 2016; Liu et al., 2018; Nayak et al., 2020; Xie et al., 2015). Therefore, WRKY from halophytic origin can be used as a candidate for improving salt tolerance in crop plants.

Transcriptomic studies in halophytes showed differential expression of genes from a diverse range of hormone signaling pathways, like the ABA, cytokinin, ethylene and auxin signaling pathways (Krishnamurthy et al., 2017; Zhang et al., 2014; Wang et al., 2015).Other key findings from transcriptomic studies in halophytes suggest that genes responsible for cell wall organization or biogenesis are upregulated under salt stress (Luo et al., 2017). Fan et al. (2013) reported enrichment of lignin biosynthetic pathways and cell wall metabolism under salt stress in *Salicornia europaea*.

The above-mentioned transcriptomics studies in halophytes have identified and isolated various salt-responsive genes, which are potential candidates to improve salinity tolerance in crop plants. Some of the potential candidate genes were functionally validated via the transgenic approach. For example, transcriptomic studies in roots of *Avicennia officinalis* revealed upregulation of cytochrome P450 gene (CYP94B subfamily) (Krishnamurthy et al., 2017). Heterogeneous expression of cytochrome P450 gene from *A. officinalis* (*AtCYP94B1*) induced salt tolerance in Arabidopsis and rice (Krishnamurthy et al., 2020). Similarly, transcriptomic studies in *Millettia pinnata* observed altered expression of some transcripts under salt stress (Huang et al., 2012). Wang et al. (2013) extended this work and performed characterization of the chalcone isomerase gene from *Millettia pinnata*. This study showed improved salt tolerance in salt-sensitive *Saccharomyces cerevisiae*. Further extension of this work by introducing the candidate gene into crop plants may result in plant salt tolerance. Some other studies were also performed that utilized key findings from transcriptomic studies in halophytes to improve salinity tolerance in plants: for example, the C2H2 zinc finger gene from zoysia japonica (Xie et al., 2015; Teng et al., 2018), phosphatidylserine synthase from *Salicornia europaea* (Fan et al., 2013; Lv et al., 2020) and the *WRKY1* gene from *Reaumuria trigyna* (Dang et al., 2013; Du et al., 2017).

Attempts have been made to perform comparative transcriptomic studies in halophytes along with their related glycophytes, which revealed that most salt-responsive genes were constitutively expressed at a higher level as compared with glycophytes (Gong et al., 2005) suggesting that an effective transcriptional regulatory network is present in salt tolerance-related genes in halophytes. Therefore, efforts were made to study *cis*-regulatory elements from the halophytic salt-responsive genes (Mishra and Tanna, 2017). *T. halophila* origin *TsVP1*gene promoter carries a 130-bp *cis*-acting element, which induces β-beta-glucuronidase (GUS) expression in transgenic Arabidopsis when exposed to salinity (Sun et al., 2010). Therefore, along with halophytic salt-responsive genes, halophytic promoters are also important candidates in developing strategies for improving salinity tolerance in crop plants.

10.7 CONCLUSIONS AND FUTURE PERSPECTIVES

Plant growth and development under stress conditions is a highly complex, dynamic system involving a multitude of mechanisms induced by both intrinsic and extrinsic signals. Plants have evolved responses to stress via a network of cellular, molecular and physiological mechanisms, and their metabolic pathways are interwoven to maintain better homeostasis. Transcriptomics approaches have played a seminal role in deciphering the defense mechanisms and the liaison among different hubs and metabolic pathways, including signal transduction. A huge amount of data has been generated in several model plants and crop plants on the genes related to stress-responsive traits and metabolites and their regulatory circuits. However, it needs to be understood how synergistically the pathways and metabolic networks coordinate under a given stress condition. The RNA-seq technology has become invaluable for obtaining molecular insights into plant stress responses globally, covering an entire spectrum of genomic resources, including lncRNAs, coding genes and alternatively spliced isoforms. Several plant stress-related databases, such as RiceSRTFDB, STIFDB2, PASmiR, QlicRice, PhytAMP, PSPDB, and PSRN, have been developed to aid researchers in their explorations of abiotic stress-responsive genomic resources. For example, the plant stress RNA-seq Nexus presents a gamut of information on stress-specific differentially expressed transcripts across various plant species and different stresses (Li et al., 2018).

Plants under realistic field conditions are exposed to more than a single stress scenario, and hence, tolerance responses to combined stresses in the natural environment and cross-interaction among them will have to be investigated. Towards this, there is a need to better comprehend the role of signaling and metabolic pathways, which are often specific and also shared by plants. Genomics advances have also played a significant role in the identification and characterization of sugar transporters, which could become targets for developing stress tolerance in plants to mitigate the effects of climate change and improve crop yield. Transcriptome studies have generated a wealth of genomic resources, including molecular markers to genetically enhance trait selection and breeding. There has also been continued interest in functional validation of candidate genes derived from the transcriptomic studies to explore their roles in diverse abiotic stresses and to improve stress tolerance using molecular breeding and biotechnological approaches. Transcriptome profiling at cellular, tissue and organ level, and also at different developmental stages, could offer valuable information on the cross-talk among the stresses and the way plants respond to stress. For example, the genes that play a crucial role during flowering under heat stress are very useful in engineering heat-tolerant cultivars. In this regard, the omics approaches have become the choice tools for unravelling such molecular mechanisms of stress tolerance. The current research achievements and advancements in transcriptomics technology, for example, single-cell transcriptomics, will speed up plant stress biology research towards designing high-yielding, stress-tolerant crops to pave the way for sustainable agriculture.

REFERENCES

Agarwal, P., Parida, S.K., Mahto, A., Das, S., Mathew, I.E., Malik, N., & Tyagi, A. K. 2014. Expanding frontiers in plant transcriptomics in aid of functional genomics and molecular breeding. *Biotechnology Journal*, *9*(12): 1480–1492. https://doi.org/10.1002/biot.201400063

Ali, A., Khan, I.U., Jan, M., Khan, H.A., Hussain, S., Nisar, M., Chung, W.S., & Yun, D.J. 2018. The high-affinity potassium transporter EpHKT1; 2 from the extremophile *Eutrema parvula* mediates salt tolerance. *Frontiers in Plant Science*, *9*(1108): 1–11. https://doi.org/10.3389/fpls.2018.01108

Aliakbari, M., Razi, H., Alemzadeh, A., & Tavakol, E. 2021. RNA-seq transcriptome profiling of the halophyte *Salicornia persica* in response to salinity. *Journal of Plant Growth Regulation*, *40*: 707–721. https://doi.org/10.1007/s00344-020-10134-z

Allen, M.R., Dube, O.P., Solecki, W., Aragón-Durand, F., Cramer, W., Humphreys, S., Kainuma, M., Kala, J., Mahowald, N., Mulugetta, Y., Perez, R., Wairiu, M., & Zickfeld, K. 2018. Framing and context. In: *Global Warming of 1.5°C. An IPCC Special Report on the Impacts of Global Warming of 1.5°C above Pre-Industrial Levels and Related Global Greenhouse Gas Emission Pathways, in the Context of Strengthening the Global Response to the Threat of Climate Change, Sustainable Development, and Efforts to Eradicate Poverty* [V. Masson-Delmotte, P. Zhai, H.-O. Pörtner, D. Roberts, J. Skea, P.R. Shukla, A. Pirani, W. Moufouma-Okia, C. Péan, R. Pidcock, S. Connors, J.B.R. Matthews, Y. Chen, X. Zhou, M.I. Gomis, E. Lonnoy, T. Maycock, M. Tignor, and T. Waterfield (eds.)]. Geneva, Switzerland: IPCC.

Amaral, M.N., Arge, L.W.P., Benitez, L.C., Danielowski, R., da Silveira Silveira, S.F., da Rosa Farias, D., de Oliveira, A.C., da Maia, L.C., & Braga, E.J.B. 2016. Comparative transcriptomics of rice plants under cold, iron, and salt stresses. *Functional & Integrative Genomics*, *16*(5): 567–579. https://doi.org/10.1007/s10142-016-0507-y

Amirbakhtiar, N., Ismaili, A., Ghaffari, M.R., Firouzabadi, F.N., & Shobbar, Z-S. 2019. Transcriptome response of roots to salt stress in a salinity-tolerant bread wheat cultivar. *PloS ONE*, *14*(3): e0213305.

Anjum, N.A., Gill, S.S., Ahmad, I., Tuteja, N., Soni, P., Pareek, A., Umar, S., Iqbal, M., Pacheco, M., Duarte, A.C., & Pereira, E. 2012. Understanding stress-responsive mechanisms in plants: an overview of transcriptomics and proteomics approaches. In *Improving Crop Resistance to Abiotic Stress*, ed. N.Tuteja, S.S.Gill, A.F.Tiburcio, and R.Tuteja, 337–355. Wiley-VCH Verlag GmbH & Co. KGaA. https://doi.org/10.1002/9783527632930.ch15

Anupama, A., Bhugra, S., Lall, B., Chaudhury, S., & Chugh, A. 2019.Morphological, transcriptomic and proteomic responses of contrasting rice genotypes towards drought stress. *Environmental and Experimental Botany*, *166*: 103795-103809. https://doi.org/10.1016/j.envexpbot.2019.06.008

Aprile, A., Mastrangelo, A.M., De Leonardis, A.M., Galiba, G., Roncaglia, E., Ferrari, F., De Bellis, L., Turchi, L., Giuliano, G., & Cattivelli, L. 2009. Transcriptional profiling in response to terminal drought stress reveals differential responses along the wheat genome. *BMC Genomics*, *10*(279): 1–18. https://doi.org/10.1186/1471-2164-10-279

Aprile, A., Havlickova, L., Panna, R., Marè, C., Borrelli, G.M., Marone, D., Perrotta, C., Rampino, P., De Bellis, L., Curn, V., & Mastrangelo, A.M. 2013. Different stress responsive strategies to drought and heat in two durum wheat cultivars with contrasting water use efficiency. *BMC Genomics*, *14*(821): 1–18. https://doi.org/10.1186/1471-2164-14-821

Arisha, M.H., Ahmad, M.Q., Tang, W., Liu, Y., Yan, H., Kou, M., Wang, X., Zhang, Y., & Li, Q. 2020. RNA-sequencing analysis revealed genes associated drought stress responses of different durations in hexaploid sweet potato. *Scientific Reports*, *10*(1): 1–17. https://doi.org/10.1038/s41598-020-69232-3

Arora, K., Panda, K.K., Mittal, S., Mallikarjuna, M.G., Rao, A.R., Dash, P.K., & Thirunavukkarasu, N. 2017. RNAseq revealed the important gene pathways controlling adaptive mechanisms under waterlogged stress in maize. *Scientific Reports*, *7*(1): 1–12. https://doi.org/10.1038/s41598-017-10561-1.

Atkinson, N.J., & Urwin, P.E. 2012. The interaction of plant biotic and abiotic stresses: from genes to the field. *Journal of Experimental Botany*, *63*(10): 3523–3543. https://doi.org/10.1093/jxb/ers100

Atkinson, N.J., Lilley, C.J., & Urwin, P.E. 2013. Identification of genes involved in the response of Arabidopsis to simultaneous biotic and abiotic stresses. *Plant Physiology*, *162*: 2028–2041.

AzzouzOlden, F., Hunt, A.G., & Dinkins, A. 2020. Transcriptome analysis of drought tolerant sorghum genotype SC56 in response to water stress reveals an oxidative stress defense strategy. *Molecular Biology Reports*, *47*: 3291–3303. https://doi.org/10.1007/s11033-020-05396-5

Bahieldin, A., Atef, A., Sabir, J.S., Gadalla, N.O., Edris, S., Alzohairy, A.M., Radhwan, N.A., Baeshen, M.N., Ramadan, A.M., Eissa, H.F., Hassan, S.M., Baeshen, N.A., Abuzinadah, O., Al-Kordy, M.A., El-Domyati, F.M., & Jansen, R.K. 2015. RNA-Seq analysis of the wild barley (*H. spontaneum*) leaf transcriptome under salt stress. *ComptesRendusBiologies*, *338*(5): 285–297. https://doi.org/10.1016/j.crvi.2015.03.010

Bainbridge, M.N., Warren, R.L., Hirst, M., Romanuik, T., Zeng, T., Go, A., Delaney, A., Griffith, M., Hickenbotham, M., Magrini, V., Mardis, E.R., Sadar, M.D., Siddiqui, A.S., Marra, M.A., & Jones, S.J.M. 2006. Analysis of the prostate cancer cell line LNCaP transcriptome using a sequencing-by-synthesis approach.*BMC Genomics*, *7*(1): 1–11. https://doi.org/10.1186/1471-2164-7-246

Bartels, D., & Sunkar, R. 2005. Drought and salt tolerance in plants. *Critical Reviews in Plant Sciences*, *24*(1): 23–58. https://doi.org/10.1080/07352680590910410

Becker-Andre, M., & Hahlbrock, K. 1989. Absolute mRNA quantification using the polymerase chain reaction (PCR). A novel approach by a PCR aided transcript titration assay (PATTY). *Nucleic Acids Research*, *17*(22): 9437–9446. https://doi.org/10.1093/nar/17.22.9437

Bedre, R., Mangu, V.R., Srivastava, S., Sanchez, L.E., & Baisakh, N. 2016. Transcriptome analysis of smooth cordgrass (*Spartina alterniflora* Loisel), a monocot halophyte, reveals candidate genes involved in its adaptation to salinity. *BMC Genomics*, *17*(1): 1–18. https://doi.org/10.1186/s12864-016-3017-3

Begcy, K., Nosenko, T., Zhou, L.Z., Fragner, L., Weckwerth, W., & Dresselhaus, T. 2019. Male sterility in maize after transient heat stress during the tetrad stage of pollen development. *Plant Physiology*, *181*: 683–700. https://doi.org/10.1104/pp.19.00707

Bowman, M.J., Park, W., Bauer, P.J., Udall, J.A., Page, J.T., Raney, J., Scheffler, B.E., Jones, D.C., & Campbell, B.T. 2013. RNA-Seq transcriptome profiling of upland cotton (*Gossypium hirsutum* L.) root tissue under water-deficit stress. *PLoS One*, *8*(12): 1–13. https://doi.org/10.1371/journal.pone.0082634

Brenner, S., Johnson, M., Bridgham, J., Golda, G., Lloyd, D.H., Johnson, D., Luo, S., McCurdy, S., Foy, M., Ewan, M., Roth, R., George, D., Eletr, S., Albrecht, G., Vermaas, E., Williams, S.R., Moon, K., Burcham, T., Pallas, M., DuBridge, R.B., Kirchner, J., Fearon, K., Mao, J., & Corcoran, K.2000. Gene expression analysis by massively parallel signature sequencing (MPSS) on microbead arrays. *Nature Biotechnology*, *18*(6): 630–634. https://doi.org/10.1038/76469

Cai, Z., He, F., Feng, X., Liang, T., Wang, H., Ding, S., & Tian, X.2020. Transcriptomic analysis reveals important roles of lignin and flavonoid biosynthetic pathways in rice thermotolerance during reproductive stage. *Frontiers in Genetics*, *11*(1120): 1–15. https://doi.org/10.3389/fgene.2020.562937

Castandet, B., Hotto, A.M., Strickler, S.R., & Stern, D.B. 2016. ChloroSeq, an optimized chloroplast RNA-Seq bioinformatic pipeline, reveals remodeling of the organellar transcriptome under heat stress. *G3: Genes, Genomes, Genetics*, *6*(9): 2817–2827. https://doi.org/10.1534/g3.116.030783

Chandran, A.K.N., Kim, J.W., Yoo, Y.H., Park, H.L., Kim, Y.J., Cho, M.H., & Jung, K.H. 2019. Transcriptome analysis of rice-seedling roots under soil–salt stress using RNA-Seq method. *Plant Biotechnology Reports*, *13*(6): 567–578. https://doi.org/10.1007/s11816-019-00550-3

ChantreNongpiur, R., Lata Singla-Pareek, S., & Pareek, A. 2016. Genomics approaches for improving salinity stress tolerance in crop plants. *Current Genomics*, *17*(4): 343–357. https://dx.doi.org/10.2174%2F1389202917666160331202517

Chen, H., Li, Y., Ma, X., Guo, L., He, Y., Ren, Z., Kuang, Z., Zhang, X., & Zhang, Z. 2019. Analysis of potential strategies for cadmium stress tolerance revealed by transcriptome analysis of upland cotton. *Scientific Reports*, *9*(1): 1–13. https://doi.org/10.1038/s41598-018-36228-z

Chen, S., & Li, H.2017.Heat stress regulates the expression of genes at transcriptional and post-transcriptional levels, revealed by RNA-seq in *Brachypodium distachyon*. *Frontiers in Plant Science*, *7*(2067): 1–13. https://doi.org/10.3389/fpls.2016.02067

Cheng, C.H. 1998. Evolution of the diverse antifreeze proteins. *Current Opinion in Genetics and Development*, *8*: 715–720. https://doi.org/10.1016/S0959-437X(98)80042-7

Choudhary, M., Jayanand, & Padaria, J.C.2015.Transcriptional profiling in pearl millet (*Pennisetum glaucum* LR Br.) for identification of differentially expressed drought responsive genes. *Physiology and Molecular Biology of Plants*, *21*(2): 187–196. https://doi.org/10.1007/s12298-015-0287-1

Collier, R.H., & Else, M. (2014).UK fruit and vegetable production – impacts of climate change and opportunities for adaptation. In *Climate Change Impact and Adaptation in Agricultural Systems*, ed. J. Fuhrer and P.J. Gregory, 88–109. Wallingford: CABI.

Dang, Z.H., Zheng, L.L., Wang, J., Gao, Z., Wu, S.B., Qi, Z., & Wang, Y.C. 2013. Transcriptomic profiling of the salt-stress response in the wild recretohalophyte *Reaumuria trigyna*. *BMC Genomics*, *14*(1): 1–18. https://doi.org/10.1186/1471-2164-14-29

Danilevskaya, O.N., Yu, G., Meng, X., Xu, J., Stephenson, E., Estrada, S., Chilakamarri, S., Zastrow-Hayes, G., & Thatcher, S.2019. Developmental and transcriptional responses of maize to drought stress under field conditions. *Plant Direct*, *3*(5): 1–20. https://doi.org/10.1002/pld3.129

Das, P., Nutan, K.K., Singla-Pareek, S.L., & Pareek, A. 2015. Understanding salinity responses and adopting "omics-based" approaches to generate salinity tolerant cultivars of rice. *Frontiers in Plant Science*, *6*(712): 1–16. https://doi.org/10.3389/fpls.2015.00712

De Oliveira, E.D., Bramley, H., Siddique, K.H., Henty, S., Berger, J., & Palta, J.A. 2013. Can elevated CO_2 combined with high temperature ameliorate the effect of terminal drought in wheat? *Functional Plant Biology*, *40*: 160–171.

Dhankher, O.P., & Foyer, C.H. 2018. Climate resilient crops for improving global food security and safety. *Plant Cell & Environment*, *41*(5): 877–884.

Diédhiou, C.J., Popova, O.V., & Golldack, D.2009. Transcript profiling of the salt-tolerant *Festuca rubra* ssp. *litoralis* reveals a regulatory network controlling salt acclimatization. *Journal of Plant Physiology*, *166*(7): 697–711. https://doi.org/10.1016/j.jplph.2008.09.015

Ding, Y., Virlouvet, L., Liu, N., Riethoven, J.J., Fromm, M., & Avramova, Z. 2014. Dehydration stress memory genes of *Zea mays*; comparison with *Arabidopsis thaliana*. *BMC Plant Biology*, *14*(1): 1–15. https://doi.org/10.1186/1471-2229-14-141

Diray-Arce, J., Clement, M., Gul, B., Khan, M.A., & Nielsen, B.L. 2015. Transcriptome assembly, profiling and differential gene expression analysis of the halophyte *Suaedafruticosa* provides insights into salt tolerance. *BMC Genomics*, *16*(1): 1–24. https://doi.org/10.1186/s12864-015-1553-x

Dobrá, J., Černý, M., Štorchová, H., Dobrev, P., Skalák, J., Jedelský, P.L., Lukšanová, H., Gaudinová, A., Pešek, B., Malbeck, J., & Vanek, T. 2015. The impact of heat stress targeting on the hormonal and transcriptomic response in Arabidopsis. *Plant Science*, *231*: 52–61. https://doi.org/10.1016/j.plantsci.2014.11.005

Dong, Z., & Chen, Y. 2013. Transcriptomics: advances and approaches. *Science China Life Sciences*, *56*(10): 960–967. https://doi.org/10.1007/s11427-013-4557-2

Du, C., Zhao, P., Zhang, H., Li, N., Zheng, L., & Wang, Y. 2017. The *Reaumuria trigyna* transcription factor RtWRKY1 confers tolerance to salt stress in transgenic Arabidopsis. *Journal of Plant Physiology*, *215*: 48–58. https://doi.org/10.1016/j.jplph.2017.05.002

Du, C., Hao, L., Chen, L., & Huaifu, F. 2021. Understanding of the postgerminative development response to salinity and drought stresses in cucumber seeds by integrated proteomics and transcriptomics analysis. *Journal of Proteomics*, *232*: 104062. https://doi.org/10.1016/j.jprot.2020.104062

Du, M., Ding, G., & Cai, Q. 2018. The transcriptomic responses of *Pinus massoniana* to drought stress. *Forests*, *9*(326): 1–15. https://doi.org/10.3390/f9060326

Duarte-Delgado, D., Dadshani, S., Schoof, H., Oyiga, B.C., Schneider, M., Mathew, B., Léon, J., & Ballvora, A. 2020. Transcriptome profiling at osmotic and ionic phases of salt stress response in bread wheat uncovers trait-specific candidate genes. *BMC Plant Biology*, *20*(1): 1–18. https://doi.org/10.1186/s12870-020-02616-9

Dudhate, A., Shinde, H., Tsugama, D., Liu, S., & Takano, T.2018.Transcriptomic analysis reveals the differentially expressed genes and pathways involved in drought tolerance in pearl millet [*Pennisetum glaucum* (L.) R. Br]. *PloS ONE*, *13*(4): 1–14. doi.org/10.1371/journal.pone.0195908

Elzanati, O., Mouzeyar, S., & Roche, J.2020.Dynamics of the transcriptome response to heat in the moss, *Physcomitrella patens*. *International Journal of Molecular Sciences*, *21*(4): 1–19. https://doi.org/10.3390/ijms21041512

Ereful, N.C., Liu, L.Y., Greenland, A., Powell, W., Mackay, I., & Leung, H. 2020. RNA-seq reveals differentially expressed genes between two indica inbred rice genotypes associated with drought-yield QTLs. *Agronomy*, *10*(5):1–19. https://doi.org/10.3390/agronomy10050621

Evers, D., Legay, S., Lamoreux, D., Hausman, J.F., Hoffmann, L., & Renaut, J. 2012. Towards a synthetic view of potato cold and salt stress response by transcriptomic and proteomic analyses. *Plant Molecular Biology*, *78*: 503–514.

Fan, P., Nie, L., Jiang, P., Feng, J., Lv, S., Chen, X., Bao, H., Guo, J., Tai, F., Wang, J., & Jia, W. 2013. Transcriptome analysis of *Salicornia europaea* under saline conditions revealed the adaptive primary metabolic pathways as early events to facilitate salt adaptation. *PLoS ONE*, *8*(11): 1–18. https://doi.org/10.1371/journal.pone.0080595

Fang, C., Dou, L., Liu, Y., Yu, J., & Tu, J. 2018. Heat stress-responsive transcriptome analysis in heat susceptible and tolerant rice by high-throughput sequencing. *Ecological Genetics and Genomics*, *6*: 33–40. https://doi.org/10.1016/j.egg.2017.12.001

FAO. 1996. *Rome Declaration and World Food Summit Plan of Action*. Food and Agricultural Organization of the United Nations (FAO) World Food Summit, Rome, Italy, 13–17 November 1996, FAO, Rome, Italy. www.fao.org/docrep/ 003/w3613e/w3613e00.HTM

FAO. 2017. *The Future of Food and Agriculture—Trends and Challenges*. Rome: Food and Agriculture Organization of the United Nations. www.fao.org/3/i6583e/i6583e.pdf

Fernando, N., Panozzo, J., Tausz, M., Norton, R.M., Fitzgerald, G.J., Myers, S., Walker, C., Stangoulis, J., & Seneweera, S. 2012. Wheat grain quality under increasing atmospheric CO_2 concentrations in a semi-arid cropping system. *Journal of Cereal Science*, *56*(3): 684–690. https://doi.org/10.1016/j.jcs.2012.07.010

Field, C.B., Barros, V., Stocker, T.F., & Dahe, Q. eds. 2012. *Managing the Risks of Extreme Events and Disasters to Advance Climate Change Adaptation: Special Report of the Intergovernmental Panel on Climate Change*. Cambridge, England: Cambridge University Press, 582 pp.

Formentin, E., Sudiro, C., Perin, G., Riccadonna, S., Barizza, E., Baldoni, E., Lavezzo, E., Stevanato, P., Sacchi, G.A., Fontana, P., Toppo, S., Morosinotto, T., Zottini, M., & Schiavo, F.L.2018. Transcriptome and cell physiological analyses in different rice cultivars provide new insights into adaptive and salinity stress responses. *Frontiers in Plant Science*, *9*(204): 1–17. https://doi.org/10.3389/fpls.2018.00204

Fracasso, A., Trindade, L.M., & Amaducci, S. 2016. Drought stress tolerance strategies revealed by RNA-Seq in two sorghum genotypes with contrasting WUE. *BMC Plant Biology*, *16*(1): 1–18. https://doi.org/10.1186/s12870-016-0800-x

Fragkostefanakis, S., Mesihovic, A., Hu, Y., & Schleiff, E. 2016. Unfolded protein response in pollen development and heat stress tolerance. *Plant Reproduction*, *29*(1): 81–91. https://doi.org/10.1007/s00497-016-0276-8

Garalde, D.R., Snell, E.A., Jachimowicz, D., Sipos, B., Lloyd, J.H., Bruce, M., Pantic, N., Admassu, T., James, P., Warland, A., Jordan, M., Ciccone, J., Serra, S., Keenan, J., Martin, S., McNeill, L., Wallace, E.J., Jayasinghe, L., Wright, C., Blasco, J., Young, S., Brocklebank, D., Juul, S., Clarke, J., Heron, A.J., & Turner, D. 2018. Highly parallel direct RNA sequencing on an array of nanopores. *Nature Methods*, *15*(3): 201–208. http://dx.doi.org/10.1038/nmeth.4577

Garg, R., Bhattacharjee, A., & Jain, M. 2015. Genome-scale transcriptomic insights into molecular aspects of abiotic stress responses in chickpea. *Plant Molecular Biology Reporter*, *33*(3): 388–400. https://doi.org/10.1007/s11105-014-0753-x

Garg, R., Shankar, R., Thakkar, B., Kudapa, H., Krishnamurthy, L., Mantri, N., Varshney, R.K., Bhatia, S., & Jain, M. 2016. Transcriptome analyses reveal genotype-and developmental stage-specific molecular responses to drought and salinity stresses in chickpea. *Scientific Reports*, *6*(1): 1–15. https://doi.org/10.1038/srep19228

Garg, R., Verma, M., Agrawal, S., Shankar, R., Majee, M., & Jain, M. 2014. Deep transcriptome sequencing of wild halophyte rice, *Porteresia coarctata*, provides novel insights into the salinity and submergence tolerance factors. *DNA Research*, *21*(1): 69–84. https://doi.org/10.1093/dnares/dst042

Gharat, S.A., Parmar, S., Tambat, S., Vasudevan, M., & Shaw, B.P. 2016. Transcriptome analysis of the response to NaCl in *Suaeda maritima* provides an insight into salt tolerance mechanisms in halophytes. *PloS ONE*, *11*(9):1–35. https://doi.org/10.1371/journal.pone.0163485

Gong, Q., Li, P., Ma, S., InduRupassara, S., & Bohnert, H.J. 2005. Salinity stress adaptation competence in the extremophile *Thellungiella halophila* in comparison with its relative *Arabidopsis thaliana*. *The Plant Journal*, *44*(5): 826–839. https://doi.org/10.1111/j.1365-313X.2005.02587.x

González-Schain, N., Dreni, L., Lawas, L.M., Galbiati, M., Colombo, L., Heuer, S., Jagadish, K.S., & Kater, M.M. 2016. Genome-wide transcriptome analysis during anthesis reveals new insights into the molecular basis of heat stress responses in tolerant and sensitive rice varieties. *Plant and Cell Physiology*, *57*(1): 57–68. https://doi.org/10.1093/pcp/pcv174

Goyal, E., Amit, S.K., Singh, R.S., Mahato, A.K., Chand, S., & Kanika, K. 2016. Transcriptome profiling of the salt-stress response in *Triticum aestivum* cv. Kharchia Local. *Scientific Reports*, *6*(1): 1–14. https://doi.org/10.1038/srep27752

Grigore, M.N., Ivanescu, L., & Toma, C. 2014. *Halophytes: An Integrative Anatomical Study*. Cham: Springer. doi: 10.1007/978-3-319-05729-3_10

Grinevich, D.O., Desai, J.S., Stroup, K.P., Duan, J., Slabaugh, E., & Doherty, C.J. 2019. Novel transcriptional responses to heat revealed by turning up the heat at night. *Plant Molecular Biology*, *101*(1): 1–19. https://doi.org/10.1007/s11103-019-00873-3

Gu, C., Xu, S., Wang, Z., Liu, L., Zhang, Y., Deng, Y., & Huang, S. 2018. De novo sequencing, assembly, and analysis of *Iris lactea* var. chinensis roots' transcriptome in response to salt stress. *Plant Physiology and Biochemistry*, *125*: 1–12. https://doi.org/10.1016/j.plaphy.2018.01.019

Guo, S.M., Tan, Y., Chu, H.J., Sun, M.X., & Xing, J.C. 2019. Transcriptome sequencing revealed molecular mechanisms underlying tolerance of *Suaeda salsa* to saline stress. *PloS ONE*, *14*(7): 1–23. https://doi.org/10.1371/journal.pone.0219979

Han, Z.J., Sun, Y., Zhang, M., & Zhai, J.T. 2020. Transcriptomic profile analysis of the halophyte *Suaeda rigida* response and tolerance under NaCl stress. *Scientific Reports*, *10*(1): 1–10. https://doi.org/10.1038/s41598-020-71529-2

Hasan, M.M.U., Ma, F., Islam, F., Sajid, M., Prodhan, Z.H., Li, F., Shen, H., Chen, Y., & Wang, X. 2019. Comparative transcriptomic analysis of biological process and key pathway in three cotton (Gossypium spp.) species under drought stress. *International Journal of Molecular Sciences*, *20*(9): 1–24. https://dx.doi.org/10.3390%2Fijms20092076

Hastilestari, B.R., Lorenz, J., Reid, S., Hofmann, J., Pscheidt, D., Sonnewald, U., & Sonnewald, S. 2018. Deciphering source and sink responses of potato plants (*Solanum tuberosum* L.) to elevated temperatures. *Plant, Cell &Environment*, *41*(11): 2600–2616. https://doi.org/10.1111/pce.13366

Hatfield, J.L., & Prueger, J.H. 2015. Temperature extremes: effect on plant growth and development. *Weather and Climate Extremes, 10*: 4–10. https://doi.org/10.1016/j.wace.2015.08.001

Hewezi, T., Léger, M., & Gentzbittel, L. 2008. A comprehensive analysis of the combined effects of high light and high temperature stresses on gene expression in sunflower. *Annals of Botany, 102*: 127–140. https://doi.org/10.1093/aob/mcn071

Huang, J., Lu, X., Yan, H., Chen, S., Zhang, W., Huang, R., &Zheng, Y. 2012. Transcriptome characterization and sequencing-based identification of salt-responsive genes in *Millettia pinnata*, a semi-mangrove plant. *DNA Research, 19*(2): 195–207. https://doi.org/10.1093/dnares/dss004

Huang, L., Kuang, L., Li, X., Wu, L., Wu, D., & Zhang, G. 2018. Metabolomic and transcriptomic analyses reveal the reasons why *Hordeum marinum* has higher salt tolerance than *Hordeum vulgare*. *Environmental and Experimental Botany, 156*: 48–61. https://doi.org/10.1016/j.envexpbot.2018.08.019

Hussain, S., Yin, H., Peng, S., Khan, F.A., Khan, F., Sameeullah, M., Hussain, H.A., Huang, J., Cui, K., & Nie, L. 2016. Comparative transcriptional profiling of primed and non-primed rice seedlings under submergence stress. *Frontiers in Plant Science, 7*(1125): 1–16. https://doi.org/10.3389/fpls.2016.01125

IPCC. 2014. *Climate Change 2014: Synthesis Report. Contribution of Working Groups I, II and III to the Fifth Assessment Report of the Intergovernmental Panel on Climate Change* [Core Writing Team, R.K. Pachauri and L.A. Meyer (eds.)]. Geneva, Switzerland: IPCC, 151 pp.

Iquebal, M.A., Sharma, P., Jasrotia, R.S., Jaiswal, S., Kaur, A., Saroha, M., Angadi, U.B., Sheoran, S., Singh, R., Singh, G.P., Rai, A., Tiwari, R., & Kumar, D. 2019. RNAseq analysis reveals drought-responsive molecular pathways with candidate genes and putative molecular markers in root tissue of wheat. *Scientific Reports, 9*(1): 1–18. https://doi.org/10.1038/s41598-019-49915-2

Ishka, M.R., Brown, E., Weigand, C., Tillett, R.L., Schlauch, K.A., Miller, G., & Harper, J.F. 2018. A comparison of heat-stress transcriptome changes between wild-type Arabidopsis pollen and a heat-sensitive mutant harboring a knockout of cyclic nucleotide-gated cation channel 16 (cngc16). *BMC Genomics, 19*(1): 1–19. https://doi.org/10.1186/s12864-018-4930-4

Jaiswal, S., Antala, T.J., Mandavia, M.K., Chopra, M., Jasrotia, R.S., Tomar, R.S., Kheni, J., Angadi, U.B., Iquebal, M.A., Golakia, B.A., Rai, A., & Kumar, D. 2018. Transcriptomic signature of drought response in pearl millet (*Pennisetum glaucum* (L.) and development of web-genomic resources. *Scientific Reports, 8*(1): 1–16. https://doi.org/10.1038/s41598-018-21560-1

Jiang, Y., & Deyholos, M.K. 2006. Comprehensive transcriptional profiling of NaCl-stressed Arabidopsis roots reveals novel classes of responsive genes. *BMC Plant Biology, 6*(1): 1–20. https://doi.org/10.1186/1471-2229-6-25

Jin, H., Dong, D., Yang, Q., & Zhu, D. 2016. Salt-responsive transcriptome profiling of *Suaeda glauca* via RNA sequencing. *PloS ONE, 11*(3): 1–14. https://doi.org/10.1371/journal.pone.0150504

Jin, Y., Yang, H., Wei, Z., Ma, H., & Ge, X. 2013. Rice male development under drought stress: phenotypic changes and stage-dependent transcriptomic reprogramming. *Molecular Plant, 6*(5):1630–1645. https://doi.org/10.1093/mp/sst067

Kaczmarek, M., Fedorowicz-Strońska, O., Głowacka, K., Waśkiewicz, A., & Sadowski, J. 2017. $CaCl_2$ treatment improves drought stress tolerance in barley (*Hordeum vulgare* L.). *Acta Physiologiae Plantarum, 39*(41): 1–11. doi:10.1007/s11738-016-2336-y

Kawaura, K., Mochida, K., & Ogihara, Y. 2008. Genome-wide analysis for identification of salt-responsive genes in common wheat. *Functional & Integrative Genomics, 8*(3): 277–286. https://doi.org/10.1007/s10142-008-0076-9

Kim, S., Na, J., Nie, H., Kim, J., Lee, J., & Kim, S. 2021. Comprehensive transcriptome profiling of *Caragana microphylla* in response to salt condition using de novo assembly. *Biotechnology Letters, 43*(1): 317–327. https://doi.org/10.1007/s10529-020-03022-9

Kjellstrom, E., Nikulin, G., Strandberg, G., Christensen, O.B., Jacob, D., Keuler, K., Lenderink, G., Van Meijgaard, E., Schar, C., Somot, S., Sorland, S.L., Teichmann, C., & Vautard, R. 2018. European climate change at global mean temperature increases of 1.5 and 2 degrees above pre-industrial conditions as simulated by the EURO-CORDEX regional climate models. *Earth System Dynamics, 9*: 459–478.

Kong, W., Zhong, H., Gong, Z., Fang, X., Sun, T., Deng, X., & Li, Y. 2019. Meta-analysis of salt stress transcriptome responses in different rice genotypes at the seedling stage. *Plants, 8*(3): 64–80.

Koussevitzky, S., Suzuki, N., Huntington, S., Armijo, L., Sha, W., Cortes, D., Shulaev, V., & Mittler, R. 2008. Ascorbate peroxidase 1 plays a key role in the response of Arabidopsis thaliana to stress combination. Journal of Biological Chemistry, 283(49): 34197–34203.

Krishnamurthy, P., Mohanty, B., Wijaya, E., Lee, D.Y., Lim, T.M., Lin, Q., Xu, J., Loh, C.S., & Kumar, P.P. 2017. Transcriptomics analysis of salt stress tolerance in the roots of the mangrove *Avicennia officinalis*. *Scientific Reports*, *7*(1): 1–19. https://doi.org/10.1038/s41598-017-10730-2

Krishnamurthy, P., Vishal, B., Ho, W.J., Lok, F.C.J., Lee, F.S.M., & Kumar, P.P. 2020. Regulation of a cytochrome P450 gene CYP94B1 by WRKY33 transcription factor controls apoplastic barrier formation in roots to confer salt tolerance. *Plant Physiology*, *184*(4): 2199–2215. https://doi.org/10.1104/pp.20.01054

Kultz, D. 2005. Molecular and evolutionary basis of the cellular stress response. *Annual Review of Physiology*, *67*: 225–257.

Kumar, K., Kumar, M., Kim, S.R., Ryu, H., & Cho, Y.G. 2013. Insights into genomics of salt stress response in rice. *Rice*, *6*(1): 1–15. https://doi.org/10.1186/1939-8433-6-27

Kumari, A., Das, P., Parida, A.K., & Agarwal, P.K. 2015. Proteomics, metabolomics, and ionomics perspectives of salinity tolerance in halophytes. *Frontiers in Plant Science*, *6*(537): 1–20. https://doi.org/10.3389/fpls.2015.00537

Kushwaha, H.R., Joshi, R., Pareek, A., & Singla-Pareek, S.L. 2016. MATH-domain family shows response toward abiotic stress in Arabidopsis and rice. *Frontiers in Plant Science*, *7*(923): 1–19. https://doi.org/10.3389/fpls.2016.00923

Lan, T., Zheng, Y., Su, Z., Yu, S., Song, H., Zheng, X., Lin, G., & Wu, W. 2019. OsSPL10, a SBP-box gene, plays a dual role in salt tolerance and trichome formation in rice (*Oryza sativa* L.). *G3: Genes, Genomes, Genetics*, *9*(12): 4107–4114. https://doi.org/10.1534/g3.119.400700

Lappalainen, T., Sammeth, M., Friedländer, M.R., Ac'tHoen, P., Monlong, J., Rivas, M.A., Gonzalez-Porta, M., Kurbatova, N., Griebel, T., Ferreira, P.G., Barann, M., Wieland, T., Greger, L., Iterson, M.V., Almlöf, J., Ribeca, P., Pulyakhina, I., Esser, D., Giger, T., Tikhonov, A., Sultan, M., Bertier, G., MacArthur, D.G., Lek, M., Lizano, E., Buermans, H.P.J., Padioleau, I., Schwarzmayr, T., Karlberg, O., Ongen, H., Kilpinen, H., Beltran, S., Gut, M., Kahlem, K., Amstislavskiy, V., Stegle, O., Pirinen, M.,Montgomery, S.B., Donnelly, P., McCarthy, M.I., Flicek, P., Strom, T.M., Geuvadis Consortium; Lehrach, H., Schreiber, S., Sudbrak, R., Carracedo, A., Antonarakis, S.E., Häsler, R., Syvänen, A.-C., vanOmmen, G.-J., Brazma, A., Meitinger, T., Rosenstiel, P., Guigó, R., Gut, I.G., Estivill, X., & Dermitzakis, E.T. 2013. Transcriptome and genome sequencing uncovers functional variation in humans. *Nature*, *501*(7468): 506–511. https://doi.org/10.1038/nature12531

Lee, J.H., Daugharthy, E.R., Scheiman, J., Kalhor, R., Yang, J.L., Ferrante, T.C., Terry, R., Jeanty, S.S., Li, C., Amamoto, R., Peters, D.T., Turczyk, B.M., Marblestone, A.H., Inverso, S.A., Bernard, A., Mali, P., Rios, X., Aach, J., & Church, G.M. 2014. Highly multiplexed subcellular RNA sequencing in situ. *Science*, *343*(6177): 1360–1363. doi: 10.1126/science.1250212

Li, J., Liu, C-C., Sun, C-H., & Chen, Y-T. 2018. Plant stress RNA-seq Nexus: a stress-specific transcriptome database in plant cells. *BMC Genomics*, *19*: 966. doi:10.1186/s12864-018-5367-5

Li, J., Gao, Z., Zhou, L., Li, L., Zhang, J., Liu, Y., & Chen, H. 2019. Comparative transcriptome analysis reveals K+ transporter gene contributing to salt tolerance in eggplant. *BMC Plant Biology*, *19*(1): 1–18. https://doi.org/10.1186/s12870-019-1663-8

Li, L., Li, M., Qi, X., Tang, X., & Zhou, Y. 2018. De novo transcriptome sequencing and analysis of genes related to salt stress response in *Glehnia littoralis*. *PeerJ*, *6*(e5681): 1–22. https://doi.org/10.7717/peerj.5681

Li, P., Cao, W., Fang, H., Xu, S., Yin, S., Zhang, Y., Lin, D., Wang, J., Chen, Y., Xu, C., & Yang, Z. 2017. Transcriptomic profiling of the maize (*Zea mays* L.) leaf response to abiotic stresses at the seedling stage. *Frontiers in Plant Science*, *8*(290): 1–13. https://doi.org/10.3389/fpls.2017.00290

Li, P., Yang, H., Wang, L., Liu, H., Huo, H., Zhang, C., Liu, A., Zhu, A., Hu, J., Lin, Y., & Liu, L. 2019. Physiological and transcriptome analyses reveal short-term responses and formation of memory under drought stress in rice. *Frontiers in Genetics*, *10*(55): 1–16. https://doi.org/10.3389/fgene.2019.00055

Li, Q., Qin, Y., Hu, X., Li, G., Ding, H., Xiong, X., & Wang, W. 2020. Transcriptome analysis uncovers the gene expression profile of salt-stressed potato (*Solanum tuberosum* L.). *Scientific Reports*, *10*(1): 1–19. https://doi.org/10.1038/s41598-020-62057-0

Li, Y.C., Meng, F.R., Zhang, C.Y., Zhang, N., Sun, M.S., Ren, J.P., Niu, H.B., Wang, X., & Yin, J. 2012. Comparative analysis of water stress-responsive transcriptomes in drought-susceptible and-tolerant wheat (*Triticum aestivum* L.). *Journal of Plant Biology*, *55*(5): 349–360. https://doi.org/10.1007/s12374-011-0032-4

Lin, J., Li, J. P., Yuan, F., Yang, Z., Wang, B. S., & Chen, M. 2018. Transcriptome profiling of genes involved in photosynthesis in *Elaeagnus angustifolia* L. under salt stress. *Photosynthetica*, *56*(4): 998–1009. https://doi.org/10.1007/s11099-018-0824-6

Liu, B., Kong, L., Zhang, Y., & Liao, Y. 2021. Gene and metabolite integration analysis through transcriptome and metabolome brings new insight into heat stress tolerance in potato (*Solanum tuberosum* L.). *Plants*, *10*(103): 1–17. https://doi.org/10.3390/plants10010103

Liu, C., Li, S., Wang, M., & Xia, G. 2012. A transcriptomic analysis reveals the nature of salinity tolerance of a wheat introgression line. *Plant Molecular Biology*, *78*(1–2): 159–169. doi:10.1007/s11103-011-9854-1

Liu, G., Zha, Z., Cai, H., Qin, D., Jia, H., Liu, C., Qiu, D., Zhang, Z., Wan, Z., Yang, Y.,Wan, B., You, A., & Jiao, C. 2020. Dynamic transcriptome analysis of anther response to heat stress during anthesis in thermotolerant rice (*Oryza sativa* L.). *International Journal of Molecular Sciences*, *21*(1155): 1–17. http://dx.doi.org/10.3390/ijms21031155

Liu, Q., Tang, J., Wang, W., Zhang, Y., Yuan, H., & Huang, S. 2018. Transcriptome analysis reveals complex response of the medicinal/ornamental halophyte *Iris halophila* Pall. to high environmental salinity. *Ecotoxicology and Environmental Safety*, *165*: 250–260. https://doi.org/10.1016/j.ecoenv.2018.09.003

Liu, S.C., Jin, J.Q., Ma, J.Q., Yao, M.Z., Ma, C.L., Li, C.F., Ding, Z.T., & Chen, L. 2016. Transcriptomic analysis of tea plant responding to drought stress and recovery. *PloS ONE*, *11*(1): 1–21. https://doi.org/10.1371/journal.pone.0147306

Liu, Z., Qin, J., Tian, X., Xu, S., Wang, Y., Li, H., Wang, X., Peng, H., Yao, Y., Hu, Z., Ni, Z., Xin, M., & Sun, Q. 2018. Global profiling of alternative splicing landscape responsive to drought, heat and their combination in wheat (*Triticum aestivum* L.). *Plant Biotechnology Journal*, *16*(3): 714–726. https://doi.org/10.1111/pbi.12822

Liu, Z., Xin, M., Qin, J., Peng, H., Ni, Z., Yao, Y., & Sun, Q. 2015. Temporal transcriptome profiling reveals expression partitioning of homeologous genes contributing to heat and drought acclimation in wheat (*Triticum aestivum* L.). *BMC Plant Biology*, *15*(1): 1–20. https://doi.org/10.1186/s12870-015-0511-8

Lobell, D.B., & Field, C.B. 2011. California perennial crops in a changing climate. *Climatic Change*, *109*(Suppl. 1): S317–S333.

Lobell, D.B., & Gourdji, S.M. 2012. The influence of climate change on global crop productivity. *Plant Physiology*, *160*(4): 1686–1697.

Lowe, R., Shirley, N., Bleackley, M., Dolan, S., & Shafee, T. 2017. Transcriptomics technologies. *PLoS Computational Biology*, *13*(5): 1–23. https://doi.org/10.1371/journal.pcbi.1005457

Lukoszek, R., Feist, P., & Ignatova, Z. 2016. Insights into the adaptive response of *Arabidopsis thaliana* to prolonged thermal stress by ribosomal profiling and RNA-Seq. *BMC Plant Biology*, *16*: 221–234. https://doi.org/10.1186/s12870-016-0915-0

Luo, Q., Teng, W., Fang, S., Li, H., Li, B., Chu, J., Li, Z., & Zheng, Q. 2019. Transcriptome analysis of salt-stress response in three seedling tissues of common wheat. *The Crop Journal*, *7*: 378–392. https://doi.org/10.1016/j.cj.2018.11.009

Luo, Q., Yu, B., & Liu, Y. 2005. Differential sensitivity to chloride and sodium ions in seedlings of *Glycine max* and *G. soja* under NaCl stress. *Journal of Plant Physiology*, *162*(9): 1003–1012. https://doi.org/10.1016/j.jplph.2004.11.008

Luo, Y., Reid, R., Freese, D., Li, C., Watkins, J., Shi, H., Zhang, H., Loraine, A., & Song, B.H. 2017. Salt tolerance response revealed by RNA-Seq in a diploid halophytic wild relative of sweet potato. *Scientific Reports*, *7*(1): 1–13. https://doi.org/10.1038/s41598-017-09241-x

Lv, S., Tai, F., Guo, J., Jiang, P., Lin, K., Wang, D., Zhang, X., & Li, Y. 2020. Phosphatidylserine synthase from *Salicornia europaea* is involved in plant salt tolerance by regulating plasma membrane stability. *Plant and Cell Physiology*, *62*(1): 66–79. https://doi.org/10.1093/pcp/pcaa141

Lwalaba, J.L.W., Zvobgo, G., Gai, Y., Issaka, J.H., Mwamba, T.M., Louis, L.T., Fu, L., Nazir, M.M., Kirika, B.A., Tshibangu, A.K. & Adil, M.F. 2021. Transcriptome analysis reveals the tolerant mechanisms to cobalt and copper in barley. *Ecotoxicology and Environmental Safety*, *209*(111761): 1–11. https://doi.org/10.1016/j.ecoenv.2020.111761

Ma, J., Li, R., Wang, H., Li, D., Wang, X., Zhang, Y., Zhen, W., Duan, H., Yan, G., & Li, Y. 2017. Transcriptomics analyses reveal wheat responses to drought stress during reproductive stages under field conditions. *Frontiers in Plant Science*, *8*(592): 1–13. https://doi.org/10.3389/fpls.2017.00592

Ma, J., Zhang, M., Xiao, X., You, J., Wang, J., Wang, T., Yao, Y., & Tian, C. 2013. Global transcriptome profiling of *Salicornia europaea* L. shoots under NaCl treatment. *PloS ONE*, *8*(6): 1–10. https://doi.org/10.1371/journal.pone.0065877

Mansouri, M., Naghavi, M.R., Alizadeh, H., Mohammadi-Nejad, G., Mousavi, S.A., Salekdeh, G.H., & Tada, Y. 2019. Transcriptomic analysis of *Aegilops tauschii* during long-term salinity stress. *Functional & Integrative Genomics*, *19*(1): 13–28. https://doi.org/10.1007/s10142-018-0623-y

Mansuri, R.M., Shobbar, Z-S., Jelodar, N.B., Ghaffari, M., Mohammadi, S.M., & Daryani, P. 2020. Salt tolerance involved candidate genes in rice: an integrative meta-analysis approach. *BMC Plant Biology*, *20*(452): 1–14. https://doi.org/10.1186/s12870-020-02679-8

Mantri, N., Patade, V., Penna, S., Ford, R., & Pang, E. 2012. Abiotic stress responses in plants: present and future. In *Abiotic Stress Responses in Plants*, ed. P. Ahmad and M. Prasad, 1–19. New York, NY: Springer. https://doi.org/10.1007/978-1-4614-0634-1_1

Marra, M.A., Hillier, L., & Waterston, R.H. 1998. Expressed sequence tags—ESTablishing bridges between genomes. *Trends* in *Genet*ics, *14*: 4–7.

Meng, X., Liu, S., Dong, T., Xu, T., Ma, D., Pan, S., Li, Z., & Zhu, M. 2020. Comparative transcriptome and proteome analysis of salt-tolerant and salt-sensitive sweet potato and overexpression of IbNAC7 confers salt tolerance in Arabidopsis. *Frontiers in Plant Science*, *11*(1342): 1–15. https://doi.org/10.3389/fpls.2020.572540

Miao, H., Sun, P., Liu, J., Wang, J., Xu, B., & Jin, Z. 2018. Overexpression of a novel ROP gene from the banana (MaROP5g) confers increased salt stress tolerance. *International Journal of Molecular Sciences*, *19*(3108): 1–19. https://doi.org/10.3390/ijms19103108

Miao, Z., Han, Z., Zhang, T., Chen, S., & Ma, C. 2017. A systems approach to a spatio-temporal understanding of the drought stress response in maize. *Scientific Reports*, *7*(1): 1–14. https://doi.org/10.1038/s41598-017-06929-y

Mishra, A., & Tanna, B. 2017. Halophytes: potential resources for salt stress tolerance genes and promoters. *Frontiers in Plant Science*, *8*(829): 1–10. https://doi.org/10.3389/fpls.2017.00829

Mishra, D.C., Arora, D., Kumar, R.R., Goswami, S., Varshney, S., Budhlakoti, N., Kumar, S., Chaturvedi, K.K., Sharma, A., Chinnusamy, V., & Rai, A. 2020. Weighted gene co-expression analysis for identification of key genes regulating heat stress in wheat. *Cereal Research Communications*, *49*: 73–81. https://doi.org/10.1007/s42976-020-00072-7

Mittler, R. 2006. Abiotic stress, the field environment and stress combination. *Trends in Plant Science*, *11*(1): 15–19. https://doi.org/10.1016/j.tplants.2005.11.002

Mortazavi, A., Williams, B.A., McCue, K., Schaeffer, L., & Wold, B. 2008. Mapping and quantifying mammalian transcriptomes by RNA-Seq. *Nature Methods*, *5*(7): 621–628. https://doi.org/10.1038/nmeth.1226

Muller, F., & Rieu, I. 2016. Acclimation to high temperature during pollen development. *Plant Reproduction*, *29*: 107–118.

Muller, R., Morant, M., Jarmer, H., Nilsson, L., & Neilson, T.H. 2007. Genome-wide analysis of the Arabidopsis leaf transcriptome reveals interaction of phosphate and sugar metabolism. *Plant Physiology*, *143*: 156–171.

Munns, R., & Tester, M. 2008. Mechanisms of salinity tolerance. *Annual Review of Plant Biology*, *59*: 651–681. https://doi.org/10.1146/annurev.arplant.59.032607.092911

Muthuramalingam, P., Krishnan, S.R., Pothiraj, R., & Ramesh, M. 2017. Global transcriptome analysis of combined abiotic stress signaling genes unravels key players in *Oryza sativa* L.: an in silico approach. *Frontiers in Plant Science*, *8*(759): 1–13. https://doi.org/10.3389/fpls.2017.00759

Muthusamy, M., Uma, S., Backiyarani, S., Saraswathi, M.S., & Chandrasekar, A. 2016. Transcriptomic changes of drought-tolerant and sensitive banana cultivars exposed to drought stress. *Frontiers in Plant Science*, *7*(1609): 1–14. https://doi.org/10.3389/fpls.2016.01609

Nayak, S.S., Pradhan, S., Sahoo, D., & Parida, A. 2020. De novo transcriptome assembly and analysis of *Phragmites karka*, an invasive halophyte, to study the mechanism of salinity stress tolerance. *Scientific Reports*, *10*(1): 1–12. https://doi.org/10.1038/s41598-020-61857-8

Nejat, N., Ramalingam, A., & Mantri, N. 2018. Advances in transcriptomics of plants. *Advances in Biochemical Engineering/Biotechnology*, *164*: 161–185.

Nelson, G., Rosegrant, M.W., Palazzo, A., Gray, I., Ingersoll, C., Robertson, R., Tokgoz, S., Zhu, T., Sulser, T., Ringler, C., Msangi, S., & You, L. 2010, *Food Security, Farming, and Climate Change to 2050: Scenarios,*

Results, Policy Options. Research reports, International Food Policy Research Institute (IFPRI). http://dx.doi.org/10.2499/9780896291867

Nikalje, G.C., & Suprasanna, P. 2018. Coping with metal toxicity–cues from halophytes. *Frontiers in Plant Science*, *9*(777): 1–11. https://doi.org/10.3389/fpls.2018.00777

Nikalje, G.C., Variyar, P.S., Joshi, M.V., Nikam, T.D., & Suprasanna, P. 2018. Temporal and spatial changes in ion homeostasis, antioxidant defense and accumulation of flavonoids and glycolipid in a halophyte *Sesuvium portulacastrum* (L.) L. *PloS ONE*, *13*(4): 1–31. https://doi.org/10.1371/journal.pone.0193394

Nishizawa, A., Yabuta, Y., Yoshida, E., Maruta, T., Yoshimura, K., & Shigeoka, S. 2006. Arabidopsis heat shock transcription factor A2 as a key regulator in response to several types of environmental stress. *Plant Journal*, *48*: 535–547.

Oh, D.H., Barkla, B.J., Vera-Estrella, R., Pantoja, O., Lee, S.Y., Bohnert, H.J., & Dassanayake, M. 2015. Cell type-specific responses to salinity – the epidermal bladder cell transcriptome of *Mesembryanthemum crystallinum*. *New Phytologist*, *207*(3): 627–644. https://doi.org/10.1111/nph.13414

Oikonomopoulos, S., Bayega, A., Fahiminiya, S., Djambazian, H., Berube, P., & Ragoussis, J. 2020. Methodologies for transcript profiling using long-read technologies. *Frontiers in Genetics*, *11*: 606. doi: 10.3389/fgene.2020.00606

Ozsolak, F., & Milos, P.M. 2011. RNA sequencing: advances, challenges and opportunities. *Nature Reviews Genetics*, *12*(2): 87–98. https://doi.org/10.1038/nrg2934

Padmalatha, K.V., Dhandapani, G., Kanakachari, M., Kumar, S., Dass, A., Patil, D.P., Rajamani, V., Kumar, K., Pathak, R., Rawat, B., Leelavathi, S., Reddy, P.S., Jain, N., Powar, K.N., Hiremath, V., Katageri, I.S., Reddy, M.K., Solanke, A.U., Reddy, V.S., & Kumar, P.A. 2012. Genome-wide transcriptomic analysis of cotton under drought stress reveal significant down-regulation of genes and pathways involved in fibre elongation and up-regulation of defense responsive genes. *Plant Molecular Biology*, *78*(3): 223–246. doi:10.1007/s11103-011-9857-y

Pan, J., Li, Z., Dai, S., Ding, H., Wang, Q., Li, X., Ding, G., Wang, P., Guan, Y., & Liu, W. 2020. Integrative analyses of transcriptomics and metabolomics upon seed germination of foxtail millet in response to salinity. *Scientific Reports*, *10*(1): 1–16. https://doi.org/10.1038/s41598-020-70520-1

Patel, P., Yadav, K., Srivastava, A.K., Suprasanna, P., & Ganapathi, T.R. 2019. Overexpression of native Musa-miR397 enhances plant biomass without compromising abiotic stress tolerance in banana. *Scientific Reports*, *9*(1): 16434. https://doi.org/10.1038/s41598-019-52858-3

Phule, A.S., Barbadikar, K.M., Maganti, S.M., Seguttuvel, P., Subrahmanyam, D., Babu, M.B.B.P., & Kumar, P.A. 2019. RNA-Seq reveals the involvement of key genes for aerobic adaptation in rice. *Scientific Reports*, *9*(1): 1–10. https://doi.org/10.1038/s41598-019-41703-2

Pierce, I., Halter, G., & Waters, E.R. 2020. Transcriptomic analysis of the heat-stress response in *Boechera depauperata* and Arabidopsis reveals a distinct and unusual heat-stress response in Boechera. *Botany*, *98*(10): 589–602. https://doi.org/10.1139/cjb-2020-0014.

Postnikova, O.A., Shao, J., & Nemchinov, L.G. 2013. Analysis of the alfalfa root transcriptome in response to salinity stress. *Plant and Cell Physiology*, *54*(7): 1041–1055. https://doi.org/10.1093/pcp/pct056

Pozhitkov, A.E., Tautz, D., & Noble, P.A. 2007. Oligonucleotide microarrays: widely applied—poorly understood. *Briefings in Functional Genomics and Proteomics*, *6*(2): 141–148. https://doi.org/10.1093/bfgp/elm014

Qian, Y., Ren, Q., Zhang, J., & Chen, L. 2019. Transcriptomic analysis of the maize (*Zea mays* L.) inbred line B73 response to heat stress at the seedling stage. *Gene*, *692*: 68–78. https://doi.org/10.1016/j.gene.2018.12.062

Qing, D. J., Lu, H. F., Li, N., Dong, H. T., Dong, D. F., & Li, Y. Z.2009.Comparative profiles of gene expression in leaves and roots of maize seedlings under conditions of salt stress and the removal of salt stress. *Plant and Cell Physiology*, *50*(4): 889–903. https://doi.org/10.1093/pcp/pcp038

Quail, M.A., Smith, M., Coupland, P., Otto, T.D., Harris, S.R., Connor, T.R., Bertoni, A., Swerdlow, H.P., & Gu, Y. 2012. A tale of three next generation sequencing platforms: comparison of Ion Torrent, Pacific Biosciences and Illumina MiSeq sequencers. *BMC Genomics*, *13*(1): 1–13. https://doi.org/10.1186/1471-2164-13-341

Rabbani, M.A., Maruyama, K., Abe, H., Khan, M.A., Katsura, K., Ito, Y., Yoshiwara, K., Seki, M., Shinozaki, K., & Yamaguchi-Shinozaki, K. 2003. Monitoring expression profiles of rice genes under cold, drought, and high-salinity stresses and abscisic acid application using cDNA microarray and RNA gel-blot analyses. *Plant Physiology*, *133*(4): 1755–1767. https://doi.org/10.1104/pp.103.025742

Rai, V., Tuteja, N., & Takabe, T. 2012. Transporters and abiotic stress tolerance in plants. In *Improving Crop Resistance to Abiotic Stress*, ed. N. Tuteja, S.S. Gill, A.F. Tiburcio, and R. Tuteja, 507–522. Weinheim: Wiley-VCH Verlag GmbH & Co. KGaA. https://doi.org/10.1002/9783527632930.ch22

Ramegowda, V., & Senthil-Kumar, M. 2015. The interactive effects of simultaneous biotic and abiotic stresses on plants: mechanistic understanding from drought and pathogen combination. *Journal of Plant Physiology*, *176*: 47–54. https://doi.org/10.1016/j.jplph.2014.11.008

Ramu, V.S., Paramanantham, A., Ramegowda, V., Mohan-Raju, B., Udayakumar, M., & Senthil-Kumar, M. 2016. Transcriptome analysis of sunflower genotypes with contrasting oxidative stress tolerance reveals individual-and combined-biotic and abiotic stress tolerance mechanisms. *PloS ONE*, *11*(6): 1–19. https://doi.org/10.1371/journal.pone.0157522

Rangan, P., Furtado, A., & Henry, R. 2020. Transcriptome profiling of wheat genotypes under heat stress during grain-filling. *Journal of Cereal Science*, *91*(102895): 1–8. https://doi.org/10.1016/j.jcs.2019.102895

Ranjan, A., & Sawant, S. 2015. Genome-wide transcriptomic comparison of cotton (*Gossypium herbaceum*) leaf and root under drought stress. *3 Biotech*, 5(4): 585–596.

Ranjan, A., Nigam, D., Asif, M.H., Singh, R., Ranjan, S., Mantri, S., Pandey, N., Trivedi, I., Rai, K.M., Jena, S.N., Koul, B., Tuli, R., Pathre, U.V., & Sawant, S.V. 2012a. Genome wide expression profiling of two accession of *G. herbaceum* L. in response to drought. *BMC Genomics*, *13*(1): 1–18. https://doi.org/10.1186/1471-2164-13-94

Ranjan, A., Pandey, N., Lakhwani, D., Dubey, N.K., Pathre, U.V., & Sawant, S.V. 2012b. Comparative transcriptomic analysis of roots of contrasting *Gossypium herbaceum* genotypes revealing adaptation to drought. *BMC Genomics*, *13*(1): 1–22. https://doi.org/10.1186/1471-2164-13-680

Rasmussen, S., Barah, P., Suarez-Rodriguez, M.C., Bressendorff, S., Friis, P., Costantino, P., Bones, A.M., Nielsen, H.B., & Mundy, J. 2013. Transcriptome responses to combinations of stresses in Arabidopsis. *Plant Physiology*, *161*(4): 1783–1794. https://doi.org/10.1104/pp.112.210773

Raza, A., Razzaq, A., Mehmood, S.S., Zou, X., Zhang, X., Lv, Y., & Xu, J. 2019. Impact of climate change on crops adaptation and strategies to tackle its outcome: a review. *Plants*, *8*(2): 34. https://doi.org/10.3390/plants8020034

Razzaque, S., Elias, S.M., Haque, T., Biswas, S., Jewel, G.N.A., Rahman, S., Weng, X., Ismail, A.M., Walia, H., Juenger, T.E., & Seraj, Z.I. 2019. Gene expression analysis associated with salt stress in a reciprocally crossed rice population. *Scientific Reports*, *9*(1): 1–17. https://doi.org/10.1038/s41598-019-44757-4

Reddy, P.S., Chakradhar, T., Reddy, R.A., Nitnavare, R.B., Mahanty, S., & Reddy, M.K. 2016. Role of heat shock proteins in improving heat stress tolerance in crop plants. In *Heat Shock Proteins and Plants. Heat Shock Proteins*, vol 10, ed. A. Asea, P. Kaur, and S. Calderwood, 283–307. Cham: Springer. https://doi.org/10.1007/978-3-319-46340-7_14

Rest, J.S., Wilkins, O., Yuan, W., Purugganan, M.D., & Gurevitch, J. 2016. Meta-analysis and meta-regression of transcriptomic responses to water stress in Arabidopsis. *The Plant Journal*, *85*(4): 548–560. https://doi.org/10.1111/tpj.13124

Rich-Griffin, C., Stechemesser, A., Finch, J., Lucas, E., Ott, S., & Schäfer, P. 2020. Single-cell transcriptomics: a high-resolution avenue for plant functional genomics. *Trends in Plant Science*, *25*(2): 186–197. https://doi.org/10.1016/j.tplants.2019.10.008

Rizhsky, L., Liang, H., Shuman, J., Shulaev, V., Davletova, S., & Mittler, R. 2004. When defense pathways collide. The response of Arabidopsis to a combination of drought and heat stress. *Plant Physiology*, *134*(4): 1683–1696.

Rosenzweig, C., Elliott, J., Deryng, D., Ruane, A.C., Muller, C., Arneth, A., Boote, K.J., Folberth, C., Glotter, M., Khabarov, N., Neumann, K., Piontek, F., Pugh, T.A.M., Schmid, E., Stehfest, E., Yang, H., & Jones, J.W. 2014. Assessing agricultural risks of climate change in the 21st century in a global gridded crop model intercomparison. *Proceedings of the National Academy of Sciences of the U.S.A.*, *111*(9): 3268–3273.

Ruiz, K.B., Maldonado, J., Biondi, S., & Silva, H. 2019. RNA-seq analysis of salt-stressed versus non salt-stressed transcriptomes of *Chenopodium quinoa* landrace R49. *Genes*, *10*(12): 1–19. https://doi.org/10.3390/genes10121042

Rutley, N., Poidevin, L., Doniger, T., Tillet, R.L., Rath, A., Forment, J., Schlauch, K., Luria, G., Ferrando, A., Harper, J.F., & Miller, G. 2021. Characterization of novel pollen-expressed transcripts reveals their potential roles in pollen heat stress response in *Arabidopsis thaliana*. *Plant Reproduction*, *34*: 61–78. https://doi.org/10.1007/s00497-020-00400-1

Saidi, Y., Finka, A., Muriset, M., Bromberg, Z., Weiss, Y.G., Maathuis, F.J., & Goloubinoff, P. 2009. The heat shock response in moss plants is regulated by specific calcium-permeable channels in the plasma membrane. *The Plant Cell*, *21*(9): 2829–2843. https://doi.org/10.1105/tpc.108.065318

Saini, N., Nikalje, G.C., Zargar, S.M., & Suprasanna, P. 2021. Molecular insights into sensing, regulation and improving of heat tolerance in plants. *Plant Cell Reports*, *21*. doi: 10.1007/s00299-021-02793-3

Samad, A.F., Sajad, M., Nazaruddin, N., Fauzi, I.A., Murad, A., Zainal, Z., & Ismail, I. 2017. MicroRNA and transcription factor: key players in plant regulatory network. *Frontiers in Plant Science*, *8*(565): 1–18. https://doi.org/10.3389/fpls.2017.00565

Sarkar, N.K., Kim, Y.K., & Grover, A. 2014. Coexpression network analysis associated with call of rice seedlings for encountering heat stress. *Plant Molecular Biology*, *84*(1–2): 125–143. https://doi.org/10.1007/s11103-013-0123-3

Sasidharan, R., Mustroph, A., Boonman, A., Akman, M., Ammerlaan, A.M., Breit, T., Schranz, M.E., Voesenek, L.A., & van Tienderen, P.H. 2013. Root transcript profiling of two Rorippa species reveals gene clusters associated with extreme submergence tolerance. *Plant Physiology*, *163*(3): 1277–1292. https://doi.org/10.1104/pp.113.222588

Schreiber, A.W., Sutton, T., Caldo, R.A., Kalashyan, E., Lovell, B., Mayo, G., Muehlbauer, G.J., Druka, A., Waugh, R., Wise, R.P., Langridge, P., & Baumann, U. 2009. Comparative transcriptomics in the Triticeae. *BMC Genomics*, *10*(1): 1–17. https://doi.org/10.1186/1471-2164-10-285

Sedeek, K.E., Mahas, A., & Mahfouz, M. 2019. Plant genome engineering for targeted improvement of crop traits. *Frontiers in Plant Science*, *10*(114): 1–16. https://doi.org/10.3389/fpls.2019.00114

Shankar, R., Bhattacharjee, A., & Jain, M. 2016. Transcriptome analysis in different rice cultivars provides novel insights into desiccation and salinity stress responses. *Scientific Reports*, *6*(1): 1–15. https://doi.org/10.1038/srep23719

Shao, H., Wang, H., & Tang, X. 2015. NAC transcription factors in plant multiple abiotic stress responses: progress and prospects. *Frontiers in Plant Science*, *6*(902): 1–8. https://doi.org/10.3389/fpls.2015.00902

Sharma, R., Singh, G., Bhattacharya, S., & Singh, A. 2018. Comparative transcriptome meta-analysis of *Arabidopsis thaliana* under drought and cold stress. *PloS ONE*, *13*(9): 1–18. https://doi.org/10.1371/journal.pone.0203266

Shaw, R., Tian, X., & Xu, J. 2021.Single-cell transcriptome analysis in plants: advances and challenges. *Molecular Plant*, *14*: 115–126. https://doi.org/10.1016/j.molp.2020.10.012

Shinde, H., Dudhate, A., Anand, L., Tsugama, D., Gupta, S.K., Liu, S., & Takano, T. 2020. Small RNA sequencing reveals the role of pearl millet miRNAs and their targets in salinity stress responses. *South African Journal of Botany*, *132*: 395–402. https://doi.org/10.1016/j.sajb.2020.06.011

Shinde, H., Dudhate, A., Tsugama, D., Gupta, S.K., Liu, S., & Takano, T. 2019. Pearl millet stress-responsive NAC transcription factor PgNAC21 enhances salinity stress tolerance in Arabidopsis. *Plant Physiology and Biochemistry*, *135*: 546–553. https://doi.org/10.1016/j.plaphy.2018.11.004

Shinde, H., Tanaka, K., Dudhate, A., Tsugama, D., Mine, Y., Kamiya, T., Gupta, S.K., Liu, S., & Takano, T. 2018. Comparative de novo transcriptomic profiling of the salinity stress responsiveness in contrasting pearl millet lines. *Environmental and Experimental Botany*, *155*: 619–627. https://doi.org/10.1016/j.envexpbot.2018.07.008

Shivhare, R., Asif, M.H., & Lata, C. 2020. Comparative transcriptome analysis reveals the genes and pathways involved in terminal drought tolerance in pearl millet. *Plant Molecular Biology*, *103*: 639–652. https://doi.org/10.1007/s11103-020-01015-w

Skorupa, M., Gołębiewski, M., Domagalski, K., Kurnik, K., Nahia, K.A., Złoch, M., Tretyn, A., & Tyburski, J. 2016. Transcriptomic profiling of the salt stress response in excised leaves of the halophyte *Beta vulgaris* ssp. *maritima*. *Plant Science*, *243*: 56–70. https://doi.org/10.1016/j.plantsci.2015.11.007

Smith, M.A., Ersavas, T., Ferguson, J.M., Liu, H., Lucas, M.C., Begik, O., Bojarski, L., Barton, K., & Novoa, E.M. 2020. Molecular barcoding of native RNAs using nanopore sequencing and deep learning. *Genome Research*, *30*(9): 1345–1353. doi:10.1101/gr.260836.120

Soeda, Y., Konings, M.C., Vorst, O., van Houwelingen, A.M., Stoopen, G.M., Maliepaard, C.A., Kodde, J., Bino, R.J., Groot, S.P., & van der Geest, A.H. 2005. Gene expression programs during *Brassica oleracea* seed maturation, osmopriming, and germination are indicators of progression of the germination process and the stress tolerance level. *Plant Physiology*, *137*(1): 354–368. https://doi.org/10.1104/pp.104.051664

Song, T., Das, D., Yang, F., Chen, M., Tian, Y., Cheng, C., Sun, C., Xu, W., & Zhang, J. 2020. Genome-wide transcriptome analysis of roots in two rice varieties in response to alternate wetting and drying irrigation. *The Crop Journal*, *8*(4): 586–601. https://doi.org/10.1016/j.cj.2020.01.007

Srivastava, A. K., Kumar, J.S., & Suprasanna, P. 2021. Seed 'primeomics': plants memorize their germination under stress. *Biological Reviews*, *96*: 1723–1743.

Stöckle, C.O., Nelson, R.L., Higgins, S., Brunner, J., Grove, G., Boydston, R., Whiting, M., & Kruger, C. (2010). Assessment of climate change impact on Eastern Washington agriculture. *ClimaticChange*, *102*: 77–102.

Su, P., Jiang, C., Qin, H., Hu, R., Feng, J., Chang, J., Yang, G., & He, G. 2019. Identification of potential genes responsible for thermotolerance in wheat under high temperature stress. *Genes*, *10*(174): 1–15. https://doi.org/10.3390/genes10020174

Sun, Q., Gao, F., Zhao, L., Li, K., & Zhang, J. 2010. Identification of a new 130 bp cis-acting element in the TsVP1 promoter involved in the salt stress response from *Thellungiella halophila*. *BMC Plant Biology*, *10*(1): 1–12. https://doi.org/10.1186/1471-2229-10-90

Sun, Z., Wang, Y., Mou, F., Tian, Y., Chen, L., Zhang, S., Jiang, Q., & Li, X. 2016. Genome-wide small RNA analysis of soybean reveals auxin-responsive microRNAs that are differentially expressed in response to salt stress in root apex. *Frontiers in Plant Science*, *6*(1273): 1–12. https://doi.org/10.3389/fpls.2015.01273

Suprasanna, P. 2020. Plant abiotic stress tolerance: Insights into resilience build-up. *Journal of Biosciences*, *45*:120. https://doi.org/10.1007/s12038-020-00088-5

Suzuki, N., Rivero, R.M., Shulaev, V., Blumwald, E., & Mittler, R. 2014. Abiotic and biotic stress combinations. *New Phytologist*, *203*: 32–43.

Tahmasebi, A., Ashrafi-Dehkordi, E., Shahriari, A.G., Mazloomi, S.M., & Ebrahimie, E. 2019. Integrative meta-analysis of transcriptomic responses to abiotic stress in cotton. *Progress in Biophysics and Molecular Biology*, *146*: 112–122. https://doi.org/10.1016/j.pbiomolbio.2019.02.005

Tan, K.C., Ipcho, S.V.S., Trengove, R.D., Oliver, R.P., & Solomon, P.S. 2009. Assessing the impact of transcriptomics, proteomics and metabolomics on fungal phytopathology. *Molecular Plant Pathology*, *10*: 703–715.

Tang, R., Gupta, S.K., Niu, S., Li, X.Q, Yang, Q., Chen, G., Zhu, W., & Haroon, M. 2020. Transcriptome analysis of heat stress response genes in potato leaves. *Molecular Biology Reports*, *47*(6): 4311–4321. doi: 10.1007/s11033-020-05485-5

Tchounwou, P.B., Yedjou, C.G., Patlolla, A.K., & Sutton, D.J. 2012. Heavy metal toxicity and the environment. In *Molecular, Clinical and Environmental Toxicology*, volume 3, Environmental Toxicology, ed. A. Luch, 133–164. Springer, Basel. https://doi.org/10.1007/978-3-7643-8340-4_6

Teng, K., Tan, P., Guo, W., Yue, Y., Fan, X., & Wu, J. 2018. Heterologous expression of a novel *Zoysia japonica* C2H2 zinc finger gene, ZjZFN1, improved salt tolerance in Arabidopsis. *Frontiers in Plant Science*, *9* (1159): 1–13. https://doi.org/10.3389/fpls.2018.01159

Tiwari, P., Srivastava, D., Chauhan, A.S., Indoliya, Y., Singh, P.K., Tiwari, S., Fatima, T., Mishra, S.K., Dwivedi, S., Agarwal, L., Singh, P.C., Asif, M.H., Tripathi, R.D., Shirke, P.A., Chakrabarty, D., Chauhan, P.S., & Nautiyal, C.S. 2021. Root system architecture, physiological analysis and dynamic transcriptomics unravel the drought-responsive traits in rice genotypes. *Ecotoxicology and Environmental Safety*, *207*(111252): 1–13. https://doi.org/10.1016/j.ecoenv.2020.111252

UNEP. 2017. www.unep.org/news-and-stories/story/high-and-dry-degraded-lands-are-driving-people-their-homes (January 12, 2022)

Van Verk, M.C., Hickman, R., Pieterse, C.M., & Van Wees, S.C. 2013. RNA-Seq: revelation of the messengers. *Trends in Plant Science*, *18*(4): 175–179. https://doi.org/10.1016/j.tplants.2013.02.001

Varier, A., Vari, A.K., & Dadlani, M. 2010. The subcellular basis of seed priming. *Current Science*, *99*(4): 450–456.

Vaughan, M.M., Block, A., Christensen, S.A., Allen, L.H., & Schmelz, E.A. 2018. The effects of climate change associated abiotic stresses on maize phytochemical defenses. *Phytochemistry Reviews*, *17*:37–49.

Velculescu, V.E., Zhang, L., Vogelstein, B., & Kinzler, K.W. 1995. Serial analysis of gene expression. *Science*, *270*(5235): 484–487. doi: 10.1126/science.270.5235.484

Virlouvet, L., & Fromm, M. 2015. Physiological and transcriptional memory in guard cells during repetitive dehydration stress. *New Phytologist*, *205*(2): 596–607. https://doi.org/10.1111/nph.13080

Voinnet, O. 2009. Origin, biogenesis, and activity of plant microRNAs. *Cell*, *136*(4): 669–687. https://doi.org/10.1016/j.cell.2009.01.046

Walia, H., Wilson, C., Condamine, P., Ismail, A.M., Xu, J., Cui, X., & Close, T.J. 2007. Array-based genotyping and expression analysis of barley cv. Maythorpe and Golden Promise. *BMC Genomics*, *8*(1): 1–14. https://doi.org/10.1186/1471-2164-8-87

Wang, X., Elling, A.A., Li, X., Li, N., Peng, Z., He, G., Sun, H., Qi, Y., Liu, X.S., & Deng, X.W. 2009. Genome-wide and organ-specific landscapes of epigenetic modifications and their relationships to mRNA and small RNA transcriptomes in maize. *Plant Cell*, *21*: 1053–1069.

Wang, B., Liu, C., Zhang, D., He, C., Zhang, J., & Li, Z. 2019. Effects of maize organ-specific drought stress response on yields from transcriptome analysis. *BMC Plant Biology*, *19*(1): 1–19. https://doi.org/10.1186/s12870-019-1941-5

Wang, H., Hu, T., Huang, J., Lu, X., Huang, B., & Zheng, Y. 2013. The expression of *Millettia pinnata* chalcone isomerase in *Saccharomyces cerevisiae* salt-sensitive mutants enhances salt-tolerance. *International Journal of Molecular Sciences*, *14*(5): 8775–8786. https://doi.org/10.3390/ijms14058775

Wang, J., Li, B., Meng, Y., Ma, X., Lai, Y., Si, E., Yang, K., Ren, P., Shang, X., & Wang, H. 2015. Transcriptomic profiling of the salt-stress response in the halophyte *Halogeton glomeratus*. *BMC Genomics*, *16*(1): 1–14. https://doi.org/10.1186/s12864-015-1373-z

Wang, J., Li, B., Yao, L., Meng, Y., Ma, X., Lai, Y., Si, E., Ren, P., Yang, K., Shang, X., & Wang, H. 2018. Comparative transcriptome analysis of genes involved in Na^+ transport in the leaves of halophyte *Halogeton glomeratus*. *Gene*, *678*: 407–416. https://doi.org/10.1016/j.gene.2018.08.025

Wang, M., Wang, Y., Zhang, Y., Li, C., Gong, S., Yan, S., Li, G., Hu, G., Ren, H., Yang, J., Yu, T., & Yang, K. 2019. Comparative transcriptome analysis of salt-sensitive and salt-tolerant maize reveals potential mechanisms to enhance salt resistance. *Genes & Genomics*, *41*(7): 781–801. https://doi.org/10.1007/s13258-019-00793-y

Wang, R., Wang, X., Liu, K., Zhang, X.J., Zhang, L.Y., & Fan, S.J. 2020. Comparative transcriptome analysis of halophyte *Zoysia macrostachya* in response to salinity stress. *Plants*, *9*(4): 1–18. https://doi.org/10.3390/plants9040458

Wang, Y., Yu, Y., Huang, M., Gao, P., Chen, H., Liu, M., Chen, Q., Yang, Z, & Sun, Q. 2020. Transcriptomic and proteomic profiles of II YOU 838 (*Oryza sativa*) provide insights into heat stress tolerance in hybrid rice. *PeerJ*, *8*(8306): 1–24. https://doi.org/10.7717/peerj.8306

Weber, A.P. 2015. Discovering new biology through sequencing of RNA. *Plant Physiology*, *169*(3): 1524–1531. https://doi.org/10.1104/pp.15.01081

Weber, A.P., Weber, K.L., Carr, K., Wilkerson, C., & Ohlrogge, J.B. 2007. Sampling the Arabidopsis transcriptome with massively parallel pyrosequencing. *Plant Physiology*, *144*(1): 32–42. https://doi.org/10.1104/pp.107.096677

Wei, H., Wang, X., He, Y., Xu, H., & Wang, L. 2021. Clock component OsPRR73 positively regulates rice salt tolerance by modulating OsHKT2; 1-mediated sodium homeostasis. *The EMBO Journal*, *40*(3): 1–17. https://doi.org/10.15252/embj.2020105086

Wilhelm, B.T., Marguerat, S., Watt, S., Schubert, F., Wood, V., Goodhead, I., Penkett, C.J., Rogers, J., & Bähler, J. 2008. Dynamic repertoire of a eukaryotic transcriptome surveyed at single-nucleotide resolution. *Nature*, *453*(7199): 1239–1243. https://doi.org/10.1038/nature07002

Wu, B., Hu, Y., Huo, P., Zhang, Q., Chen, X., & Zhang, Z. 2017. Transcriptome analysis of hexaploid hulless oat in response to salinity stress. *PloS ONE*, *12*(2): 1–16. https://doi.org/10.1371/journal.pone.0171451

Wu, Y., Deng, Z., Lai, J., Zhang, Y., Yang, C., Yin, B., Zhao, Q., Zhang, L., Li, Y., Yang, C., & Xie, Q. 2009. Dual function of Arabidopsis ATAF1 in abiotic and biotic stress responses. *Cell Research*, *19*(11): 1279–1290. https://doi.org/10.1038/cr.2009.108

Xia, H., Ma, X., Xu, K., Wang, L., Liu, H., Chen, L., & Luo, L. (2020). Temporal transcriptomic differences between tolerant and susceptible genotypes contribute to rice drought tolerance. *BMC Genomics*, *21*(1): 1–18. https://doi.org/10.1186/s12864-020-07193-7

Xie, Q., Niu, J., Xu, X., Xu, L., Zhang, Y., Fan, B., Liang, X., Zhang, L., Yin, S., & Han, L. 2015. De novo assembly of the Japanese lawngrass (*Zoysia japonica* Steud.) root transcriptome and identification of candidate unigenes related to early responses under salt stress. *Frontiers in Plant Science*, *6*(610): 1–14. https://doi.org/10.3389/fpls.2015.00610

Xiong, H., Guo, H., Xie, Y., Zhao, L., Gu, J., Zhao, S., Li, J., & Liu, L. 2017. RNAseq analysis reveals pathways and candidate genes associated with salinity tolerance in a spaceflight-induced wheat mutant. *Scientific Reports*, *7*(1): 1–13. https://doi.org/10.1038/s41598-017-03024-0

Xue, G.P., Way, H.M., Richardson, T., Drenth, J., Joyce, P.A., & McIntyre, C.L. 2011. Overexpression of TaNAC69 leads to enhanced transcript levels of stress up-regulated genes and dehydration tolerance in bread wheat. *Molecular Plant*, *4*(4): 697–712. https://doi.org/10.1093/mp/ssr013

Yamamoto, N., Takano, T., Tanaka, K., Ishige, T., Terashima, S., Endo, C., Kurusu, T., Yajima, S., Yano, K., & Tada, Y. 2015. Comprehensive analysis of transcriptome response to salinity stress in the halophytic turf grass *Sporobolus virginicus*. *Frontiers in Plant Science*, *6*(241): 1–14. https://doi.org/10.3389/fpls.2015.00241

Yang, L., Jin, Y., Huang, W., Sun, Q., Liu, F., & Huang, X. 2018. Full-length transcriptome sequences of ephemeral plant *Arabidopsis pumila* provides insight into gene expression dynamics during continuous salt stress. *BMC Genomics*, *19*(1): 1–14. https://doi.org/10.1186/s12864-018-5106-y

Yang, Y., & Guo, Y. 2018. Elucidating the molecular mechanisms mediating plant salt-stress responses. *New Phytologist*, *217*(2): 523–539. https://doi.org/10.1111/nph.14920

Yao, L., Wang, J., Li, B., Meng, Y., Ma, X., Si, E., Ren, P., Yang, K., Shang, X., & Wang, H. 2018. Transcriptome sequencing and comparative analysis of differentially-expressed isoforms in the roots of *Halogeton glomeratus* under salt stress. *Gene*, *646*: 159–168. https://doi.org/10.1016/j.gene.2017.12.058

Yao, Y., Zhang, X., Wang, N., Cui, Y., Zhang, L., & Fan, S. 2020. Transcriptome analysis of salt stress response in halophyte *Atriplex centralasiatica* leaves. *Acta Physiologiae Plantarum*, *42*(1): 1–15. https://doi.org/10.1007/s11738-019-2989-4

Ye, W., Wang, T., Wei, W., Lou, S., Lan, F., Zhu, S., Li, Q., Ji, G., Lin, C., Wu, X., & Ma, L. 2020. The full-length transcriptome of *Spartina alterniflora* reveals the complexity of high salt tolerance in monocotyledonous halophyte. *Plant and Cell Physiology*, *61*(5): 882–896. https://doi.org/10.1093/pcp/pcaa013

You, J., Zhang, Y., Liu, A., Li, D., Wang, X., Dossa, K., Zhou, R., Yu, J., Zhang, Y., Wang, L., & Zhang, X. 2019. Transcriptomic and metabolomic profiling of drought-tolerant and susceptible sesame genotypes in response to drought stress. *BMC Plant Biology*, *19*(1): 1–16. https://doi.org/10.1186/s12870-019-1880-1

Younesi-Melerdi, E., Nematzadeh, G.A., Pakdin-Parizi, A., Bakhtiarizadeh, M.R., & Motahari, S.A. 2020. De novo RNA sequencing analysis of *Aeluropus littoralis* halophyte plant under salinity stress. *Scientific Reports*, *10*(1): 1–14. https://doi.org/10.1038/s41598-020-65947-5

Yuan, F., Lyu, M.J.A., Leng, B.Y., Zhu, X.G., & Wang, B.S. 2016. The transcriptome of NaCl-treated *Limonium bicolor* leaves reveals the genes controlling salt secretion of salt gland. *Plant Molecular Biology*, *91*(3): 241–256. https://doi.org/10.1007/s11103-016-0460-0

Zahaf, O., Blanchet, S., de Zélicourt, A., Alunni, B., Plet, J., Laffont, C., de Lorenzo, L., Imbeaud, S., Ichanté, J.L., Diet, A., Badri, M., Zabalza, A., González, E.M., Delacroix, H., Gruber, V., Frugier, F., & Crespi, M. 2012. Comparative transcriptomic analysis of salt adaptation in roots of contrasting *Medicago truncatula* genotypes. *Molecular Plant*, *5*(5): 1068–1081. https://doi.org/10.1093/mp/sss009

Zapata, J.M., Martínez-García, V., & Lefebvre, S. 2007. Phylogeny of the TRAF/MATH domain. In *TNF Receptor Associated Factors (TRAFs). Advances in Experimental Medicine and Biology*, ed. H. Wu, volume 597, 1–24. Springer, New York, NY. https://doi.org/10.1007/978-0-387-70630-6_1

Zhang, A., Han, D., Wang, Y., Mu, H., Zhang, T., Yan, X., & Pang, Q. 2018. Transcriptomic and proteomic feature of salt stress-regulated network in Jerusalem artichoke (*Helianthus tuberosus* L.) root based on de novo assembly sequencing analysis. *Planta*, *247*(3): 715–732. https://doi.org/10.1007/s00425-017-2818-1

Zhang, D.F., Zeng, T.R., Liu, X.Y., Gao, C.X., Li, Y.X., Li, C.H., Song, Y.C., Shi, Y.S., Wang, T.Y., & Yu, L.I. 2019. Transcriptomic profiling of sorghum leaves and roots responsive to drought stress at the seedling stage. *Journal of Integrative Agriculture*, *18*(9): 1980–1995. https://doi.org/10.1016/S2095-3119(18)62119-7

Zhang, S.S., Yang, H., Ding, L., Song, Z.T., Ma, H., Chang, F., & Liu, J.X. 2017. Tissue-specific transcriptomics reveals an important role of the unfolded protein response in maintaining fertility upon heat stress in Arabidopsis. *The Plant Cell*, *29*(5): 1007–1023. https://doi.org/10.1105/tpc.16.00916

Zhang, X., & Cai, X. 2011. Climate change impacts on global agricultural land availability. *Environmental Research Letters*, *6*(1): 1–9. 10.1088/1748-9326/6/1/014014

Zhang, X., Liao, M., Chang, D., & Zhang, F. 2014. Comparative transcriptome analysis of the Asteraceae halophyte *Karelinia caspica* under salt stress. *BMC Research Notes*, *7*(1): 1–9. https://doi.org/10.1186/1756-0500-7-927

Zhang, Y., Liu, Z., Khan, A.A., Lin, Q., Han, Y., Mu, P., Liu, Y., Zhang, H., Li, L., Meng, X., Ni, Z., & Xin, M. 2016. Expression partitioning of homeologs and tandem duplications contribute to salt tolerance in wheat (*Triticum aestivum* L.). *Scientific Reports*, *6*(1): 1–10. https://doi.org/10.1038/srep21476

Zhao, L., Yang, Z., Guo, Q., Mao, S., Li, S., Sun, F., Wang, H., & Yang, C. 2017. Transcriptomic profiling and physiological responses of halophyte *Kochia sieversiana* provide insights into salt tolerance. *Frontiers in Plant Science*, *8*(1985): 1–13. https://doi.org/10.3389/fpls.2017.01985

Zhou, J., Huang, J., Tian, X., Zheng, J., & He, X. 2020 Transcriptome analysis reveals dynamic changes in the salt stress response in Salix. *Journal of Forestry Research*, *31*: 1851–1862. https://doi.org/10.1007/s11676-019-00941-w

Zhou, Y., Yang, P., Cui, F., Zhang, F., Luo, X., & Xie, J. 2016. Transcriptome analysis of salt stress responsiveness in the seedlings of Dongxiang wild rice (*Oryza rufipogon* Griff.). *PloS ONE*, *11*(1): 1–25. https://doi.org/10.1371/journal.pone.0146242

Zhu, T., and Wang, X. 2000. Large-scale profiling of the Arabidopsis transcriptome. *Plant Physiology*, *124*: 1472–1476. https://doi.org/10.1104/pp.124.4.1472

Zhu, Q., Zhang, J., Gao, X., Tong, J., Xiao, L., Li, W., & Zhang, H. 2010. The Arabidopsis AP2/ERF transcription factor RAP2. 6 participates in ABA, salt and osmotic stress responses. *Gene*, *457*(1–2): 1–12. https://doi.org/10.1016/j.gene.2010.02.011

Ziemann, M., Kamboj, A., Hove, R.M., Loveridge, S., El-Osta, A., & Bhave, M. 2013. Analysis of the barley leaf transcriptome under salinity stress using mRNA-Seq. *Acta Physiologiae Plantarum*, *35*(6): 1915–1924. https://doi.org/10.1007/s11738-013-1230-0

11 Deciphering the Molecular Mechanism of Salinity Tolerance in Halophytes Using Transcriptome Analysis

Tejas C. Bosamia[1], *Doddabhimappa R. Gangapur*[1,2], *Parinita Agarwal*[1], *and Pradeep K. Agarwal**[1,2]
[1]Division of Plant Omics, CSIR-Central Salt and Marine Chemicals Research Institute (CSIR-CSMCRI), Council of Scientific & Industrial Research (CSIR), Gijubhai Badheka Marg, Bhavnagar (Gujarat), India
[2]Academy of Scientific and Innovative Research, CSIR-CSMCRI, Gijubhai Badheka Marg, Bhavnagar (Gujarat), India
*Corresponding author: E-mail: pagarwal@csmcri.res.in

CONTENTS

Abbreviations: ABA: abscisic acid; APX: ascorbate peroxidase; AREB: ABA-responsive elements; BR: brassinosteroids; CaM: calmodulin; CAT: catalase; cGMP: cyclic guanosine monophosphate; CK: cytokinins; DHAR: dehydroascorbate reductase; EST: Express Sequence Tag; GO: gene ontology; GR: glutathione reductase; HKT: histidine kinase transporter; IAA: indole-3-acetic acid; KAT: potassium transporter; KEGG: Kyoto Encyclopedia of Genes and Genomes; MAPKs: mitogen-activated protein kinases; MDA: malondialdehyde; MDHAR: monodehydroascorbate reductase;

DOI: 10.1201/9781003159636-11

NCC: nitrogen-containing compounds; NHX: Na^+/H^+ exchanger; NSCC: nonselective cation channels; PA: phosphatidic acid; POX: peroxidase; PPO: polyphenol oxidase; QACs: quaternary ammonium compounds; ROS: reactive oxygen species; SAGE: serial analysis of gene expression; SnRKs: sucrose non-fermenting 1-related protein kinases; SOD: superoxide dismutase; SOS: salt overly sensitive; TFs: transcription factors

11.1 INTRODUCTION

Plants are exposed to different abiotic stresses, such as high and low temperature, high salinity, drought and heavy metal, which adversely affect their growth and productivity. Among the different abiotic stresses, salinity is one of the most severe global problems, as the high salt concentration hampers the growth, development, and productivity of the crop (Chen *et al.*, 2018). A significant reduction in yield and quality of crops occurs due to salinity because of severe ion toxicity and water deficiency (physiological drought) (Agarwal *et al.*, 2020; Flowers and Colmer, 2015). Approximately 20% of the world's cultivated lands and half of irrigated lands are threatened by salinity (Shabala, 2013; Dang *et al.*, 2013). At global level, 831.4 mha of land is affected by soil salinity (FAO, 2014), and in India alone, approximately 6.74 mha of land is saline (Kumar and Sharma, 2020). Furthermore, increasing salinity of irrigation water in dry areas and rising sea water level due to climate change may lead to the situation becoming more pronounced in the future. Soil with salinity in excess of 4 ds/m is considered saline soil. Saline soils fall into the following categories: (i) sodic (or alkaline) soils, (ii) saline soils and (iii) saline/sodic soils. Sodic soils are found in arid and semi-arid regions and contain a crust of insoluble sodium carbonate and bicarbonate. Sodic soils with low soluble salt content have an electrical conductivity (EC) <4 dS/m, an exchangeable sodium percentage (ESP) >15% and a pH >8.5. Saline soils, found in arid regions, estuaries and coastal fringes, have a high concentration of water-soluble salts with an EC >4 dS/m, an ESP <15% and a pH <8.5. Saline–sodic soils are found in arid and semi-arid regions and possess an EC >4 dS/m, ESP >15% and pH <8.5 (Rengasamy, 2002).

Plants follow different molecular, biochemical and physiological mechanisms for adaptation in a highly saline environment. They try to avoid salt by selective uptake of ions, or by localizing them in the vacuoles or extruding them from the cell to maintain low toxicity in the cytoplasm. However, the efficacy of salinity tolerance varies between different types of plants. Most crop plants (glycophytes) are susceptible to high salinity and their metabolism is seriously affected by the saline condition (Bedre *et al.*, 2016).

In contrast, halophytes are able to grow and complete their life cycle under high salt concentrations (>200 mM NaCl). Halophytes represent only 1% of world flora and are rich in diversity (Nikalje *et al.*, 2018). Different halophytes show a different response towards abiotic stresses. In nature, halophytes grow in salt water areas like marine lands and desert regions. Halophytes have adopted specific salt-tolerant mechanisms over glycophytes, including osmotic adjustment, ion uptake and transport, salt accumulation and cellular compartmentalization (Agarwal *et al.*, 2020). The study of halophytes may unravel the molecular networks of genes and their regulatory sequences involved in adaptive mechanisms. Several salt stress-responsive promoters and genes have been isolated from halophyte species and transformed into crop plants through genetic engineering approaches to obtain abiotic stress tolerance (Afzal *et al.*, 2020; Kotula *et al.*, 2020).

11.2 MOLECULAR MECHANISMS OF SALINITY TOLERANCE

11.2.1 Perception of Salt Stress and Early Response

Two processes are thought to operate during salt stress in plants: rapid reduction of water availability due to high concentration of sodium ion in the soil, and accumulation of sodium ion in the shoot, which severely hampers photosynthesis in the plant (Van Zelm *et al.*, 2020). Osmotic or drought stress due to the unavailability of water is imposed as the early response of salinity stress, whereas the sodium-specific responses are induced later (Shavrukov, 2013). After NaCl application, plants achieve osmotic homeostasis relatively quickly or very early. Sodium stress comes later, after approximately 1–3 days, which triggers the different genes responsible for salinity tolerance.

11.2.2 Maintaining Sodium/Potassium Homeostasis and Role of Ion Transporter

As plants are exposed to high salinity, the cellular concentration of the ions increases gradually, which is harmful to both glycophytes and halophytes. To counter this problem, halophytes have developed special adaptive features like compartmentalizing the excess salts into different tissues or moving them into the vacuole (Zhu, 2001; Manchanda and Garg, 2008). Membrane transporter proteins have a pivotal role in the movement of the Na^+ in different tissues (Apse *et al.*, 1999). To maintain cellular ion homeostasis, coordinated action of membrane proteins, various pumps, Ca^{2+} sensors and their downstream interacting partners is required. Different Na^+/H^+ antiporters help in ion homeostasis and reduce cellular toxicity. The Na^+/H^+ exchanger (NHX) present at the vacuole membrane helps to eliminate excess salts from the cytoplasm. H^+-ATPase permits the NHX to pair with the passive movement of H^+ inside the vacuole and maintain the electrochemical gradient. The SOS (salt overly sensitive) pathway, involved in ion homeostasis, is operated by three major proteins: SOS1, SOS2, and SOS3 (Liu *et al.,* 1998; Liu *et al.*, 2000; Shi *et al.*, 2000). SOS1 encodes a membrane-integrated Na^+/H^+ antiporter, which is essential for the regulation of Na^+ efflux in the apoplast. A low-affinity Na^+ transporter like the histidine kinase transporter (HKT) blocks the entry of Na^+ into the cytosol. Based on the affinity for either sodium or potassium, HKT protein gene families are divided into two families (Platten *et al.*, 2006). The members of subfamily 1 contain serine residue at the first pore-loop domain, and they show higher selectivity for Na^+ than for K^+, whereas the members of subfamily 2 have a glycine residue in place of serine and function as Na^+/K^+ co-transporters. Under salt stress, most halophytes accumulate more Na^+ in their shoots than in their roots while retaining higher levels of K^+ than do glycophytes and, thus, a more optimal K^+/Na^+ ratio.

11.2.3 Production of Reactive Oxygen Species (ROS)

The salinity-induced high osmotic pressure inside the cells is responsible for the generation of ROS, including superoxide ($O_2^{-\bullet}$), hydrogen peroxide (H_2O_2), hydroxyl radical ($^{\bullet}OH$) and singlet oxygen ($^1O_2$) (Imlay, 2003; Khedia *et al.*, 2019). The high concentration of ROS has harmful effects in the cell by oxidation of membranes, proteins and nucleic acids, and causes lipid peroxidation, protein denaturation and DNA mutation. Plants have developed and adopted the enzymatic and non-enzymatic antioxidant system in response to salinity-induced osmotic stress. The enzymatic component consists of enzymes such as catalase (CAT), superoxide dismutase (SOD), ascorbate peroxidase (APX), peroxidase (POX), dehydroascorbate reductase (DHAR), glutathione reductase (GR), ascorbate-glutathione cycle enzyme and monodehydroascorbate reductase (MDHAR). Because plant cells do not have a special strategy to detoxify $^{\bullet}OH$ enzymatically, plants can diminish the $^{\bullet}OH$-induced damage to cell membranes by avoiding its formation or detoxifying it using low-molecular-weight non-enzymatic antioxidant molecules, for example, ascorbate, tocopherols, carotenoids, glutathione and polyphenols (Sharma *et al.*, 2012). The comparison between two halophytic grass species upon salinity stress showed a substantial increase in SOD and CAT activities in *Halopyrum mucronatum* compared with *Cenchrus ciliaris.* Similarly, a study was carried out to compare salt tolerance ability between halophyte and glycophyte plant species. The accumulation of a high level of ROS in the halophyte *Sesuvium portulacastrum* was only observed after a substantial time, while in the case of the glycophyte *Brassica juncea*, a high level of ROS and malondialdehyde (MDA) accumulation occurred as early as two days. A better antioxidant defence mechanism to prevent oxidative damage and proper energy utilization are the key determinant factors for salinity tolerance (Srivastava *et al.*, 2015). In the halophyte species *S. portulacastrum*, the activities of antioxidative enzymes such as POX, polyphenol oxidase (PPO), CAT, and SOD increased upon salt stress up to the level of 600 mM NaCl concentration (Rajaravindran and Natarajan, 2012).

11.2.4 Production and Transport of Salt Stress-Responsive Hormones

Plant hormones are organic endogenous signalling compounds, coordinating both structural and physiological activities of a plant under optimal conditions and during stress conditions. Abscisic acid (ABA) has key roles in maturation, senescence, regulation of root growth and stomata closing during abiotic stress. Apart from ABA, other hormones, including cytokinins (CK), brassinosteroids (BR), and auxins, also have important roles in abiotic stress tolerance (Verma *et al.*, 2016). The plant response to water deficit and physiological drought induced by salinity is mediated by ABA signalling. ABA biosynthesis genes, including *NINECIS-EPOXYCAROTENOID DIOXYGENASES* (*NCEDs*), *ABA DEFICIENTS* (*ABAs*) and *ALDEHYDEOXIDASE3*, are induced rapidly during the stress (Ruiz-Sola *et al.*, 2014) reducing water loss via transpiration by triggering stomatal closure (Wilkinson and Davies, 2010). The perception of ABA leads to activation of phosphorylation activity of sucrose non-fermenting 1-related protein kinases (SnRKs) (Van Zelm *et al.*, 2020). SnRKs are involved in the downstream regulation of various processes, including ion transport, ROS production, gene transcription, and closing of stomata. SnRKs also act upon the transcription factors that target the genes containing *ABRE*s in their promoters. NAC-, MYB-, HD-Zip, AP2/ERF- and WRKY-type are several types of *cis*-regulatory elements that induce ABA-responsive genes upon salt stress. Several phytohormones other than ABA, including CK, ethylene, BR, jasmonate, salicylate, and nitric oxide, also have an influence on the regulation of stomata (Acharya and Assmann, 2009). The role of indole-3-acetic acid (IAA) in drought tolerance has been reported by Zhang *et al.* (2009). CK plays a role as a positive regulator of auxin biosynthesis and is involved in the homeostatic feedback regulatory loop with IAA signalling, which maintains CK and IAA concentrations in developing root and shoot tissues (Jones *et al.*, 2010).

11.2.5 Osmotic Adjustment and Other Cellular Components

The excess ion concentration in the cytoplasm imparts osmotic stress in the cells, which can also be managed by accumulating low-molecular-mass compounds. These uncharged, polar and soluble compounds do not interfere with the plant's normal biochemical reactions. These compounds include simple sugars, sugar alcohols, complex sugars, nitrogen-containing compounds (NCC), quaternary ammonium compounds (QACs), polyamines, and sulfonium compounds (Parvaiz and Satyawati, 2008). These compounds have several roles in maintaining the osmotic pressure of the cell, including osmotic adjustment, osmoprotection, carbon storage, and radical scavenging. Polyols like mannitol act as low-molecular-weight chaperones and improve stress tolerance by scavenging hydroxyl radicals and stabilizing macromolecular structures (Bohnert *et al.*, 1995). Transgenic tobacco and *Arabidopsis* expressing mannitol showed better salt stress tolerance and improved growth attributes (Thomas *et al.*, 1995). Many amino acids, including proline, alanine, arginine, glycine, serine, leucine, valine, glutamine, asparagine, and the non-protein amino acids (citrulline and ornithine), accumulate in plants during salt stress. In *Suaeda altissima* (seablite), the concentration of proline increased two-fold in the shoot and five-fold in the root upon 750 mM NaCl-induced salinity stress (Meychik *et al.*, 2013). One of the QACs, glycine betaine (GB), is most abundantly present in the plant during salt and drought stress (Mansour, 2000). The GB is primarily localized in the chloroplast and plays a crucial role in the protection of chloroplast and thylakoid membrane, thus maintaining photosynthesis processes and the integrity of the plasma membrane.

11.2.6 Expression of Signalling Molecules and Role of Phospholipids and Protein Kinases

Plants use calcium (Ca^{2+}) as a secondary messenger and signalling molecule that regulates many developmental processes and stress adaptation (Kudla *et al.*, 2010). When plants experience biotic and abiotic stresses, cellular Ca^{2+} levels rise temporarily, which transduces the signal for the expression of downstream target genes (Galon *et al.*, 2008; Reddy *et al.*, 2011). The cytoplasmic Ca^{2+} signals can regulate the transcription of the nuclear genes by phosphorylation and dephosphorylation of transcription factors (TFs) (Galon *et al.*, 2008). Changes in cytosolic Ca^{2+} concentrations may cause either activation or suppression of the Ca^{2+}-responsive proteins and sensors such as calmodulin (CaM). Salinity and osmotic stresses cause modification of phospholipids present in the membrane that also act as signalling molecules. Several phospholipid signals, including polyphosphoinositides and phosphatidic acid (PA), are produced upon short exposure to salt stress (Van Zelm *et al.*, 2020). PA can bind and influence various downstream proteins of the ABA signalling pathway. The PA activates the MAPKs, which have a downstream target of the SOS pathway/NHX. The plant produces diverse MAPKs during the development stages and environmental changes. The MAPK signalling pathway is composed of three main cascading kinases: MAPK (MPK), MAPK kinase (MKK), and MAPKK kinases (MAPKKK or MEKK). Proteins in another protein kinase family, SnRK2 subclass 1, are activated in an ABA-independent manner during salinity and osmotic stress. Subclasses 2 and 3 are required for ABA responses (Kulik *et al.*, 2011). The ABA-independent SnRks influence the transcription of aquaporins and cytochrome P450, which affect auxin biosynthesis and transport as well as having a major role in root growth and branching during salt stress (Kawa *et al.*, 2020).

11.2.7 Transcription Factors

TFs are proteins that identify and bind to specific upstream promoter elements and control the expression of particular genes (Agarwal and Jha, 2010). TFs are the key modulators in cellular functions and have various roles in gene expression, including chromatin remodelling and recruitment/stabilization of the Pol II transcription-initiation complex (Singh, 1998). The role of TFs in salinity tolerance is to regulate the salinity-responsive gene and restore cellular ion homeostasis. Plant hormones act as a key integrator, which induces a cascade of signalling events that subsequently activate the TFs that increase expression of stress-responsive genes. The stress-responsive proteins are subsequently involved in various cellular responses, including detoxification of ROS, biosynthesis of cell walls, regulation of stomatal aperture, and maintaining the integrity of plasma membranes. The major plant-specific stress-responsive TFs are MYB, WRKY, bZIP, ZFP, HD-ZIP, NAC, etc. Many studies have shown that overexpression of different TFs mitigates abiotic stress in plants (Agarwal and Jha, 2010; Baillo *et al.*, 2019; Wang *et al.*, 2016a). ABA-dependent stress response pathways are mainly regulated by the ABA-responsive element binding factor (AREB/ABF) family of TFs, while ABA-independent pathways are regulated by the dehydration responsive element binding (DREB) TF family. ABF is the subfamily of bZIP that binds to conserved ABA-responsive elements (ABRE). The transcriptional activation of ABFs is mediated by the phosphorylation activities of kinases. DREB2, belonging to the APETLA2/ethylene responsive element binding protein family, plays an important role in salt, drought and heat shock response and induces salt and osmotic stress tolerance genes. Similarly, MYB TFs are one of the largest families of TFs identified in plants. Several MYB TFs function as regulators of biotic and abiotic stress tolerance in plants. WRKY TF families are characterized by a highly conserved N-terminus DNA-binding domain. WRKYs have multiple roles in plant immune response to biotic stress and osmotic stress induced during salinity and drought. Qin *et al.* (2013) confirmed salinity tolerance in the *Arabidopsis* plant by overexpressing wheat WRKY79. The transgenic plant exhibited more oxidative stress tolerance by developing more lateral roots and longer primary roots, producing more proline, and increased expression of salt-responsive genes.

In recent years, applications of RNA-seq using next-generation sequencing have gained access to the comprehensive transcriptome of halophytes, leading to the identification of novel and unique salt-responsive genes. In this chapter, we explain the transcriptome analyses and their role in finding new clues linked to the salinity tolerance mechanism in halophytes. Further, we discuss the representation of different salt-responsive genes or TFs in different halophytes and how the salt stress response is perceived, the role of ion transporters in maintain sodium/potassium homeostasis, and downstream regulation of TFs in the context of halophytes.

11.3 ERA OF TRANSCRIPTOME STUDIES

11.3.1 From EST to Whole Transcriptome

The transcriptome is comprised of all transcripts, including coding and noncoding, in an individual or a population of cells. The term "transcriptome" was defined by Velculescu *et al.* (1997). Transcriptome analysis began with the primitive technique called Express Sequence Tag (EST), a subsequence of a cDNA sequence generated with earlier-generation sequencers (Sanger sequencing method). EST does not require prior knowledge of the origin of the samples and is useful for novel gene identification, but cannot provide the quantitation of expression of the gene. It is also a time-consuming, low-throughput and costly method. The automation of Sanger sequencing technology leads to the development of a method called Serial/Cap Analysis of Gene Expression (SAGE/ CAGE), which was one of the first attempts to quantify gene expression on a global basis. Almost at the same time, an array-based technology called microarray emerged. It is a complementary probe hybridization-based technology in which probes are arrayed into glass slides and fluorescent labelled target transcripts are added. Because of their high throughput and lower cost, microarrays were widely used throughout the 2000s. However, unlike SAGE, they are limited to probing with an array of the genes that are already known, so a reference genome or transcriptome is a must for microarrays. The recent advent of sequencing technology (next-generation sequencing technology (second-generation sequencer)) led to intense development in transcriptome analysis and supplanted microarray-based approaches as the technology of choice for gene expression analysis. RNA-seq supports both the discovery and the quantification of transcripts using a single high-throughput sequencing assay. A reference genome or transcriptome is used for read alignment, but if a reference sequence is not available, the transcriptome can be assembled de novo using the reads and subsequently used for read alignment. In addition to gene expression analysis, RNA-seq is quite effective in detecting alternative splicing events. As a result, it has grown to be a more popular transcript profiling approach over the last decade (Wang *et al.*, 2019). In 2006, the first RNA-seq paper was published using the 454-pyrosequencing technology (Bainbridge *et al.*, 2006). However, the dominance of RNA-seq application in transcriptome study started with the invention of a new short-read sequencing method developed by Solexa (now Illumina) (Mortazavi *et al.*, 2008). The short-read RNA-seq, such as 454-pyrosequencer, Illumina and Ion torrent, has limitations regarding the accuracy of identifying multiple full-length transcripts (Wang *et al.*, 2016b). With long-read sequencing technology, including Oxford Nanopore and PacBio single-molecule sequencing (third-generation sequencer), it is now possible to generate single read for each transcript. Each transcript can be accurately captured and studied individually, since it directly provides full-length cDNA sequences (Wang *et al.*, 2016b). It is more suitable for detection of alternative splice events and novel isoforms, and to develop a gene model for orphan species, which have an incomplete or missing reference genome (Wang *et al.*, 2016b).

11.3.2 Data Analysis

RNA-seq experiments produce a huge number of raw sequence datasets, which are subjected to downstream processing and data analysis to obtain fruitful information. Data analysis is usually performed on a high-configuration computer having bioinformatics software. Most popular RNA-seq programs are run from a command-line interface, either in a Unix environment or within the R/Bioconductor statistical environment (Huber *et al.*, 2015). The analysis of RNA-seq data comprises four steps: quality control, alignment, quantification and differential expression (Van Verk *et al.*, 2013). Raw reads obtained from the sequencer are not perfect, and downstream analyses are required, including low-quality and complexity sequence trimming, adaptor sequence trimming, GC content distribution and removal of duplicate reads. Only quality control-checked reads are subjected to use for the alignment process. To determine the expression of a particular gene, the sequence reads abundance is estimated using reads alignment to a reference genome or de novo alignment to one another if no reference is available. There are many bioinformatics tools available to perform the short-read alignment. It is important to check the quality of mapping by the percentage of read mapped. The alignment tool also permits the identification of splice junctions, intron splice site information and mRNA isoforms. Identification of splice junctions prevents misalignment and improves the accuracy of gene expression. The complete construction of a transcriptome using short reads is challenging because of the multi-location nature of genes and genome complexity. Long-read sequencing technology provides a long enough sequence to cover a complete transcript of the gene and is a promising alternative. The next step of data processing after alignment is quantification, which may be performed at the gene, exon and transcript level. The quantification is carried out by counting the number of reads that map with the reference genome or de novo assembly. The raw read counts are affected by factors including the transcript length and the total number of reads. The differentially expressed genes are obtained by normalizing the raw read counts. A statistical test is applied to normalized read counts to decide whether or not, for a given gene, observed differences in expression are statistically significant. The final output of the data analyses will be gene lists, which are significantly differentially regulated between experimental groups. A brief bioinformatics pipeline to analyse RNA-seq data is visualized in Figure 11.1. Many software and web-based tools are available to visualize and interpret the differentially expressed gene data. Heatmap is one of the most common methods to represent gene expression data across the various treatment conditions. For interpretation, gene set enrichment analysis is used, which focuses on the functional annotation of differentially expressed genes. The annotation results are categorized as differentially expressed genes that are associated with a certain biological process or molecular function. The pathway enrichment analysis gives clues about gene lists from genome-scale data that are enriched in biological pathways. The procedure of pathway enrichment analysis comprises identification of the gene list, determination of statistically enriched pathways, and visualization and interpretation of results.

11.4 RNA-SEQ-BASED APPROACHES TO STUDY HALOPHYTES

The transcriptome sequencing of halophytes has recently been exploited to determine the molecular basis of the salt tolerance mechanism. A number of genes and networks are involved in the regulation of salinity stress in plants. These genes are briefly grouped into the following categories: (i) genes that encode TFs, hormones and general metabolism, (ii) genes that encode for the transporter proteins and antiporter protein that are involved in ion homeostasis, (iii) protein activity regulation genes, encoding for kinases and phosphatases, and (iv) genes encoding proteins that protect the cell from osmotic stress, including osmoprotectants. Several studies have been reported that include the use of 454 pyrosequencer, Illumina HiSeq and PacBio Iso-Seq platforms to study halophytes. The details of the RNA-seq-based approaches to study the molecular mechanism are demonstrated in Table 11.1.

Most transcriptome studies are carried out in *Suaeda* species (Diray-Arce *et al.*, 2015; Jin *et al.*, 2016; Gharat *et al.*, 2016; Xu *et al.*, 2017; Guo et al., 2019b; Han *et al.*, 2020; Zhang *et al.*, 2020b). This perennial halophyte has succulent leaves and a strong ability to accumulate and sequester Na^+ and Cl^- without the aid of salt glands, bladder or trichomes. The salt tolerance capacity of

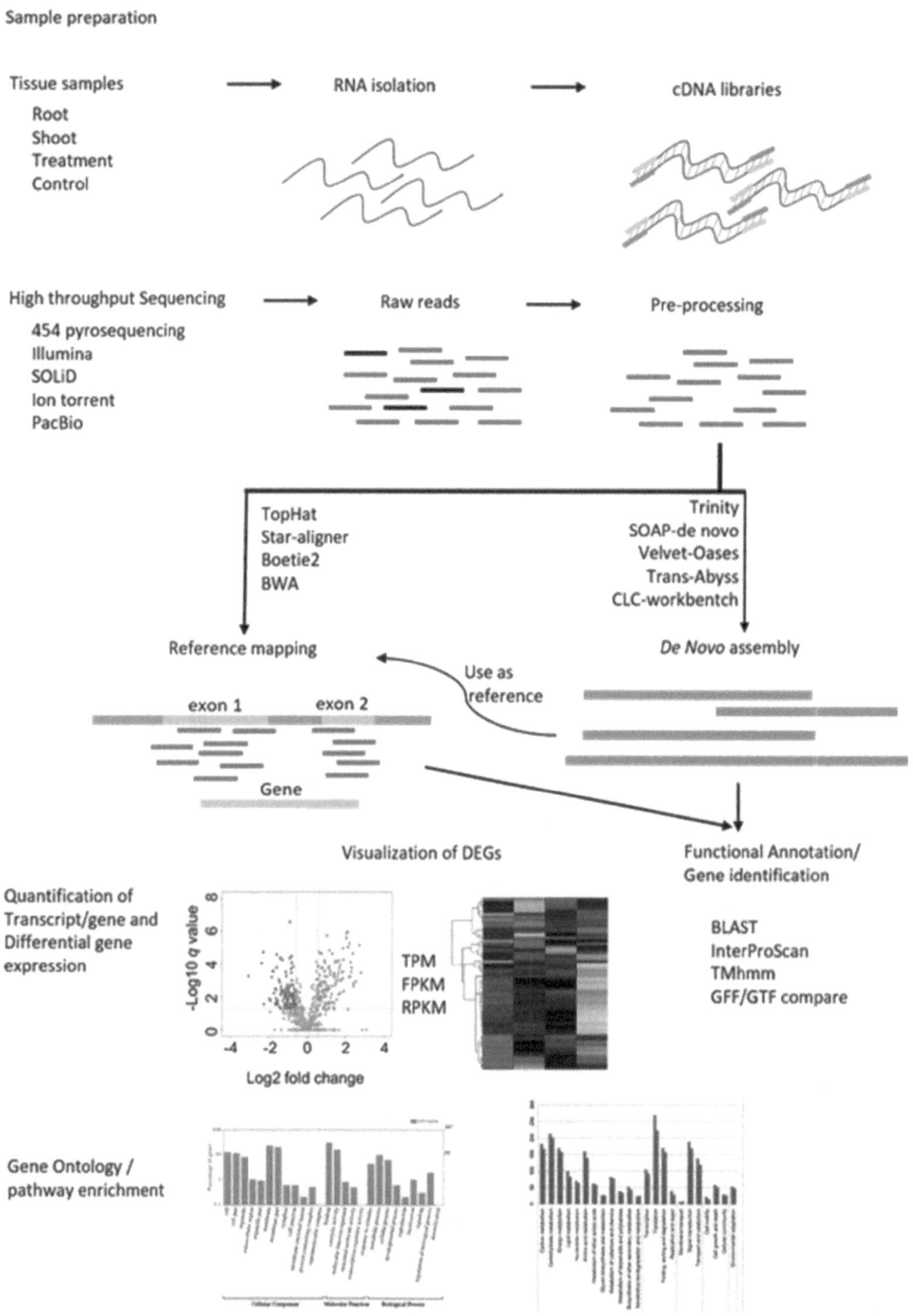

FIGURE 11.1 A schematic workflow of RNA-seq data analyses. Typical transcriptome sequencing of plant tissue comprises three major steps: sample preparation, transcriptome sequencing and in silico analyses. Tissue samples are subjected to RNA isolation, followed by sequencer-dependent cDNA library preparation. The high-throughput sequencing is carried out in various next-generation sequencing platforms. The raw reads generated are subjected to downstream analyses of pre-processing (quality check), de novo assembly and/or reference mapping, functional annotation, quantification of differential gene expression and visualization of analysis data.

TABLE 11.1
RNA-Seq Studies of Known Halophytes under Salt Stress

Sl. No	Halophytes (Platform for RNA-Seq)	High-Quality Reads Generated through Transcriptome (Data in Gb)	Time Point Interval and Salinity Treatments	DEGs across Treatment	GO Term Enriched	Pathway Enriched	Software Used	Reference
1	*Aeluropus littoralis* (HiSeq2500)	364,937,505	NaCl (200 and 400 mM) for 72 h	1861	120	2	Trinity, BLAST, HMMER, TransDecoder, TMHMM v2.0, SignalP v4.1, KEGG, RSEM tool	Younesi-Melerdi *et al.* (2020)
2	*Atriplex canescens* (Hiseq2000)	26.78 Gb	100 mM NaCl for 6 and 24 h	14,686 in leaves, 16,306 in root	52	–	Trinity, TGICL, Blast2GO, KEGG	Guo *et al.* (2019a)
3	*Atriplex centralasiatica* (HiSeqX)	331,486,650	300 mM NaCl for 7 d	9144	240	101	Trinity, BLAST, KEGG RSEM, Mapman	Yao *et al.* (2020)
4	*Beta vulgaris* ssp. *Maritime* (Miseq)	35,370,704	0 mM, 150 mM and 300 mM NaCl for 48 h	1246	–	–	Trinity, Blast2GO, Tophat2, Cufflink2	Skorupa *et al.* (2016)
5	*Betula halophile* (Hiseq2500)	451,940,854 (67.79 Gb)	0 mM, 200 mM NaCl for 24 h	–	56	129	Trinity, BLAST, BLAST2GO, EMBOSS, KOBAS	Shao *et al.* (2018)
6	*Chenopodium quinoa* (Hiseq2000)	142,935,936 (14.4 Gb)	300 mM NaCl for 1 h and 5 d	2416	–	–	CLC Genomics, Blast2GO	Ruiz *et al.* (2019)
7	*Clerodendrum inerme* (Hiseq2500)	683,948,928 (101.5 Gb)	400 mM NaCl for 0, 6, 24 and 72 h	4186 across all samples	52	31	Trinity, BlastX, KEGG, RSEM	Xiong *et al.* (2019)
8	*Halogeton glomeratus* (Hiseq2000)	265,300,480 (23.9 Gb)	200 mM NaCl for 6, 12, 24 and 72 h	17,410	44	128	Trinity, BLAST, Blast2GO, ERANGE, KEGG, WEGO	Wang *et al.* (2015)
9	*Halogeton glomeratus* (PacBio Iso-Seq)	433,420	0, 100, 200, 400 mM NaCl for 3 d	–	–	–	BLAST, SMRT and BLAST2GO	Wang *et al.* (2017)
10	*Helianthus tuberosus* (Hiseq4000)	60,069,922 (18 Gb)	50, 150 or 250 mM NaCl for 15 d	5432	102	19	Trinity, BLASTX, Blast2GO, KEGG, DESeq	Zhang *et al.* (2018)

(continued)

TABLE 11.1 (Continued)
RNA-Seq Studies of Known Halophytes under Salt Stress

Sl. No	Halophytes (Platform for RNA-Seq)	High-Quality Reads Generated through Transcriptome (Data in Gb)	Time Point Interval and Salinity Treatments	DEGs across Treatment	GO Term Enriched	Pathway Enriched	Software Used	Reference
11	*Hordeum brevisubulatum* (Hiseq2500)	72,066,413 (9 Gb)	350 mM NaCl/ 20 µM ABA for 0, 1, 6 and 24 h	47,605	–	–	Trinity, BLAST, KEGG, STEM	Zhang *et al.* (2020a)
12	*Iris halophile* (Hiseq)	380,000,000 (57 Gb)	150, 300 mM NaCl for 7 d	1120 in shoot and 100 in root	38	19	Trinity, BLAST, RSEM, DESeq	Liu *et al.* (2018)
13	*Karelinia capsica* (Hiseq2000)	101,930,000	300 mM NaCl for 3, 6, 12, 24, 36 and 48 h	60,127	69	202	Trinity, BLAST, Glimmer, EMBOSS, KEGG, WEGO	Zhang *et al.* (2014)
14	*Kochia sieversiana* (Hiseq2000)	502,491,351	NaCl (0, 80, 160, 240, 320, 400, 480, 600, 800 and 1000 mM) for 48 h	3012 in shoot and 3621 in root	55 in shoot and 50 in root	142 in shoot, 161 in root	Trinity, TransDecoder, BLAST, KEGG, RSEM	Zhao *et al.* (2017)
15	*Leymus chinensis* (PacBio Iso-Seq and Nova Seq 6000)	45,037 from PacBio, ~10 Gb data for each sample from NovaSeq6000	$NaHCO_3$ and Na_2CO_3, pH 8.8 as the alkaline-stress treatment with total salinity of 200 and 300 mM for 2–30 d	2216	–	9	SMRTlink, isoseq3, TransDecoder, HISAT2, RSEM, KEGG	Wang *et al.* (2020a)
16	*Lycium ruthenicum* (Hiseq)	208,873,796 (62.13 Gb)	75 mM NaCl for 48 h	199	Top 10	8	Trinity, BLAST, RSEM, DESeq, KEGG	Wang *et al.* (2020b)
17	*Phragmites karka* (Hiseq2500)	165,021,860	150 mM NaCl for 48 and 72 h	954 in leaves and 1097 in root	24	20	BinPacker, rnaSpades, CD-HIT-EST, CAP3, BLASTX, Bowtie2, RSEM, edgeR	Nayak *et al.* (2020)
18	*Porteresia coarctata* (GA II)	374,759,624	450 and 700 mM NaCl for 12 h	15,158	60	19	Velvet, Oases, ABySS, Trinity, CLC Genomics, CDHIT, BLASTX	Garg *et al.* (2014)

19	*Puccinellia distans* (Hiseq)	4,540,000,000	0, 20, 40, 80 and 120 µM of Na_2SeO_4 for 7 d	3525	51	121	CLC Genomics, Blast2GO	Kök *et al.* (2020)
20	*Salicornia europaea* (Hiseq2000)	2000 Mb	200 mM NaCl for 3, 12 and 24 h /3, 7 and 21 d	11,706 across all samples	–	17	SOAPdenovo, BLASTE, STScan, MapMan	Fan *et al.* (2013)
21	*Salicornia europaea* (Hiseq2000)	80,000,000 (7.21 Gb)	0, 10, 200, 400, 800 mM NaCl for 72 h	3979	199	120	Trinity, Blast2GO, KEGG	Ma *et al.* (2013)
22	*Salicornia europaea* (Miseq)	~108,000,000 (~16.2 Gb)	Salt affected site S1 and S2 season: fall [F] and spring [S]	38,384	–	–	Trinity, GeneMark S-T BLAST, DESeq	Furtado *et al.* (2019)
23	*Salicornia persica* (Hiseq2500)	133,872,558 (13.5 Gb)	300 mM NaCl for 72 h	1595	24	–	CLC Genomics, Blast2GO	Aliakbari et al. (2021)
24	*Sedum lineare* (Hiseq2500)	233,544,300	50, 100, 150, 200 and 250 mM NaCl for 7 d	584	31	–	Trinity, BLAST, KEGG, NGSQCToolkit, DEGseq	Song *et al.* (2019)
25	*Spartina alterniflora* (454 GS FLX)	770,690 (249.6 Mb)	500 mM NaCl for 6, 12, 24 and 72 h	–	–	–	Bowtie, BLAST, BinGO tool, HMMER, DEGseq	Bedre *et al.* (2016)
26	*Spartina alterniflora* and *Spartina maritima* (454 GS FLX)	1,114,825	Samples collected from two sites of Atlantic coast	409	–	–	GS Assembler v 2.3	De Carvalho *et al.* (2013)
27	*Spartina alterniflora* (Hiseq2500 and PacBio Iso-Seq)	410,233 insert reads from PacBio Iso-Seq, 298,143,652 (7.4 Gb) from Hiseq2500	0, 350, 500 and 800 mM NaCl for 24 h	2380	16	–	CDHIT-EST, GMAP, Blast2GO TransDecode, BinGO RSEM	Ye *et al.* (2020)
28	*Sporobolus virginicus* (Hiseq)	270,558,254	500 mM NaCl for 48 h	3748 in shoot and 4594 in root	147 in shoot and 263 in root	17 in shoot and 57 in root	Trinity, SOAPdenovo-Trans, Velvet, Oases, BLASTX, KEGG	Yamamoto *et al.* (2015)
29	*Suaeda fruticose* (Hiseq 2000)	283,587,292 (56 Gb)	0 and 300 mM NaCl	519	–	–	Oases, Velvet, CDHIT-EST, Blast2Go, EdgeR	Diray-Arce *et al.* (2015)

(continued)

TABLE 11.1 (Continued)
RNA-Seq Studies of Known Halophytes under Salt Stress

Sl. No	Halophytes (Platform for RNA-Seq)	High-Quality Reads Generated through Transcriptome (Data in Gb)	Time Point Interval and Salinity Treatments	DEGs across Treatment	GO Term Enriched	Pathway Enriched	Software Used	Reference
30	*Suaeda glauca* (Hiseq 2500)	179,289,824 (45.1 Gb)	0, 100, 200, 300, 400 and 1000 mM NaCl for 24 h	231	84	116	Trinity, Blast2GO, DESeq	Jin *et al.* (2016)
31	*Suaeda maritima* (Hiseq 2000)	300,430,008 (30.6 Gb)	100 mL 2.0% NaCl for 9 h	1382	Top 20	35	Trinity, CD-HIT-EST, BLAST, Bowtie 2, DESeq	Gharat *et al.* (2016)
32	*Suaeda rigida* (Hiseq 4000)	2212 million (~343.9 Gb)	100, 300, 500 mM NaCl for 3 h and 5 d	66 for 3 h and 726 for 5 days across all samples	–	–	Trinity, BLAST, GATK3, RSEM, KEGG	Han *et al.* (2020)
33	*Suaeda salsa* (Hiseq 2500)	137.5 million (13.75 Gb)	30% solution of sea salts for 10 d	68,599 in leaves and 77,250 in root	Top 20 in leaf and root	24 in leaves and 16 in root	Trinity, HTSwq, BLAST, KEGG	Guo *et al.* (2019b)
34	*Suaeda salsa* (Hiseq 2500)	194.87 million	Plants randomly selected from the saline inland	5825	–	111	Trinity, BLAST, DESeq	Xu *et al.* (2017)
35	*Suaeda salsa* (Hiseq X)	272,740,498	300 mM NaCl for 7 d	7753	171	55	Trinity, Blast2GO, RESM, Mapman	Zhang *et al.* (2020b)
36	*Reaumuria trigyna* (Hiseq 2000)	54.68 million	100 mM to 400 mM NaCl for 0.5, 1, 2, 4 and 8 h	5032	27	119	SOAPdenovo, Blast2GO	Dang *et al.* (2013)
37	*Zoysia macrostachya* (Hiseq)	326,357,784	300 mM NaCl for 24 h	8703	486	–	Trinity, RSEM, Goseq	Wang *et al.* (2020c)
38	*Zygophyllum xanthoxylum* (Hiseq 2000)	107.09 million	50 mM NaCl and –0.5 MPa osmotic stress for 6 and 24 h	6258 in root and 2486 in leaves	–	–	BLAST2GO	Ma *et al.* (2016)

Suaeda species is linked to its capability to retain higher levels of K^+ and so a higher K^+/Na^+ ratio in the shoot. Zhang *et al.* (2020b) reported that the higher expression of the potassium transporter (KEA3) is responsible for the sequestration of Na^+ and Cl^- into vacuoles. The GO enrichment analysis of *Suaeda salsa* transcriptome revealed that the genes related to ion transport, ROS scavengers, and TFs (such as HB-7, MYB, NAC, and HD-Zip) were highly expressed upon salt stress (Gharat *et al.*, 2016; Zhang *et al.*, 2020b). Similarly, upon salinity treatment, higher expression of the Na^+ /H^+ exchanger, V-H^+ ATPase, choline monooxygenase, and potassium and chloride channels was reported, which could reduce the overaccumulation of Na^+ and Cl^- in shoot and transport into vacuoles (Guo *et al.*, 2019b). Besides the ion transporter, glycine betaine, glutathione S-transferase, L-ascorbate, Fe-SOD and flavonoids function as antioxidants in *Suaeda* species (Gharat *et al.*, 2016; Guo *et al.*, 2019b). The halophyte *Salicornia* is being studied in detail by several researchers in order to understand the salinity tolerance mechanism. *Salicornia* species are small annual branched herbs with a high salt accumulation capability. The gene network analysis using transcriptome sequencing of *Salicornia persica* upon salt stress indicated that Na^+ compartmentalization is a major component of salinity tolerance, which is regulated by ABA and calcium signalling (Aliakbari *et al.*, 2021). Similarly, Furtado *et al.* (2019) mentioned the higher accumulation of Na^+ in shoot and K^+ and Ca^{2+} in root. The transcriptome analysis and pathway enrichment analysis of *Salicornia europaea* showed the differential expression of the genes related to cell wall metabolism and lignin biosynthesis to promote the development of xylem for increased sodium uptake upon salinity stress (Fan *et al.*, 2013). Furthermore, a large number of differentially expressed genes are identified in *Salicornia europaea* that are involved in ion homeostasis and osmotic adjustment during salinity, including cation transporters and proteins for the synthesis of low-molecular-weight compounds (Ma *et al.*, 2013). A long-read technology (Iso-Seq) for transcriptome sequencing has been employed to capture a full catalogue of transcriptomes and their variants in *Halogeton glomeratus* upon salinity stress. *Halogeton* species are annual/perennial herbs with succulent leaves. A total of 433,420 reads were generated, which were further assembled into 54,835 high-quality non-redundant consensus isoforms, and 53,230 (97.07%) were annotated. The data generated using Iso-Seq are more contiguous, contain a higher proportion of the intact coding sequences and can be used as a reference transcriptome for the study of the salt tolerance mechanism in *Halogeton glomeratus* (Wang *et al.*, 2017). Similarly, Illumina transcriptome sequencing of *Halogeton glomeratus* upon salt stress indicated upregulation of several genes, including V-ATPase (V-type H^+ -transporting ATPase), glutathione S-transferases, peroxidases and serine hydroxymethyltransferase (Wang *et al.*, 2015). In contrast to the above-mentioned halophytes, *Atriplex* species are typical secretohalophytes with salt bladders. The genus *Atriplex* is a C4 annual/perennial woody shrub with excellent salt tolerance adaptability. Guo *et al.* (2019a) mentioned that Na^+ transporters in leaves (such as AcSOS1, AcHKT1, and AcNHX) play crucial roles in the excretion of salt via epidermis bladder cells. The organic osmolytes in leaves were significantly upregulated by NaCl treatment. The aquaporin proteins play an important role in the regulation of water balance during salinity stress. A similar study of the transcriptome of *Atriplex centralasiatica* leaves indicated the role of cation potassium channel (KAT) and potassium transporter (KT), including potassium channel protein (KAT1), sodium transporter (HKT1) and cyclic nucleotide-gated channel protein (CNGC), in sequestering excessive Na^+ into bladder cells (Yao *et al.*, 2020). The GO enrichment analysis also showed the upregulation of genes associated with the ABA-dependent signalling pathway and TFs. A grass family halophyte, *Spartina alterniflora*, exhibited strong saline environmental adaptation, including greater ion selectivity to uptake potassium and exclude sodium, resulting in a lower Na^+/K^+ ratio and development of specialized salt glands to secrete ions. The gene network analysis of *S. alterniflora* indicated that the protein kinase-encoding genes (*SaOST1*, *SaCIPK10* and *SaLRRs*) have a crucial role in the regulation of the salt tolerance gene network (Ye *et al.*, 2020). The differential gene expression analysis of *S. alterniflora* upon salt stress showed upregulation of TFs (ARF, AUX/IAA, CCAAT, E2F-DP, and squamosa-promoter binding protein), ion transporters

(Na^+/H^+-antiporters and high- and low-affinity K^+ channels) and protein kinases (CDPK/SnRK family of protein kinases, casein kinase (CK) family, cyclin-dependent kinase family, serine-threonine protein kinase family and MAPK family) (Bedre *et al.*, 2016). The transcriptome analysis of *Puccinellia distans* was carried out under selenium stress and identified eight enzymes in sulphur metabolism based on KEGG pathway enrichment, and JA induces upregulation of antioxidant and defence-related genes to accumulate Se and cope with Se stress (Kök *et al.*, 2020). On a similar note, a study on *Iris halophila* under salinity stress identified key genes that encode for flavonoid and lignin biosynthetic pathways, hormone signalling and sodium/potassium ion transporter, and are distributed in significantly enriched 19 KEGG pathways (Liu *et al.*, 2018). *Aeluropus littoralis* is an annual grass family halophyte plant with salt glands and bladders. The transcriptome study of *A. littoralis* identified vacuoles and PRC1 complex enriched in the cell component, and the MAPK signalling and phytohormone signal transduction significantly enriched in the KEGG pathway (Younesi-Melerdi *et al.*, 2020). The whole-transcriptome profiling of leaf and root tissue of *Phragmites karka* under salinity treatment showed the upregulation of different TFs (including WRKY, MYB, CCCH, and NAC), caffeolyl shikimate esterase, alanine glyoxylate aminotransferase, and pheophorbide-a-oxygenase (Nayak *et al.*, 2020). Transcriptome sequencing of an endangered halophyte species, *Betula halophila*, under salt stress showed that four pathways were enriched significantly: fatty acid elongation, ribosome, sphingolipid metabolism, and flavonoid biosynthesis (Shao *et al.*, 2018). In the case of salt-tolerant *Beta vulgaris*, transcriptome sequencing revealed the differential expression of genes that encode for membrane transport proteins such as PIP (Plasma membrane intrinsic proteins) and TIP (Plant tonoplast intrinsic proteins), aquaporins, H^+-PPase, ion channels, ABC transporters or low-molecular-weight metabolite transporters (Skorupa *et al.*, 2016). A time-course transcriptome study of *Clerodendrum inerme* was carried out upon salt treatment for 0 h, 6 h, 24 h, and 72 h. The GO enrichment analysis in *C. inerme* indicated total 52 GO term and 40, 23 and 22 enriched pathways for the 6 h, 72 h, and 24 h treatments, respectively (Xiong *et al.*, 2019).

11.5 CONCLUSIONS

The recent advent of sequencing technology has allowed halophyte species to be exploited for revealing the molecular basis of salinity tolerance. To sum up, the ABA-dependent signalling pathway and calcium signalling are the key regulators of salinity stress in halophytes. The downstream regulation of ABA-mediated process includes ion transport, ROS production, gene transcription and closing of stomata during salinity stress. Ion transporters, including KAT, HKT1, CNGC, NHX, V-ATPase, and Na^+/H^+ ion exchangers, are highly upregulated upon salt stress in halophytes. Regardless of whether halophytes contain salt glands and bladder or not, ion transporters have a major role in maintaining the ion and water balance in the metabolically active cells of the plant. Apart from Na^+ compartmentalization into vacuole and salt bladder as adaptation, a higher level of small organic molecules as osmoprotectants plays an important role in preventing damage caused by salinity-induced osmotic stress in plants.

ACKNOWLEDGEMENTS

CSIR-CSMCRI Communication No 39/2021. P.A. acknowledges the financial support from DSTWOS-A scheme. All the authors acknowledge CSIR, New Delhi for financial support.

REFERENCES

Acharya BR, Assmann SM (2009) Hormone interactions in stomatal function. *Plant Molecular Biology*, 69(4): 451–462

Afzal MZ, Jia Q, Ibrahim AK, Niyitanga S, Zhang L (2020) Mechanisms and signaling pathways of salt tolerance in crops: understanding from the transgenic plants. *Tropical Plant Biology*: 13(4), 297–320

Agarwal P, Dabi M, Kinhekar K, Gangapur DR, Agarwal PK (2020) Special adaptive features of plant species in response to salinity. In: Hasanuzzaman M, Tanveer M (ed) *Salt and Drought Stress Tolerance in Plants*. Springer, Cham, pp. 53–76

Agarwal PK, Jha B (2010) Transcription factors in plants and ABA dependent and independent abiotic stress signalling. *Biologia Plantarum*, 54(2): 201–212

Aliakbari M, Razi H, Alemzadeh A, Tavakol E (2021) RNA-seq Ttranscriptome profiling of the halophyte *Salicornia persica* in response to salinity. *Journal of Plant Growth Regulation*, 40(2):707–721

Apse MP, Aharon GS, Snedden WA, Blumwald E (1999) Salt tolerance conferred by overexpression of a vacuolar Na+/H+ antiport in *Arabidopsis*. *Science* 285(5431): 1256–1258

Baillo EH, Kimotho RN, Zhang Z, Xu P (2019) Transcription factors associated with abiotic and biotic stress tolerance and their potential for crops improvement. *Genes* 10(10): 771

Bainbridge MN, Warren RL, Hirst M, Romanuik T, Zeng T, Go A, Delaney A, Griffith M, Hickenbotham M, Magrini V, Mardis ER, Sadar MD, Siddiqui AS, Marra MA, Jones SJ (2006) Analysis of the prostate cancer cell line LNCaP transcriptome using a sequencing-by-synthesis approach. *BMC Genomics* 7(1): 1–11

Bedre R, Mangu VR, Srivastava S, Sanchez LE, Baisakh N (2016) Transcriptome analysis of smooth cordgrass (*Spartina alterniflora* Loisel), a monocot halophyte, reveals candidate genes involved in its adaptation to salinity. *BMC Genomics* 17(1): 1–18

Bohnert HJ, Nelson DE, Jensen RG (1995) Adaptations to environmental stresses. *The Plant Cell* 7(7): 1099

Chen M, Yang Z, Liu J, Zhu T, Wei X, Fan H, Wang B (2018) Adaptation mechanism of salt excluders under saline conditions and its applications. *International Journal of Molecular Sciences* 19(11): 3668

Dang ZH, Zheng LL, Wang J, Gao Z, Wu SB, Qi Z, Wang YC (2013) Transcriptomic profiling of the salt-stress response in the wild recretohalophyte *Reaumuria trigyna*. *BMC Genomics* 14(1): 1–18

De Carvalho JF, Poulain J, Da Silva C, Wincker P, Michon-Coudouel S, Dheilly A, Naquin D, Boutte J, Salmon A, Ainouche M (2013) Transcriptome *de novo* assembly from next-generation sequencing and comparative analyses in the hexaploid salt marsh species *Spartina maritima* and *Spartina alterniflora* (Poaceae). *Heredity* 110(2): 181–193

Diray-Arce J, Clement M, Gul B, Khan MA, Nielsen BL (2015) Transcriptome assembly, profiling and differential gene expression analysis of the halophyte *Suaeda fruticosa* provides insights into salt tolerance. *BMC Genomics* 16(1): 1–24

Fan P, Nie L, Jiang P, Feng J, Lv S, Chen X, Bao H, Guo J, Tai F, Wang J, Jia W, Li Y (2013) Transcriptome analysis of *Salicornia europaea* under saline conditions revealed the adaptive primary metabolic pathways as early events to facilitate salt adaptation. *PLoS ONE* 8(11): e80595

FAO (2014) (www.fao.org/faostat/en/)

Flowers TJ, Colmer TD (2015) Plant salt tolerance: adaptations in halophytes. *Annals of Botany* 115: 327–331

Furtado BU, Nagy I, Asp T, Tyburski J, Skorupa M, Gołębiewski M, Hulisz P, Hrynkiewicz K (2019) Transcriptome profiling and environmental linkage to salinity across *Salicornia europaea* vegetation. *BMC Plant Biology* 19(1): 1–14

Galon Y, Nave R, Boyce JM, Nachmias D, Knight MR, Fromm H (2008) Calmodulin-binding transcription activator (CAMTA) 3 mediates biotic defense responses in *Arabidopsis*. *FEBS Letters* 582(6): 943–948

Garg R, Verma M, Agrawal S, Shankar R, Majee M, Jain M (2014) Deep transcriptome sequencing of wild halophyte rice, *Porteresia coarctata*, provides novel insights into the salinity and submergence tolerance factors. *DNA Research* 21(1): 69–84

Gharat SA, Parmar S, Tambat S, Vasudevan M, Shaw BP (2016) Transcriptome analysis of the response to NaCl in *Suaeda maritima* provides an insight into salt tolerance mechanisms in halophytes. *PloS ONE* 11(9): e0163485

Guo H, Zhang L, Cui YN, Wang SM, Bao AK (2019a) Identification of candidate genes related to salt tolerance of the secretohalophyte *Atriplex canescens* by transcriptomic analysis. *BMC Plant Biology* 19(1): 1–17

Guo SM, Tan Y, Chu HJ, Sun MX, Xing JC (2019b) Transcriptome sequencing revealed molecular mechanisms underlying tolerance of *Suaeda salsa* to saline stress. *PloS ONE* 14(7): e0219979

Han ZJ, Sun Y, Zhang M, Zhai JT (2020) Transcriptomic profile analysis of the halophyte *Suaeda rigida* response and tolerance under NaCl stress. *Scientific Reports* 10(1): 1–10

Huber W, Carey VJ, Gentleman R, Anders S, Carlson M, Carvalho BS, Bravo HC, Davis S, Gatto L, Girke T, Gottardo R, Hahne F, Hansen KD, Irizarry RA, Lawrence M, Love MI, MacDonald J, Obenchain V, Oleś AK, Pagès H, Reyes A, Shannon P, Smyth GK, Tenenbaum D, Waldron L, Morgan M (2015) Orchestrating high-throughput genomic analysis with Bioconductor. *Nature Methods* 12(2): 115

Imlay JA (2003) Pathways of oxidative damage. *Annual Reviews in Microbiology* 57(1): 395–418

Jin H, Dong D, Yang Q, Zhu D (2016) Salt-responsive transcriptome profiling of *Suaeda glauca* via RNA sequencing. *PloS ONE* 11(3): e0150504

Jones B, Gunnerås SA, Petersson SV, Tarkowski P, Graham N, May S, Dolezal K, Sandberg G, Ljung K (2010) Cytokinin regulation of auxin synthesis in *Arabidopsis* involves a homeostatic feedback loop regulated via auxin and cytokinin signal transduction. *The Plant Cell* 22(9): 2956–2969

Kawa D, Meyer AJ, Dekker HL, Abd-El-Haliem AM, Gevaert K, Van De Slijke E, Maszkowska J, Bucholc M, Dobrowolska G, De Jaeger G, Schuurink RC, Haring MA, Testerink C (2020) SnRK2 protein kinases and mRNA decapping machinery control root development and response to salt. *Plant Physiology* 182(1): 361–377

Khedia J, Agarwal P, Agarwal PK (2019) Deciphering hydrogen peroxide-induced signalling towards stress tolerance in plants. *3 Biotech* 9(11): 1–13

Kök AB, Mungan MD, Doğanlar S, Frary A. (2020) Transcriptomic analysis of selenium accumulation in *Puccinellia distans* (Jacq.) Parl., a boron hyperaccumulator. *Chemosphere* 245: 125665

Kotula L, Garcia Caparros P, Zörb C, Colmer TD, Flowers TJ (2020) Improving crop salt tolerance using transgenic approaches: an update and physiological analysis. *Plant, Cell & Environment* 43(12): 2932–2956

Kudla J, Batistič O, Hashimoto K (2010) Calcium signals: the lead currency of plant information processing. *The Plant Cell* 22(3): 541–563

Kulik A, Wawer I, Krzywińska E, Bucholc M, Dobrowolska G (2011) SnRK2 protein kinases—key regulators of plant response to abiotic stresses. *Omics: a Journal of Integrative Biology* 15(12): 859–872

Kumar P, Sharma PK (2020) Soil salinity and food security in India. Frontiers in Sustainable Food Systems 4: 174.

Liu J, Zhu JK (1998) A calcium sensor homolog required for plant salt tolerance. *Science* 280: 1943–1945

Liu JP, Ishitani M, Halfter U, Kim CS, Zhu JK (2000) The *Arabidopsis thaliana* SOS2 gene encodes a protein kinase that is required for salt tolerance. *Proceedings of the National Academy of Sciences of the United States of America* 97: 3730–3734

Liu Q, Tang J, Wang W, Zhang Y, Yuan H, Huang S (2018) Transcriptome analysis reveals complex response of the medicinal/ornamental halophyte *Iris halophila* Pall. to high environmental salinity. *Ecotoxicology and Environmental Safety* 165: 250–260

Ma J, Zhang M, Xiao X, You J, Wang J, Wang T, Yao Y, Tian C (2013) Global transcriptome profiling of *Salicornia europaea* L. shoots under NaCl treatment. *PloS ONE* 8(6): e65877

Ma Q, Bao AK, Chai WW, Wang WY, Zhang JL, Li YX, Wang SM (2016) Transcriptomic analysis of the succulent xerophyte *Zygophyllum xanthoxylum* in response to salt treatment and osmotic stress. *Plant and Soil* 402(1): 343–361

Manchanda G, Garg N (2008) Salinity and its effects on the functional biology of legumes. *Acta Physiologiae Plantarum* 30(5): 595–618

Mansour MMF (2000) Nitrogen containing compounds and adaptation of plants to salinity stress. *Biologia Plantarum* 43(4): 491–500

Meychik NR, Nikolaeva YI, Yermakov IP (2013) Physiological response of halophyte (*Suaeda altissima* (L.) Pall.) and glycophyte (*Spinacia oleracea* L.) to salinity. *American Journal of Plant Sciences* 4: 427–435

Mortazavi A, Williams BA, McCue K, Schaeffer L, Wold B (2008) Mapping and quantifying mammalian transcriptomes by RNA-Seq. *Nature Methods* 5(7): 621–628

Nayak SS, Pradhan S, Sahoo D, Parida A (2020) De novo transcriptome assembly and analysis of *Phragmites karka*, an invasive halophyte, to study the mechanism of salinity stress tolerance. *Scientific Reports* 10(1): 1–12

Nikalje GC, Srivastava AK, Pandey GK, Suprasanna P (2018) Halophytes in biosaline agriculture: mechanism, utilization, and value addition. *Land Degradation & Development* 29(4): 1081–1095

Parvaiz A, Satyawati S (2008) Salt stress and phyto-biochemical responses of plants – a review. *Plant Soil and Environment* 54(3): 89

Platten JD, Cotsaftis O, Berthomieu P, Bohnert H, Davenport RJ, Fairbairn DJ, Horie T, Leigh RA, Lin HX, Luan S, Mäser P, Pantoja O, Rodríguez-Navarro A, Schachtman DP, Schroeder JI, Sentenac H, Uozumi N, Véry AA, Zhu JK, Dennis ES, Tester M (2006) Nomenclature for HKT transporters, key determinants of plant salinity tolerance. *Trends in Plant Science* 11(8): 372–374

Qin Z, Lv H, Zhu X, Meng C, Quan T, Wang M, Xia G (2013) Ectopic expression of a wheat WRKY transcription factor gene TaWRKY71-1 results in hyponastic leaves in *Arabidopsis thaliana*. *PLoS ONE* 8(5): e63033

Rajaravindran M, Natarajan S (2012) Effects of salinity stress on growth and antioxidant enzymes of the halophyte *Sesuvium portulacastrum*. *International Journal of Research in Plant Science* 2(1): 23–28

Reddy AS, Ali GS, Celesnik H, Day IS (2011) Coping with stresses: roles of calcium-and calcium/calmodulin-regulated gene expression. *The Plant Cell* 23(6): 2010–2032

Rengasamy P (2002) Transient salinity and subsoil constraints to dryland farming in Australian sodic soils: an overview. Australian Journal of Experimental Agriculture 42(3): 351–361

Ruiz KB, Maldonado J, Biondi S, Silva H (2019) RNA-seq analysis of salt-stressed versus non salt-stressed transcriptomes of *Chenopodium quinoa* landrace R49. *Genes* 10(12): 1042

Ruiz-Sola MÁ, Arbona V, Gómez-Cadenas A, Rodríguez-Concepción M, Rodríguez-Villalón A (2014) A root specific induction of carotenoid biosynthesis contributes to ABA production upon salt stress in *Arabidopsis*. *PLoS ONE* 9(3): e90765

Shabala S (2013) Learning from halophytes: physiological basis and strategies to improve abiotic stress tolerance in crops. *Annals of Botany* 112: 1209–1221 .

Shao F, Zhang L, Wilson IW, Qiu D (2018) Transcriptomic analysis of *Betula halophila* in response to salt stress. *International Journal of Molecular Sciences* 19(11): 3412

Sharma P, Jha AB, Dubey RS, Pessarakli M (2012) Reactive oxygen species, oxidative damage, and antioxidative defense mechanism in plants under stressful conditions. *Journal of Botany* 2012: 217037

Shavrukov Y (2013) Salt stress or salt shock: which genes are we studying? *Journal of Experimental Botany* 64(1): 119–127

Shi H, Ishitani M, Kim C, Zhu JK (2000) The *Arabidopsis thaliana* salt tolerance gene SOS1 encodes a putative Naþ/Hþ antiporter. *Proceedings of the National Academy of Sciences of the United States of America* 97: 6896–6901

Singh KB (1998) Transcriptional regulation in plants: the importance of combinatorial control. *Plant Physiology* 118(4): 1111–1120

Skorupa M, Gołębiewski M, Domagalski K, Kurnik K, Nahia KA, Złoch M, Tretyn A, Tyburski J (2016) Transcriptomic profiling of the salt stress response in excised leaves of the halophyte *Beta vulgaris* ssp. maritima. *Plant Science* 243: 56–70.

Song Y, Yang X, Yang S, Wang J. (2019) Transcriptome sequencing and functional analysis of *Sedum lineare* Thunb. upon salt stress. *Molecular Genetics and Genomics* 294(6): 1441–1453

Srivastava AK, Srivastava S, Lokhande VH, D'Souza SF, Suprasanna P (2015) Salt stress reveals differential antioxidant and energetics responses in glycophyte (*Brassica juncea* L.) and halophyte (*Sesuvium portulacastrum* L.). *Frontiers in Environmental Science* 3: 19

Thomas JC, Sepahi M, Arendall B, Bohnert HJ (1995) Enhancement of seed germination in high salinity by engineering mannitol expression in *Arabidopsis thaliana*. *Plant, Cell & Environment* 18(7): 801–806

Van Verk MC, Hickman R, Pieterse CM, Van Wees SC (2013) RNA-Seq: revelation of the messengers. *Trends in Plant Science* 18(4): 175–179

Van Zelm E, Zhang Y, Testerink C (2020) Salt tolerance mechanisms of plants. *Annual Review of Plant Biology* 71: 403–433

Velculescu VE, Zhang L, Zhou W, Vogelstein J, Basrai MA, Bassett Jr DE, Hieter P, Vogelstein B, Kinzler KW (1997) Characterization of the yeast transcriptome. *Cell* 88(2): 243–251

Verma V, Ravindran P, Kumar PP (2016) Plant hormone-mediated regulation of stress responses. *BMC Plant Biology* 16(1): 1–10

Wang B, Kumar V, Olson A, Ware D (2019) Reviving the transcriptome studies: an insight into the emergence of single-molecule transcriptome sequencing. *Frontiers in Genetics* 10: 384

Wang B, Tseng E, Regulski M, Clark TA, Hon T, Jiao Y, Lu Z, Olson A, Stein AC, Ware D (2016b) Unveiling the complexity of the maize transcriptome by single-molecule long-read sequencing. *Nature Communications* 7(1): 1–13

Wang H, Guo J, Tian Z, Li J, Deng L, Zheng Y, Yuan Y (2020b) Transcriptome profiling of mild-salt responses in *Lycium ruthenicum* early seedlings to reveal salinity-adaptive strategies. *Acta Physiologiae Plantarum* 42(4): 1–17

Wang H, Wang H, Shao H, Tang X (2016a) Recent advances in utilizing transcription factors to improve plant abiotic stress tolerance by transgenic technology. *Frontiers in Plant Science* 7: 67

Wang H, Xiang Y, Li LH, Bhanbhro N, Yang CW, Zhang Z (2020a) Photosynthetic response and transcriptomic profiling provide insights into the alkali tolerance of clone halophyte *Leymus chinensis*. *Photosynthetica* 58(3): 780–789

Wang J, Li B, Meng Y, Ma X, Lai Y, Si E, Yang K, Ren P, Shang X, Wang H (2015) Transcriptomic profiling of the salt-stress response in the halophyte *Halogeton glomeratus*. *BMC Genomics* 16(1): 1–14

Wang J, Yao L, Li B, Meng Y, Ma X, Wang H. (2017) Single-molecule long-read transcriptome dataset of halophyte *Halogeton glomeratus*. *Frontiers in Genetics* 8: 197

Wang R, Wang X, Liu K, Zhang XJ, Zhang LY, Fan SJ (2020c) Comparative transcriptome analysis of halophyte *Zoysia macrostachya* in response to salinity stress. *Plants* 9(4): 458

Wilkinson S, Davies WJ (2010) Drought, ozone, ABA and ethylene: new insights from cell to plant to community. *Plant, Cell & Environment* 33(4): 510–525

Xiong Y, Yan H, Liang H, Zhang Y, Guo B, Niu M, Jian S, Ren H, Zhang X, Li Y, Zeng S, Wu K, Zheng F, Teixeira da Silva JA, Ma G (2019) RNA-Seq analysis of *Clerodendrum inerme* (L.) roots in response to salt stress. *BMC Genomics* 20(1): 1–18

Xu Y, Zhao Y, Duan H, Sui N, Yuan F, Song J (2017) Transcriptomic profiling of genes in matured dimorphic seeds of euhalophyte *Suaeda salsa*. *BMC Genomics* 18(1): 1–14

Yamamoto N, Takano T, Tanaka K, Ishige T, Terashima S, Endo C, Kurusu T, Yajima S, Yano K, Tada Y (2015) Comprehensive analysis of transcriptome response to salinity stress in the halophytic turf grass *Sporobolus virginicus*. *Frontiers in Plant Science* 6: 241

Yao Y, Zhang X, Wang N, Cui Y, Zhang L, Fan S (2020) Transcriptome analysis of salt stress response in halophyte *Atriplex centralasiatica* leaves. *Acta Physiologiae Plantarum* 42(1): 1–15

Ye W, Wang T, Wei W, Lou S, Lan F, Zhu S, Li Q, Ji G, Lin C, Wu X, Ma L (2020) The full-length transcriptome of *Spartina alterniflora* reveals the complexity of high salt tolerance in monocotyledonous halophyte. *Plant and Cell Physiology* 61(5): 882–896

Younesi-Melerdi E, Nematzadeh GA, Pakdin-Parizi A, Bakhtiarizadeh MR, Motahari SA (2020). *De novo* RNA sequencing analysis of *Aeluropus littoralis* halophyte plant under salinity stress. *Scientific Reports* 10(1): 1–14

Zhang A, Han D, Wang Y, Mu H, Zhang T, Yan X, Pang Q (2018) Transcriptomic and proteomic feature of salt stress-regulated network in Jerusalem artichoke (*Helianthus tuberosus* L.) root based on de novo assembly sequencing analysis. *Planta* 247(3): 715–732

Zhang L, Wang Y, Zhang Q, Jiang Y, Zhang H, Li R (2020a) Overexpression of HbMBF1a, encoding multiprotein bridging factor 1 from the halophyte *Hordeum brevisubulatum*, confers salinity tolerance and ABA insensitivity to transgenic *Arabidopsis thaliana*. *Plant Molecular Biology* 102(1): 1–17

Zhang X, Ervin EH, Evanylo GK, Haering KC (2009) Impact of biosolids on hormone metabolism in drought-stressed tall fescue. *Crop Science* 49(5): 1893–1901

Zhang X, Liao M, Chang D, Zhang F (2014) Comparative transcriptome analysis of the Asteraceae halophyte *Karelinia caspica* under salt stress. *BMC Research Notes* 7(1): 1–9

Zhang X, Yao Y, Li X, Zhang L, Fan S (2020b) Transcriptomic analysis identifies novel genes and pathways for salt stress responses in *Suaeda salsa* leaves. *Scientific Reports* 10(1): 1–12

Zhao L, Yang Z, Guo Q, Mao S, Li S, Sun F, Wang H, Yang C (2017) Transcriptomic profiling and physiological responses of halophyte *Kochia sieversiana* provide insights into salt tolerance. *Frontiers in Plant Science* 8: 1985

Zhu JK (2001) Plant salt tolerance. *Trends in Plant Science* 6(2): 66–71

12 Seed Aging in Crops
A Proteomics Perspective

*Truong Van Nguyen[1], Ravi Gupta[2], Sun Tae Kim[1], and Cheol Woo Min[1,3]**
[1]Department of Plant Bioscience, Life and Industry Convergence Research Institute, Pusan National University, Miryang, South Korea
[2]College of General Education, Kookmin University, Seoul, South Korea
[3]Department of Plant Bioscience, Pusan National University, Miryang, South Korea
Corresponding author: E-mail: min0685@pusan.ac.kr

CONTENTS

12.1 INTRODUCTION

Seed aging is an undesirable event that leads to a reduction in agricultural productivity because of the reduced longevity and loss of seed viability (Anderson and Baker 1983). Surprisingly, more than 25% of seed productivity is lost due to deteriorative stress during post-harvest storage (Shelar, Shaikh, and Nikam 2008). Representatively, the temperature and moisture content are negatively correlated with the germination ability of the seeds during pre- and post-harvesting storage. During aging, physiological, biological, and biochemical alterations take place in the seeds, resulting in loss of enzymatic activities, increased protein degradation, decreased respiratory system, and elevated lipid peroxidation, among others (Murata, Tsuchiya, and Roos 1984; Bailly et al. 1996). Therefore, the maintenance of seeds for sustaining their viability and productivity during storage is one of the essential prerequisites for the potential agricultural quality of seeds (Finch-Savage and Bassel 2016).

The food reserves of seeds, including storage proteins, oils, carbohydrates, and a variety of secondary metabolites, are rapidly mobilized during germination to provide the energy necessary for seedling emergence and initial seedling growth (Rajjou et al. 2012). In healthy and vital seeds,

DOI: 10.1201/9781003159636-12

reactive oxygen species (ROS) are present in optimal concentrations and play pivotal roles in cell signaling and regulation of seed germination and dormancy (Kurek, Plitta-Michalak, and Ratajczak 2019; Wang et al. 2012a). However, aging results in an accumulation of ROS, which cause oxidation of seed food reserves and other cellular contents, including DNA, RNA, and membrane lipids including phosphatidylcholine, phosphatidylethanolamine, and phosphatidylglycerol (Chhabra, Shabnam, and Singh 2019). In the last three decades, there have been multiple efforts to identify the major factors associated with seed aging; however, the exact mechanisms and downstream effects of seed aging remain largely unknown. The recent advancements in proteomics allow a detailed characterization of the seed proteome of many plants, leading to a significant improvement in the understanding of the complex mechanism of seed aging (Min et al. 2019).

12.2 CROP SEEDS AND AGING

The initial aging of seeds primarily depends on the moisture content, temperature, and biophysical conditions during the storage of seeds (Ratajczak et al. 2019). Based on the moisture content of the mature seeds, these are mainly divided into three categories: orthodox, recalcitrant, and intermediate seeds (Linington and Pritchard 2001). Of these, orthodox seeds have a moisture content lower than 5–10% and can be stored for relatively longer periods at subfreezing temperatures. Several reports with proteome and genome analysis showed that the orthodox seeds exhibit tolerance to desiccation stress because of their limited metabolism and extensive antioxidant system (Kurek, Plitta-Michalak, and Ratajczak 2019). Most species of annual and biennial crops, including most of the legumes, produce orthodox seeds (Tazi et al. 2018). It was reported that the seed aging-induced rapid increase of ROS mainly takes place in mitochondria, and thus, mitochondrial proteins are at high risk of oxidation (Walters, Ballesteros, and Vertucci 2010). Since mitochondrial proteins are particularly associated with energy metabolism and the tricarboxylic acid (TCA) cycle, ROS-induced dysfunction of these enzymes results in overall downregulation of energy production during aging (Rhoads et al. 2006; Winger et al. 2007). On the other hand, recalcitrant seeds are relatively sensitive to desiccation and have a moisture content between 25% and 45%; because of this, these seeds need to be stored in relatively higher-temperature conditions. Moreover, due to their relatively higher moisture content, recalcitrant seeds are more sensitive to deterioration stress than orthodox seeds; therefore, the longevity of recalcitrant seeds is lower than that of orthodox seeds. Most recalcitrant seeds have a relatively larger size, of which they include cacao, avocado, and seeds of horticultural trees. In contrast, intermediate seeds show relatively high desiccation tolerance and are suitable for storage at lower temperatures than recalcitrant seeds (Linington and Pritchard 2001).

12.3 PHYSIOLOGICAL AND BIOCHEMICAL CHANGES DURING SEED AGING

The seed aging induced by abnormal storage conditions results in membrane disruption, DNA damage, a decrease of protein synthesis, and lipid peroxidation (Mahjabin, Bilal, and Abidi 2015). Moreover, these changes are accompanied by several other physiological alterations, such as decreased seed germination and decreased shoot and root lengths (Min et al. 2016, 2017). In order to understand the mechanism of seed aging, soybean seeds were subjected to artificial aging treatment at 42 °C with 99% relative humidity conditions. Aged seeds showed only 63% and 37% seed germination after three and seven days of artificial aging treatment, respectively. Moreover, a reduction of 72.8% and 77.1% in the shoot and root lengths, respectively, was also observed as compared with control, while the moisture content of deteriorated seeds had increased significantly (Min et al. 2017; Kapoor et al. 2010). Besides these physiological alterations, changes in biochemical parameters, including decrease of total sugar, protein, and oil content, activation of various proteases, and ROS content were also observed in the deteriorated seeds (Mahjabin, Bilal, and Abidi 2015). Based on these results, many research groups suggested that accumulation of ROS, including hydrogen peroxide (H_2O_2) and MDA, is mainly associated with the loss of seed viability (Kibinza et al. 2006; Murthy, Kumar, and Sun 2003; Parkhey, Naithani, and Keshavkant 2012). Particularly, the rapid increase of MDA content, which is generated because of lipid peroxidation in seeds, has been suggested as one of the major markers for seed deterioration in *Triticum aestivum* (Lv et al. 2016), *Brassica napus* (Yin et al. 2015), *Oryza sativa* (Gao et al. 2016), and *Glycine max* (Min et al. 2017). Furthermore, peroxidation of membrane lipids in deteriorated seeds is mainly associated with the leakage of ions, amino acids, and sugars (López-Fernández et al. 2018; Murthy, Kumar, and Sun 2003). A recent study demonstrated that MDA and H_2O_2 levels rapidly increased in deteriorated seeds, whereas the activities of antioxidant enzymes, including catalase, superoxide dismutase, and peroxidase, were greatly reduced (Min et al. 2017; Gao et al. 2016). Besides, the content of abscisic acid (ABA) in deteriorated seeds was higher compared with control, and treatment with gibberellic acid (GA), which has ABA-antagonistic roles, showed recovery of seeds from deteriorated status (Yin et al. 2015). Moreover, major/minor metabolism-related proteins showed differential abundance profiles together with differential modulation of various proteins involved in cellular functions, including protein degradation, protein synthesis, stress, and development, as observed in wheat, rice, soybean seeds, and rapeseed (Figure 12.1A and B) (Lv et al. 2016; Yin et al. 2015; Gao et al. 2016; Min et al. 2017).

12.4 PROTEOMICS OF SOYBEAN SEED DURING AGING

Recently, a few proteomics-based studies have been carried out to understand the physiological and biochemical changes at the proteome level in soybean seeds. A high proportion of the seed storage proteins (SSPs), also referred to as high-abundant proteins (HAPs), in soybean seeds is one of the major hurdles in the analysis and identification of the low-abundant proteins (LAPs), which are mainly associated with signaling and various metabolic pathways (Min et al. 2020). Therefore, a few research groups have developed novel techniques using isopropanol (Natarajan et al. 2009), calcium (Krishnan, Oehrle, and Natarajan 2009), and protamine sulfate (Kim et al. 2015) for the depletion of HAPs and subsequent enrichment of LAPs in soybean seeds. A number of recent studies have been conducted to identify the major changes during deterioration of soybean seeds at the proteome level (Min et al. 2016, 2017).

12.4.1 Enrichment of Low-Abundance Proteins

Soybean seeds are a rich source of proteins and oils; therefore, several studies have used transcriptomic, metabolomic, and proteomic approaches to investigate the changes in gene, metabolite, and protein profiles under specific stress conditions. However, the presence of major SSPs, including β-conglycinin and glycinin, which account for approximately up to 70% of total proteins, is a major bottleneck for high-throughput proteome analysis of soybean seeds (Min et al. 2019). Therefore, Kim and co-workers developed a reproducible and efficient method for the enrichment of LAPs using protamine sulfate (PS) (Kim et al. 2015). The efficacy of the PS-precipitation (PSP) method in the enrichment of LAPs was initially shown on two-dimensional gel electrophoresis (2-DGE) analysis, which led to the identification of potential factors related to protein accumulation in soybean seeds (Min et al. 2015). A comparative study showed that the PSP method led to the depletion of up to 20% of the SSPs as compared with the non-depleted studies (Min et al. 2019). Using a label-free quantitative proteomic approach, another similar study showed that most of the identified proteins in a depleted fraction of soybean seeds were associated with major carbohydrate (CHO) metabolism, secondary metabolism, and amino acid metabolism, among others (Min et al. 2019).

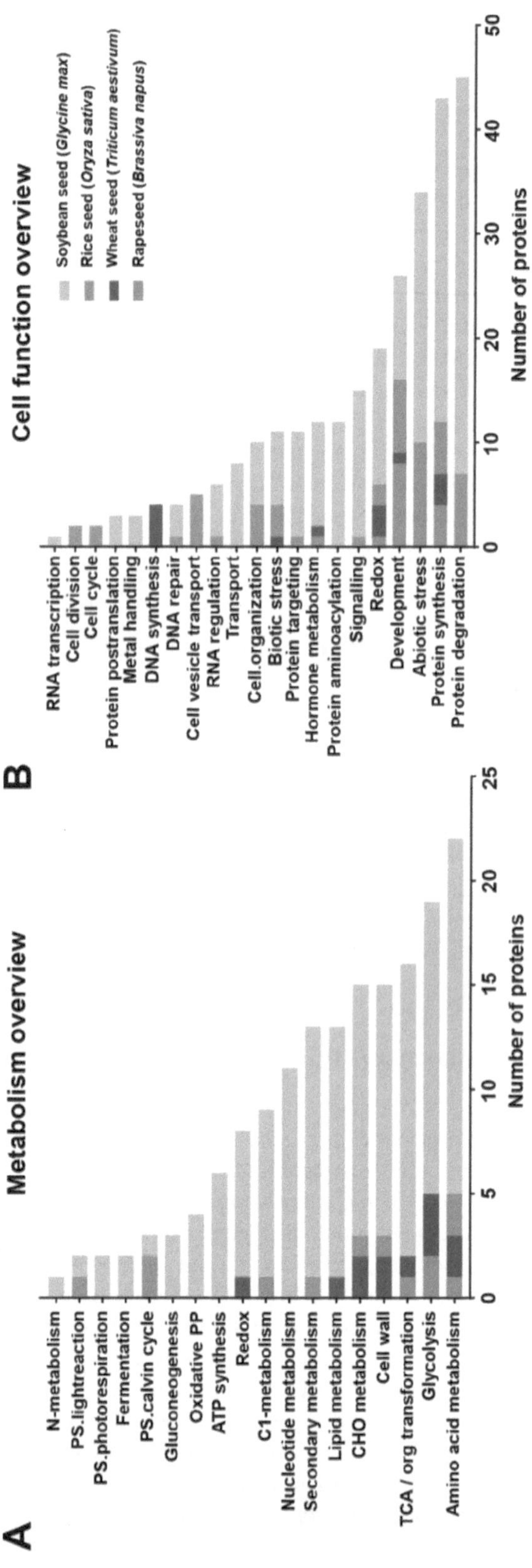

FIGURE 12.1 Functional classification of differentially modulated proteins related to aging in wheat (*Triticum aestivum*), rice (*Oryza sativa*), soybean seeds (*Glycine max*), and rapeseed (*Brassica napus*) using MapMan software. The bar chart shows the corresponding metabolism (A) and cell function (B) overview in each crop.

12.4.2 Proteomic Analysis Using LAPs Extracted from Aged Seeds

Min and co-workers analyzed the seed aging proteome of soybean using 2-DGE and high-throughput label-free quantitative approaches. The LAPs were isolated from aged soybean seeds harvested after zero, one, and two day at 42 °C with 99% relative humidity and recovery at 37 °C for 3 h. A total of 825, 683, 785, and 660 protein spots were observed on 2-DGE from the samples harvested at zero, one, and two days and the recovery sample, respectively (Min et al. 2016). Among these, a total of 33 protein spots showed significant differential abundance, of which 31 showed decreased and two showed increased abundance during seed aging. Differential proteins were associated with diverse metabolic pathways and included an isoform of alpha-1,2-glucan phosphorylase, urease, heat shock protein (HSP), alpha-D-phosphoglucomutase, UDP-glucose pyrophosphorylase, calreticulin, and superoxide dismutase, among others. Moreover, the functional analysis of proteins with decreased abundance showed that these were mainly associated with primary metabolic processes, including CHO metabolism, protein metabolism, nucleobase-containing compound metabolism, and cellular amino acid metabolism. In addition, proteins with decreased abundance were mainly associated with catalytic activity, including oxidoreductase activity, isomerase activity, hydrolase activity, and transferase activity the in molecular function category of gene ontology (GO) enrichment analysis. In particular, four proteins with decreased abundance in aged seeds were related to starch phosphorylation or degradation. Since these proteins are directly or indirectly related to the early development stage of seeds, seed dehydration, and desiccation tolerance during the late stage of seed development (da Silva et al. 1997), the degradation of these proteins was suggested to be the potential reason for the loss of viability or reduced germination of soybean seeds during aging.

To extend this observation, label-free quantitative proteomic analysis was carried out, leading to the identification of 1,626 proteins in the LAP-enriched fraction of the soybean seed proteome (Min et al. 2017). A sequential statistical analysis resulted in the identification of 146 significantly modulated proteins, with 352 and 35 aging-degraded and induced proteins, respectively. The hierarchical clustering analysis showed two major abundance profiles divided into 141 and 5 proteins with decreased and increased abundance, respectively. The functional classification of the significant proteins showed that these were mainly involved in primary metabolism, protein degradation, protein folding, and ROS detoxification. In particular, 40 and 116 proteins showing decreased abundance and aging-induced degradation, respectively, were related to various metabolic pathways associated with energy production, including CHO metabolism, amino acid metabolism, and lipid metabolism. Moreover, decreased abundance of the five deubiquitinating enzymes, which have pivotal roles in cell cycle regulation, protein degradation, gene expression, the DNA repair system, and plant development/growth, was also observed during aging treatment (Reyes-turcu, Ventii, and Wilkinson 2009; Isono and Nagel 2014). These overall results indicate that the degradation of protein synthesis and various energy generation-related proteins affects seed longevity and germination during aging in soybean seeds (Min et al. 2017).

12.4.3 Proteomic Analysis of Development Stages of Soybean Seed after Aging Treatment

In addition to the soybean seed proteomic analysis using matured seeds during aging, attempts have also been made to understand the proteome profile of soybean seeds at the R7 growth stage (Wang et al. 2012b). Using 2-DGE-based proteomic analysis coupled with matrix-assisted laser desorption ionization-time of flight tandem mass spectrometry (MALDI-TOF/TOF MS/MS), a total of 42 differentially modulated proteins were identified, of which 22 and 20 protein spots showed increased and decreased abundance, respectively. Functional classification indicated that the identified proteins were mainly associated with CHO metabolism, signal transduction, protein biosynthesis, photosynthesis, protein folding and assembly, and nitrogen metabolism, among others. Particularly, this study suggested that the developing seed (R7 stage) showed enhanced resistance to deteriorative stress through various metabolic processes, including the reduction of carbon loss, photo inhibition damage, ROS, and DNA damage, while storage of arginine and amino acid synthesis, among others, were increased for further absorption of nitrogen compounds against deterioration stress. Taken together, these findings suggested key involvement of deterioration stress during the soybean seed development stage (Wang et al. 2012b).

12.5 PROTEOMIC ANALYSIS OF OTHER CROPS OF AGED SEEDS

12.5.1 Proteomic Analysis of Rapeseed

Brassica napus (rapeseed) is one of the most important oil-producing crops in the world. However, the storage of rapeseeds in abnormal conditions leads to the loss of seed viability, especially during the summer season, because rapeseeds are always harvested in the late spring season (Cartea et al. 2019; Yin et al. 2015). Therefore, to understand the seed aging mechanism in rapeseeds, Yin and co-workers carried out proteomic analysis using a 2-DGE-based approach (Yin et al. 2015). For determination of aging-related proteins, an artificial aging treatment at 40 °C with 90% of relative humidity was given to the seeds. The seeds artificially aged by this method showed approximately 1.8-fold higher ion leakage and MDA content, along with a higher concentration of ROS, as compared with control seeds. Further comparative proteomic analysis led to the identification of 81 significantly modulated proteins between the healthy and artificially aged seeds. Functional annotation of the differential proteins showed that these were mainly proteins related to protein modification and destination, development, and cell structure. Consequently, the delay of germination in artificially aged rapeseeds was mainly associated with the inhibition of basic cellular metabolism, such as protein metabolism and protein destination, together with an increase in ABA hormone content, which is a known inhibitor of seed germination, by activation of ABA biosynthesis enzymes (Gubler, Millar, and Jacobsen 2005; Yin et al. 2015). Therefore, it was suggested that an overall decrease in the major metabolism related-proteins and activation of ABA biosynthesis enzymes are among the important factors involved in the seed aging process in rapeseeds.

12.5.2 Proteomic Analysis of Wheat Seed

Wheat (*Triticum aestivum*) is one of the most important and widely cultivated crops due to its high gluten and storage protein content (Shewry and Hey 2015). In order to understand the seed aging mechanisms in wheat seeds, an integrated biochemical and tandem mass tag (TMT)labeling-based quantitative proteome analysis was carried out on aged wheat embryos (Lv et al. 2016). Seed vigor test and embryo structure analysis showed a gradual decrease in germination rate and degradation of granules during seed aging. Proteome analysis led to the identification of 6,281 proteins, of which 162 showed significant modulation among the sample sets, with 36 and 126 proteins exhibiting increased and decreased abundance profiles, respectively. Functional classification of the differential proteins showed that proteins related to protein destination and storage, metabolism and energy supply, and stress defense were decreased during the artificial aging process. Particularly, proteins related to the degradation of storage reserves, such as alpha-amylase inhibitors and oleosin proteins, and those related to the response to oxidative, osmotic, and temperature pressures, including catalase, UPD-glucose-6-dehydrogenase, translationally-controlled tumor proteins (TCTP), and multiprotein bridging factor, and the response to biotic stress, with agglutinin isolectin, non-specific lipid-transfer proteins, ribosomal protein, and asparagine synthetase, were found to be decreased during aging. On the basis of these results, it was proposed that the reduction of proteins related to various storage components led to the increased activity of amylase and protease, which are associated with decomposition of storage substances, possibly playing an important role during wheat seed aging and potentially being considered as a new marker for evaluation of seed vigor in wheat (Lv et al. 2016).

12.5.3 Proteomic Analysis of Rice Seed

Rice is the most important cereal crop worldwide, and particularly, hybrid rice cultivars are widely adopted in Asian countries due to their high yield and resistance to various stressors. However, the storability of these rice cultivars is comparatively poorer than for inbred varieties (Gao et al. 2016). A proteomics approach was used to identify the aging-associated proteins for further understanding of the aging mechanism in the hybrid rice cultivars. Proteins were isolated from embryos of control and aged seeds of two different varieties of hybrid rice (IIY998 and BY998) having a similar genetic background and resolved on 2-DGE. This approach identified a total of 110 protein spots, of which 78 differential protein spots were successfully identified using liquid chromatography-tandem mass spectrometry (LC-MS/MS). The functional classification of the identified proteins showed that these were mainly involved in storage, protein synthesis and destination, metabolism, cell structure/cytoskeleton, redox regulation, and disease/defense. Interestingly, it was revealed that significant modulation of the two types of globulin proteins associated with assimilation of nitrogen and carbon sources for germination and seedling growth in the BY998 cultivar resulted in comparatively higher seed viability than in the IIY998 cultivar. On the other hand, the glutelin content of rice seed embryo in the aging-sensitive cultivar (IIY998) was found to be higher than in the BY998 cultivar. This indicated that the glutelin content in rice seed may have a negative correlation with rice seed viability. The identification and functional classification of seed storability-related proteins related to the various physiological and biological changes, including reduction of energy and increase of oxidative damage, may provide a possible explanation of why rice seed viability is reduced during long-term storage in abnormal conditions (Gao et al. 2016).

12.6 CONCLUSION

To better understand the physiological alteration, particularly the seed viability mechanism, induced by deterioration stress in plant seeds, several studies have been conducted using physiological, biochemical, and high-throughput proteomic tools. These analyses collectively suggest an overall decrease of the proteins involved in CHO metabolism, signal transduction, protein synthesis/destination, photosynthesis, protein folding and assembly, and nitrogen metabolism with higher concentrations of ROS and MDA in aged seeds across species. In particular, most of the studies suggested that an optimum concentration of storage proteins is crucial to maintain the germination ability of seeds, and the inhibition or degradation of these proteins, caused by activation of various proteases, could be one of the major factors affecting seed viability during long-term (or abnormal) storage of seeds in deteriorative conditions (Figure 12.2). In addition, the increase of ABA content by activation of ABA biosynthesis enzymes showed negative correlation with seed vigor in rapeseeds. Taken together, these proteomics studies have provided a better understanding of the complex mechanism of seed aging. Furthermore, the list of identified proteins can serve as potential targets for manipulation to control the negative effects of aging-induced changes in crop seeds in future.

ACKNOWLEDGEMENT

This study was financially supported by the two-year Program of the Pusan National University.

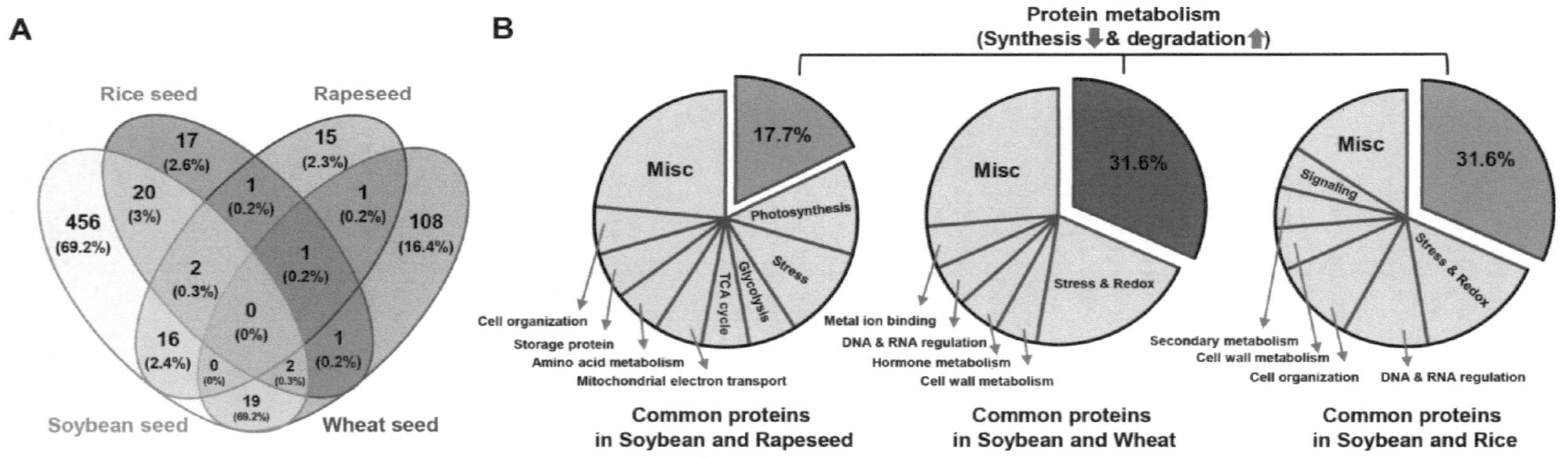

FIGURE 12.2 Identification of common proteins by label-free quantitative and 2-DGE approaches across the species of plant seeds and annotation of their functions. (A) Venn diagram showing the distribution of the common and specific proteins in wheat (*Triticum aestivum*), rice (*Oryza sativa*), soybean (*Glycine max*) seeds, and rapeseed (*Brassica napus*). (B) Pie charts showing the common functions and different functions in comparisons of soybean seed and rapeseed (green, left on B), soybean and wheat seeds (blue, middle of B), and soybean and rice seeds (31.6%, red, right on B), respectively.

REFERENCES

Anderson, J. D., and Baker, J. E. 1983. Deterioration of seeds during aging. *Phytopathology* 73:321–325.

Bailly, C., Benamar, A., Corbineau, F., and Côme, D. 1996. Changes in malondialdehyde content and in superoxide dismutase, catalase and glutathione reductase activities in sunflower seeds as related to deterioration during accelerated aging. *Physiologia Plantarum* 97:104–110.

Cartea, E., Haro-Bailón, A. D., Padilla, G., et al. 2019. Seed oil quality of *Brassica napus* and *Brassica rapa* germplasm from northwestern Spain. *Foods* 8:1–10.

Chhabra, R., Ansari, S., and Singh, T. 2019. Seed aging, storage and deterioration: An irresistible physiological phenomenon. *Agricultural Reviews* 40:1–6.

da Silva, P. M. F. R., Eastmond, P. J., Hill, L. M., Smith, A. M., and Rawsthorne. S. 1997. Starch metabolism in developing embryos of oilseed rape. *Planta* 203: 480–487.

Finch-Savage, W. E., and Bassel, G. W. 2016. Seed vigour and crop establishment: Extending performance beyond adaptation. *Journal of Experimental Botany* 67:567–591.

Gao, J., Fu, H., Zhou, X., et al. 2016. Comparative proteomic analysis of seed embryo proteins associated with seed storability in rice (*Oryza sativa* L) during natural aging. *Plant Physiology and Biochemistry* 103:31–44.

Gubler, F., Millar, A. A., and Jacobsen, J. V. 2005. Dormancy release, ABA and pre-harvest sprouting. *Plant Biotechnology* 8:183–187.

Isono, E., and Nagel, M. 2014. Deubiquitylating enzymes and their emerging role in plant biology. *Plant Science* 5:1–6.

Kapoor, N., Arya, A., Siddiqui, M. A., Amir, A., and Kumar, H. 2010. Seed deterioration in chickpea (*Cicer arietinum* L.) under accelerated ageing. *Asian Journal of Plant Sciences* 9:158–162.

Kibinza, S., Vinel, D., Côme, D., Bailly, C., and Corbineau, F. 2006. Sunflower seed deterioration as related to moisture content during ageing, energy metabolism and active oxygen species scavenging. *Physiologia Plantarum* 128:496–506.

Kim, Y. J., Wang, Y., Gupta, R., et al. 2015. Protamine sulfate precipitation method depletes abundant plant seed-storage proteins: A case study on legume plants. *Proteomics* 15:1760–1764.

Krishnan, H. B., Oehrle, N. W., and Natarajan, S. S. 2009. A rapid and simple procedure for the depletion of abundant storage proteins from legume seeds to advance proteome analysis: A case study using *Glycine max*. *Proteomics* 9:3174–3188.

Kurek, K., Plitta-Michalak, B., and Ratajczak, E. 2019. Reactive oxygen species as potential drivers of the seed aging process. *Plants* 8:1–13.

Linington, S. H., and Pritchard, H. W. 2001. Gene banks. *Encyclopedia of Biodiversity* 3:165–181.

López-Fernández, M. P., Moyano, L., Correa, M. D., et al. 2018. Deterioration of willow seeds during storage. *Scientific Reports* 8:1–11.

Lv, Y., Zhang, S., Wang, J., and Hu, Y. 2016. Quantitative proteomic analysis of wheat seeds during artificial ageing and priming using the isobaric tandem mass tag labeling. *PloS ONE* 11:1–31.

Mahjabin, Bilal, S., and Abidi, A. B. 2015. Physiological and biochemical changes during seed deterioration: A review. *International Journal of Recent Scientific Research* 6:3416–3422.

Min, C. W., Gupta, R., Agrawal, G. K., Rakwal, R., and Kim, S. T. 2019. Concepts and strategies of soybean seed proteomics using the shotgun proteomics approach. *Expert Review of Proteomics* 16:795–804.

Min, C. W., Kim, Y. J., Gupta, R., et al. 2016. High-throughput proteome analysis reveals changes of primary metabolism and energy production under artificial aging treatment in *Glycine max* seeds. *Applied Biological Chemistry* 59:841–853.

Min, C. W., Lee, S. H., Cheon, Y. E., et al. 2017. In-depth proteomic analysis of *Glycine max* seeds during controlled deterioration treatment reveals a shift in seed metabolism. *Journal of Proteomics* 169:125–135.

Min, C. W., Park, J., Bae, J. W., et al. 2020. In-depth investigation of low-abundance proteins in matured and filling stages seeds of *Glycine max* employing a combination of protamine sulfate precipitation and TMT-based quantitative proteomic analysis. *Cells* 9:1–20.

Min, C. W., Gupta, R., Kim, S. W., et al. 2015. Comparative biochemical and proteomic analyses of soybean seed cultivars differing in protein and oil content. *Journal of Agricultural and Food Chemistry* 63:7134–7142.

Murata, M., Tsuchiya, T., and Roos, E. E. 1984. Chromosome damage induced by artificial seed aging in barley – 3. Behavior of chromosomal aberrations during plant growth. *Theoretical and Applied Genetics* 67:161–170.

Murthy, U. M. N., Kumar, P. P., and Sun, W. Q. 2003. Mechanisms of seed ageing under different storage conditions for *Vigna radiata* (L.) Wilczek: Lipid peroxidation, sugar hydrolysis, Maillard reactions and their relationship to glass state transition. *Journal of Experimental Botany* 54:1057–1067.

Natarajan, S. S., Krishnan, H. B., Lakshman, S., and Garrett, W. M. 2009. An efficient extraction method to enhance analysis of low abundant proteins from soybean seed. *Analytical Biochemistry* 394:259–268.

Parkhey, S., Naithani, S. C., and Keshavkant, S. 2012. ROS production and lipid catabolism in desiccating *Shorea robusta* seeds during aging. *Plant Physiology and Biochemistry* 57:261–267.

Rajjou, L., Duval, M., Gallardo, K., et al. 2012. Seed germination and vigor. *Annual Review of Plant Biology* 63:507–533.

Ratajczak, E., Małecka, A., Ciereszko, I., and Staszak, A. M. 2019. Mitochondria are important determinants of the aging of seeds. *International Journal of Molecular Sciences* 20:1–12.

Reyes-turcu, F. E., Ventii, K. H., and Wilkinson, K. D. 2009. Regulation and cellular roles of ubiquitin-specific deubiquitinating enzymes. *Annual Review of Biochemistry* 78:363–397.

Rhoads, D. M., Umbach, A. L., Subbaiah, C. C., and Siedow, Z. N. 2006. Mitochondrial reactive oxygen species. Contribution to oxidative stress and interorganellar signaling. *Plant Physiology* 141:357–366.

Shelar, V., Shaikh, R., and Nikam, A. S. 2008. Soybean seed quality during storage: A review. *Agricultural Reviews* 29:125–131.

Shewry, P. R., and Hey, S. J. 2015. The contribution of wheat to human diet and health. *Food and Energy Security* 4:178–202.

Tazi, M., Hall, G., Sika, G., and Kugbei, S. 2018. Seed quality assurance. In *Seeds Toolkit*, ed. H. Bahri, D. G. Méndez, R. Duffy, S. Stevens, and D. Moretti, 1–109. Rome: The Food and Agriculture Organization of the United Nations and AfricaSeeds Press.

Walters, C., Ballesteros, D., and Vertucci, V. A. 2010. Structural mechanics of seed deterioration: Standing the test of time. *Plant Science* 179:565–573.

Wang, F., Wang, R., Jing, W., and Zhang, W. 2012a. Quantitative dissection of lipid degradation in rice seeds during accelerated aging. *Plant Growth Regulation* 66:49–58.

Wang, L., Ma, H., Song, L., Shu, Y., and Gu, W. 2012b. Comparative proteomics analysis reveals the mechanism of pre-harvest seed deterioration of soybean under high temperature and humidity stress. *Journal of Proteomics* 75:2109–2127.

Winger, A. M., Taylor, N. L., Heazlewood, J. L., Day, D. A., and Millar, A. H. 2007. The cytotoxic lipid peroxidation product 4-hydroxy-2-nonenal covalently modifies a selective range of proteins linked to respiratory function in plant mitochondria. *Journal of Biological Chemistry* 282:37436–37447.

Yin, X., He, D., Gupta, R., and Yang, P. 2015. Physiological and proteomic analyses on artificially aged *Brassica napus* seed. *Frontiers in Plant Science* 6:1–11.

13 Crop Proteomics
Towards Systemic Analysis of Abiotic Stress Responses

*Asmat Farooq[1,2], Rakeeb Ahmad Mir[3], Vikas Sharma[2], Mohammad Maqbool Pakhtoon[1], Kaiser Ahmad Bhat[1,3], Ali Asgar Shah[3], and Sajad Majeed Zargar[1]**
[1]Proteomics Laboratory, Division of Plant Biotechnology, Sher-e-Kashmir University of Agricultural Sciences & Technology of Kashmir, Srinagar, J&K, India
[2]Division of Biochemistry, Faculty of Basic Sciences, Sher-e-Kashmir University of Agricultural Sciences and Technology of Jammu, Main Campus Chatha, Jammu, J&K, India
[3]Department of Biotechnology, BGSB University, Rajouri, J&K, India
*Corresponding author: E-mail: smzargar@skuastkashmir.ac.in

CONTENTS

DOI: 10.1201/9781003159636-13

13.1 INTRODUCTION

Proteomics is a set of advanced techniques employed to study genetic functions through the interface of total proteins (Brandes et al., 2020). In other words, "proteomics" is an applied field, which is used to study and categorize comprehensive deposits of proteins inside the cells at a particular instance (Gupta et al.,2015). Proteomics techniques can be used to study protein expression profiling, protein modifications, protein–protein interactions, and cell–environment interactions. In-plant biology, functional and structural proteomics, in addition to expression analysis, are involved in characterizing the proteins in total (Hubbard, 2002; Sali et al., 2003; Aslam et al. 2017). The various stress responses in different crop plants are characterized by extremely complicated and cross-linked defense/stress signaling pathways for establishing unique biomarkers using -omics technologies such as genomics (transcriptomics), proteomics, and metabolomics approaches (Zargar et al., 2011). The main requirements of plants for proper growth, development, and reproduction are water, light, and mineral nutrition. A wide range of abiotic as well as biotic factors limit the growth and development of plants. Water scarcity or drought, temperature stress, salinity, heavy metal toxicity, and nutrient deficiencies are various abiotic stresses that retard the growth and metabolism of plants (Chaves and Oliveira 2004; Zargar et al. 2017). These abiotic stresses severely influence plant growth and can decrease the yield (Kumar et al. 2019). When plants experience nutrition stress, they become unable to complete their normal growth and developmental stages. For example, salt stress causes ion-intolerance (Munns and Tester 2008),while drought results in low germination rates and consequent impairment of photosynthesis and depletion of membrane complexity, therefore, enhancing the production of oxygen free radicals like ROS (reactive oxygen species) (Greenberg et al. 2008; Khan et al. 2021). Persistent drought results in loss of leaf water potential, impairment in the opening of stomata, and reduction in the size of leaves; inhibits root development; decreases seed quality and quantity, plant size, and functionality; slows flower formation and fruit formation; and therefore, inhibits plant development and yield (Xu et al. 2016). Salt stress and water scarcity are the chief causes of absorption pressures in plants. Moreover, heat stress causes widespread precipitation and accumulation of cell proteins, whose aggregation can stimulate cellular death (Bakthisaran et al. 2015). Lower temperature interferes with metabolism by altering membrane characteristics, changing protein structure, and inhibiting catalytic reactions (Sehrawat et al. 2014). Likewise, metal stress interferes in various necessary metabolic processes, such as photosynthesis, respiration, nitrogen and protein metabolism, and nutrient absorption (Singh et al. 2013). Therefore, a standard modification is desirable in the production strategy to fortify conventional crop production methods. An alternative is to make use of various -omics techniques/strategies to identify advantageous genotypes for raising transgenic crops. In that case, the highly productive traits that exist may be needed as a basic substance for incorporation of some constructive attributes from otherwise unaffected traits, races, or uncultivated relatives. The modern observation of research is that the success of conservative breeding is appreciably enhanced by utilizing various proteomics approaches. Nowadays, crop proteomics is utilized globally in all main cereals because of the available amount of essential heritable and genomic assets. The success of proteomics techniques lies in efficiently assisting the screening and development of stress-resilient cereal crops such as wheat, rice, and maize (Abreu et al. 2013).

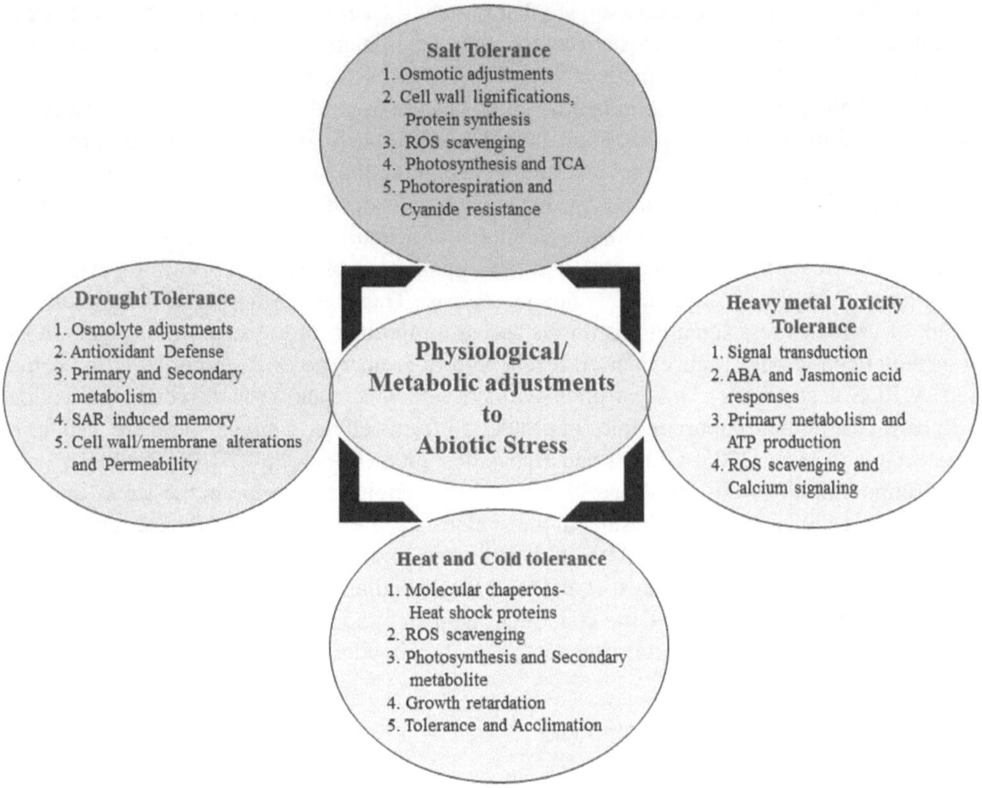

FIGURE 13.1 Insights into different abiotic stress responses: analysis.

13.2 BASIC PLANT RESPONSES AGAINST ABIOTIC STRESSES

Extended abiotic stress causes the generation of ROS in various cellular organelles, such as chloroplasts, mitochondria, and peroxisomes (Riyazuddin et al. 2020). ROS attack DNA, lipids, proteins, and carbohydrates and disturb the common functions of cells, which in extreme stress situations leads to apoptosis and other types of cell death, such as ferroptosis (Riyazuddin and Gupta 2021). Therefore, in order to limit these free radicals, plants have developed different enzymatic and non-enzymatic defensive proteins (Qureshi et al. 2018). These proteins function as antioxidants and include superoxide dismutase, catalase, peroxidases, and glutathione reductase, while non-enzymatic antioxidants include glutathione, which works in a collaborative manner to protect the cells against ROS (Hasanuzzaman et al., 2019). To lessen the consequence of environmental constraints, the accumulation of compatible solutes like proline and glycine betaine occurs in the cells. Such solutes help in osmotic alteration and defend against different cellular components. In the case of increased CO_2 in the environment, this stimulates the antioxidant defense in response to damage caused by oxygen radicals inside the cells, causing accumulation of solutes, and therefore, can help against salinity and water stress in plants and maintain their integrity and better crops for a sustainable future against abiotic stresses. The responses of plants against different stress conditions are presented in Figure 13.1.

13.2.1 Solute Accumulation

The generation of large quantities of low-molecular-weight organic solutes, called osmolytes, is one of the important metabolic responses of plant cells to abiotic stresses. Osmolytes are hydrophilic molecules, which aid in carrying out important physiological (catalytic) processes (Rhodes et al. 2002). Various types of organic solutes that protect the cells from harmful effects of stresses are reported, including sugars (glucose, fructose, sucrose, trehalose, raffinose, and fructans), sugar alcohols (mannitol, glycerol, and methylated inositols), amino acid-derived solutes (proline, glycine betaine, β-alanine, and proline betaine), amines (1,4,5,6-tetrahydro-2-methyl-4-carboxyl pyrimidine), and sulfur-containing biomolecules (choline-O-sulphate, dimethyl sulphonio propionate) (Flowers and Colmer 2008). The synthesis and its accumulation are among the early responses in plants when exposed to different abiotic stresses such as water stress, heat stress, salinity stress, and osmotic and oxidative stress. Various research has shown that the metabolic pathway genes are regulated by different phytohormones, calcium ion signaling, and kinases such as mitogen-activated protein kinases (MAPKs) to combat these abiotic stresses. Thus, the synthesis and accumulation are the result of various stress signaling pathways and are important for the survival of plants in such environmental conditions. Their enhanced levels help to protect the cells, minimizing the damage caused by ROS accumulation, loss of photosynthetic efficiency, and growth reduction (Golldack et al., 2014). Due to their important roles in protection from cellular damage, these are also termed cryoprotectants (Yancey 2005; Groppa and Benavides 2008; Khan et al. 2010). Proline is one of the most important and well-characterized amino acids playing a role in abiotic stress tolerance. The increase in cellular osmolarity results in the synthesis and accumulation of these active solutes, which is followed by water influx into or reduced efflux from the cells, which causes turgor and increases the cellular expansion (Hayat et al. 2012; Roychoudhury et al. 2015). Moreover, they also protect various macromolecules of the cell, such as nucleic acids, proteins, lipids, and enzymes (Rodríguez et al. 2005; Groppa and Benavides 2008; Roychoudhury et al. 2015; Paul et al. 2018).

13.2.2 Antioxidative Defense Responses

Various stresses are responsible for generating the production of ROS, including $O2^{\bullet}$, H_2O, H_2O_2, $OH^{\bullet}$ alkyl oxides, hydro-peroxides, and carbonyl oxides. These have harmful effects on biomolecules such as proteins, lipids, carbohydrates, and nucleic acids, finally leading to the apoptosis of cells (Foyer and Noctor 2005). The generation and elimination of ROS usually occur in the cells. In the mitochondria, the generation of ROS and hydrogen peroxides occurs because of excess reduction/oxidation of the electron transport chain (ETC). Similarly, chloroplasts are responsible for the major production of O_2 and H_2O_2 (Davletova et al. 2005), because the oxygen concentration in the chloroplast is greater than in other organelles. The light-induced reduction of O_2 to $O_2{\bullet}-$ in the photosynthetic ETC is known as the Mehler reaction. The superoxides generated are converted to H_2O_2either naturally or through the catalytic activity of the enzymatic antioxidant superoxide dismutase. Hydrogen peroxide is also converted to hydroxyl radical (OH•). The main producers of H_2O_2 and superoxides in plant cells are peroxisomes, which include both the peroxisomal matrix and the peroxisomal membrane. In the matrix, xanthine oxidoreductase catalyzes the oxidation of xanthine or hypoxanthine to uric acid, which is further acted upon by urate oxidase, giving rise to allantoin (Werner and Witte 2011). In the peroxisomal membranes, transmembrane or integral membrane proteins of molecular mass 18, 29, and 32 kDa involved in the ETC, consisting of flavoproteins, NADH, and cytochrome b, are found to be responsible for the generation of oxygen free radicals (Corpas et al. 2015). Hydrogen peroxide generation also occurs to a great extent through photorespiration due to the oxidation of glycolate by the catalytic action of the enzyme glycolate oxidase, beta-oxidation of fatty acid, and flavin oxidase-catalyzed processes (Baker and Graham 2013).

13.2.3 Antioxidant Enzyme Generation

13.2.3.1 Superoxide Dismutase

Superoxide dismutase is the most powerful and the first endogenous antioxidant enzyme that plays an important role against ROS to protect cells against oxidative damage (Ighodaro and Akinloye

2018). This enzyme is the primary response of cells against free radical species and functions by modifying reactive superoxides and hydrogen peroxides to reduce their harmful effects, such as membrane disruption, protein changes, and genetic mutation (Ahmad et al. 2012). Dismutase was primarily observed by Cannon et al. (1987) in crops like maize and is responsible for the conversion of O_2•– into $H_2O_2$and O_2.

13.2.3.2 Catalase

Catalases are the most important antioxidative enzymes found in peroxisomes and glyoxysomes and are involved in the elimination of H_2O_2, which is produced in response to various stresses in plants (Ahmad et al. 2012). Catalase converts harmful peroxides to H_2O and O_2 and does not use a reducing agent; therefore, it helps cells against the effects of H_2O_2 (Ahmad et al. 2012). It is ubiquitous in glyoxysome organelles during germination in barley and eliminates hydrogen peroxides generated through the β-oxidation of lipids (Jiang and Zhang 2002). In addition, it is found that peroxisomes of the leaves of C3 plants produce peroxides in photorespiration (Kiani et al. 2015), and it has also been reported that more peroxisomes are formed under abiotic stress to prevent the diffusion of peroxides from the cytoplasm of the cells (Lopez-Huertas et al. 2000).

13.2.3.3 Ascorbate Peroxidase

Ascorbate peroxidase (APX) is another antioxidant enzyme mostly observed in higher plants and algae (Raven 2003). This enzyme uses ascorbic acid to convert peroxides to water and mono-dehydroascorbate in the ascorbate–glutathione reaction. Its isoforms are observed as an important stress response in the peroxide family under abiotic stress (Mittler and Poulos 2005). Plant APXs have been observed in various cellular compartments, and eight isozymes have been identified in *Arabidopsis thaliana*. Of these, some are soluble cytosolic (APX1, APX2, APX6), and some are bound to the microsome membrane (APX3, APX4, APX5) and the chloroplast sAPX and tAPX (Panchuk et al. 2002). Increased expression of cytosolic as well as peroxisomal APXs has been observed in sensitive cultivars. Similarly, the expression of chloroplastic APXs was found to be stimulated in response to drought stress. These observations highlight the ability of these enzymes to effectively scavenge ROS at their generation site (D'Arcy-Lameta et al. 2006). The thylakoid APX expression levels were found to be enhanced in drought-tolerant cultivars, and stromal APX2 was found to be higher in drought-sensitive cultivars (Secenji et al. 2010). The APX2 gene was increased especially during light stress or heat stress conditions (Mullineaux and Karpinski 2002). Lu et al. (2007) confirmed that cytosolic APX 1 causes salinity tolerance in hybrid Arabidopsis.

13.2.3.4 Glutathione Reductase

It is an oxidoreductase enzyme present in both prokaryotes as well as eukaryotes (Romero-Puertas et al. 2006). Glutathione reductase enzymes function in the ascorbate–glutathione complex and help in maintaining the equilibrium between reduced glutathione and the ascorbate pool (Ahmad et al. 2012). Meldrum and Tarr (1935) first reported reductase enzyme in yeast, followed by its discovery in plants in 1951 (Mapson and Goddard 1951). In crops such as wheat (Selote and Khanna Chopra 2006) and rice (Selote and Khanna-Chopra 2004), enhanced activity of glutathione reductase enzyme was observed in drought stress, while rice (Guo et al. 2006), maize (Hodges et al. 1997) and tomato (Walker and McKersie 1993) showed increased reductase activity under low-temperature and metal toxicity stress.

13.2.4 Non-Enzymatic Antioxidant Generation

13.2.4.1 Ascorbic Acid

Ascorbic acid is an important antioxidant utilized by APX to inhibit the adverse effects of peroxides by converting them into H_2O and generating mono-dehydroascorbate (Ahmad et al. 2009).

13.2.4.2 α-Tocopherol

Another non-enzymatic antioxidant, α-tocopherol or Vitamin E, is lipid-soluble and is synthesized in plants under abiotic stress. Along with antioxidative proteins, it is responsible for the scavenging of free radicals (Munne-Bosch and Algere 2003). α-tocopherol is responsible for helping in the protection of photosystem II (PSII) against the free radicals generated during metabolic reactions in chloroplasts (Lopez-Huertas et al. 2000). Munne-Bosch and Algere (2003) observed that it helps to protect the membrane integrity and enhances the resistance in plants under abiotic stresses. Falk et al. (2003) observed the enhancement of expression of α-tocopherol genes under oxidative and water stress (Zhang and Schmidt 2000).

13.2.4.3 Reduced Glutathione

Reduced glutathione, an SH-containing antioxidant which comprises L-glutamic acid, L-cysteine, and glycine amino acids, is ubiquitous in plants, animals, and microorganisms and acts as an important protectant under various abiotic stresses, particularly during oxidative stress (Asthir et al. 2020). It takes part in important biological processes such as nucleic acid biosynthesis, protein biosynthesis, and membrane stabilization. Meyer et al. (2005) described how peroxide concentrations are regulated by the activity of glutathione, and reduced/oxidized glutathione is important in balancing peroxide concentrations in cells; therefore, it plays an important part in ROS generation pathways (Pastori and Foyer 2002). Tripeptide unreduced glutathione occurs in different compartments of the cell such as cytoplasm, endoplasmic reticulum, vacuoles, mitochondria, and chloroplasts (Sankar et al. 2007). Glutathione is an important non-protein thiol that is nucleophilic, is necessary for bonding with metals (Rodriguez et al. 2005), and acts as a vital non-enzymatic antioxidant response against various oxidative abiotic stresses (Samarah 2005).

13.3 MAJOR ABIOTIC STRESS RESPONSES: MOLECULAR PERSPECTIVE

13.3.1 Drought Stress Response

Plants respond to water stress with composite biochemical, physiological, morphological, and anatomical changes through small and large developmental and growth-associated metabolism. The usual drought response mechanism consists of various characteristics: (1) escape by completing their life cycle rapidly (like premature flowering); (2) prevention through increasing water uptake capacity (like alteration in root architecture or conservation of water by decreasing water loss, such as closing/reducing stomata, leaf area); (3) tolerance by improvement in osmotic alteration and enhancing cell wall flexibility to uphold tissue pressure; (4) resistance by shifting metabolic pathways under harsh water stress conditions; and (5) drought-level metabolic behaviors to enable plants to resist prolonged stress (such as DNA/RNA alterations). These reactions can occur concurrently in a plant's response to water stress and cease after rehydration (Mitra 2001). Researchers have observed important conserved water stress-responsive proteins, such as membrane-stabilizing proteins and late embryogenic abundant proteins (LEA), which possess water binding ability and are known as dehydrins (Riyazuddin et al. 2021). These stress-responsive proteins are water-soluble and therefore, contribute to tolerance responses through modification of cellular composition and cellular metabolism (Osakabe et al. 2014). Various heat shock proteins are reported to play critical functions in maintaining the conformation of proteins. These heat shock proteins are low in molecular weight and are produced in response to abiotic stresses, especially during heat stress (Bakthisaran et al. 2015). These proteins are also known as molecular chaperones, as they are involved in protein folding processes and avoiding denaturation under stress conditions. In addition, various transcription factors, such as MYB, MYC, DREB/CBF, ABF/AREB, NAC, WRK-Y, and SnRK2, are synthesized to control and establish the adaptive response under water stress (Sakuma et al. 2006; Nakashima et al. 2009; Tran et al., 2010; Umezawa et al. 2014; Saruhashi et al. 2015).

13.3.2 Salinity Stress Responses

Salinity stress is regarded as one of the major abiotic stresses that limit crop production in dry and low-rainfall areas (Ashraf and Harris 2004; Hussain et al. 2009; Riyazuddin et al. 2020). It has affected nearly 80 million hectares of cultivable soil globally (Munns and Tester 2008). Salinity is primarily apparent in the roots, which affects plant development through the induction of pressures because of decreased water accessibility and also due to the ionic stress caused by solute inequality in the cell cytoplasm (Ahmed et al. 2012; Munns 2005; Conde et al. 2011). Therefore, ionic balance is a crucial feature of salinity stress adjustment in resistant crops. Ionic balance involves multiple responses, depending on the resistance of the particular crop and the level of the salinity stress. It was observed in different plant species that the concentration and absorption of K^+/Na^+ and Ca^{2+}/Na^+ selectively manage inequality in salinity-tolerant and salinity-sensitive crops (Phang et al. 2008; Sanchez et al. 2008). In addition, the crop homology of ion transporters is related to salinity stress responses and involves Na^+ transporters and Na^+/H^+ antiporters (Phang et al. 2008; Benedito et al. 2010; Teakle et al. 2010). Amongst these, the vacuolar Na^+/H^+ antiporter NHX1 (for Na^+/H^+ exchanger) is responsible for Na^+ compartmentation (Zahran et al. 2007; Teakle et al. 2010). Other major responses of salinity stress tolerance include accumulation of osmoprotectants and stress proteins, and maintenance of ROS homeostasis and ionic balance. The transcriptome analysis of crops leads to the identification of stress-regulating genes overexpressed in roots in comparison to leaves. Major groups expressed in response to the constraints are genes and transcription factors linked to stress-related signaling pathways (Fan et al. 2013).

13.3.3 Heat Stress Responses

Among the major effects of heat stress is the increased production of ROS, which causes oxidative stress inside the whole plant body (Hasanuzzaman et al. 2013). Plants are able to cope with heat stress through morphological, physiological, and biochemical adaptations inside the system and often by developing signal transduction for altering metabolism. Plants change their metabolism in response to high-temperature stress specifically through the accumulation of compatible solutes or osmolytes, which help in maintaining the proteins and cellular structures, maintenance of cell turgidity by osmotic adjustments, and modifying the antioxidant defense responses and re-forming the redox balance and homeostasis inside the cells (Tiwari and Yadav 2019). In the chloroplast, heat stress causes alterations in grana pile, changes in the thylakoid structure, and a decrease of PS II antenna complex (Shaheen et al. 2016). The PS II and oxygen-evolving complexes are prime targets, as are thylakoid membranes, carbon fixation, and stroma photochemical reactions (Gerganova et al. 2016). The inhibition of Rubisco activase activity and S-adenosyl-L-homocysteine hydrolase activity has been observed in heat stress, leading to variation in the entire proteomics of leaves of a heat-resistant tomato variety (Zhou et al., 2011). Moreover, altered expression of the genes associated with the glyoxylate pathway, carbohydrate metabolic reactions, photosynthetic reactions, and cellular protective responses have also been reported in response to heat stress in various plants. Geneticists have traced molecular mechanisms to develop novel crops resilient to heat stress through controlled variations in temperature conditions. Similarly, Rubisco enzyme activity was found to be decreased in tomatoes infected with viruses such as cucumber mosaic virus (Di Carli et al. 2010), and at the same time, it was found to be enhanced in water stress, salt stress, and nutritional stress (Salekdeh and Komatsu 2007). Similarly, S-adenosyl-L-homocysteine hydrolase enzyme activity was also observed to be enhanced in the tolerant cultivars of wild tomato immunized with *Clavibacter michiganensis* (Afroz et al. 2011). Protein analysis was carried out by Muneer et al. (2016) on the grafting union of three tomato genotypes that showed higher activity of antioxidant enzymes such as SOD, APX, and CAT. Approximately 40 genes were observed to be differentially expressed in these genotypes (Super Sunload, B-blocking, and Super Doterang) when exposed to high- or low-temperature stress conditions (Muneer et al. 2016). The characterization of the stress-resistant proteins through matrix-assisted laser desorption ionization-time of flight (MALDI-TOF) was established by using immune-blotting techniques.

13.3.4 Heavy Metal Toxicity Responses

The term "heavy metal toxicity" typically refers to the accumulation of metals such as cadmium (Cd), copper (Cu), chromium (Cr), lead (Pb), and zinc (Zn) and harmful metalloids such as arsenic (As) and boron (B), which causes harmful consequences for plant growth and development (Hossain et al. 2012). Heavy metal toxicity causes different harms, and they are constantly increasing in the environment due to industries, mining, and metropolitan activities (Breckle 1991). They act as essential elements, but when their concentration exceeds a particular level, they inhibit plant growth instead of promoting plant growth and development (Clemens 2001). These heavy metals accumulate primarily in the plant roots through particular/general ion carriers or channels (Bubb and Lester 1991). The deficiency of specific transporters, which are mainly involved in the uptake of essential elements such as Zn^{2+}, Fe^{2+}, and Ca^{2+}, also allows the uptake of Cd^{2+} and Pb^{2+} (Perfus-Barbeoch et al. 2002). The harmful effects of heavy metals involve attaching ions to sulfhydryl functional sites of proteins; the substitution of important cations from functional sites, most importantly enzyme deactivation; and the generation of ROS, causing oxidation of macromolecules or biomolecules (Sharma and Dietz 2009). From the very beginning, widespread investigation of plants' responses to heavy metal stress has been performed to unravel the pathways of tolerance. Genomic techniques are practical in dealing with plant environmental stress reactions, including heavy metal stress (Bohnert et al. 2006). However, alteration of genetic expression at the transcription level has not always been reflected at the protein stage. Profound proteome observation is of great importance in identifying stress proteins that take part in heavy metal detoxification pathway(s). With time, higher plants have developed complicated processes to adjust the absorption, transport, and quantity of different heavy metal ions. Various responses against heavy metal stress inside cells to protect the cells from harmful effects involve structural and functional modifications of the plasma membrane and the biosynthesis of different transporters and SH-containing chelating complexes for vacuole separation. Besides, the enhanced abundance of defensive proteins for useful ROS scavenging and heat shock proteins to restore the usual protein configuration also help heavy metal-stressed plants to maintain the reduction/oxidation balance. Balancing important metabolic pathways – photosynthesis and mitochondrial respiration – assists stressed plants to generate higher reduction capability to balance the high ATP demands of heavy metal-stressed cells (Nadarajah et al. 2020).

13.3.5 Nutrition Deficiency Responses

The deficiency of mineral nutrients has been reported to influence plant growth by inducing changes in growth pattern, chemical composition, and antioxidant defense capacity, ultimately rendering the plant susceptible to various stress factors (Hajiboland 2012; Urwat et al. 2019). Plants growing on acidic soils have less access to several essential mineral nutrients, such as P, K, Ca, Mg, Cu, and Zn (Muthu Kumar et al. 2014). Isolation and characterization of genes involved in the absorption of minerals and their interaction within the various plant tissues will undoubtedly throw light on the mechanism operating in the optimum utilization of minerals for developing healthy and productive plants. The trace element-detoxifying systems and their manipulation in various transgenic crops for increased tolerance have been observed (Cherian and Oliveira 2005). These were classified and categorized into two groups based upon their functions. One group included genes that synthesize the metallothioneins, phytochelatins, glutathione, and various classes of peptides that control and maintain heavy metal concentrations in crops. These responsive proteins help in detoxifying heavy metals by decreasing their free concentrations, chelating heavy metal ions, and acting as precursors of phytochelatins (Cobbett and Goldsbrough 2002; Choppala et al. 2014). Another group includes the transporters that help in controlling the transport of toxic cations. For example, overexpression of yeast transport YCF1 provided enhanced tolerance and accumulation of Cd and Pb through increased transport capacity to the transgenic plants (Song et al. 2003). Similarly, other transporters such as calcium vacuolar transporter CAX2, Zn transporter (ZAT), and ferric reductase (FRE1, FRE2) were also found to provide tolerance to and accumulation of heavy metals in plants (Xia et al. 2018).

13.4 PROTEOMIC APPROACHES: TOWARDS UNDERSTANDING ABIOTIC STRESSES

A proteomic approach is carried out using two different methods: gel-based approaches and gel-free approaches (Zargar et al. 2016). Gel-based approaches include protein separation by gel electrophoresis and then their identification and characterization by mass spectrometry, while the gel-free method involves protease digestion and then liquid chromatographic separation and mass spectrometric characterization without resolving the proteins on polyacrylamide gels (Champagne et al. 2013). One- or two-dimensional gel electrophoresis and difference in gel electrophoresis (DIGE) are some gel-based approaches, while MudPIT, isotope-coded affinity tag (ICAT), tandem mass tags (TMT), and iTRAQ are some of the gel-free approaches that have been used to carry out abiotic stress response analysis (Washburn et al. 2001).

13.4.1 Gel-Based Proteomics Approach

These types of approaches are widely utilized for analyzing the global expression of proteins and mainly include two-dimensional electrophoresis (2D) followed by identification of differentially modulated proteins by mass spectrophotometry (Kłodzińska and Buszewski 2013). In the separation by 2D, the proteins are separated based upon their isoelectric point (charge) in the first dimension followed by a resolution based upon their molecular mass in the second dimension (Roy et al. 2011). Separated proteins appear as spots after staining with a dye such as CBB, $AgNO_3$, or SYPRO (Robinson et al. 2011). This step is followed by the identification of the proteins appearing as spots by mass spectrometry approaches (Magdeldin et al. 2014). In order to overcome the reproducibility issues of the 2D system, a modified variant, DIGE, is used, in which the samples are labeled with the particular type of fluorescent dye used earlier to run on the same gel. The various dyes commonly used are cyanine based, such as Cy3-NHS and Cy5-NHS, which are of the same molecular mass, amine-reactive, and positively charged. Since different samples are run on the same gel, DIGE is more reproducible and allows even minor detection of differences between similar proteins in different samples. Thus, DIGE is a sensitive technique that allows detection of very small levels, down to 0.5 fmoles, of proteins, and this detection system is linear over a million-fold concentration range (Blundon and Ganesan 2019; Minden et al. 2012; Saia-cereda et al. 2017). The gel-based approach has been greatly advantageous in carrying out the important basic aim of proteomic analysis through identification, characterization, and quantification of a large number of target proteins under different abiotic stresses with the help of mass spectrometry.

13.4.2 Gel-Free Proteomic Approach

This approach to abiotic stress analysis has been widely advantageous over gel-based proteomic studies (Zhou et al. 2012). It involves quantification of proteins without label methods or tag-based labeling in which different mass tags, for example ICAT, isobaric tags for relative and absolute quantification (iTRAQ), or TMT, are commonly utilized. Similarly, dimethyl labels and 180 labels have recently been added to these approaches to identify different proteins or peptides involved in the underlying stress response. For stable isotope labeling by amino acids in cell culture, 15N labels are used for metabolite labeling of proteins *in vivo* in cells (El-khateeb et al. 2019). The labeling-based method involves differentially labeling samples, then combining them and subjecting them to mass spectrometry analysis and observations. The quantification of proteins is based upon the light and heavy peptide intensities. These label-based quantitation methods are also known as multiplex proteomics, as different peptide fragments from different conditions are labeled with isobaric tags and are then combined and ionized. Upon their separation, the reporter ions are released and encode different conditions. The relative signals can be used to quantify the protein abundance in the sample (Sonnett et al. 2018; Stadlmeier et al. 2018). The multiplex proteomic approach is therefore highly reproducible, and the throughput increases, as all the samples can be targeted in a single run. This limits the costs and analysis of pre-fractionated samples as compared with label-free quantification. This approach allows quantifications of more protein, and therefore, these benefits make the multiplex proteomic approach an attractive alternative for the relative quantification of the whole proteome(Connell et al. 2018). On the contrary, recently, the label-free quantitative technique has attained significant importance for identifying a wide range of proteins. In this approach, the samples to be analyzed are directly injected into a mass spectroscope for comparison (Abdallah et al. 2012; Megger et al. 2013). This label-free method as several advantages over label-based proteomics: it is cost-effective, less time consuming as compared with labeling techniques that require various labeling steps, and more sensitive than label-based methods. Label-free proteomic analysis may be divided into two quantification strategies. One is called spectral counting, in which the number of fragment ion spectra (tandem mass spectrometry (MS/MS)) for peptides of a particular protein are counted and compared, while another approach measures the mass spectrometric signal intensity of peptide ion precursors or measures the chromatographic peak regions. In gel-free proteomics, trypsin-digested peptides are first resolved mainly by reverse-phase liquid chromatography, where peptides are separated through various properties such as charge, hydrophobicity, etc., and are afterward ionized through an ion source and detected by mass spectroscopy. Each peptide of any specific charge and mass forms a particular single isotopic mass peak. The peak is a function of retention time, and the area under the curve can be established in the extracted ion chromatogram (Wasinger et al. 2013; Al Shweiki et al. 2017; Mergner et al. 2020).

13.5 CROP PROTEOMICS: AN UPDATE

13.5.1 Wheat

Wheat (*Triticum* sp.) acts as an important food crop globally, providing an important source of gluten and different storage proteins (Gill et al. 2004). *Triticum* sp. is usually grown in water stress areas, and hence it is necessary to study the defense signaling pathways under these conditions. Various proteomic approaches have become promising tools in identifying signaling molecules under water stress conditions (Komatsu et al. 2014). A number of resistant and stressed varieties of wheat under water stress conditions have been observed and studied for ROS-regulated responses (Hajheidari et al. 2007). Abscisic acid-induced defense proteins, late embryogenesis abundant proteins, CDKS such as zinc finger proteins, MYB transcriptional factors, and WRKY domain-containing proteins are expressed in wheat cultivars under water deficit/stress conditions as observed by quantitative proteomic analyses (Kamal et al. 2010). A wide range of different proteomic approaches was utilized for the identification and characterization of various salt-induced changes in protein networks and therefore, provided insights into the wheat salt stress responses (Wang et al. 2008; Gao et al. 2011; Mahajan et al. 2017). The effects of salinity stress on the pattern of protein expression in the chloroplast of wheat were monitored using a 2D approach coupled with LTQ-Fourier transform ion cyclotron resonance hybrid mass spectroscopy (Kamal et al. 2012). Salt stress decreased the rate of photosynthesis, the rate of transcription, stomatal efficiency, water potential, and the chlorophyll A and B levels while increasing the proline levels. The various electron transport complex components, such as reaction center of photosystem I, oxygen-evolving complex, and cytochrome b6f complex, were also observed to be reduced due to salinity stress in wheat. A 2D-based analysis of leaf proteome in wheat coupled with MALDI-TOF identified various important salinity stress-responsive proteins highly involved in membrane transport, ROS scavenging, energy synthesis, carbon metabolism, and proper protein folding (Caruso et al. 2008; Gao et al. 2011). H^+-ATPases, glutathione S-transferase, ferritin, and triose-phosphate isomerase were some of the highly increased proteins identified in the salinity-tolerant response in the different wheat cultivars. An iTRAQ-based proteome analysis of wheat flag leaves during high-temperature stress identified 3499 proteins, of which 258 were heat-responsive, including 155 up-regulated and 105 down-regulated proteins (Lu et al. 2007). Similarly, proteomic analysis using a 2D with MALDI/TOF mass spectrometry-based system led to the identification of cadmium (heavy metal stress)-responsive proteins in wheat roots. These cadmium-responsive proteins consisted of two major groups: those involved in the chelation of metal by glutathione in the cytosol and others involved in the efflux of cadmium into the vacuole, a scavenging system by the ascorbate–glutathione cycle (Wang et al. 2011).

13.5.2 Rice

Oryza sativa is one of the major food crops grown globally as an important staple food. It is widely used as a model crop for carrying out molecular biology and genome studies (Meng et al. 2019). International rice genome sequencing under the rice research project in the year 2005 gave a complete genetic draft of the rice genome and therefore, assisted in carrying out various proteomic studies on rice in different stress conditions. Proteome profiles of rice seedlings, androecium, shoot, roots, and leaves of different cultivars and at different levels have been investigated in water stress conditions (Koller et al. 2002; Shu et al. 2011; Liu and Bennett 2011; Muthurajan et al. 2011; Xiong et al. 2010; Ji et al. 2012; Mirzaei et al. 2012). A label-free proteomic approach led to the identification of 1487 drought-responsive proteins in rice seedlings. Different ATPases, GTPases, and water channels like aquaporins were differentially modulated under different water conditions (Mirzaei et al. 2012). Research associated with proteomic studies on rice crops was excellently reviewed by Agrawal et al. (2006), Komatsu and Yano (2006), He and Yang (2013), and Kim et al. (2014) under different stress conditions.

13.5.3 Maize

Zea mays is also another important crop acting as a staple food globally along with other food crops, and the United States Department of Agriculture reported about 839 million tons of maize production in the year 2012. Besides acting as a food crop, it also provides an important feed for animals (Strable and Scanlon 2009). Maize is one of the most important crops, having a variable level of genetic changes because of cross-pollination, and thus, it is being widely used for observing different processes such as variation, mutations, recombinations, imprinting behavior, and different expression changes in nucleic acids and proteins during abiotic stresses such as water stress through different proteomic approaches (Doebley 2004; Benešová et al. 2012; Yang et al. 2014; Vu et al. 2016; Zhao et al. 2016a, b). A 2D-based analysis of dehydration-responsive proteins in *Zea mays* at various stages of growth and germination stages led to the identification of about 11 proteins, such as LEA proteins, EMB 564, GLOBULIN 2, CYSTATIN, CLASS I HSP, and NBS-LRR resistant proteins. These proteins accumulated in embryos upon maturation and decreased during germination (Huang et al. 2012). Various studies were carried out for this crop under drought and well-watered conditions to characterize quantitative traits; for grain yield and anthesis-silking interval were observed (Li et al. 2016). Another abiotic stress that adversely affects the production and its economy is cold/low temperature. Studies on different cultivars have led to the identification of various quantitative traits for germination stages and early growth in the field environment (Allam et al. 2016). The Plant Proteome Database (http://ppdb.tc.cornell.edu), TIGR Maize Database (http://maize.jcvi.org), and Maize Assembled Genomic Island (http://magi.plantgenomics.iastate.edu/) have been developed and are available for use. The information obtained by researching the maize proteome has led to enhancing its nutritional values, resistance to different stresses, improving crop yields, and adaptation to the environment. Further, because of widespread co-linearity between cereal genomics of maize, sorghum, rice, wheat, and barley (Bennetzen et al. 2003), the information may be applied to related crops, and new, advanced varieties could be developed for meeting the globally increasing demands for food and feed. The maize proteome under abiotic stresses is found to respond strongly through alterations of various metabolic processes, such as carbohydrate metabolism, as well as the activation of antioxidant/ROS scavenging responses, such as glutathione S-transferase and APX (He et al. 2013).

13.5.4 Barley

Hordeum vulgare, another important food crop that usually grows in arid and semi-arid soils, is reported to possess several resistant responses against different environmental stresses, such as drought, salinity, high temperature, etc. Therefore, it is greatly utilized to develop different resistant cultivars in different parts of the varied climates across the world. In 2012, this cereal crop acted as an important model for carrying out different studies, such as proteomic analysis, genetic studies, and different stress responses, when its genome was completely sequenced by the international barley sequencing association (Shakhatreh et al. 2010). Responsive proteins in leaves and roots of different cultivars of barley were reported using mass spectrometry techniques-based proteomics by Rollins et al. (2013). Through differential gene expression analysis in leaves as well as in roots under drought stress and well-watered conditions, nearly 66 defensive proteins were identified (Wendel boe-Nelson et al. 2012). Similarly, the relative proteome analysis of stress-tolerant and sensitive cultivars has led to the characterization of proteins expressed during dehydration stress (Ashoub et al. 2013). Results obtained from this study showed up-regulation of various proteins, such as NADP-dependent malic acid enzyme, lipoxygenase, sucrose synthase, and betaine aldehyde dehydrogenase. The detoxifying cellular enzymes and photorespiration-controlling enzymes were found to be increased in tolerant cultivars but not in sensitive ones (Ashoub et al. 2013).

13.5.5 Sorghum

Sorghum is a C4 cereal food crop, belonging to the family Poaceae, with multiple uses, high metabolic efficiency, and high production efficacy (Alkaraki et al. 1995). It acts as a staple food for the people of arid and semi-arid regions in Asia and Africa and is a rich source of proteins. It has been widely utilized for studying responses to various abiotic stresses, such as dehydration, high temperature, etc., with the help of the 2D-based proteomic approaches. For example, observations under dehydration have led to the identification of different dehydration-responsive proteins in the leaves of sorghum (Jedmowski et al. 2014). The whole-genome database network of sorghum is made available on the website http://structuralbiology.cau.edu.cn/sorghum/index.html (Tian et al. 2016). Various metabolic proteins, enzymes, and chaperones have been identified using a gel-based approach of proteomics under different abiotic stresses and after their recovery from such stresses. Methionine synthase, SAM synthase, and nitrile lyase were different metabolic proteins identified and were found to be increased under stress conditions. Similarly, pyruvate kinase enzyme was also seen to be up-regulated using proteomic analysis (Jedmowski et al. 2014). In low-temperature-tolerant cultivars, different transcription factors such as DREB, C-repeat binding factor, and ethylene-responsive factors were observed to be up-regulated. Various proteins such as heat shock proteins, glutathione S-transferase, cytochromes, etc. were observed to be differentially expressed during low-temperature stress in tolerant cultivars (Chopra et al. 2015).

13.5.6 Common Bean

Phaseolus vulgaris L., a major food crop grown worldwide, faces various abiotic stresses, particularly dehydration stress. The proteome analysis of common bean under different abiotic stresses has led to the identification and characterization of various proteins that are differentially expressed and were found to be highly involved in amino acid and carbohydrate metabolism (Yang et al. 2013; Zargar et al. 2017). 2D was utilized to identify the different stress-responsive proteins in the common bean leaves of various tolerant and sensitive cultivars, and the proteins spotted were mainly characterized into various groups that were involved in energy metabolism, photosynthesis, protein biosynthesis, proteolytic degradation, and response to stress(Zadražnik et al. 2013; Manzoor et al. 2021). Dehydrins are a class of stress-responsive proteins that protect membrane integrity and macromolecular functions, therefore increasing water stress tolerance in common bean. The common bean root proteome is widely altered by exposure to low-temperature stress, depending upon the length and the time of exposure to stress conditions (Badowiec et al. 2014). Exposure to stress for a short time causes the activation of protection mechanisms and expression of defensive proteins, the abundance of proteins that are related to energy metabolism, and the generation of various defense response phytohormones. The common bean response to any abiotic stress is strongly dependent upon the length and time of exposure, which leads to dynamic alterations to the whole bean proteome.

13.6 CONCLUSION AND FUTURE PERSPECTIVES

With the expanding advancements in technologies in the field of agriculture, the breeding of specific cultivars has shown promising results in major food crop production and quality enhancement when facing different abiotic stresses while taking into account the growing population. Crop proteomics, analyzing the different signaling pathways of different abiotic stress responses, has greatly helped in the selection and breeding of resistant and tolerant varieties for sustainable development. Crop proteomics has been widely applied to crops such as wheat, rice, and barley under various abiotic stresses and is a promising tool for observation of quantitative traits that are responsible for resistance and tolerance under various abiotic stresses. One of the major constraints to -omics techniques, including genomics and proteomics, is high cost, but if this is efficiently regulated, such approaches can help develop crops of higher potential with specific traits.

REFERENCES

Abdallah, Cosette, Eliane Dumas-Gaudot, Jenny Renaut, and Kjell Sergeant. "Gel-based and gel-free quantitative proteomics approaches at a glance." *International Journal of Plant Genomics* 2012 (2012): 494572.

Abreu, Isabel A., Ana Paula Farinha, Sónia Negrão, Nuno Gonçalves, Cátia Fonseca, Mafalda Rodrigues, Rita Batista, Nelson JM Saibo, and M. Margarida Oliveira. "Coping with abiotic stress: proteome changes for crop improvement." *Journal of Proteomics* 93 (2013): 145–168.

Afroz, Amber, Ghulam Muhammad Ali, Asif Mir, and Setsuko Komatsu. "Application of proteomics to investigate stress-induced proteins for improvement in crop protection." *Plant Cell Reports* 30, no. 5 (2011): 745–763.

Agrawal, Ganesh Kumar, and Randeep Rakwal. "Rice proteomics: a cornerstone for cereal food crop proteomes." *Mass Spectrometry Reviews* 25, no. 1 (2006): 1–53.

Ahmad, Parvaiz, Ashwani Kumar, Muhammad Ashraf, and Nudrat Aisha Akram. "Salt-induced changes in photosynthetic activity and oxidative defense system of three cultivars of mustard (*Brassica juncea* L.)." *African Journal of Biotechnology* 11, no. 11 (2012): 2694–2703.

Ahmad, Parvaiz, Cheruth Abdul Jaleel, M. M. Azooz, and Gowher Nabi. "Generation of ROS and non-enzymatic antioxidants during abiotic stress in plants." *Botany Research International* 2, no. 1 (2009): 11–20.

Al-Karaki, G. N., R. B. Clark, and C. Y. Sullivan. "Effects of phosphorus and water stress levels on growth and phosphorus uptake of bean and sorghum cultivars." *Journal of Plant Nutrition* 18, no. 3 (1995): 563–578.

Allam, Mohamed, Pedro Revilla, Abderrahmane Djemel, William F. Tracy, and Bernardo Ordás."Identification of QTLs involved in cold tolerance in sweetx field corn." *Euphytica* 208, no. 2 (2016): 353–365.

AlShweiki, Mhd Rami, Susann Mönchgesang, Petra Majovsky, Domenika Thieme, Diana Trutschel, and Wolfgang Hoehenwarter. "Assessment of label-free quantification in discovery proteomics and impact of technological factors and natural variability of protein abundance." *Journal of Proteome Research* 16, no. 4 (2017): 1410–1424.

Ashoub, Ahmed, Tobias Beckhaus, Thomas Berberich, Michael Karas, and Wolfgang Brüggemann. "Comparative analysis of barley leaf proteome as affected by drought stress." *Planta* 237, no. 3 (2013): 771–781.

Ashraf, M. P. J. C., and P. J. C. Harris. "Potential biochemical indicators of salinity tolerance in plants." *Plant Science* 166, no. 1 (2004): 3–16.

Aslam, Bilal, Madiha Basit, Muhammad Atif Nisar, Mohsin Khurshid, and Muhammad Hidayat Rasool. "Proteomics: technologies and their applications." *Journal of Chromatographic Science* 55, no. 2 (2017): 182–196.

Asthir, Bavita, Gurpreet Kaur, and Balraj Kaur. "Convergence of pathways towards ascorbate–glutathione for stress mitigation." *Journal of Plant Biology* 63 (2020): 243–257.

Badowiec, Anna, and Stanisław Weidner. "Proteomic changes in the roots of germinating *Phaseolus vulgaris* seeds in response to chilling stress and post-stress recovery." *Journal of Plant Physiology* 171, no. 6 (2014): 389–398.

Baker, Alison, and Ian A. Graham, eds. *Plant Peroxisomes: Biochemistry, Cell Biology and Biotechnological Applications*. Springer Science & Business Media, 2013.

Bakthisaran, Raman, Ramakrishna Tangirala, and Ch Mohan Rao. "Small heat shock proteins: role in cellular functions and pathology." *Biochimica et Biophysica Acta (BBA)-Proteins and Proteomics* 1854, no. 4 (2015): 291–319.

Benedito, Vagner A., Haiquan Li, Xinbin Dai, Maren Wandrey, Ji He, Rakesh Kaundal, Ivone Torres-Jerez et al. "Genomic inventory and transcriptional analysis of *Medicago truncatula* transporters." *Plant Physiology* 152, no. 3 (2010): 1716–1730.

Bennetzen, Jeffrey L., and Jianxin Ma. "The genetic colinearity of rice and other cereals on the basis of genomic sequence analysis." *Current Opinion in Plant Biology* 6, no. 2 (2003): 128–133.

Benešová, Monika, Dana Hola, Lukáš Fischer, Petr L. Jedelský, František Hnilička, Naďa Wilhelmová, Olga Rothova et al. "The physiology and proteomics of drought tolerance in maize: early stomatal closure as a cause of lower tolerance to short-term dehydration?" *PLoS ONE* 7, no. 6 (2012): e38017.

Blundon, Malachi, Vinitha Ganesan, Brendan Redler, Phu T. Van, and Jonathan S. Minden. "Two-dimensional difference gel electrophoresis." In *Electrophoretic Separation of Proteins*, pp. 229–247. Humana Press, New York, NY, 2019.

Bohnert, Hans J., Qingqiu Gong, Pinghua Li, and Shisong Ma. "Unraveling abiotic stress tolerance mechanisms–getting genomics going." *Current Opinion in Plant Biology* 9, no. 2 (2006): 180–188.

Brandes, Nadav, Nathan Linial, and Michal Linial. "PWAS: proteome-wide association study—linking genes and phenotypes by functional variation in proteins." *Genome Biology* 21, no. 1 (2020): 1–22.

Breckle, S. W. "Growth under stress: heavy metals." In: *Plant Roots: The Hidden Half* (Y. Waisel, A. Eshel and U. Kafkafi, eds.), Marcel Dekker (1991): 351–373.

Bubb, J. M., and J. N. Lester. "The impact of heavy metals on lowland rivers and the implications for man and the environment." *Science of the Total Environment* 100 (1991): 207–233.

Cannon, R. E., J. A. White, and J. G. Scandslios. *Proceedings of the National Academy of Sciences of the United States of America* 84 (1987): 179–183.

Caruso, Giuseppe, Chiara Cavaliere, Chiara Guarino, Riccardo Gubbiotti, Patrizia Foglia, and Aldo Laganà. "Identification of changes in *Triticum durum* L. leaf proteome in response to salt stress by two-dimensional electrophoresis and MALDI-TOF mass spectrometry." *Analytical and Bioanalytical Chemistry* 391, no. 1 (2008): 381–390.

Champagne, Antoine, and Marc Boutry. "Proteomics of nonmodel plant species." *Proteomics* 13, no. 3–4 (2013): 663–673.

Chaves, Maria Manuela, and Margarida M. Oliveira. "Mechanisms underlying plant resilience to water deficits: prospects for water-saving agriculture." *Journal of Experimental Botany* 55, no. 407 (2004): 2365–2384.

Cherian, Sam, and M. Margarida Oliveira. "Transgenic plants in phytoremediation: recent advances and new possibilities." *Environmental Science & Technology* 39, no. 24 (2005): 9377–9390.

Choppala, Girish, Saifullah, Nanthi Bolan, Sadia Bibi, Muhammad Iqbal, Zed Rengel, Anitha Kunhikrishnan, Nanjappa Ashwath, and Yong Sik Ok. "Cellular mechanisms in higher plants governing tolerance to cadmium toxicity." *Critical Reviews in Plant Sciences* 33, no. 5 (2014): 374–391.

Chopra, Ratan, Gloria Burow, Chad Hayes, Yves Emendack, Zhanguo Xin, and John Burke. "Transcriptome profiling and validation of gene based single nucleotide polymorphisms (SNPs) in sorghum genotypes with contrasting responses to cold stress." *BMC Genomics* 16, no. 1 (2015): 1–11.

Clemens, Stephan. "Molecular mechanisms of plant metal tolerance and homeostasis." *Planta* 212, no. 4 (2001): 475–486.

Cobbett, Christopher, and Peter Goldsbrough. "Phytochelatins and metallothioneins: roles in heavy metal detoxification and homeostasis." *Annual Review of Plant Biology* 53, no. 1 (2002): 159–182.

Conde, Artur, M. Manuela Chaves, and Hernâni Gerós. "Membrane transport, sensing and signaling in plant adaptation to environmental stress." *Plant and Cell Physiology* 52, no. 9 (2011): 1583–1602.

Corpas, Francisco J., Dharmendra K. Gupta, and José M. Palma. "Production sites of reactive oxygen species (ROS) in organelles from plant cells." In *Reactive Oxygen Species and Oxidative Damage in Plants under Stress*, pp. 1–22. Springer, Cham, 2015.

D'Arcy-Lameta, Agnes, Roselyne Ferrari-Iliou, Dominique Contour-Ansel, Anh-Thu Pham-Thi, and Yasmine Zuily-Fodil. "Isolation and characterization of four ascorbate peroxidase cDNAs responsive to water deficit in cowpea leaves." *Annals of Botany* 97, no. 1 (2006): 133–140.

Davletova, Sholpan, Karen Schlauch, Jesse Coutu, and Ron Mittler. "The zinc-finger protein Zat12 plays a central role in reactive oxygen and abiotic stress signaling in Arabidopsis." *Plant Physiology* 139, no. 2 (2005): 847–856.

DiCarli, Mariasole, Maria Elena Villani, Linda Bianco, Raffaele Lombardi, Gaetano Perrotta, Eugenio Benvenuto, and Marcello Donini. "Proteomic analysis of the plant–virus interaction in cucumber mosaic virus (CMV) resistant transgenic tomato." *Journal of Proteome Research* 9, no. 11 (2010): 5684–5697.

Doebley, John. "The genetics of maize evolution." *Annual Review Genetics* 38 (2004): 37–59.

El-Khateeb, Eman, Areti-Maria Vasilogianni, Sarah Alrubia, Zubida M. Al-Majdoub, Narciso Couto, Martyn Howard, Jill Barber, Amin Rostami-Hodjegan, and Brahim Achour. "Quantitative mass spectrometry-based proteomics in the era of model-informed drug development: applications in translational pharmacology and recommendations for best practice." *Pharmacology & Therapeutics* 203 (2019): 107397.

Falk, Jon, Gaby Andersen, Birgit Kernebeck, and Karin Krupinska. "Constitutive overexpression of barley 4-hydroxyphenylpyruvate dioxygenase in tobacco results in elevation of the vitamin E content in seeds but not in leaves." *FEBS Letters* 540, no. 1–3 (2003): 35–40.

Fan, Xiu-Duo, Jia-Qi Wang, Na Yang, Yuan-Yuan Dong, Liang Liu, Fa-Wei Wang, Nan Wang et al. "Gene expression profiling of soybean leaves and roots under salt, saline–alkali and drought stress by high-throughput Illumina sequencing." *Gene* 512, no. 2 (2013): 392–402.

Flowers, Timothy J., and Timothy D. Colmer. "Salinity tolerance in halophytes." *New Phytologist* 179, no. 4 (2008): 945–963.

Foyer, Christine H., and Graham Noctor. "Redox homeostasis and antioxidant signaling: a metabolic interface between stress perception and physiological responses." *The Plant Cell* 17, no. 7 (2005): 1866–1875.

Gao, L., X. Yan, X. Li, G. Guo, Y. Hu, W. Ma, and Y. Yan. "Proteome analysis of wheat leaf under salt stress by two-dimensional difference gel electrophoresis (2D-DIGE)." *Phytochemistry* 72, no. 10 (2011): 1180–1191.

Gerganova, Milena, Antoaneta V. Popova, Daniela Stanoeva, and Maya Velitchkova. "Tomato plants acclimate better to elevated temperature and high light than to treatment with each factor separately." *Plant Physiology and Biochemistry* 104 (2016): 234–241.

Gill, Bikram S., Rudi Appels, Anna-Maria Botha-Oberholster, C. Robin Buell, Jeffrey L. Bennetzen, Boulos Chalhoub, Forrest Chumley et al. "A workshop report on wheat genome sequencing: International Genome Research on Wheat Consortium." *Genetics* 168, no. 2 (2004): 1087–1096.

Golldack, Dortje, Chao Li, Harikrishnan Mohan, and Nina Probst. "Tolerance to drought and salt stress in plants: unraveling the signaling networks." *Frontiers in Plant Science* 5 (2014): 151.

Greenberg, B. M., X. D. Huang, P. Gerwing, X. M. Yu, P. Chang, S. S. Wu, K. Gerhardt, J. Nykamp, X. Lu, and B. Glick. "Phytoremediation of salt impacted soils: greenhouse and the field trials of plant growth promoting rhizobacteria (PGPR) to improve plant growth and salt phytoaccumulation." In *Proceeding of the 33rd AMOP technical seminar on environmental contamination and response.* Environment Canada, Ottawa, pp. 627–637. 2008.

Groppa, M. D., and M. P. Benavides. "Polyamines and abiotic stress: recent advances." *Amino Acids* 34, no. 1 (2008): 35.

Guo, Zhenfei, Weichao Ou, Shao-yun Lu, and Q. Zhong. "Differential responses of antioxidative system to chilling and drought in four rice cultivars differing in sensitivity." *Plant Physiology and Biochemistry* 44, no. 11–12 (2006): 828–836.

Gupta, Ravi, Yiming Wang, Ganesh K. Agrawal, Randeep Rakwal, Ick H. Jo, Kyong H. Bang, and Sun T. Kim. "Time to dig deep into the plant proteome: a hunt for low-abundance proteins." *Frontiers in Plant Science* 6(2015): 1–3. doi:10.3389/fpls.2015.00022.

Hajheidari, Mohsen, Alireza Eivazi, Bob B. Buchanan, Joshua H. Wong, Islam Majidi, and Ghasem Hosseini Salekdeh. "Proteomics uncovers a role for redox in drought tolerance in wheat." *Journal of Proteome Research* 6, no. 4 (2007): 1451–1460.

Hajiboland, R. "Effect of micronutrient deficiencies on plants stress responses." In *Abiotic Stress Responses in Plants*, pp. 283–329. Springer, New York, NY, 2012.

Hasanuzzaman, Mirza, Kamrun Nahar, Md Alam, Rajib Roychowdhury, and Masayuki Fujita. "Physiological, biochemical, and molecular mechanisms of heat stress tolerance in plants." *International Journal of Molecular Sciences* 14, no. 5 (2013): 9643–9684.

Hayat, Shamsul, Qaiser Hayat, Mohammed Nasser Alyemeni, Arif Shafi Wani, John Pichtel, and Aqil Ahmad. "Role of proline under changing environments: a review." *Plant Signaling &Behavior* 7, no. 11 (2012): 1456–1466.

He, Dongli, and Pingfang Yang. "Proteomics of rice seed germination." *Frontiers in Plant Science* 4 (2013): 246.

Hodges, D. Mark, Christopher J. Andrews, Douglas A. Johnson, and Robert I. Hamilton. "Antioxidant enzyme responses to chilling stress in differentially sensitive inbred maize lines." *Journal of Experimental Botany* 48, no. 5 (1997): 1105–1113.

Hossain, Zahed, Mohammad-Zaman Nouri, and Setsuko Komatsu. "Plant cell organelle proteomics in response to abiotic stress." *Journal of Proteome Research* 11, no. 1 (2012): 37–48.

Huang, Hui, Ian Max Møller, and Song-Quan Song. "Proteomics of desiccation tolerance during development and germination of maize embryos." *Journal of Proteomics* 75, no. 4 (2012): 1247–1262.

Hubbard, Michael J. "Functional proteomics: the goalposts are moving." *PROTEOMICS: International Edition* 2, no. 9 (2002): 1069–1078.

Hussain, Khalid, Abdul Majeed, Khalid Nawaz, and M. Farrukh Nisar. "Effect of different levels of salinity on growth and ion contents of black seeds (*Nigellasativa* L.)." *Current Research Journal of Biological Sciences* 1, no. 3 (2009): 135–138.

Ighodaro, O. M., and O. A. Akinloye. "First line defence antioxidants – superoxide dismutase (SOD), catalase (CAT) and glutathione peroxidase (GPX): their fundamental role in the entire antioxidant defence grid." *Alexandria Journal of Medicine* 54, no. 4 (2018): 287–293.

Jedmowski, Christoph, Ahmed Ashoub, Tobias Beckhaus, Thomas Berberich, Michael Karas, and Wolfgang Brüggemann. "Comparative analysis of *Sorghum bicolor* proteome in response to drought stress and following recovery." *International Journal of Proteomics* 2014 (2014): 395905.

Ji, Kuixian, Yangyang Wang, Weining Sun, Qiaojun Lou, Hanwei Mei, Shihua Shen, and Hui Chen. "Drought-responsive mechanisms in rice genotypes with contrasting drought tolerance during reproductive stage." *Journal of Plant Physiology* 169, no. 4 (2012): 336–344.

Jiang, Mingyi, and Jianhua Zhang. "Role of abscisic acid in water stress-induced antioxidant defense in leaves of maize seedlings." *Free Radical Research* 36, no. 9 (2002): 1001–1015.

Kamal, Abu Hena Mostafa, Kun Cho, Da-Eun Kim, Nobuyuki Uozumi, Keun-Yook Chung, Sang Young Lee, Jong-Soon Choi, Seong-Woo Cho, Chang-Seob Shin, and Sun Hee Woo. "Changes in physiology and protein abundance in salt-stressed wheat chloroplasts." *Molecular Biology Reports* 39, no. 9 (2012): 9059–9074.

Kamal, Abu Hena Mostafa, Ki-Hyun Kim, Kwang-Hyun Shin, Jong-Soon Choi, Byung-Kee Baik, Hisashi Tsujimoto, Hwa Young Heo, Chul-Soo Park, and Sun-Hee Woo. "Abiotic stress responsive proteins of wheat grain determined using proteomics technique." *Australian Journal of Crop Science* 4, no. 3 (2010): 196.

Khan, Imran, Samrah Afzal Awan, Rizwana Ikram, Muhammad Rizwan, Nosheen Akhtar, Humaira Yasmin, Riyaz Z. Sayyed, Shafaqat Ali and Noshin Ilyas. "Effects of 24-epibrassinolide on plant growth, antioxidants defense system, and endogenous hormones in two wheat varieties under drought stress." *Physiologia Plantarum* 172, no. 2 (2021): 696–706.

Kiani-Pouya, Ali. "Changes in activities of antioxidant enzymes and photosynthetic attributes in triticale (× Triticosecale Wittmack) genotypes in response to long-term salt stress at two distinct growth stages." *Acta Physiologiae Plantarum* 37, no. 4 (2015): 72.

Kim, Ilgook, Bomi Lee, Dongsu Song, and Jong-In Han. "Effects of ammonium carbonate pretreatment on the enzymatic digestibility and structural features of rice straw." *Bioresource Technology* 166 (2014): 353–357.

Kłodzińska, Ewa, and Bogusław Buszewski. "Two-dimensional gel electrophoresis (2DE)." In *Electromigration Techniques*, pp. 133–158. Springer, Berlin, Heidelberg, 2013.

Koller, Antonius, Michael P. Washburn, B. Markus Lange, Nancy L. Andon, Cosmin Deciu, Paul A. Haynes, Lara Hays et al. "Proteomic survey of metabolic pathways in rice." *Proceedings of the National Academy of Sciences* 99, no. 18 (2002): 11969–11974.

Komatsu, Setsuko, Abu H. M. Kamal, and Zahed Hossain. "Wheat proteomics: proteome modulation and abiotic stress acclimation." *Frontiers in Plant Science* 5 (2014): 684.

Komatsu, Setsuko, and Hiroyuki Yano. "Update and challenges on proteomics in rice." *Proteomics* 6, no. 14 (2006): 4057–4068.

Kumar, A., J. S. Patel, Vijay Singh Meena, and P. W. Ramteke. "Plant growth-promoting rhizobacteria: strategies to improve abiotic stresses under sustainable agriculture." *Journal of Plant Nutrition* 42, no. 11–12 (2019): 1402–1415.

Liu, Jian-Xiang, and John Bennett. "Reversible and irreversible drought-induced changes in the anther proteome of rice (*Oryza sativa* L.) genotypes IR64 and Moroberekan." *Molecular Plant* 4, no. 1 (2011): 59–69.

Li, Xiangnan, and Fulai Liu. "Drought stress memory and drought stress tolerance in plants: biochemical and molecular basis." In *Drought Stress Tolerance in Plants*, Vol 1, pp. 17–44. Springer, Cham, 2016.

Lopez-Huertas, Eduardo, Wayne L. Charlton, Barbara Johnson, Ian A. Graham, and Alison Baker. "Stress induces peroxisome biogenesis genes." *The EMBO Journal* 19, no. 24 (2000): 6770–6777.

Lu, Zhenqiang, Dali Liu, and Shenkui Liu. "Two rice cytosolic ascorbate peroxidases differentially improve salt tolerance in transgenic Arabidopsis." *Plant Cell Reports* 26, no. 10 (2007): 1909–1917.

Magdeldin, Sameh, Shymaa Enany, Yutaka Yoshida, Bo Xu, Ying Zhang, Zam Zureena, Ilambarthi Lokamani, Eishin Yaoita, and Tadashi Yamamoto. "Basics and recent advances of two dimensional-polyacrylamide gel electrophoresis." *Clinical Proteomics* 11, no. 1 (2014): 16.

Mahajan, Reetika, Muslima Nazir, Nusrat Sayeed, Vandna Rai, Roohi Mushtaq, Khalid Z. Masoodi, Shafiq A. Wani et al. "Unraveling the Abiotic Stress Tolerance in Common Bean Through Omics." In *Plant Omics and Crop Breeding*, pp. 311–365. Apple Academic Press, 2017.

Manzoor, Madhiya, Sajad Majeed Zargar, Parveen Akhter, Uneeb Urwat, Reetika Mahajan, Sajad Ahmad Bhat, Tanveer Ali Dar, and Imran Khan. "Morphological, biochemical, and proteomic studies revealed impact of Fe and P crosstalk on root development in *Phaseolus vulgaris* L." *Applied Biochemistry and Biotechnology* 193, no. 12 (2021): 3898–3914.

Mapson, L. W., and D. R. Goddard. "The reduction of glutathione by plant tissues." *Biochemical Journal* 49, no. 5 (1951): 592–601.

Megger, Dominik A., Thilo Bracht, Helmut E. Meyer, and Barbara Sitek. "Label-free quantification in clinical proteomics." *Biochimica et Biophysica Acta (BBA)-Proteins and Proteomics* 1834, no. 8 (2013): 1581–1590.

Meldrum, Norman Urquhart, and Hugh Lewis Aubrey Tarr. "The reduction of glutathione by the Warburg-Christian system." *Biochemical Journal* 29, no. 1 (1935): 108.

Meng, Qingfeng, Ravi Gupta, Chul Woo Min, Jongyun Kim, Katharina Kramer, Yiming Wang, Sang Ryeol Park, Iris Finkemeier, and Sun Tae Kim. "A proteomic insight into the MSP1 and Flg22 induced signaling in *Oryza sativa* leaves." *Journal of Proteomics* 196 (2019): 120–130. doi:10.1016/j.jprot.2018.04.015.

Mergner, Julia, Martin Frejno, Markus List, Michael Papacek, Xia Chen, Ajeet Chaudhary, Patroklos Samaras et al. "Mass-spectrometry-based draft of the Arabidopsis proteome." *Nature* 579, no. 7799 (2020): 409–414.

Meyer, Andrea, Dennis B. Hansen, Cláudia S.G. Gomes, Timothy J. Hobley, Owen R.T. Thomas, and Matthias Franzreb. "Demonstration of a strategy for product purification by high-gradient magnetic fishing: recovery of superoxide dismutase from unconditioned whey." *Biotechnology Progress* 21, no. 1 (2005): 244–254.

Mirzaei, Mehdi, Dana Pascovici, Brian J. Atwell, and Paul A. Haynes. "Differential regulation of aquaporins, small GTPases and V-ATPases proteins in rice leaves subjected to drought stress and recovery." *Proteomics* 12, no. 6 (2012): 864–877.

Mitra, Jiban. "Genetics and genetic improvement of drought resistance in crop plants." *Current Science* 80, no. 6(2001): 758–763.

Mittler, Ron, and Thomas L. Poulos. "Ascorbate peroxidase." In *Antioxidants and Reactive Oxygen Species in Plants* (Smirnoff, N., ed.), Blackwell Publishing (2005): 101–140.

Mullineaux, Philip, and Stanislaw Karpinski. "Signal transduction in response to excess light: getting out of the chloroplast." *Current Opinion in Plant Biology* 5, no. 1 (2002): 43–48.

Muneer, Sowbiya, Chung Ho Ko, Hao Wei, Yuze Chen, and Byoung Ryong Jeong."Physiological and proteomic investigations to study the response of tomato graft unions under temperature stress." *PLoS ONE* 11, no. 6 (2016): e0157439.

Munné-Bosch, Sergi, and Leonor Alegre. "Drought-induced changes in the redox state of α-tocopherol, ascorbate, and the diterpene carnosic acid in chloroplasts of Labiatae species differing in carnosic acid contents." *Plant Physiology* 131, no. 4 (2003): 1816–1825.

Munns, Rana, and Mark Tester. "Mechanisms of salinity tolerance." *Annual Review of Plant Biology* 59 (2008): 651–681.

Munns, Rana. "Genes and salt tolerance: bringing them together." *New Phytologist* 167, no. 3 (2005): 645–663.

Muthukumar, Thangavelu, Perumalsamy Priyadharsini, Eswaranpillai Uma, Sarah Jaison, and Radha Raman Pandey. "Role of arbuscular mycorrhizal fungi in alleviation of acidity stress on plant growth." In *Use of Microbes for the Alleviation of Soil Stresses*, volume 1, pp. 43–71. Springer, New York, NY, 2014.

Muthurajan, Raveendran, Zahra-Sadat Shobbar, S. V. K. Jagadish, Richard Bruskiewich, Abdelbagi Ismail, Hei Leung, and John Bennett. "Physiological and proteomic responses of rice peduncles to drought stress." *Molecular Biotechnology* 48, no. 2 (2011): 173–182.

Nadarajah, Kalaivani K. "ROS homeostasis in abiotic stress tolerance in plants. "*International Journal of Molecular Sciences* 21, no. 15 (2020): 5208.

Nakashima, Kazuo, Yusuke Ito, and Kazuko Yamaguchi-Shinozaki. "Transcriptional regulatory networks in response to abiotic stresses in Arabidopsis and grasses." *Plant Physiology* 149, no. 1 (2009): 88–95.

O'Connell, Jeremy D., Joao A. Paulo, Jonathon J. O'Brien, and Steven P. Gygi. "Proteome-wide evaluation of two common protein quantification methods." *Journal of Proteome Research* 17, no. 5 (2018): 1934–1942.

Osakabe, Yuriko, Keishi Osakabe, Kazuo Shinozaki, and Lam-Son Phan Tran. "Response of plants to water stress." *Frontiers in Plant Science* 5 (2014): 86.

Panchuk, Irina I., Roman A. Volkov, and Friedrich Schöffl. "Heat stress-and heat shock transcription factor-dependent expression and activity of ascorbate peroxidase in Arabidopsis." *Plant Physiology* 129, no. 2 (2002): 838–853.

Pastori, Gabriela M., and Christine H. Foyer. "Common components, networks, and pathways of cross-tolerance to stress. The central role of 'redox' and abscisic acid-mediated controls." *Plant Physiology* 129, no. 2 (2002): 460–468.

Perfus-Barbeoch, Laetitia, Nathalie Leonhardt, Alain Vavasseur, and Cyrille Forestier. "Heavy metal toxicity: cadmium permeates through calcium channels and disturbs the plant water status." *The Plant Journal* 32, no. 4 (2002): 539–548.

Phang, Tsui-Hung, Guihua Shao, and Hon-Ming Lam. "Salt tolerance in soybean." *Journal of Integrative Plant Biology* 50, no. 10 (2008): 1196–1212.

Qureshi, Muhammad Kamran, Sana Munir, Ahmad Naeem Shahzad, Sumaira Rasul, Wasif Nouman, and Kashif Aslam. "Role of reactive oxygen species and contribution of new players in defense mechanism under drought stress in rice." *International Journal of Agriculture and Biology* 20, no. 6 (2018): 1339–1352.

Raven, Emma Lloyd. "Understanding functional diversity and substrate specificity in haem peroxidases: what can we learn from ascorbate peroxidase?" *Natural Product Reports* 20, no. 4 (2003): 367–381.

Rhodes, David, A. Nadolska-Orczyk, and P. J. Rich. "Salinity, osmolytes and compatible solutes." In *Salinity: Environment–Plants–Molecules*, pp. 181–204. Springer, Dordrecht, 2002.

Riyazuddin, Riyazuddin, Nisha Nisha, Kalpita Singh, Radhika Verma, and Ravi Gupta. "Involvement of dehydrin proteins in mitigating the negative effects of drought stress in plants." *Plant Cell Reports*, 2021. doi:10.1007/s00299-021-02720-6.

Riyazuddin, Riyazuddin, Radhika Verma, Kalpita Singh, Nisha Nisha, Monika Keisham, KaushalKumar Bhati, Sun Tae Kim, and Ravi Gupta. "Ethylene: a master regulator of salinity stress tolerance in plants."*Biomolecules* 10, no. 6 (2020): 1–22. doi:10.3390/biom10060959.

Robinson, Aisling A., Ciara A. McManus, and Michael J. Dunn. "Two-dimensional polyacrylamide gel electrophoresis." In *Sample Preparation in Biological Mass Spectrometry*, pp. 217–242. Springer, Dordrecht, 2011.

Rodrıguez, P., A. Torrecillas, M. A. Morales, M. F. Ortuno, and M. J. Sánchez-Blanco. "Effects of NaCl salinity and water stress on growth and leaf water relations of *Asteriscus maritimus* plants." *Environmental and Experimental Botany* 53, no. 2 (2005): 113–123.

Rollins, J. A., E. Habte, S. E. Templer, T. Colby, J. Schmidt, and M. VonKorff. "Leaf proteome alterations in the context of physiological and morphological responses to drought and heat stress in barley (*Hordeum vulgare* L.)." *Journal of Experimental Botany* 64, no. 11 (2013): 3201–3212.

Romero-Puertas, M. C., F. J. Corpas, L. M. Sandalio, M. Leterrier, M. Rodríguez-Serrano, L. A. del Río, and J. M. Palma. "Glutathione reductase from pea leaves: response to abiotic stress and characterization of the peroxisomal isozyme." *New Phytologist* 170 (2006): 43–52.

Roy, Ansuman, Paul J. Rushton, and Jai S. Rohila. "The potential of proteomics technologies for crop improvement under drought conditions." *Critical Reviews in Plant Sciences* 30, no. 5 (2011): 471–490.

Roychoudhury, Aryadeep, Aditya Banerjee, and Vikramjit Lahiri. "Metabolic and molecular-genetic regulation of proline signaling and its cross-talk with major effectors mediates abiotic stress tolerance in plants." *Turkish Journal of Botany* 39, no. 6 (2015): 887–910.

Saia-Cereda, Veronica M., Adriano Aquino, Paul C. Guest, and Daniel Martins-de-Souza. Two-dimensional gel electrophoresis: A reference protocol. In *Proteomic Methods in Neuropsychiatric Research*, pp. 175–182. Springer, Cham, 2017.

Sakuma, Yoh, Kyonoshin Maruyama, Yuriko Osakabe, Feng Qin, Motoaki Seki, Kazuo Shinozaki, and Kazuko Yamaguchi-Shinozaki. "Functional analysis of an Arabidopsis transcription factor, DREB2A, involved in drought-responsive gene expression." *The Plant Cell* 18, no. 5 (2006): 1292–1309.

Salekdeh, Ghasem Hosseini, and Setsuko Komatsu. "Crop proteomics: aim at sustainable agriculture of tomorrow." *Proteomics* 7, no. 16 (2007): 2976–2996.

Sali, Andrej, Robert Glaeser, Thomas Earnest, and Wolfgang Baumeister. "From words to literature in structural proteomics." *Nature* 422, no. 6928 (2003): 216–225.

Samarah, Nezar H. "Effects of drought stress on growth and yield of barley." *Agronomy for Sustainable Development* 25, no. 1 (2005): 145–149.

Sanchez, Diego H., Felix Lippold, Henning Redestig, Matthew A. Hannah, Alexander Erban, Ute Krämer, Joachim Kopka, and Michael K. Udvardi. "Integrative functional genomics of salt acclimatization in the model legume *Lotusjaponicus*." *The Plant Journal* 53, no. 6 (2008): 973–987.

Sankar, B., C. Abdul Jaleel, P. Manivannan, A. Kishorekumar, R. Somasundaram, and R. Panneerselvam. "Effect of paclobutrazol on water stress amelioration through antioxidants and free radical scavenging enzymes in *Arachis hypogaea* L." *Colloids and Surfaces B: Biointerfaces* 60, no. 2 (2007): 229–235.

Saruhashi, Masashi, Totan Kumar Ghosh, Kenta Arai, Yumiko Ishizaki, Kazuya Hagiwara, Kenji Komatsu, Yuh Shiwa et al. "Plant Raf-like kinase integrates abscisic acid and hyperosmotic stress signaling upstream of SNF1-related protein kinase2." *Proceedings of the National Academy of Sciences* 112, no. 46 (2015): E6388–E6396.

Saxena, Raghvendra, Rajesh Singh Tomar, and Manish Kumar. "Exploring nanobiotechnology to mitigate abiotic stress in crop plants." *Journal of Pharmaceutical Sciences and Research* 8, no. 9 (2016): 974.

Sečenji, Maria, Eva Hideg, Attila Bebes, and János Györgyey. "Transcriptional differences in gene families of the ascorbate–glutathione cycle in wheat during mild water deficit." *Plant Cell Reports* 29, no. 1 (2010): 37–50.

Sehrawat, Ankita, and Renu Deswal. "S-nitrosylation analysis in *Brassica juncea* apoplast highlights the importance of nitric oxide in cold-stress signaling." *Journal of Proteome Research* 13, no. 5 (2014): 2599–2619.

Selote, Devarshi S., and Renu Khanna-Chopra. "Drought-induced spikelet sterility is associated with an inefficient antioxidant defence in rice panicles." *Physiologia Plantarum* 121, no. 3 (2004): 462–471.

Selote, Devarshi S., and Renu Khanna-Chopra. "Drought acclimation confers oxidative stress tolerance by inducing co-ordinated antioxidant defense at cellular and subcellular level in leaves of wheat seedlings." *Physiologia Plantarum* 127, no. 3 (2006): 494–506.

Shaheen, Muhammad R., Choudhary M. Ayyub, Muhammad Amjad, and Ejaz A. Waraich. "Morpho-physiological evaluation of tomato genotypes under high temperature stress conditions." *Journal of the Science of Food and Agriculture* 96, no. 8 (2016): 2698–2704.

Shakhatreh, Y., N. Haddad, M. Alrababah, S. Grando, and S. Ceccarelli. "Phenotypic diversity in wild barley (*Hordeum vulgare* L. ssp. spontaneum (C. Koch) Thell.) accessions collected in Jordan." *Genetic Resources and Crop Evolution* 57, no. 1 (2010): 131–146.

Sharma, Shanti S., and Karl-Josef Dietz. "The relationship between metal toxicity and cellular redox imbalance." *Trends in Plant Science* 14, no. 1 (2009): 43–50.

Shu, Liebo, Qiaojun Lou, Chenfei Ma, Wei Ding, Jia Zhou, Jinhong Wu, Fangjun Feng, et al. "Genetic, proteomic and metabolic analysis of the regulation of energy storage in rice seedlings in response to drought." *Proteomics* 11, no. 21 (2011): 4122–4138.

Singh, Harminder Pal, Priyanka Mahajan, Shalinder Kaur, Daizy R. Batish, and Ravinder K. Kohli. "Chromium toxicity and tolerance in plants." *Environmental Chemistry Letters* 11, no. 3 (2013): 229–254.

Song, Won-Yong, Eun Ju Sohn, Enrico Martinoia, Yong Jik Lee, Young-Yell Yang, Michal Jasinski, Cyrille Forestier, Inwhan Hwang, and Youngsook Lee. "Engineering tolerance and accumulation of lead and cadmium in transgenic plants." *Nature Biotechnology* 21, no. 8 (2003): 914–919.

Sonnett, Matthew, Eyan Yeung, and Martin Wuhr. "Accurate, sensitive, and precise multiplexed proteomics using the complement reporter ion cluster." *Analytical Chemistry* 90, no. 8 (2018): 5032–5039.

Stadlmeier, Michael, Jana Bogena, Miriam Wallner, Martin Wühr, and Thomas Carell. "A sulfoxide-based isobaric labelling reagent for accurate quantitative mass spectrometry." *Angewandte Chemie International Edition* 57, no. 11 (2018): 2958–2962.

Strable, Josh, and Michael J. Scanlon. "Maize (*Zea mays*): a model organism for basic and applied research in plant biology." *Cold Spring Harbor Protocols* 2009, no. 10 (2009): pdb-emo132.

Tian, Tian, Qi You, Liwei Zhang, Xin Yi, Hengyu Yan, Wenying Xu, and Zhen Su. "SorghumFDB: sorghum functional genomics database with multidimensional network analysis." *Database* 2016 (2016): baw099.

Tiwari, Yogesh Kumar, and Sushil Kumar Yadav. "High temperature stress tolerance in maize (*Zea mays* L.): physiological and molecular mechanisms." *Journal of Plant Biology* 62, no. 2 (2019): 93–102.

Tran, Lam-Son Phan, and Keiichi Mochida. "Identification and prediction of abiotic stress responsive transcription factors involved in abiotic stress signaling in soybean." *Plant Signaling & Behavior* 5, no. 3 (2010): 255–257.

Umezawa, Taishi, Fuminori Takahashi, and Kazuo Shinozaki. "Phosphorylation networks in the abscisic acid signaling pathway." In *The Enzymes*, vol. 35, pp. 27–56. Academic Press, 2014.

Urwat, Uneeb, Sajad Majeed Zargar, Madhiya Manzoor, Syed Mudasir Ahmad, Nazir Ahmad Ganai, Imtiyaz Murtaza, Imran Khan, and F. A. Nehvi. "Morphological and biochemical responses of *Phaseolus vulgaris* L. to mineral stress under in vitro conditions." *Vegetos* 32, no. 3 (2019): 431–438.

Vu, Lam Dai, Elisabeth Stes, Michiel VanBel, Hilde Nelissen, Davy Maddelein, Dirk Inzé, Frederik Coppens, Lennart Martens, Kris Gevaert, and Ive De Smet. "Up-to-date workflow for plant (phospho) proteomics identifies differential drought-responsive phosphorylation events in maize leaves." *Journal of Proteome Research* 15, no. 12 (2016): 4304–4317.

Walker, Mark A., and Bryan D. Mckersie. "Role of the ascorbate-glutathione antioxidant system in chilling resistance of tomato." *Journal of Plant Physiology* 141, no. 2 (1993): 234–239.

Wang, Meng-Cheng, Zhen-Ying Peng, Cui-Ling Li, Fei Li, Chun Liu, and Guang-Min Xia. "Proteomic analysis on a high salt tolerance introgression strain of *Triticum aestivum/Thinopyrum ponticum*." *Proteomics* 8, no. 7 (2008): 1470–1489.

Wang, Yun, Yaru Qian, Hao Hu, Yan Xu, and Haijun Zhang. "Comparative proteomic analysis of Cd-responsive proteins in wheat roots." *Acta Physiologiae Plantarum* 33, no. 2 (2011): 349–357.

Washburn, Michael P., Dirk Wolters, and John R. Yates. "Large-scale analysis of the yeast proteome by multidimensional protein identification technology." *Nature Biotechnology* 19, no. 3 (2001): 242–247.

Wasinger, Valerie C., Ming Zeng, and Yunki Yau. "Current status and advances in quantitative proteomic mass spectrometry." *International Journal of Proteomics* 2013 (2013): 180605.

Wendelboe-Nelson, Charlotte, and Peter C. Morris. "Proteins linked to drought tolerance revealed by DIGE analysis of drought resistant and susceptible barley varieties." *Proteomics* 12, no. 22 (2012): 3374–3385.

Werner, Andrea K., and Claus-Peter Witte. "The biochemistry of nitrogen mobilization: purine ring catabolism." *Trends in Plant Science* 16, no. 7 (2011): 381–387.

Xia, Yan, Juan Liu, Ying Wang, Xingxing Zhang, Zhenguo Shen, and Zhubing Hu. "Ectopic expression of *Vicia sativa* caffeoyl-CoA O-methyltransferase (VsCCoAOMT) increases the uptake and tolerance of cadmium in Arabidopsis." *Environmental and Experimental Botany* 145 (2018): 47–53.

Xiong, Jian-Hua, Bin-Ying Fu, Hua-Xue Xu, and Yang-Sheng Li. "Proteomic analysis of PEG-simulated drought stress responsive proteins of rice leaves using a pyramiding rice line at the seedling stage." *Botanical Studies* 51, no. 2 (2010): 137–145.

Xu, Zhenzhu, Yanling Jiang, Bingrui Jia, and Guangsheng Zhou. "Elevated-CO_2 response of stomata and its dependence on environmental factors." *Frontiers in Plant Science* 7 (2016): 657.

Yang, Liming, Tingbo Jiang, Jake C. Fountain, Brian T. Scully, Robert D. Lee, Robert C. Kemerait, Sixue Chen, and Baozhu Guo. "Protein profiles reveal diverse responsive signaling pathways in kernels of two maize inbred lines with contrasting drought sensitivity." *International Journal of Molecular Sciences* 15, no. 10 (2014): 18892–18918.

Yang, Zhong-Bao, Dejene Eticha, Hendrik Führs, Dimitri Heintz, Daniel Ayoub, Alain Van Dorsselaer, Barbara Schlingmann, Idupulapati Madhusudana Rao, Hans-Peter Braun, and Walter Johannes Horst. "Proteomic and phosphoproteomic analysis of polyethylene glycol-induced osmotic stress in root tips of common bean (*Phaseolus vulgaris* L.)." *Journal of Experimental Botany* 64, no. 18 (2013): 5569–5586.

Yancey, Paul H. "Organic osmolytes as compatible, metabolic and counteracting cytoprotectants in high osmolarity and other stresses." *Journal of Experimental Biology* 208, no. 15 (2005): 2819–2830.

Zadražnik, Tanja, Kristin Hollung, Wolfgang Egge-Jacobsen, Vladimir Meglič, and Jelka Šuštar-Vozlič. "Differential proteomic analysis of drought stress response in leaves of common bean (*Phaseolus vulgaris* L.)." *Journal of Proteomics* 78 (2013): 254–272.

Zahran, Hamdi H., M. Carmen Marín-Manzano, A. Juan Sánchez-Raya, Eulogio J. Bedmar, Kees Venema, and M. Pilar Rodríguez-Rosales. "Effect of salt stress on the expression of NHX-type ion transporters in *Medicago intertexta* and *Melilotus indicus* plants." *Physiologia Plantarum* 131, no. 1 (2007): 122–130.

Zargar, Sajad Majeed, Muslima Nazir, Ganesh Kumar Agarwal, and Randeep Rakwal. "OMICS based interventions for climate proof crops." *Genomics and Applied Biology* 2 (2011): 24–28.

Zargar, S., N. Gupta, R. Mir, and V. Rai. "Shift from gel based to gel free proteomics to unlock unknown regulatory network in plants: a comprehensive review." *Journal of Advanced Research in Biotechnology* 1, no. 19.10 (2016): 15226.

Zargar, Sajad Majeed, Nancy Gupta, Muslima Nazir, Reetika Mahajan, Firdose A. Malik, Nageebul R. Sofi, Asif B. Shikari, and R. K. Salgotra. "Impact of drought on photosynthesis: Molecular perspective." *Plant Gene* 11 (2017): 154–159.

Zargar, Sajad Majeed, Reetika Mahajan, Muslima Nazir, Preeti Nagar, Sun Tae Kim, Vandna Rai, Antonio Masi, et al. "Common bean proteomics: Present status and future strategies." *Journal of Proteomics* 169 (2017): 239–248.

Zhao, Feiyun, Dayong Zhang, Yulong Zhao, Wei Wang, Hao Yang, Fuju Tai, Chaohai Li, and Xiuli Hu. "The difference of physiological and proteomic changes in maize leaves adaptation to drought, heat, and combined both stresses." *Frontiers in Plant Science* 7 (2016): 1471.

Zhou, Hu, Zhibing Ning, Amanda E. Starr, Mohamed Abu-Farha, and Daniel Figeys."Advancements in top-down proteomics." *Analytical Chemistry* 84, no. 2 (2012): 720–734.

Zhou, Suping, Roger J. Sauvé, Zong Liu, Sasikiran Reddy, Sarabjit Bhatti, Simon D. Hucko, Yang Yong, Tara Fish, and Theodore W. Thannhauser. "Heat-induced proteome changes in tomato leaves." *Journal of the American Society for Horticultural Science* 136, no. 3 (2011): 219–226.

14 Metabolites and Abiotic Stress Tolerance in Plants

Radha Mishra[1] and Indresh Kumar Pandey[1,2,]*
[1]Department of Botany, University of Allahabad, Prayagraj, India
[2]Govt. PG College, Karnprayag, Chamoli, India
*Corresponding author: E-mail: pandey197@gmail.com

CONTENTS

14.1 INTRODUCTION

Plants exist in ever-changing environmental conditions that often impede their growth and physiological actions. The development and yield of plants are firmly interlinked with the change in external climatic factors. Plants are generally exposed to numerous kinds of abiotic stress, like drought, salinity, temperature, chilling, nutrient deficiency, and heavy metals, which obstruct their growth and productivity worldwide (Boyer, 1982; Araus et al., 2002). Due to the multiple abiotic stresses, it has been estimated that there is a potential yield loss of more than 50% (Bray et al., 2000), which puts pressure on the sustainability of the agriculture sector. These abiotic stressors are consequences of climate change that put pressure on plants to modulate themselves so as to

DOI: 10.1201/9781003159636-14

withstand the environmental influences and also induce various kinds of adaptation approaches to combat the environmental stresses. There are different ways that are used to improve stress tolerance in plants, but the understanding of the basis of abiotic stress tolerance requires significant effort. Metabolites act as biostimulators or bio-effectors and can be used to gain a better perception of their complex regulation and metabolism. Nowadays, "Omics" techniques are useful in understanding the abiotic stress tolerance mechanism in different plants. In plants, the production of metabolites is directly proportional to physiological and developmental conditions. It has been reported that abiotic stress conditions alter the expression of certain groups of genes, which may be directly involved in the metabolites biosynthesis pathways (Tuteja, 2007). In this chapter, we describe some of the most important metabolites that can influence plant abiotic stress tolerance, and a few special metabolites that are accumulated in plants during stress conditions, and attempt to figure out the effect of abiotic stresses on metabolite production.

14.2 TYPES OF METABOLITES

Metabolites can be divided into two types, primary and secondary.

14.2.1 Primary Metabolites

The term "metabolites" refers to the small molecules that are the intermediate or final products of metabolism. A primary metabolite is a key component in maintaining normal physiological processes, and it is usually referred to as a "central metabolite" (Aharoni and Galili, 2011). Primary metabolites are essential for the survival and existence of an organism, as these are involved in the growth, development, and reproduction of the organism (Crueger and Crueger, 1990). Carbohydrates, amino acids, and lipids are considered primary metabolites and could also be involved in the abiotic stress tolerance of plants. Of these, several sugars function as compatible solutes and participate in stress tolerance, including drought (Moles et al., 2018; Shanazari et al., 2018) and salinity (Annunziata et al., 2017), by osmotic adjustment, maintaining membrane stability and oxidative stress protection (Figure 14.1). In addition to the sugars, amino acids such as glycine and proline

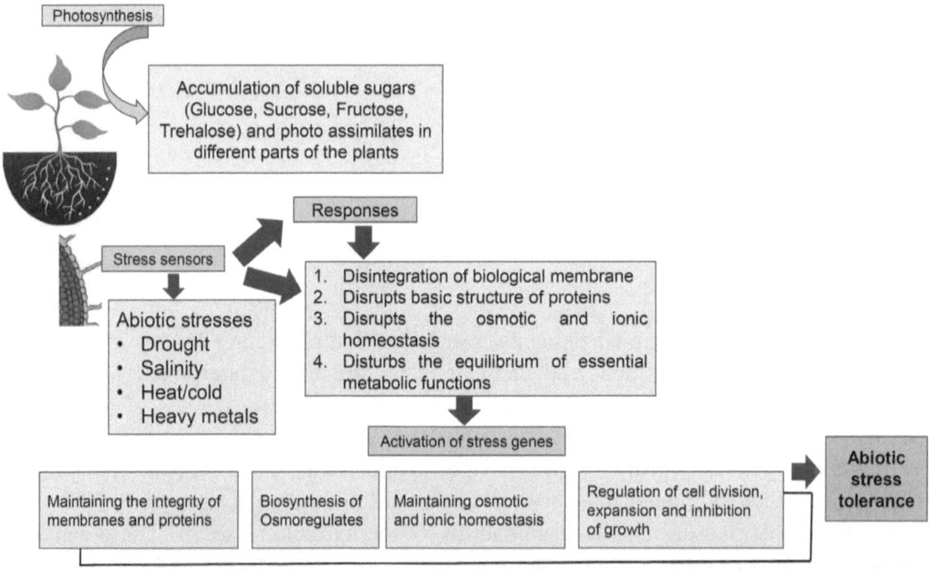

FIGURE 14.1 Overall view of primary metabolites' response in abiotic stress mechanism.

also function as compatible solutes. Moreover, amino acids also act as precursors for the synthesis of various secondary metabolites that participate in plant defense mechanisms (Griffiths et al., 2016). The term "secondary" was introduced by A. Kossel in 1891 to classify the primary metabolites that are present in every living cell that is capable of dividing. The secondary metabolites, on the other hand, are found only coincidentally and are not of important significance for the organism's life.

14.2.1.1 Carbohydrates

The accumulation of carbohydrates in higher plants is accompanied by their remarkable ability to tolerate unfavorable environmental conditions. The modification in carbohydrate homeostasis permits plants to counter the situations of abiotic stress (Pommerrenig et al., 2018). Zhang et al. (2018b) demonstrated the desiccation tolerance of *Oropetium thomaeum* grass during different environmental stresses. According to the overall picture from preliminary studies on the protective mechanism of carbohydrates, it is not amazing that deficiency in starch, sucrose, or carbohydrate affects the metabolic ability of plants to combat stress-inducing situations (Wanner and Junttila, 1999, Haisler et al., 2014, Schmitz et al., 2014; Le Hir et al., 2015) and also affects the turnover of carbohydrate by lowering and enhancing sugar content in the plant. Due to the typical accumulation of soluble sugars in leaves (Wulff-zottele et al., 2010; Hedrich et al., 2015), this causes a specific problem for plants. High cytosolic sugar content usually signals to tune down photosynthesis, which generally lowers the expression of a plethora of genes with relevance for CO_2 fixation (Eveland and Jackson, 2012). Trehalose, starch, fructans, and raffinose family oligosaccharides are the sub-types of carbohydrates and compatible metabolites that show influential effects on the plant under abiotic stress. Table 14.1 describes the role of different metabolites against abiotic stresses and their influence on the plant parts.

TABLE 14.1
Effect of Primary Metabolites and Their Counter Effect towards Abiotic Stress on Plants

Primary Metabolites	Abiotic Stress	Affected Phyto-Parts	Significance
		Carbohydrates	
Trehalose	Cold and salinity	Photosynthetic apparatus, reproductive tissue	Abnormal embryogenesis, leaf growth, increment in plant height, delayed germination (Wang et al., 2016), activation of defense-related enzymes (Govind et al., 2016)
Starch	All abiotic stresses	The stem of the plant	Modified granules in plant and enhanced starch content in the plant (Huang et al., 2018)
Fructans	Salt stress	Whole plant	Nutritional improved variety and develop crop resilient variety (Veenstra et al., 2017)
Raffinose	Drought and temperature stress	Seed part	Promote plant germination (Sengupta et al., 2015)
		Amino acids	
Proline	Salt stress and drought stress	The whole plant (mainly leaf)	Limiting the yield loss in agriculture (Semida et al., 2018)
γ-amino N-butyric acid	Under salinity stress and drought stress	Plant tissue	Develop drought and salt resistance variety (Ji et al., 2018; Li et al., 2018)
Glycine betaine	Drought stress, salt stress	Whole plant	Maintenance of cell osmotic pressure, protection of protein (Castiglioni et al., 2018) and develop stress resistance variety.

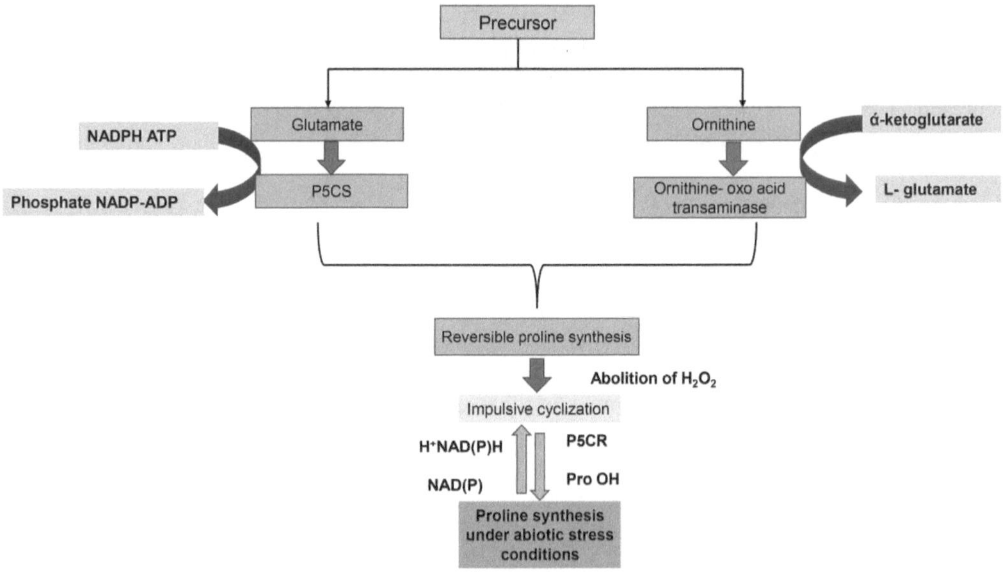

FIGURE 14.2 Proline metabolism pathways and their influences on plants.

14.2.1.2 Amino Acids

Proline (Pro) is one of the extensively studied amino acids in plant abiotic stress tolerance (Fichman et al., 2015). Singh et al. (2017) described the role of proline as an osmolyte in plants as a sensor of environmental stress. External application of proline treatment provokes salt stress tolerance in maize by reducing oxidative damage and maintaining ionic homeostasis (de Freitas et al., 2018). Moreover, proline metabolism also provides stress defense by helping to maintain $NADPH/NADP^+$ balance and driving the oxidative burst of the hypersensitive response (Figure 14.2) (Miller et al., 2009; Ben Rejeb et al., 2014). In addition, proline metabolism also influences the ROS signal pathway in plant parts (Shinde et al., 2016).

14.2.1.3 Polyamines

Polyamines are cationic compounds having two or more amine groups. They are low-molecular-weight organic molecules present in most organisms with a multitude of activities due to their diversity and position of amino groups (Morgen, 1999). Due to the cationic nature of polyamines, they bind with DNA, RNA, and proteins via electrostatic linkages, resulting in either stabilization or destabilization (Flink and Pettijohn,1975, Kusano et al., 2008). Polyamines are known to be the regulators of diverse developmental processes, including survival of plant embryos and translocation in eukaryotes (Takahashi and Kakehi, 2010); cell signaling and membrane stabilization (Tober and Tober,1984); cell proliferation and gene expression modulation (Igarashi and Kashiwagi, 2000); and apoptosis and programmed cell death (Thomas and Thomas, 2001, Seiler and Raul, 2005). Polyamines and abiotic stress are correlated with each other, as they alter a plant's response to a wide range of environmental stresses.

Modulated amounts of polyamines may act as a signal to alter the structural changes in RNA (m, r, and t) and to induce post-translational modifications of proteins (Agostinelli, 2016; Singh, 2017). Diverse abiotic stimuli under which polyamine accumulation has been observed include mineral deficiency (Richards and Coleman, 1952), metal toxicity (Chaudhary et al., 2012), salinity (Lefevre, 2001; Humel et al., 2004), and high and low temperature (Oshima, 2007). Polyamines alter protein synthesis; stimulate assembly of 30S ribosomal subunits; and stimulate Ile-tRNA formation

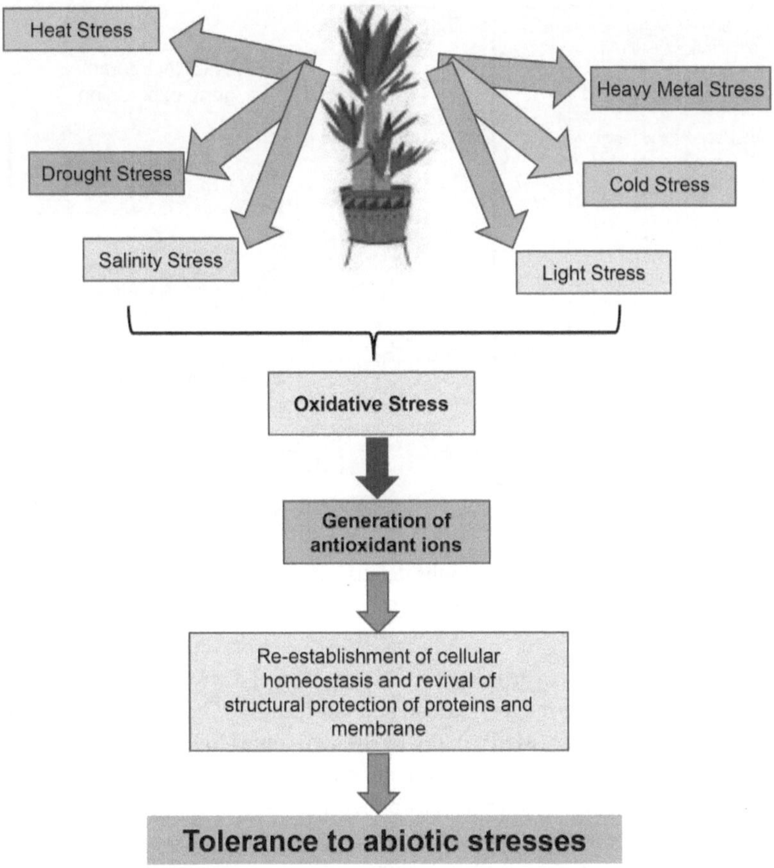

FIGURE 14.3 Schematic representation of polyamines (PAs)-mediated abiotic stress tolerance in plants.

(Igarashi et al., 2000). Also, modulated titers of polyamines in combination with epibrassinolides, an active form of brassinosteroids, were reported to regulate abscisic acid (ABA) and indole-3-acetic acid (IAA) pathways, which in turn enhances tolerance to metal toxicity in order to tolerate abiotic stresses (Chaudhary et al., 2012). A brief overview of polyamines in abiotic stresses is illustrated in Figure 14.3.

14.2.2 Secondary Metabolites

Secondary metabolites are a heterogeneous group of natural metabolic products that are not essential for vegetative growth of the producing organisms; they are considered differentiation compounds, playing adaptive roles in plants (Seigler et al., 1998). These secondary metabolites are mainly synthesized through the phenylpropanoid pathway. Environmental stresses increase the accumulation of phenylpropanoids and have a marked effect on phenolic levels in plant tissue (Dixon et al., 1995; Chalker et al, 1989). The concentrations of various secondary plant products are strongly dependent on the growing conditions and have an impact on the metabolic pathways responsible for the accumulation of the related natural products. Secondary metabolites are involved in protective activities in abiotic stress conditions (Figure 14.4). Plants produce various kinds of secondary metabolites, which may be categorized into three major groups based on their origin.

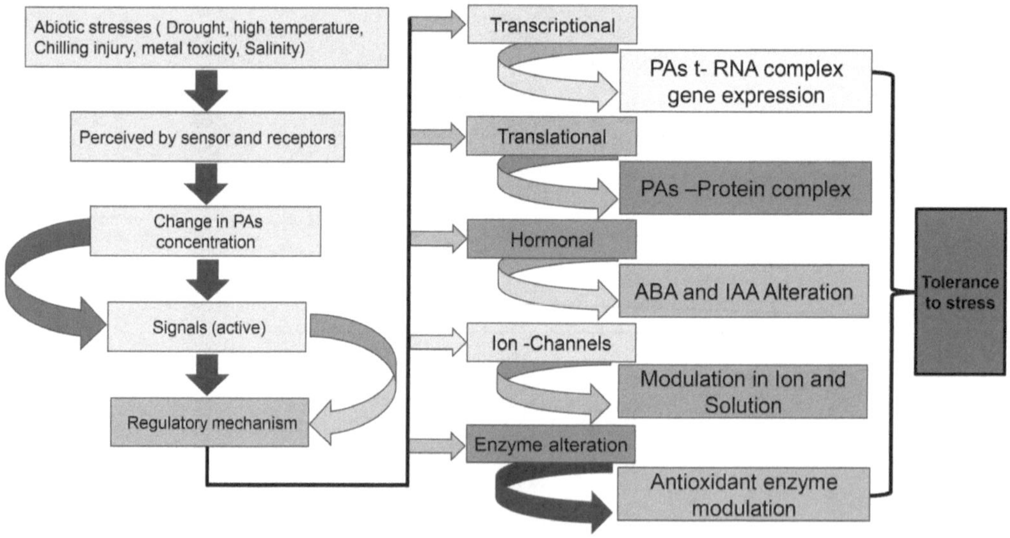

FIGURE 14.4 Role of secondary metabolites in the defense mechanism of plants.

14.2.2.1 Terpenoids

Aromatic plants have several properties due to the presence of various terpenoids. All terpenoids are obtained from isoprene units arranged in head and tail fashion. The mevalonate (MVA) pathway and 2-methylerythritol 4-phosphate (MEP) pathway are two routes to synthesize terpenoids in plants (Degenhardt and Lincoln, 2006) in the cytoplasm and chloroplast, respectively. Various components of terpenoids, such as monoterpenes, diterpenes, and tetrapenes, are synthesized in plastids, whereas sesquiterpenes and triterpenes are synthesized in the cytosol. Monoterpenes include menthol, citral, geraniol, linalool, camphor, and many others having antimicrobial and antioxidant activities. The excretion of these compounds depends upon different environmental stresses.

14.2.2.2 Phenolic Compounds

Phenolics are another salient category of secondary metabolites produced in response to various abiotic stressors. Phenolics include flavonoids, coumarins, lignins, and tannins (Gumul et al., 2007). The shikimic acid and malonic acid pathways are responsible for the synthesis of all phenolic compounds. Phenolic compounds have the ability to curtail the formation of free radicals. These compounds are useful in maintaining plant defense strategies.

14.2.2.3 Nitrogen Group

These are nitrogen-containing secondary metabolites. Nitrogen-accommodating secondary metabolites are mainly divided into two types, alkaloids and glycosides, which are non-protein amino acids. Alkaloids are known as the third largest group of secondary metabolites, which carry heterocyclic nitrogen atoms. These are derived from amino acids such as ornithine, lysine, phenylalanine, tryptophan, tyrosine, histidine, and aspartic acid. During abiotic stress conditions, concentrations of alkaloids increase in plants. Other than alkaloids, two important groups of N-containing secondary metabolites are also found; these are cyanogenic glycosides and glucosinolates released as by-products of plants in stressed conditions.

14.3 ROLE OF ABIOTIC STRESS TOLERANCE AND METABOLITES

The huge metabolic variety observed in plants is an immediate aftereffect of continuous developmental cycles. There are more than 200,000 known plant secondary metabolites, addressing a tremendous repository of assorted functions. At the point when the climate is unfavorable and plant growth is influenced, metabolism is significantly engaged in signaling, physiological regulation, and defense responses. Simultaneously, abiotic stresses affect the biosynthesis, concentration, transport, and storage of primary and secondary metabolites. Metabolic adjustments in response to abiotic stressors involve fine adjustments in amino acid, carbohydrate, and amine metabolic pathways. Proper activation of early metabolic responses helps cells restore chemical and energetic imbalances imposed by the stress and is crucial to acclimation and survival. Time-series experiments have revealed that metabolic activities respond to stress more quickly than transcriptional activities do. In order to study and map all the simultaneous metabolic responses and, more importantly, to link these responses to a specific abiotic stress, integrative and comprehensive analyses are required. Metabolomics is the systematic approach through which qualitative and quantitative analysis of a large number of metabolites is increasing our knowledge of how complex metabolic networks interact and how they are dynamically modified under stress adaptation and tolerance processes.

14.3.1 Salinity-Induced Adaptation in Metabolites

Salt stress is a major hurdle to increasing the growth and production of plants throughout the world and is a major concern to global food security (Jamil et al., 2006). Salinity appeared because of the use of excess inorganic salts as fertilizers and the deficient quality of irrigation water. Plants counter these stresses through various biochemical and physiological processes, like decreased stomatal conductance, carbon fixation and efficiency of light-harvesting mechanisms, suppression of cell growth, accelerated respiration, and enhancement of osmolytes and proteins involved in stress tolerance. Plant growth in saline soil is restricted due to a higher concentration of soluble salts and cytosolic osmotic pressure. Because of salt stress, the basic requirements (water and nutrients) of the plant are not fulfilled, which promotes physiological and metabolic disturbances in the plant. Plants growing in saline soils accumulate various compatible osmoprotectants such as proline, glycine, betaine, and sugar to manage the adverse effect of salt stress (Said-Al Ahl and Omer, 2011; Mahajan et al., 2020; Szabados and Savource, 2010, Chen and Murata, 2002). To prevent the adverse effects of salinity stress, plants have managed stress development strategies involving the activity of antioxidants, like ascorbic acid, glutathione, tocopherol, flavonoids, and carotenoids (Ahmed, 2009), and antioxidant enzymes, such as superoxide dismutase, catalase, guaiacol peroxidase, and ascorbate peroxidase (Arora et al., 2002). The production of these osmolytes, enzymes, and metabolites is either primary or secondary in nature. They are all responsible for protection and provide a defense mechanism for plant metabolism (Figure 14.5).

14.3.2 Drought-Induced Changes in Metabolome

Drought stress is one of the foremost abiotic stresses and causes substantial modification in plant growth and yield. The situation of drought arrives when the level of water in the soil decreases. It also occurs due to insufficient rainfall and water loss continuously through the process of transpiration and evaporation (Jaleel et al., 2007). Water stress is an important factor that obstructs the growth of plants at an initial stage. It has adverse effects on plant morphology. To endure drought, tolerant plants start protection approaches against water scarcity (Chaves and Oliveira, 2004). Plants demonstrate a diversity of physiological and biochemical reactions at the cellular and whole-organism level induced by drought stress.

Under drought conditions, the adjustment of photosynthesis and osmoregulation is one of the earliest plant strategies (Slama et al., 2015). Metabolomic studies identified the accumulation of osmolytes, such as some carbohydrates (e.g., glucose, sucrose, trehalose, and raffinose), polyols (e.g., mannitol and sorbitol), amino acids (e.g., proline, betaine, valine, leucine, and isoleucine),

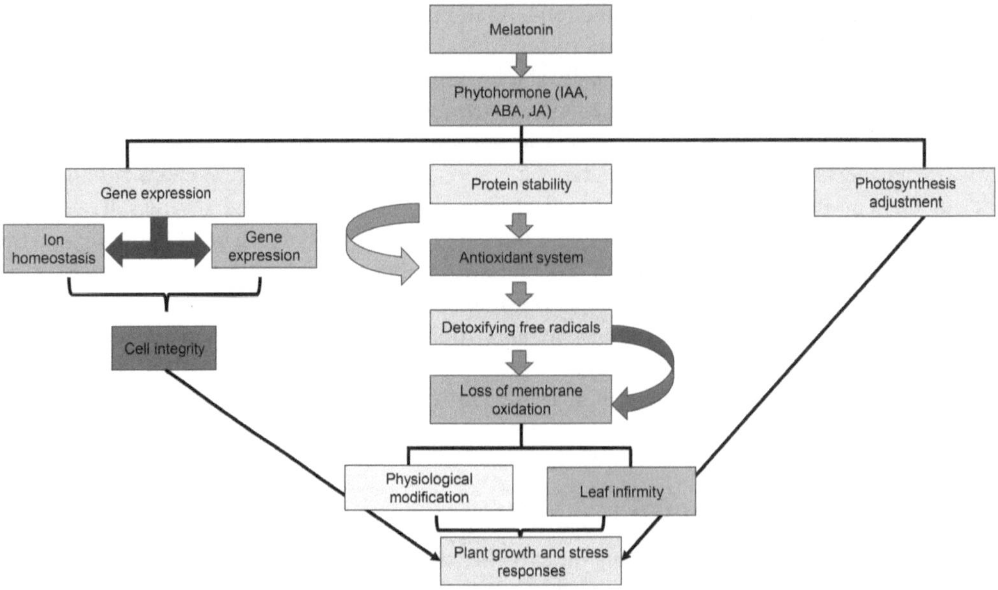

FIGURE 14.5 Mode of action of salt stress tolerance mechanism.

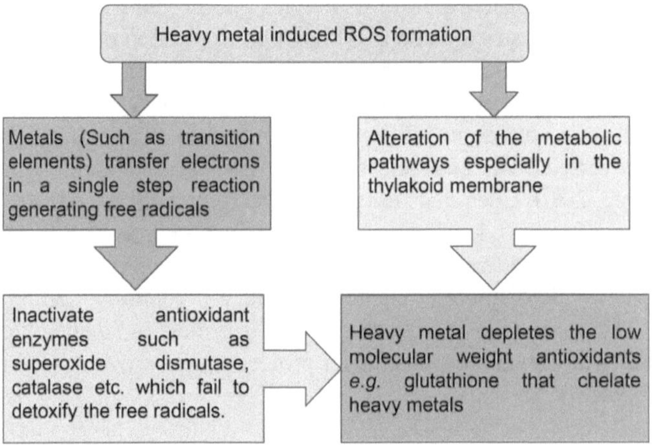

FIGURE 14.6 Mode of action of drought stress tolerance mechanism.

quaternary ammonium compounds (e.g., glycine betaine, *b*-alanine betaine, and proline betaine), and polyamines (e.g., putrescine, spermidine, and spermine), which have an important role in the decrease of the osmotic potential and maintenance of cell turgor pressure (as the cell takes up water) and contribute to the stabilization of membranes, enzymes, and proteins (Jorge and António, 2018; Sharma et al., 2019). Also, the accumulation of osmolytes helps to control ROS levels, provides energy to cope with stress, contributes to repair processes, and supports further growth (Silva et al., 2018; Fàbregas and Fernie, 2019).

From extensive series of investigations, it has been suggested that drought stress increases the amount of metabolite excretion in the plants, which reduces the adverse effect of drought stress in plant species (Figure 14.6).

14.3.3 Waterlogging Conditions Induce Modification in Various Metabolites

Constant flooding or waterlogging across the globe impacts soil quality and reduces crop yield. During the most commonly occurring temporary waterlogging, water-saturated soil not only reduces oxygen and availability of nutrients but also changes the microbial atmosphere, which results in the formation of toxic elements and finally reduces the crop yield. O_2 impairment is the most significant biochemical factor during waterlogging (Jackson and Colmer, 2005). When air spaces in the soil are acquired by water, root surroundings have hypoxia due to O_2 consumption by respiring roots. In waterlogged stress conditions, plants sometimes show organ-specific metabolic effects. An accumulation of starch in leaves of the waterlogged plant is generally noticed (Irfan et al., 2010). The waterlogged condition leads to an increase in alanine, gamma-aminobutyric acid (GABA), and polyols and induces changes in organic acids and a decrease in many amino acids, limiting the catabolism of the plant, which is responsible for plant growth. From extensive studies, it has been observed that waterlogging has important consequences for leaf metabolism, which changes the metabolic activity of plants by exerting feedback on sugar export and metabolism. In order to cope, plants develop certain strategies and produce protective enzymes, hormonal regulation, and upregulation of certain genes, which are all ways to protect plants up to a certain level of waterlogging, but in very waterlogged conditions, plants do not show avoidance mechanisms, and these conditions are lethal for plant survival. Taken as a whole, waterlogging conditions show a distinct metabolic response in roots and leaves. A study by Diab and Limami (2016) suggests that oxygen shortage (in root) and hypoxia result in a decline in unloading efficiency in roots, causing altered sugar translocation, and this exerts feedback on leaf metabolism so as to redirect sugar to other metabolic pathways. Enhancement of polyamine metabolism in root is a good example, and its repression in shoot could be associated with a change in polyamine circulation in both xylem and phloem sap. It reflects the strategy to export redox power when oxygen limitation caused by waterlogging impedes NADH reoxidation in roots. These strategies help plants to adapt to waterlogging stress conditions.

14.3.4 Modification of Metabolites in Respect to Temperature Extremes

The temperature stress of plants can be divided into the influence of temperature that causes a high temperature and chilling injuries (Hale and Orcutt, 1987). Temperature stress is one of the important physical factors that potentially affect the ontology and developmental rates of plants. Elevated (heat stress) and low temperatures (cold stress) cause modification in various physiological, biochemical, and molecular processes. Temperature stress produces leaf senescence, membrane damage, degradation of chlorophyll, and denaturation of proteins (Pradhan et al., 2017). Increased heat stress leads to the overproduction and accumulation of various organic and inorganic osmolytes (Kim et al., 2019). These osmolytes protect plants against stresses by cellular osmotic acclimatization, detoxification of ROS, shielding of biological membranes, and equilibration of enzymes and proteins (Bohnert and Jensen, 1996; Verbruggen and Hermans, 2008). One chief effect of heat stress is the production of ROS, causing oxidative damage to the cells and tissues. High-temperature stress induces the production of phenolic components like flavonoids and phenylpropanoids (Morrison and Stewart, 2002). Accumulation of soluble sugars under heat stress has been reported in sugarcane, which has considerable heat tolerance (Wahid and Close, 2007). Heat shock proteins (HSPs) are completely involved in the heat stress response. Responses involving stress proteins are a significant strategic adaptation mechanism to overcome environmental stresses. Chilling temperatures are another kind of extreme temperature that affects plant growth and development (Gupta and Deswal, 2014; Sehrawat et al., 2013). Along with the ultrastructural modifications, chilling also affects an array of physiological, biochemical, and molecular alterations, such as photo-inhibition of photosystem-I (Kudoh and Sonoike, 2002) and increased hydrogen peroxide (H_2O_2) accumulation in chilled leaves.

Under chilling stress, plants enhance the production of metabolites such as terpenes and phenolic components, which modulate cell organelles' function. In low-temperature conditions, these stabilize thylakoids of chloroplast and stimulate antioxidant production, which is a kind of tolerance mechanism of plants to cope with chilling stress. Effects of cold stress on polyamine accumulation have also been reported (Posmyk et al., 2009). Hummel et al (2004) reported that cold tolerance is associated with an increased level of polyamines, metabolites that act as a marker for chilling tolerance. From this evidence, it can be concluded that plants show quiescent adaptation strategies to protect themselves against chilling stress.

14.3.5 High Light Intensity Induces Metabolite Alteration

It is well known that light is a physical or environmental factor that can influence metabolite production. Modulation in light conditions induces many physiological activities in plants. Light determines the growth and development of plants because it influences metabolism. High light intensity enhances the production of proline, aspartate, and phenolic compounds in the plant as an avoidance mechanism of plants towards high light intensity. High light intensity promotes the accumulation of osmolytes and generates ROS mechanisms in the plant to acclimatize towards the light intensity. High light intensity induces the production of flavonoids, sinapate ester, iso-flavonoids, and psoralens (Hahlbrock, 1981; Lois, 1994). These compounds absorb light and protect against its harmful effects. Too much high light irradiance acts as an abiotic factor for plants. Wulff-zottele et al. (2010) conducted metabolite profiling of *Arabidopsis* leaves and suggested that most of the metabolites of glycolysis, the TCA cycle, and the oxidative pentose phosphate pathway were altered, which indicated that plants exposed to high light undergo a metabolic shift and enhance the Calvin Benson cycle to fix more carbon. In addition, the elevation of glycine indicated the activation of the photorespiration pathway. It shows the accumulation of the photorespiratory intermediates glycine and glycolate in the early phase (50–60 min after transition). Interestingly, the response during the mid-phase (80–360 min) shares the same properties with low-temperature treatment, which includes the accumulation of shikimate, phenylalanine, and fructose and the decrease of succinate. Further, metabolic changes clearly suggest that the responses to adapt to the enhanced intensity of light, which is part of the perception signaling relay, alert the plant cell that it needs to respond to the stress (Caldana et al., 2011).

14.3.6 Nutrient Deficiency Acts as Abiotic Stress and Induces Metabolite Production

Plant growth and development firmly rely on the ample amount of nutrients present in the soil. Nutrients play a crucial role in metabolic regulation, production, and development of new tissue and structural components of plants. Soil texture and structure as well as nutrients play an important role in the development and growth of plants (Adelusi and Ailema, 2006). Nutrient stress arises due to a combination of other stresses, such as water and salinity stress. These stresses disrupt the availability, portioning, and transport of nutrients. Nutrient inadequacy also takes place due to the competition of Na^+ and Cl^- with other nutrients such as Ca^+, K^+, and NO_3 with an increase in salinity stress (Said-Al Ahl and Omer, 2011). A plethora of nutrients and a lack of nutrients in the soil are both conditions that influence metabolite production. It has been reported that nitrogen deficiency enhances the accumulation and synthesis of phenolic compounds and flavonoids, and overproduction of secondary metabolites leads to osmolyte generation, which affects plant defense mechanisms. The deprivation of nutrients in soil degrades the cell membrane of plant cells but compensates with greater emission of isoprene and a family of different metabolites, which helps the plant to survive despite the nutrient deficiency, but acute deficiency of nutrients causes limitation of plant growth, which may be lethal in plants (Table 14.2).

TABLE 14.2
Limitation of Nutrients Affects the Metabolites in Stress Conditions

Nutrient Deficiency	Affected Metabolites	Modulation	References
Carbon	Carbohydrates, organic acids, and other C-containing metabolites like myo-inositol, raffinose, fatty acids, central amino acids (glutamine, glutamate, aspartate)	Decreased	Osuna et al. (2007); Usadel et al. (2009).
Nitrogen (useful in the synthesis of phospholipids and many secondary metabolites that have diverse roles in signaling, structure, and adaptation)	Amino acid and TCA intermediates including citrate, succinate, and malate	Decreased	Urbanczyk-Wochniak and Femie (2005); Stitt et al. (2003).
P (Phosphorus), an essential component of intermediates in central and energy metabolism	Starch, sucrose, and amino acids	Increased	Hernandez et al. (2007); Hernandez et al. (2009)
Sulfur (another macronutrient essential for S-containing amino acids	Organic sulfur-containing compounds (cysteine and glutathione) also decrease lipid content	Decreased	Hirai et al. (2004); Hirai et al. (2005).

14.3.7 Heavy Metals Induce Modification in Different Metabolites

Metals having a density higher than 5 gcm^{-3} are considered heavy metals. They are rare naturally occurring elements found in the soil, originating from weathering of parent rocks, pedogenesis, and anthropogenic activities. Heavy metals such as zinc, copper, molybdenum, manganese, and nickel are essential trace elements that play a significant role in various biological processes. These heavy metals cause soil quality to deteriorate and suppress plant growth and development. Heavy metal stress causes contamination in soil, which further induces several undesirable changes in the metabolic activity of plants. The uptake of heavy metals by plants can affect photosynthetic pigments, sugar, proteins, and non-protein production, leading to plant death during stress. The carbon fixed is predominantly allocated to secondary metabolite synthesis as the best option to enhance plant growth.

Heavy metal stress stimulates the synthesis of glutathione-metabolizing enzymes, polyamines, and phenolics. The increased phenolics possess antioxidant properties that chelate metals (Diaz et al., 2001). Phenolics, being antioxidants, readily scavenge free radicals such as hydrogen peroxide and superoxide ions by donating atoms and thus help plants to adapt to heavy metal stress (Figure 14.7).

14.4 NOVEL APPROACHES IN ABIOTIC STRESS TOLERANCE: MELATONIN AND SEROTONIN

14.4.1 Melatonin

Melatonin, obtained from tryptophan, was investigated in plants about 20 years ago, and nowadays there is expanding verification that melatonin is a major regulatory molecule involved in growth and development as well as in stress responses in the plant (Wang et al., 2018). Erland

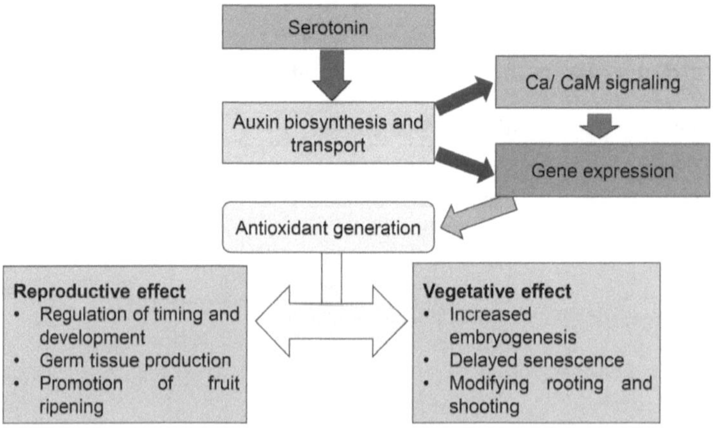

FIGURE 14.7 Heavy metal-induced oxidative stress by different mechanisms.

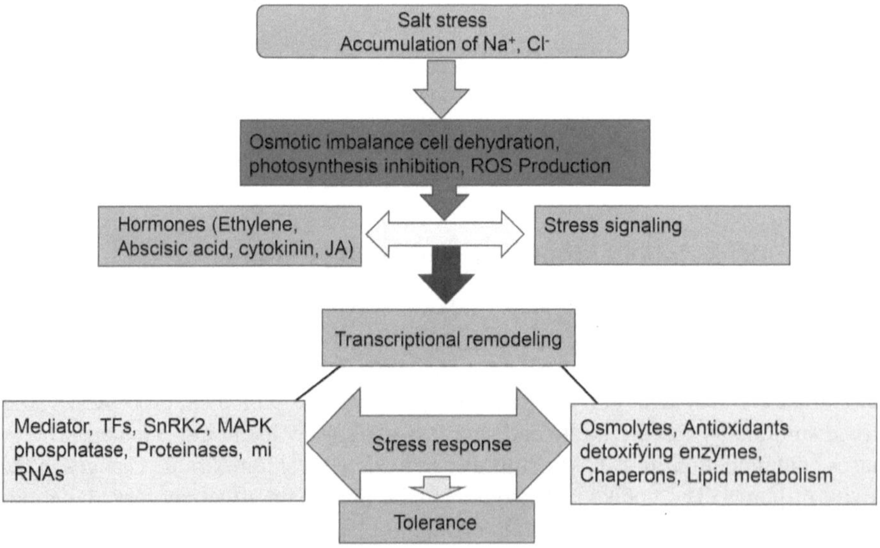

FIGURE 14.8 Role of melatonin in plant responses and stress tolerance.

et al. (2018) investigated systematically the metabolic background of melatonin and serotonin in plant morphogenesis. From the studies, it can be considered that melatonin functions as a potent antioxidant to detoxify free radicals and therefore improves oxidative stress tolerance in plants. Melatonin functions as a universal defensive signal to alleviate damage induced by different stressful conditions. Figure 14.5 shows the powerful roles of melatonin and its possible interactions with phytohormones to regulate gene expression, protein stabilization, and modification of genes, which protects the plant in extreme environmental stress conditions. Various downstream stress-responsive pathways are activated, enhancing the antioxidant system, decreasing free radicals, increasing osmoprotectants, correcting ionic disorders, alleviating membrane oxidation, and delaying leaf senescence (Wang et al., 2013; Shi et al., 2015). Stress-inhibited photosynthesis is partially recovered in the presence of melatonin. The plant thus displays better growth and increased stress tolerance after melatonin production, so it can be concluded from the studies that melatonin is a novel metabolite

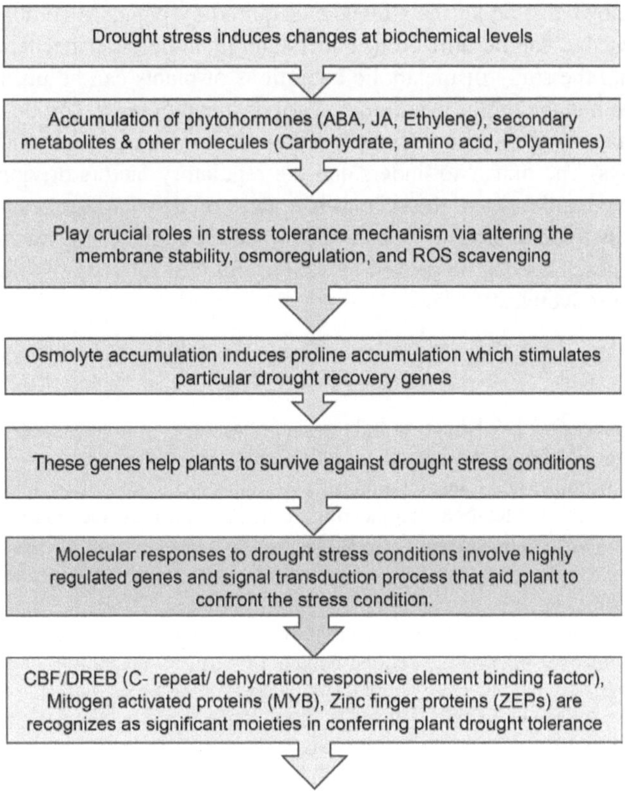

FIGURE 14.9 Summary of the mechanism of serotonin's effect on reproductive and vegetative growth and development.

and an important stress signaling molecule that functions as a general abiotic stress regulator (Wang et al., 2018).

14.4.2 Serotonin

Melatonin and serotonin have overlapping regulatory characteristics in plant abiotic stress tolerance (Kaur et al., 2015). Erland et al. (2016) reported that serotonin is an ancient and important regulatory molecule in plant developmental processes. Serotonin induces alteration in plant growth under abiotic stress (Figure 14.9). By analyzing metabolites of rice plants, Gupta and De (2017) reported elevated serotonin levels in tolerant rice varieties during salt stress, acting as a significant molecule providing tolerance to plants against adverse conditions. The exploration of regulatory roles of serotonin is an emerging area of research and requires a huge attempt to interpret the action and approaches of this new compound in plant abiotic stress avoidance.

14.5 CONCLUSION AND FUTURE PERSPECTIVE

Plants are highly adaptative to the varied environmental conditions existing on the planet. They have developed their own specific defensive traits for combating environmental factors. Since the beginning of life on this planet, life has been dependent on plant survival, so it is necessary to understand the crucial aspects of environmental changes with reference to plant adaptation. The phytoconstituents are a key feature for maintaining plant architecture and productivity. This chapter provides abundant knowledge about the strategies acquired by plants to combat stress conditions. This insightful knowledge can be utilized to understand plant defense mechanisms against environmental stresses, and the study of metabolic alterations of plants can be utilized in metabolome analysis via advanced metabolomics techniques. Various metabolites are produced in abiotic stress conditions, and the identification of metabolites acts as a biomarker, which can be helpful to plant biologists in two ways. The first is to understand the regulatory factors driving plant metabolism under diverse abiotic conditions, and the second is to develop new approaches by using this knowledge to improve stress-tolerant crops by metabolic or genetic engineering via emerging techniques related to metabolomics. This can be proved to be a useful tool for society to create resistant varieties with enhanced productivity.

REFERENCES

Adelusi, A.A., and Aileme, J.D. 2006. Effects of light and nutrient stress on some growth parameters of cowpea (*Vigna unguiculata* (L.) Walp). *Research Journal Botany* 1: 95–103.

Agostinelli, E. 2016. Polyamines and transglutaminases: future perspectives. *Amino Acids* 48: 2273–2281.

Aharoni, A., and Galili, G. 2011. Metabolic engineering of the plant primary-secondary metabolism interface. *Current Opinion in Biotechnology* 22(2): 239–244.

Ahmed, S. 2009. Effect of soil salinity on the yield and yield components of mungbean. *Pakistan Journal of Botany* 41: 263–268.

Annunziata, M.G., Ciarmiello, L.F., Woodrow, P., Maximova, E., Fuggi, A., and Carillo, P. 2017. Durum wheat roots adapt to salinity remodeling the cellular content of nitrogen metabolites and sucrose. *Frontiers in Plant Science* 7: 2035.

Araus, J.L., Slafer, G.A., Reynolds, M.P., and Royo, C. 2002. Plant breeding and drought in C3 cereals: what should we breed for? *Annals of Botany* 89: 925–940.

Arora, A., Sairam, R.K., and Srivastava, G.C. 2002. Oxidative stress and antioxidative system in plants. *Current Science* 82(10): 1227–1238.

Ben Rejeb, K., Abdelly, C., and Savoure, A. 2014. How reactive oxygen species and proline face stress together. Plant Physiology and Biochemistry 80: 278–284.

Bohnert, H.J., and Jensen, R.G. 1996. Strategies for engineering water-stress tolerance in plants. *Trends in Biotechnology* 14: 89–97.

Boyer, J.S. 1982. Plant productivity and environment. *Science* 218: 443–448.

Bray, E.A., Bailey-Serres, J., and Weretilnyk, E. 2000. Response to abiotic stress. In Gruissem, W. and Jones, R. (Eds), *Biochemistry and Molecular Biology of Plants.* American Society of Plant Physiologists, Rockville, USA, 1158–1203.

Caldana, C., Degenkolbe, T., Cuadros-Inostroza, A., Klie, S., Sulpice, R., Leisse, A., Steinhauser, D., Fernie, A.R., Willmitzer, L., and Hannah, M.A. 2011. High-density kinetic analysis of the metabolomic and transcriptomic response of Arabidopsis to eight environmental conditions. *The Plant Journal* 67: 869–884.

Castiglioni, P., Bell, E., Lund, A., et al. 2018. Identification of GB1, a gene whose constitutive overexpression increases glycinebetaine content in maize and soybean. Plant Direct 2: e00040.

Chalker-Scott, L., and Fuchigami, L.H. 1989. The role of phenolic compounds in plant stress responses. In Paul H. Li (Ed.), *Low Temperature Stress Physiology in Crops*. CRC Press Inc., Boca Raton, Florida, 67–79.

Chaves, M.M., and Oliveira, M.M. 2004. Mechanisms underlying plant resilience to water deficits: prospects for water-saving agriculture. *Journal of Experimental Botany* 55: 2365–2384.

Chen, T.H., and Murata, N. 2002. Enhancement of tolerance of abiotic stress by metabolic engineering of betaines and other compatible solutes. *Current Opinion in Plant Biology* 5: 250–257.

Choudhary, S.P., Kanwar, M., Bhardwaj, R., Yu, J.Q., and Tran, L.S.P. 2012. Chromium stress mitigation by polyamine-brassinosteroid application involves phytohormonal and physiological strategies in *Raphanus sativus* L. *PLoS ONE* 7(3): e33210.

Choudhary, S.P., Oral, H.V., Bhardwaj, R., Yu, J.Q., and Tran, L.S.P. 2012. Interaction of brassinosteroids and polyamines enhances copper stress tolerance in *Raphanus sativus*. *Journal of Experimental Botany* 63: 5659–5675.

Crueger, W., and Crueger, A. 1990. *Biotechnology: A Textbook of Industrial Microbiology*. In Brook, T.D. (Ed.), *Technology & Engineering*. Sinauer Associates, Sunderland, MA.

de Freitas, P.A.F., de Souza Miranda, R., Marques, E.C., et al. 2018. Salt tolerance induced by exogenous proline in maize is related to low oxidative damage and favourable ionic homeostasis. *Journal of Plant Growth Regulation* 37: 911–924.

Degenhardt, D.C., and Lincoln, D.E. 2006. Volatile emissions from an odorous plant in response to herbivory and methyl jasmonate exposure. *Journal of Chemical Ecology* 32: 725–743.

Diab, H. and Limami, A.M. 2016. Reconfiguration of N metabolism upon hypoxia stress and recovery: roles of alanine aminotransferase (AlaAT) and glutamate dehydrogenase (GDH). *Plants* 5(2): 25.

Díaz, J., Bernal, A., Pomar, F., and Merino, F. 2001. Induction of shikimate dehydrogenase and peroxidase in pepper (*Capsicum annuum* L.) seedlings in response to copper stress and its relation to lignification. *Plant Science* 161: 179–188.

Dixon, R.A. and Paiva, N. 1995. Stressed induced phenyl propanoid metabolism. *Plant Cell* 7: 1085–1097.

Erland, L.A., Shukla, M.R., Singh, A.S., Murch, S.J., and Saxena, P.K. 2018. Melatonin and serotonin: mediators in the symphony of plant morphogenesis. *Journal of Pineal Research* 64: e12452.

Erland, L.A., Turi, C.E., and Saxena, P.K. 2016. Serotonin: an ancient molecule and an important regulator of plant processes. *Biotechnology Advances* 34: 1347–1361.

Eveland, A.L., and Jackson, D.P. 2012. Sugars, signalling, and plant development. *Journal of Experimental Botany* 63: 3367–3377.

Fàbregas, N., and Fernie, A.R. 2019. The metabolic response to drought. *Journal of Experimental Botany* 70: 1077–1085.

Fichman, Y., Gerdes, S.Y., Kovács, H., Szabados, L., Zilberstein, A., and Csonka, L.N. 2015. Evolution of proline biosynthesis: enzymology, bioinformatics, genetics, and transcriptional regulation. *Biological Reviews* 90: 1065–1099.

Flink, I., and Pettijohn, D.E. 1975. Polyamines stabilise DNA folds. *Nature* 253: 62–63.

Govind, S.R., Jogaiah, S., Abdelrahman, M., Shetty, H.S., and Tran, L.S.P. 2016. Exogenous trehalose treatment enhances the activities of defense-related enzymes and triggers resistance against downy mildew disease of pearl millet. *Frontiers in Plant Science* 7: 1593.

Griffiths, C.A., Paul, M.J., and Foyer, C.H. 2016. Metabolite transport and associated sugar signalling systems underpinning source/sink interactions. *Biochimica et Biophysica Acta (BBA)-Bioenergetics* 1857: 1715–1725.

Gumul, D., Korus, J., and Achremowicz, B. 2007. The influence of extrusion on the content of polyphenols and antioxidant/antiradical activity of rye grains (*Secale cereale* L.). *Acta Scientiarum Polonorum Technologia Alimentaria* 6: 103–111.

Gupta, P., and De, B. 2017. Metabolomics analysis of rice responses to salinity stress revealed elevation of serotonin, and gentisic acid levels in leaves of tolerant varieties. *Plant Signaling & Behavior* 12: e1335845.

Gupta, R., and Deswal, R. 2014. Antifreeze proteins enable plants to survive in freezing conditions. *Journal of Biosciences* 39: 931–944.

Hahlbrock, K. 1981. Flavonoids. In Stumpf, P.K. and Conn, E.E. (Eds), *Biochemistry of Plants*, Vol. 7, Academic Press, New York, pp. 425–456.

Hale, M.G., and Orcutt, D.M. 1987. *The Physiology of Plants under Stress*. John Wiley & Sons, New York.

Häusler, R.E., Heinrichs, L., Schmitz, J., and Flügge, U.I. 2014. How sugars might coordinate chloroplast and nuclear gene expression during acclimation to high light intensities. *Molecular Plant* 7: 1121–1137.

Hedrich, R., Sauer, N., and Neuhaus, H.E. 2015. Sugar transport across the plant vacuolar membrane: nature and regulation of carrier proteins. *Current Opinion in Plant Biology* 25: 63–70.

Hernández, G., Ramírez, M., Valdés-López, O., et al. 2007. Phosphorus stress in common bean: root transcript and metabolic responses. *Plant Physiology* 144: 752–767.

Hernández, G., Valdés-López., O, Ramírez, M., et al. 2009. Global changes in the transcript and metabolic profiles during symbiotic nitrogen fixation in phosphorus-stressed common bean plants. *Plant Physiology* 151: 1221–1238.

Hirai, M.Y., Klein, M., Fujikawa, Y., et al. 2005. Elucidation of gene-to-gene and metabolite-to-gene networks in *Arabidopsis* by integration of metabolomics and transcriptomics. *Journal of Biological Chemistry* 280: 25590–25595.

Hirai, M.Y., Yano, M., Goodenowe, D.B., et al. 2004. Integration of transcriptomics and metabolomics for understanding of global responses to nutritional stresses in *Arabidopsis thaliana*. *Proceedings of the National Academy of Sciences* 101: 10205–10210.

Huang, X.F., Vincken, J.P., Visser, R.G., and Trindade, L.M. 2018. Production of heterologous storage polysaccharides in potato plants. *Annual Plant Reviews* 41: 389–408.

Hummel, I., Gouesbet, G., El Amrani, A., Aïnouche, A., and Couée, I. 2004. Characterization of the two arginine decarboxylase (polyamine biosynthesis) paralogues of the endemic subantarctic cruciferous species *Pringlea antiscorbutica* and analysis of their differential expression during development and response to environmental stress. *Gene* 342: 199–209.

Igarashi, K., and Kashiwagi, K. 2000. Polyamines: mysterious modulators of cellular functions. *Biochemical and Biophysical Research Communications* 271: 559–564.

Irfan, M., Hayat, S., Hayat, Q., Afroz, S., and Ahmad, A. 2010. Physiological and biochemical changes in plants under waterlogging. *Protoplasma* 241: 3–17.

Jackson, M.B., and Colmer, T.D. 2005. Response and adaptation by plants to flooding stress. *Annals of Botany* 96: 501–505.

Jaleel, C.A., Manivannan, P., Sankar, B., Kishorekumar, A., Gopi, R., Somasundaram, R., and Panneerselvam, R. 2007. Induction of drought stress tolerance by ketoconazole in *Catharanthus roseus* is mediated by enhanced antioxidant potentials and secondary metabolite accumulation. *Colloids and Surfaces B, Biointerfaces* 60: 201–206.

Jamil, M., Deog Bae, L., Kwang Yong, J., et al. 2006. Effect of salt (NaCl) stress on germination and early seedling growth of four vegetables species. *Journal of Central European Agriculture* 7(2): 273–282.

Ji, T., Li, S., Li, L., et al. 2018. Cucumber Phospholipase D alpha gene overexpression in tobacco enhanced drought stress tolerance by regulating stomatal closure and lipid peroxidation. *BMC Plant Biology* 18: 355.

Jorge, T.F., and António, C. 2018. Plant metabolomics in a changing world: metabolite responses to abiotic stress combinations. In V. Andjelkovic (Ed.), *Plant, Abiotic Stress and Responses to Climate Change*, IntechOpen, London, UK, 111–132.

Le Hir, R., Spinner, L., Klemens, P.A.W., et al. 2015. Disruption of the sugar transporters AtSWEET11 and AtSWEET12 affects vascular development and freezing tolerance in Arabidopsis. *Molecular Plant* 8: 1687–1690.

Li, J., Guo, X., Zhang, M., et al. 2018. OsERF71 confers drought tolerance via modulating ABA signaling and proline biosynthesis. *Plant Science* 270: 131–139.

Kaur, H., Mukherjee, S., Baluska, F., and Bhatla, S.C. 2015. Regulatory roles of serotonin and melatonin in abiotic stress tolerance in plants. *Plant Signaling & Behavior* 10: e1049788.

Kim, S.W., Gupta, R., Min, C.W., et al. 2019. Label-free quantitative proteomic analysis of *Panax ginseng* leaves upon exposure to heat stress. *Journal of Ginseng Research* 43: 143–153.

Kudoh, H. and Sonoike, K. 2002. Irreversible damage to photosystem I by chilling in the light: cause of the degradation of chlorophyll after returning to normal growth temperature. Planta 215: 541–548.

Kusano, T., Berberich, T., Tateda, C., and Takahashi, Y. 2008. Polyamines: essential factors for growth and survival. *Planta* 228: 367–381.

Lefevre, I., Gratia, E., and Lutts, S. 2001. Discrimination between the ionic and osmotic components of salt stress in relation to free polyamine level in rice (*Oryza sativa*). *Plant Science* 161: 943–952.

Lois, R. 1994. Accumulation of UV-absorbing flavonoids induced by UV-B radiation in *Arabidopsis thaliana* L. *Planta* 194: 498–503.

Mahajan, M., Kuiry, R., and Pal, P.K. 2020. Understanding the consequence of environmental stress for accumulation of secondary metabolites in medicinal and aromatic plants. *Journal of Applied Research on Medicinal and Aromatic Plants* 18: 100255.

Mahajan, M., Sharma, S., Kumar, P., and Pal, P.K. 2020. Foliar application of KNO_3 modulates the biomass yield, nutrient uptake and accumulation of secondary metabolites of *Stevia rebaudiana* under saline conditions. *Industrial Crops and Products* 145: 112102.

Miller, G., Honig, A., Stein, H., Suzuki, N., Mittler, R., and Zilberstein, A. 2009. Unraveling Δ1-pyrroline-5-carboxylate-proline cycle in plants by uncoupled expression of proline oxidation enzymes. *Journal of Biological Chemistry* 284: 26482–26492.

Moles, T.M., Mariotti, L., De Pedro, L.F., Guglielminetti, L., Picciarelli, P., and Scartazza, A. 2018. Drought induced changes of leaf-to-root relationships in two tomato genotypes. *Plant Physiology and Biochemistry* 128: 24–31.

Morgan, D.M. 1999. Polyamines. *Molecular Biotechnology* 11: 229–250.

Morrison, M.J., and Stewart, D.W. 2002. Heat stress during flowering in summer Brassica. *Crop Science* 42: 797–803.

Oshima, T. 2007. Unique polyamines produced by an extreme thermophile, *Thermus thermophilus*. *Amino Acids* 33: 367–372.

Osuna, D., Usadel, B., Morcuende, R., et al. 2007. Temporal responses of transcripts, enzyme activities and metabolites after adding sucrose to carbon-deprived Arabidopsis seedlings. *The Plant Journal* 49: 463–491.

Pommerrenig, B., Ludewig, F., Cvetkovic, J., Trentmann, O., Klemens, P.A., and Neuhaus, H.E. 2018. In concert: orchestrated changes in carbohydrate homeostasis are critical for plant abiotic stress tolerance. *Plant and Cell Physiology* 59: 1290–1299.

Posmyk, M.M., Bałabusta, M., Wieczorek, M., Sliwinska, E., and Janas, K.M. 2009. Melatonin applied to cucumber (*Cucumis sativus* L.) seeds improves germination during chilling stress. *Journal of Pineal Research* 46: 214–223.

Pradhan, J., Sahoo, S.K., Lalotra, S., and Sarma, R.S. 2017. Positive impact of abiotic stress on medicinal and aromatic plants. *International Journal of Plant Sciences (Muzaffarnagar)* 12 (2): 309–313.

Rejeb, K.B., Abdelly, C., and Savouré, A. 2014. How reactive oxygen species and proline face stress together. *Plant Physiology and Biochemistry* 80: 278–284.

Richards, F.J., and Coleman, R.G. 1952. Occurrence of putrescine in potassium-deficient barley. *Nature* 170: 460–460.

Said-Al Ahl, H.A.H., and Omer, E.A. 2011. Medicinal and aromatic plants production under salt stress. A review. *Herba Polonica* 57: 72–87.

Schmitz, J., Heinrichs, L., Scossa, F., et al. 2014. The essential role of sugar metabolism in the acclimation response of *Arabidopsis thaliana* to high light intensities. *Journal of Experimental Botany* 65: 1619–1636.

Sehrawat, A., Gupta, R., and Deswal, R. 2013. Nitric oxide-cold stress signalling cross-talk, evolution of a novel regulatory mechanism. Proteomics 13: 1816–1835.

Seiler, N., and Raul, F. 2005. Polyamines and apoptosis. *Journal of Cellular and Molecular Medicine* 9: 623–642.

Semida, W.M., Hemida, K.A., and Rady, M.M. 2018. Sequenced ascorbate–proline–glutathione seed treatment elevates cadmium tolerance in cucumber transplants. *Ecotoxicology and Environmental Safety* 154: 171–179.

Sengupta, S., Mukherjee, S., Basak, P., and Majumder, A.L. 2015. Significance of galactinol and raffinose family oligosaccharide synthesis in plants. *Frontiers in Plant Science* 6: 656.

Shanazari, M., Golkar, P., and Maibody, A.M.M. 2018. Effects of drought stress on some agronomic and biophysiological traits of *Trititicum aestivum*, *Triticale*, and *Tritipyrum* genotypes. *Archives of Agronomy and Soil Science* 64: 2005–2018.

Sharma, A., Shahzad, B., Rehman, A., Bhardwaj, R., Landi, M., and Zheng, B. 2019. Response of phenylpropanoid pathway and the role of polyphenols in plants under abiotic stress. *Molecules* 24: 2452.

Shi, H., Jiang, C., Ye, T., et al. 2015. Comparative physiological, metabolomic, and transcriptomic analyses reveal mechanisms of improved abiotic stress resistance in bermudagrass [*Cynodondactylon* (L). Pers.] by exogenous melatonin. *Journal of Experimental Botany* 66: 681–694.

Shinde, S., Villamor, J.G., Lin, W., Sharma, S., and Verslues, P.E., 2016. Proline coordination with fatty acid synthesis and redox metabolism of chloroplast and mitochondria. *Plant Physiology* 172: 1074–1088.

Silva, S., Santos, C., Serodio, J., et al. 2018. Physiological performance of drought-stressed olive plants when exposed to a combined heat-UV-B shock and after stress relief. *Functional Plant Biology* 45: 1233–1240.

Singh, A., Sharma, M.K., and Sengar, R.S. 2017. Osmolytes: proline metabolism in plants as sensors of abiotic stress. *Journal of Applied and Natural Science* 9: 2079–2092.

Slama, I., Abdelly, C., Bouchereau, A., Flowers, T., and Savouré, A. 2015. Diversity, distribution and roles of osmoprotective compounds accumulated in halophytes under abiotic stress. *Annals of Botany* 115: 433–447.

Stitt, M., and Fernie, A.R. 2003. From measurements of metabolites to metabolomics: an "on the fly" perspective illustrated by recent studies of carbon–nitrogen interactions. *Current Opinion in Biotechnology* 14: 136–144.

Szabados, L., and Savouré, A. 2010. Proline: a multifunctional amino acid. *Trends in Plant Science* 15: 89–97.

Tabor, C.W., and Tabor, H. 1984. Polyamines. *Annual Review of Biochemistry* 53: 749–790.

Takahashi, T., and Kakehi, J.I. 2010. Polyamines: ubiquitous polycations with unique roles in growth and stress responses. *Annals of Botany* 105: 1–6.

Thomas, T., and Thomas, T.J. 2001. Polyamines in cell growth and cell death: molecular mechanisms and therapeutic applications. *Cellular and Molecular Life Sciences CMLS* 58: 244–258.

Tuteja, N. 2007. Mechanisms of high salinity tolerance in plants. *Methods in Enzymology* 428: 419–438.

Urbanczyk-Wochniak, E., and Fernie, A.R. 2005. Metabolic profiling reveals altered nitrogen nutrient regimes have diverse effects on the metabolism of hydroponically-grown tomato (*Solanum lycopersicum*) plants. *Journal of Experimental Botany* 56: 309–321.

Usadel, B., Poree, F., Nagel, A., et al. 2009. A guide to using MapMan to visualize and compare Omics data in plants: a case study in the crop species, Maize. *Plant Cell & Environment* 32: 1211–1229.

Veenstra, L.D., Jannink, J.L., and Sorrells, M.E. 2017. Wheat fructans: a potential breeding target for nutritionally improved, climate-resilient varieties. *Crop Science* 57: 1624–1640.

Verbruggen, N., and Hermans, C. 2008. Proline accumulation in plants: a review. *Amino Acids* 35: 753–759.

Wahid, A., and Close, T.J. 2007. Expression of dehydrins under heat stress and their relationship with water relations of sugarcane leaves. *Biologia Plantarum* 51: 104–109.

Wang, C.L., Zhang, S.C., Qi, S.D., Zheng, C.C., and Wu, C.A. 2016. Delayed germination of *Arabidopsis* seeds under chilling stress by overexpressing an abiotic stress inducible GhTPS11. *Gene* 575: 206–212.

Wang, P., Sun, X., Chang, C., Feng, F., Liang, D., Cheng, L., and Ma, F. 2013. Delay in leaf senescence of *Malus hupehensis* by long-term melatonin application is associated with its regulation of metabolic status and protein degradation. *Journal of Pineal Research* 55: 424–434.

Wang, Y., Reiter, R.J., and Chan, Z. 2018. Phytomelatonin: a universal abiotic stress regulator. *Journal of Experimental Botany* 69: 963–974.

Wanner, L.A., and Junttila, O. 1999. Cold-induced freezing tolerance in *Arabidopsis*. *Plant Physiology* 120: 391–400.

Wulff-zottele, C., Gatzke, N., Kopka, J., Orellana, A., Hoefgen, R., Fisahn, J., and Hesse, H. 2010. Photosynthesis and metabolism interact during acclimation of *Arabidopsis thaliana* to high irradiance and sulphur depletion. *Plant, Cell & Environment* 33: 1974–1988.

Zhang, Y., Swart, C., Alseekh, S., Scossa, F., Jiang, L., Obata, T., Graf, A., and Fernie, A.R. 2018. The extra-pathway interactome of the TCA cycle: expected and unexpected metabolic interactions. *Plant Physiology* 177: 966–979.

15 Genome Editing for Developing Abiotic Stress-Resilient Plants

Debajit Das[1], Sanjeev Kumar[2], Saradia Kar[3], Channakeshavaiah Chikkaputtaiah[4], and Dhanawantari L. Singha[4]**

[1]Department of Agricultural Biotechnology, Assam Agricultural University, Jorhat, Assam, India
[2]Department of Biosciences & Bioengineering, Indian Institute of Technology Guwahati, India
[3]Plant Molecular Biotechnology and Functional Genomics Laboratory, Assam University, Silchar, India
[4]Biological Sciences and Technology Division, CSIR-North East Institute of Science and Technology, Jorhat, Assam, India
*Corresponding authors: channakeshav@neist.res.in; dhanawantarisingha@gmail.com

CONTENTS

DOI: 10.1201/9781003159636-15

15.1 INTRODUCTION

Being sessile, crop plants are frequently exposed to several abiotic stresses leading to severe yield reduction of more than 50% on average, seriously disrupting world agricultural productivity. In this respect, advances in plant breeding strategies are being implicated in developing abiotic stress-tolerant crop species; however, these interventions are insufficient to provide a significant improvement of the trait and guarantee global future food sustainability, considering the complex inheritance of abiotic stress tolerance. Along with conventional plant breeding approaches, strategies such as genetic manipulation through modern plant biotechnological interventions are proved to be among the solutions. The foundation of plant biotechnology research is to bring about genetic modifications technically for crop improvement, yield, and other purposeful variations. To address some of the critical drawbacks of genetic modification, like that of the incorporation of foreign DNA, the focus nowadays is turning to genome editing technology (GET) due to its versatility, greater efficiency, and accurate targeting for developing abiotic stress-tolerant crops. In the last decade, the alteration in the genome by inserting or deleting specific sequences has revolutionized plant science. Reproducible genome editing (GE) techniques have been invented to make changes in the genomic sequences by which sequence-specific modifications can be carried out in broad-spectrum cell types (Gaj et al. 2016). Such methods lead to the functional analysis of genes and help in discovering novel traits in plants as well as animals. The concept of GET was adopted from directed double-stranded breaks (DSBs) in sister chromatids during meiosis, which later undergo recombination (Rudin et al. 1989). Model experiments in yeasts showed that specific nucleases led to homologous repair by carrying out double-stranded (ds) cuts of unwanted mutated regions (Carroll 2017). Site-specific endonucleases (SSEs) play the main role in carrying out GET, as these enzymes make ds cuts with many fewer non-specific cleavages to facilitate the desired modification. To date, four SSE cuts have been reported: clustered regularly interspaced short palindromic repeats (CRISPR)/CRISPR associated protein 9 (Cas9), zinc finger nucleases (ZFNs), transcription activator-like effector nucleases (TALENs), and meganucleases. These genome modifying enzymes act as the foundation for transcription factors (TFs), which are the key players of any gene expression. The ds cuts on the DNA sequences can be repaired by homologous recombination (HR) and non-homologous end-joining (NHEJ) (Mushtaq et al. 2018). Gene-edited crops are different from transgenic crops because the latter have a foreign DNA sequence inserted. However, a gene-edited crop is a defined mutant with its own edited part, which makes the crops more acceptable for crop breeding than the controversial transgenics (Jaganathan et al. 2018).

The principal aim of the present chapter is to focus on the different GETs that have been utilized to bring about a ground-breaking change in genetic engineering. In addition to the fundamental principles ingrained in the functioning of each technique, emphasis is given to the implementation of these strategies in plant biology over the last five years, demonstrating their significant role in plant breeding programs. Largely based on the targeted nucleases used for GE, ZFNs, TALENs, and CRISPR/Cas9 are indeed the three key strategies.

15.2 APPLICATION OF GENOME EDITING TECHNIQUES IN PLANTS

Customized GE strategies provide scientists with the potential to introduce beneficial traits succinctly and efficiently instead of relying on traditional breeding. The available GE techniques are ZFNs, TALENs, Sequence-Specific Nucleases (SSNs), meganucleases, and CRISPR/Cas. ZFNs, TALENs, SSNs, and meganucleases are considered first-generation GET, whereas CRISPR/Cas is considered the most advanced second-generation GET. The second-generation GET is widely used for the development of abiotic stress-resilient plants due to the simplicity, low cost, and ease of GE (Surabhi et al. 2019). All the techniques depend on the cellular DNA repair mechanism. Both the first- and the second-generation GE techniques modify the target sequence specifically by nucleases (SSN). The resulting DSB in the genome are repaired by the plant through insertion and deletion (INDELS) by using NHEJ and homology-directed recombination (HDR) pathways (Jaganathan et al. 2018).

15.3 FIRST-GENERATION GETS

15.3.1 Meganucleases

Meganucleases are a cluster of naturally occurring rare-cutting endonucleases. They are considered as naturally occurring endonucleases or homing endonucleases with high specificity and low toxicity. The existence of meganucleases was first reported during the 1990s. Since then, they have been used as tools for inducing mutations through HR. They create ds DNA breaks through site-specific DNA cleavage. They have been used for gene targeting in plant cells (Puchta et al. 1993) as well as insects (Rong and Golic, 2000), and mammalian cells (Donoho et al. 1998). In nature, meganucleases are present in phages, bacteria, archaebacteria, and various eukaryotes. Among the five families of meganucleases, the LAGLIDADG protein family is well studied and generally encodes within introns (Silva et al. 2011). Their target recognition site is around 12–45 nt (Thierry et al. 1992). This large recognition site is the major limitation to meganuclease-induced gene targeting, despite their high specificity (Epinat et al. 2003). For its efficient application, first, these recognition sites need to be introduced. Apart from these, modification in the naturally existing meganucleases is mandatory to increase its efficiency further. One such modification is the generation of hybrid endonucleases derived from existing endonucleases through domain swapping (Epinat et al. 2003). The availability of such variants increases the possibility of using meganucleases in GE.

15.3.2 Zinc Finger Nucleases

Zinc finger nuclease is the first synthetic GET to make DSB. It is made up of two monomers of 6 bp flanking the specific spacer sequence (Figure 1a). The dimer is responsible for making DSB in between the spacer sequence. The cut leads to either insertion of new sequences or deletion of the particular site, creating a frameshift mutation in the region. The joining of the broken ends is stimulated by HR. If there are no homologous sequences present, the cell shifts to NHEJ (Mushtaq et al. 2018). The ZFN is mainly derived from *Flavobacterium okeanokoites* as a type II restriction enzyme called FokI (Li et al. 1992). The cleaving site of the enzyme is made up of Cys-His zinc fingers (ZFs), which are constituted of 30 amino acids. This cleavage site attaches to 3 bp DNA (Carroll 2011). Two sets of ZFs are responsible for making the cleavage with monomeric domains. The first de novo ZFN-based mutagenesis was carried out in Drosophila and targeted a gene replacement in DNA from the germline cells (Bibikova et al. 2002). The GET carried out by ZFN requires a TF for specific cell types and a carrier for genomic integration (Carroll 2011). Successful efficiencies in inserting both homologous and non-homologous inserts have also been reported by injecting ZFN mRNAs and donor DNA into embryos. Most of the GE work using ZFNs had started in both plants and animals in early 2000. An important concern regarding ZFNs is their non-specific mutations in DNA sequences, which hampers the efficiency of their editing techniques. Steps like the development of artificial obligate heterodimer ZFN architecture have been adopted based on charge-to-charge repulsion to reduce the homodimerization of ZFN domains at the wrong sites (Doyon et al. 2011).

15.3.3 Transcription Activator-Like Effector Nuclease

Transcription activator-like effector (TALE) proteins are bacterial effectors, and TALENs, which are TALE proteins fused with nuclease, are TFs translocated from bacteria like *Xanthomonas* sp. to the plant cells (Boch and Bonus 2010). The DNA binding domain fuses with the carboxy-terminal FokI cleavage domain (Gaj et al. 2016), dimerizing the TALEN proteins, which cut two ends of a 12–19-bp spacer sequence (Miller et al. 2011). The TALE binding domains have repeating sequences with 34 residues, which was a breakthrough finding to create an SSN binding site (Li et al. 2011). Such engineered TALE proteins fused with nuclease, better known as TALENs, can modify nearly any type of gene (Figure 15.1b). Therefore, unlike ZFNs, TALENs have an increased affinity to target DNA. The latter have many fewer base mismatches and target overlaps with the adjacent sequences. The TALENs have been reported to exhibit low nuclease-associated cytotoxicity due to large DNA target sites and a thymine-rich required 5′ target site end. Both the characteristics contribute to their specificity (Wright et al. 2014). TALEN-mediated targeted mutagenesis of endogenous genes has been carried out in most model plants and crops, such as *Arabidopsis*, rice, tobacco, wheat, barley, etc. (Forner et al. 2015, Zhang et al. 2013, Wang et al. 2014). Christian et al. 2013 reported a poor transmission rate in the first generation of *Arabidopsis*, while Li et al. (2012) asserted over 50% transfer of TALEN-induced mutations to the T1 rice generation.

15.4 SECOND-GENERATION GETS

15.4.1 CRISPR/Cas9: A Recently Developed Genome Editing Technology

CRISPR is a recent novel GE technique that is successfully used in improved crop production. The CRISPR system plays a major role in providing adaptive immunity to bacteria. The palindromic repeats with the help of guided RNA and Cas (CRISPR associated) proteins were first discovered in *Streptococcus pyogenes*, which could cleave an invading viral genome. The CRISPR-containing loci are typically flanked by Cas protein sequences. The sequences are reported to be homologous to foreign pathogenic elements of the virus (Barrangou and Marraffini 2014). The short segment of foreign DNA is incorporated within the CRISPR sequence, followed by transcription, generating a CRISPR RNA (crRNA). The crRNA anneal with the trans-activating crRNA (tracrRNA) forms a single guide RNA (SgRNA). The latter is utilized to target the pathogenic dsDNA and cleave it by associating itself with Cas9 protein (Gaj et al. 2016). The Cas9 protein requires two elements for its attachment: a seed sequence of 17–20 bp within crRNA and a proto-spacer adjacent motif (PAM) (Jain et al. 2019). Each two-nuclease domain of Cas9 cleaves one of the two strands of DNA at three bases upstream of PAM, forming a blunt end (Wu et al. 2014) (Figure 15.1c). The DSBs are repaired by NHEJ and HDR. HDR can be used to introduce exogenous DNA at the repairing site of the cleaved DNA. This process is applied in developing a novel GE tool that can be accessed in both plant and animal cells.

If we compare it with ZFNs and chimeric TALENs targeting the promoters of TFs, CRISPR/Cas9 aims at the target sequence through complementary sequence-based interaction between gRNA and target DNA. The system has advantages based on design, specificity, cost-effectiveness, and specificity, as it eliminates the step of making new proteins to recognize target sites (Vats et al. 2019). The first CRISPR work to manipulate the mammalian genome was performed in the year 2013. The required protein deactivation was carried out by producing indels (insertion and deletion) in the target loci (Cong et al. 2013). Simultaneously, model plants like *Arabidopsis*, tobacco, rice,

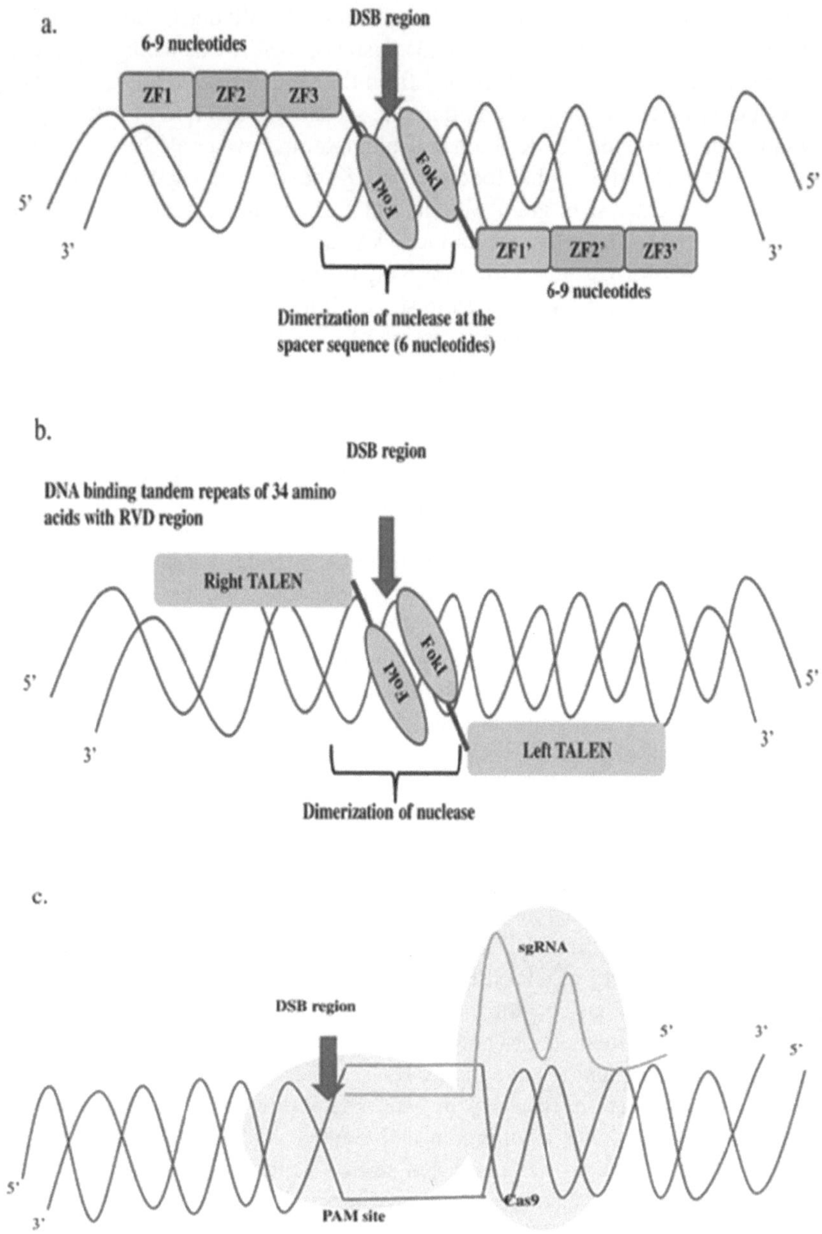

FIGURE 15.1 Designer nucleases for genome editing. (a) ZFN technique has three zinc finger proteins (ZF1, ZF2, ZF3, and ZF1′, ZF2′, ZF3′) on opposite sides of the FokI nuclease dimer, providing nine-nucleotide specificity in the two flanking regions with three-nucleotide specificity each. The FokI is bound to opposite strands with nuclease domain 5 situated at the C-terminal of ZFNs. The dimer is responsible for making DSB in the 6-bp spacer sequence. (b) TALEN is an engineered fused TALE protein with FokI nuclease at the carboxy-terminal. TALEN is used for the recognition of tandem repeats of 16 identical protein domains made up of 34 amino acids each. Within these repeats, at the 12th and 13th positions, the amino acid residues are more variable, and the region is known as repeat-variable diresidues (RVD). (c) CRISPR/Cas9 is composed of CRISPR-containing loci typically flanked by Cas protein sequences on each side. The short segment of foreign DNA gets incorporated within the CRISPR sequence followed by transcription generating CRISPR RNA (crRNA). The crRNA anneals with the trans-activating crRNA (tracrRNA) forming a single guide RNA (sgRNA). The sgRNA associates with Cas9 protein and hybridizes with the target DNA by bringing conformational changes. This change induces the subsequent cut as DSBs.

and wheat were the plants in which GE was first attempted utilizing CRISPR/Cas9 technology in the year 2013 (Li et al. 2013, Shan et al. 2013). Using CRISPR/Cas technology, attempts have been made to produce biofuels and "biotech oil" from the seeds of *Camelina sativa* by improving its natural fatty acid production. On consumption, the seeds have resulted in better options against cell oxidation and hence, provided better human health (Kamburova et al. 2017). Gene editing is a promising approach for achieving global food security. Knocking out negative regulating genes for yield-related factors like grain size, grain weight, and grain number created expected phenotypic losses in crops like rice and wheat (Nadolska-Orczyk et al. 2017). This loss could easily be overcome by utilizing the CRISPR/Cas9 tool for improving the quality of yield traits (Chen et al. 2019). The CRISPR/Cas9 tool has also been reported to successfully target viral genomes and pathogenic bacterial plasmids. Therefore, stable over-expression of gRNA and Cas9 protein targeting viruses like geminiviruses has been produced to develop antiviral breeding in plants (Ali et al. 2015, Baltes et al. 2015).

15.5 MAJOR GENOME EDITING TECHNIQUES USED FOR DEVELOPING ABIOTIC STRESS TOLERANCE IN PLANTS

Because of so many ecological changes in abiotic constraints, particularly salt concentration, drought, temperature, and heavy metals, global agricultural production volumes are greatly diminishing. Among the most feasible and environmentally friendly techniques to deal with this problem and generate abiotic stress-resistant crop plants is traditional breeding. Selective breeding has indeed led to considerable production improvement in different crops for environmental stress tolerance. Nevertheless, in addition to the traditional solution to producing durable crop plants for abiotic stress, more efficacious and elegant techniques with profound impacts are expected to counter all such obstacles (Driedonks et al. 2016).

Latest revelations in GE approaches offer us new possibilities for improving existing crops through systematic genetic manipulation of specific crop attributes. GE is described as a range of innovative tools in molecular biology that permits accurate, reliable, and coordinated alterations at a specific genomic location (Chen and Gao 2014). For the last 20 years, genetic manipulations employing ZFNs (Kim et al. 1996) and TALENs (Christian et al. 2010) have been the only choices available; however, the discovery of CRISPR/Cas (Jinek et al. 2012) mechanisms that provide versatility and ease of tailored GE has become the center of attention. Many such technologies leverage standard SSNs, which can be mediated to pinpoint specific genes and to create DSBs. The intrinsic repairing mechanisms of the plant resolve the DSBs by either NHEJ, which might eventually lead to nucleotide addition or deletion, resulting in gene knock-downs, or by HR, which can induce substitution and insertion of genes (Symington and Gautier 2011). The application of GE technology in a large number of plants has generated several gene knock-out variants and numerous gene replacement and insertion mutants, and most of these mutants are proven to be valuable for crop improvements.

15.5.1 Application of ZFNs for Trait Improvement in Plants

ZFNs are recognized as the first generation of GE techniques based on active Cys2-His2 (ZF) domain working criteria (Kim et al. 1996, Gaj et al. 2013, Palpant and Dudzinski, 2013, Pabo et al. 2001).

The application of ZFNs relies on optimizing zinc fingers against target DNA sequences, trailed by binding of unique ZFs to broader segments. It has been extensively used in numerous organisms, notably plants, ever since the discovery of ZFNs in 1996 (Gaj et al. 2013). To stimulate disruptions in an AP2/ERF family TF gene, *ABA-INSENSITIVE 4*, implicated in environmental stress tolerance in *Arabidopsis*, a tailored ZFN was utilized together with a heat-shock promoter (Osakabe et al. 2010). A greater percentage (up to 3%) of genetic variations occurring in the target mutants was recorded. Analogously, the larger number of ZFN-mediated genes targeting the endogenous tobacco *acetolactate synthase* genes (*ALS SuRA* and *SuRB*) has been revealed to provide resistance to imidazolinone and sulfonylurea herbicides (Townsend et al. 2009). The implication of ZFNs for gene stacking in maize (*Zea mays*) is also being documented (Ainley et al. 2013, Petolino et al. 2010). Moreover, ZFN was employed to manipulate an existing plant gene for malate dehydrogenase (MDH), leading to improved MDH-carrying plants with augmented yield (Shukla et al. 2016).

In current history, by adeptly showing their efficiency to manipulate gene expression, ZFNs have modernized GE strategies and developed alternative ideas for fundamental and applied research in this perspective. In line with greater precision and reliability with a negligible off-target negative effect, they have provided many advantages over other existing GE techniques. Present attempts are now more centered on developing the techniques of design and implementation to extend their uses across a range of crop plants.

15.5.2 Application of TALEN for Trait Improvement in Plants

TALEN-mediated GE has proven to be an effective method for incredibly efficient, sensitive, and scalable GE techniques with targeted genomic alterations in plants (Jankele and Svoboda 2014). Considering the convenience of rapid site-directed modification via the TALEN system, different genes are being customized and effectively used in a wide range of plant species, including *Arabidopsis*, tomato, rice, potato, wheat, soybean, and maize (Xiong et al. 2015).

In the latest study, *Arabidopsis* plants were designed employing TALENs to reconfigure five intrinsic genes, *ADH1*, *TT4*, *MAPKKK1*, *DSK2B*, and *NATA2*, culminating in 2–15% somatic mutagenesis in the *Arabidopsis* seedlings (Christian et al. 2013). Likewise, a TALEN technique has been utilized in wheat (*Triticum aestivum*) to create specific modifications in the three homoalleles encoding MILDEW-RESISTANCE LOCUS (MLO) to develop resistance to powdery mildew, a feature rarely present in native populations (Wang et al. 2014). Nishizawa-Yokoi et al. (2016) used a TALEN-mediated somatic mutagenesis in rice calli in its waxy locus. When the homologous joining pathway was studied for the TALEN target sites, it showed alt-NHEJ-induced deletions in the lig4 gene. TALENs have been used to develop herbicide-resistant crops. Li et al. (2016) produced TALEN-mediated *OsALS* gene targeting cassettes to create double point mutations in crops. The *OsALS* gene is targeted by herbicides like chlorsulfuron and bi-spyribac sodium (BS), inhibiting the enzyme activity. The two point mutations confer tolerance to the BS effect. The change in the gene was brought about by HR-mediated mutagenesis, and the change was successfully transferred to the T1 generation. Hence, the TALEN-mediated GE technique plays a part in developing herbicide-resistant crops. In a recent study, efforts have been made to curing plants of cytoplasmic male sterility (CMS) using TALEN-mediated mitochondrial GE. The defect causes failed pollen development or absence of seed setting, which is used to produce hybrid seeds from the F1 generation. Kazamath et al. (2019) attempted to knock out orf79 and orf125, the CMS-causative gene, with mito-TALENs. Their work could cure male sterility in rice and rapeseeds. DSBs made by mito-TALENs were repaired by HR deleting the target genes. Soybean plants have been configured through TALEN technology to eliminate polyunsaturated fats by inducing alterations in the fatty acid pathway genes *FAD2-1A* and *FAD2-1B*, required to convert monounsaturated fat (oleic acid) to polyunsaturated fat (linoleic acid) by seed metabolism. Subsequent *FAD2-1A* and *FAD2-1B* mutated soybean lines have been shown to increase the concentration of oleic acid from 20% to 80% and to decrease the concentration of linoleic acid from around 50% to less than 4% (Haun et al. 2014). Clasen and colleagues have reported that the cold storage and processing properties for potato (*Solanum tuberosum*) have been prolonged using TALENs (Clasen et al. 2016). Nowadays, due to certain drawbacks, especially additional cost and intensive protein engineering, GE methods, such as ZFNs and TALENs, are being overlooked following the introduction of ground-breaking GE technologies such as CRISPR/

Cas9, CRISPR/Cpf1, etc. Practically speaking, a very limited number of studies have so far been conducted using ZFNs and TALENs to enhance abiotic stress tolerance in crops.

15.5.3 Applications of CRISPR/Cas Technology for Improving Abiotic Stress Tolerance in Crop Plants

Plants respond to environmental cues by activating various pathways premised on stress-inducing gene expression (Golldack et al. 2014). A substantial number of dynamic interconnected network genes (regulatory, metabolic, and signaling) that contribute to abiotic stress responses are being used in the genetic engineering scheme to ameliorate abiotic stress and to modulate resilience in both models and crops (Mickelbart et al. 2015). GE has made great strides ever since the breakthrough of the CRISPR/Cas9 magic tool and has accrued phenomenal success. Over the past decades, CRISPR/Cas9 RNA-guided techniques have been expeditiously used for crop improvement, although their use in climate-smart crop development has only been discussed in a very few studies. To date, due to the complex nature of abiotic stress, minimal GE studies have been performed to establish climate-resilient crops as compared with pathogen resistance. The implications of CRISPR technology to impart abiotic stress tolerance in different crops under oscillating external environmental conditions are briefly illustrated in the following and presented in Table 15.1.

15.5.3.1 CRISPR/Cas-Mediated Drought Stress Tolerance

The most debilitating stress hampering plant growth and yield is water deficit or drought, contrary to other abiotic stresses (Zhang et al. 2018). It often leads to severe crop failures annually and is likely to be further exacerbated with rising global temperatures. Plants exhibit different phenotypic (decreased leaf area), physiological (controlled stomatal opening and closing and water-use efficiency), and biochemical (enzymatic and non-enzymatic antioxidants and osmolyte accumulation) manifestations during water deficit conditions (Fang and Xiong 2015). Due to the drawbacks of greenhouse experiments and additional expense, the use of the transgenic-based strategy by introgression of different structural (reactive oxygen species (ROS) scavengers) and regulatory genes (TFs, phosphatases, kinases, etc.) to develop genetically modified plants has attracted stereotypical views. Nowadays, GE strategy is the preferred method for plant breeders to overcome such issues.

For practical study contributing to a better interpretation of the intrinsic networks involved with *SlMAPK3*-mediated drought stress tolerance in tomato plants, the CRISPR/Cas9 platform was adopted for knock-down of a tomato mitogen-activated protein kinase 3 (*slmapk3*) gene (Wang et al. 2017). DuPont biologists adeptly edited a gene encoding maize negative ethylene response regulator, *ARGOS8*, using CRISPR/Cas9 technology (Shi et al. 2017). In that research, they observed that even under drought stress conditions, CRISPR/Cas9-driven *ARGOS8* mutants exhibited a higher maize grain yield. Two genes pertaining to abiotic stress, *TaDREB3* and *TaDREB2*, were investigated using a CRISPR/Cas9 approach in wheat protoplast. The expression of mutant genes in about 70% of transfected protoplasts has been demonstrated with a T7 endonuclease assay. Compared with wild cultivars, the mutant plants displayed elevated resistance to drought (Kim et al. 2018). To provoke mutation in the *Arabidopsis OST2* gene, Osakabe et al. (2016) used CRISPR/Cas9 technology; the alteration culminated in a modified stomatal opening and closing trend in response to perturbations in environmental parameters, thereby augmenting the dehydration stress tolerance of the plants. Likewise, in *Arabidopsis*, knock-out of *MIR169a* by a CRISPR/Cas9 GE tool imparted drought resilience: 54–57% of mutant plants survived even after 12 days of drought stress treatment and re-watering, although no recovery was recorded in wild-type plants (Zhao et al. 2016). The stress/ABA-activated protein kinase 2 (*SAPK2*) gene, a principal abscisic acid (ABA) signaling mediator, was successfully engineered to develop loss-of-function mutant rice plants through the CRISPR/Cas9 device (Lou et al. 2017). The *sapk2* rice variants tend to be more vulnerable to drought and oxidative stress than wild-type plants, indicating that *SAPK2* is critical to rice drought tolerance

TABLE 15.1
List of Genes Targeted through CRISPR/Cas Strategy in Model and Crop Plants for Improving Abiotic Stress Tolerance

Model Plants/ Crops Name	Gene(s) Targeted	Gene Function(s)	DSB Repair Mechanism	Traits/Stress Response	References
Arabidopsis (*Arabidopsis thaliana*)	*OST2 (OPEN STOMATA 2) (AHA1)*	Proton pumping	NHEJ	Tolerance to drought stress	Osakabe et al. (2016)
	MIR169a	Regulation of gene expression	HDR	Tolerance to drought stress	Zhao et al. (2016)
	AITR	ABA-induced transcriptional repressor	NHEJ	Tolerance to salinity stress	Chen et al. (2021)
	CBF	C-repeat binding factor	NHEJ	Tolerance to salinity stress	Zhao et al. (2016)
	SIZ1	C2H2 type zinc finger protein	NHEJ and HDR	Tolerance to salinity stress	Han et al. (2019)
	UGT79B2, UGT79B3	Regulation of metabolic process	NHEJ	Double mutants showed susceptibility responses to cold, drought, salinity stress	Li et al. (2017)
Tomato (*Solanum lycopersicum*)	*SlHyPRP1*	Multi-stress tolerance	NHEJ	Tolerance to salinity stress	Tran et al. (2021), Saikia et al. (2020)
	SlMPK3	Regulation of gene expression, protection from oxidative damage	NHEJ	Tolerance to drought stress	Wang et al. (2017)
	SlARF4	Auxin signaling pathway	NHEJ	Tolerance to salinity stress	Bouzroud et al. (2020)
	SP5G, SP	Regulators of day length sensitivity	NHEJ	Tolerance to salinity stress	Li et al. (2018), Zhang et al. (2018)
	SlCBF1	Transcription factor	NHEJ	Tolerance to cold stress	Li et al. (2018)
	WUS	Transcriptional activator/repressor of genes in the shoot apical meristem	NHEJ	Tolerance to salinity stress	Li et al. (2018)
	GGP1	Involved in biosynthesis of vitamin C	NHEJ	Tolerance to salinity stress	Li et al. (2018)
	HKT1;2	High affinity potassium transporter	HDR	Tolerance to salinity stress	Vu et al. (2020)
	SlNPR1	Phytohormone signaling	NHEJ	Tolerance to drought stress	Li et al. (2019)

(*continued*)

TABLE 15.1 (Continued)
List of Genes Targeted through CRISPR/Cas Strategy in Model and Crop Plants for Improving Abiotic Stress Tolerance

Model Plants/ Crops Name	Gene(s) Targeted	Gene Function(s)	DSB Repair Mechanism	Traits/Stress Response	References
	CLV3	Regulates shoot and floral meristem development	NHEJ	Tolerance to salinity stress	Li et al. (2018), Van Eck et al. (2019)
Rice (*Oryza sativa*)	*OsMPK2, OsDEP1*	Multiple abiotic stress tolerance	NHEJ, HDR	Yield under stress	Shan et al. (2014)
	OsBEL, OsAOX1a, OsAOX1b, OsAOX1c	Regulation of transcription (*BEL*), cold-responsive gene family (*AOX1a, AOX1b, AOX1c*)	NHEJ	Multiple abiotic stress tolerance response	Xu et al. (2015)
	DST	Zinc finger transcription factor	NHEJ	Tolerance to salinity stress	Kumar et al. (2020)
	RR9, RR10	Participates in cytokinin signaling pathway	NHEJ	Tolerance to salinity stress	Wang et al. (2019)
	RR22	Transcription factor	NHEJ	Tolerance to salinity stress	Zhang et al. (2019)
	OsSAPK2	Regulator of hyper-osmotic stress signaling and ABA-dependent plant development	NHEJ	Tolerance to drought stress	Lou et al. (2017)
	BGE3	Cytokinin transporter	NHEJ	Tolerance to salinity stress	Yin et al. (2020)
	SPL10	Transcription factor	NHEJ	Tolerance to salinity stress	Lan et al. (2019)
	OsRAV2	AP2/ERF responsive transcription factor	NHEJ	Tolerance to salinity stress	Duan et al. (2016)
	OsAnn3	Act as Ca^{2+}-dependent phospholipid binding protein	NHEJ	Tolerance to cold stress	Shen et al. (2017)
	OsMPK5	Participates in stress-responsive signaling pathway	NHEJ	Multiple abiotic stress tolerance response	Xie and Yang (2013)
	FLN2	Involved in sucrose metabolism	NHEJ	Tolerance to salinity stress	Chen et al. (2019)
	MIR528	Salt stress response regulator	NHEJ	Tolerance to salinity stress	Zhou et al. (2017)
	BBS1	Involved in chaperone-mediated signaling	NHEJ	Tolerance to salinity stress	Zeng et al. (2018)
	NAC041	Transcription factor	NHEJ	Tolerance to salinity stress	Bo et al. (2019)

	OsPDS, OsPMS3, OsEPSPS, OsDERF1, OsMSH1, OsMYB1, OsROC5, OsSPP, OsMYB5, OsYSA	Pigment synthesis (*OsPDS*), aromatic amino acid synthesis (*OsEPSPS*), DNA mismatch repair protein (*OsMSH1*), Rice Outermost Cell-specific gene5 (*OsROC5*), AP2 domain containing protein (*OsDERF1*), pentatricopeptide repeat domain containing protein (*OsYSA*), fertility-related gene (*OsPMS3*), transcription factors (*OsMYB1, OsMYB5*)	NHEJ	Drought and other multiple abiotic stress tolerance responses	Zhang et al. (2016), Zhang et al. (2014)
	OsNAC14	Transcription factor	HDR	Tolerance to drought stress	Shim et al. (2018)
Maize (*Zea mays*)	*ARGOS8*	Aluminium-induced protein superfamily pseudogene	HDR	Tolerance to drought stress	Shi et al. (2017)
	ZmHKT1	Sodium transporter HKT1	NHEJ	Tolerance to salinity stress	Zhang et al. (2018)
Wheat (*Triticum aestivum*)	*TaDREB2, TaDREB3*	Wheat dehydration responsive element binding transcription factor	NHEJ	Tolerance to drought stress	Kim et al. (2018)
Soybean (*Glycine max*)	*DREB2A, DREB2B*	Dehydration responsive element binding transcription factor	NHEJ	Tolerance to drought and salinity stress	Curtin et al. (2018)

and possibly a candidate gene for crop improvement (Lou et al. 2017). The receptor component of salicylic acid phytohormone is nonexpressor of pathogenesis-related 1(*NPR1*) (Wu et al. 2012). The mutant tomato (*SlNPR1*) resulting from the CRISPR technique showed elevated MDA and H_2O_2 levels and decreased levels of enzymatic antioxidants such as catalase, peroxidase, ascorbate peroxidase, and superoxide dismutase. In addition, other primary genes associated with drought, such as *SlGST*, *SlDHN*, and *SlDREB*, were silenced. These findings showed that the *SlNPR1* gene might potentially withstand drought conditions in tomato breeding. Overall, experimental studies have shown that there is spectacular potential to boost drought tolerance in several agriculturally important crop plants by using CRISPR/Cas9 technologies.

15.5.3.2 CRISPR/Cas-Mediated Salt Stress Tolerance

One of the most devastating variables affecting agricultural production is salinity. Literally 800 million hectares of farmland are seriously affected by salinity globally, as per the 2008 FAO (Food and Agriculture Organization) survey report (FAO 2008). Countless physiochemical modifications are induced in crop plants by salinity stress, which include high osmotic stress, ionic disturbances, and secondary stress (Isayenkov and Maathuis 2019). Apart from serious yield penalties, salinity stress can impact numerous growth characteristics in crop plants, such as plant height, fresh weight output, and plant biomass production (Semiz et al. 2012).

In salinity and drought stress conditions, the NAC TF family plays a central role in curtailing diverse plant metabolic pathways (Shen et al. 2017). Via utilizing the CRISPR/Cas9 paradigm to verify the crucial role of the NAC rice TF coding gene *OsNAC041* during salt stress treatments, Bo et al. (2019) created the *OsNAC041* mutant. Wild-type seedling shoots were taller relative to the mutant plants under 150 mmol/l NaCl treatment. The wild seedling type remained intact, while almost all mutant seedlings died, implying that *OsNAC041* is an important player of salinity stress tolerance in rice. Similarly, Curtin and co-workers performed knock-out mutations of two *Drb2a* and *Drb2b* genes applying the CRISPR/Cas9 strategy and observed that these genes control salt and drought resistance in soybeans (Curtin et al. 2018). The *OsRR22* gene encoding the rice type-B response regulator was knocked down via the CRISPR/Cas9 approach to significantly boost rice salinity tolerance (Zhang et al. 2019). The SQUAMOSA promoter binding like protein 10 (*OsSPL10*), a negative regulatory salt stress-responsive gene in rice, was knocked out using the CRISPR/Cas9 strategy. Interestingly, knock-out mutant lines exhibited greater tolerance to salinity stress (Lan et al. 2019). The role of the protein OsRAV2 TF in salinity stress is excellently defined. The researchers highlighted the significance of the promoter region of this gene, GT-1, known for salt induction, in a study by Li et al. (2020). Under salinity stress, homozygous mutant lines developed by CRISPR/Cas-driven editing of the *GT-1* promoter region could not express OsRAV2 TF. This shows the significance of this promoter GT-1 region in imparting salt stress tolerance in rice. Furthermore, CRISPR/Cas9 based SlHyPRP1 multiplex editing resulted in the precise deletion and salt stress tolerance in cultivated tomatoes (Tran et al. 2021; Saikia et al. 2020). *SlARF4*, an auxin response factor in tomato that is critical for salinity and osmotic stress tolerance through the CRISPR/Cas strategy, has been successfully mutated by Bouzroud and colleagues (Bouzroud et al. 2020). Under normal and stressful situations, the *SlARF4* mutant plants exhibited better salinity and osmotic stress tolerance by decreased stomatal conductance combined with increased leaf relative water content and ABA content. To date, significantly less research has been performed on crop plants using GE strategies to enhance salinity stress tolerance. CRISPR/Cas9 can be exploited by molecular biologists to strengthen tolerance to salinity stress and to simulate the physiological effects of plant growth and development.

15.5.3.3 CRISPR/Cas-Mediated Heat Stress Tolerance

As a consequence of global warming, high temperatures, which create heat stress, have become an important agricultural issue in several parts of the world (Tubiello et al. 2007). Under heat stress, plant cells often exhibit membrane disruption, ROS accumulation, and metabolic disruptions that further reduce crop productivity and quality (Iba 2002).

Nowadays, among all the other techniques used for gene editing, CRISPR/Cas9 technology has been a pivotal tool for successful GE to study the molecular mechanism of heat stress and improve thermo-tolerance in crop plants (Nguyen et al. 2018). CRISPR/Cas9-steered GE for thermo-tolerance has been achieved by modifying the SlaGAMOUS-LIKE 6 (*SIAGL6*) gene in tomato, which improved the fruit setting during heat exposure (Klap et al. 2017). Analogously, Yu and his colleagues reported in a study that *SlMAPK3* knock-out induced by the CRISPR/Cas9 device improves heat stress tolerance by ROS homeostasis in tomato plants (Yu et al. 2019). Presently, attempts to enhance heat tolerance exploit a number of gene(s) mainly associated with ethylene-responsive TFs through genetic manipulation, mostly with the overarching intention of maximizing productivity during environmental stresses such as heat stress (Li et al. 2016, Shi et al. 2017a, Kim et al. 2018). Nevertheless, given the volume of distinct genomic information presented by CRISPR, a handful of field research initiatives have been undertaken by breeding programs.

15.5.3.4 CRISPR/Cas-Mediated Cold Stress Tolerance

Cold stress in different regions of the world has impacted agricultural production and crop quality and distribution. Leaf photosynthesis and biomass production, which are often the key determinants of yield components, are primarily influenced by cold stress (Liu et al. 2019). Low-temperature yield losses are a critical constraint in rice production, not just in high-latitude or high-altitude regions but also in nations like Thailand and the Philippines. Multiple transgenic approaches have been developed continuously to ramp up cold resistance in different crops. The prime focus of many such interventions is to specify the unique genes capable of upping cold tolerance (Shen et al. 2017a, b).

In contrast to cold-stressed wild-type plants, CRISPR/Cas9 was exploited to create knock-down variants for the japonica rice *annexin* gene (*OsANN3*), leading to a drastic increase in relative ionic conductivity and a noticeable decline in the viability level of T1 mutant groups. This finding implies that *OsANN3* plays a significant role in adaptation to cold exposure by rice; hence, the gene could perhaps be a viable target for engineering rice with better cold tolerance (Shen et al. 2017a). Using CRISPR/Cas9 and RNAi technologies, Li et al. 2016 reported that two *Arabidopsis* glycosyltransferase genes, *UG79B2* and *UGT79B3*, had a positive impact on hypothermia, primarily controlled by CBF1. Wild-type 12-day-old *Arabidopsis*, over-expressed *UGT79B2/B3OE* plants, and *ugt79b2/b3* Cas9 variant lines were subjected to cold conditions at 4 °C. Both *ugt79b2/b3* mutated lines and the wild type entirely whitened at a lower temperature, −12 °C, with severe ion loss and limited survival, although the overexpressed *UGT79B2/B3OE* plants surprisingly exhibited 25% survivability, coinciding with little or no ion leakage, elevated antioxidant capacity, and ROS-scavenging anthocyanin, allowing increased tolerance to cold stress (Li et al. 2016). Zhao et al. (2016) adopted the CRISPR/Cas strategy to create single, double, and triple *Arabidopsis* mutants and observed that CBF triple (*CBF1*, *CBF2*, and *CBF3*) mutants are extremely vulnerable to cold exposure compared with wild-type plants, implying that all three CBF genes are central to cold acclimation. Via over-expression and mutation, CRISPR/Cas GE strategies have identified various cold stress-related genes in plants such as rice and *A. thaliana*.

15.6 EXTENSION AND LATEST GE TECHNOLOGY INCARNATIONS FOR ABIOTIC STRESS TOLERANCE IN CROPS

In agriculture, the CRISPR/Cas9 technology has acquired significant implications; nonetheless, certain limitations restrict its use in plant GE. Independent variables impacting the CRISPR/Cas9 system, including Cas9 cleavage activity, off-target effects and sgRNA configuration, vector delivery methods, selection of target sites, and HDR frequency, have been explored in various studies. To address these problems with the advancement of plant biotechnology, many unique innovative improvements are routinely incorporated into the GE strategy.

15.6.1 Vector Delivery

There are some drawbacks to the traditional cargo-vectors for GE that impede the precise editing process in plants. Vector delivery based on CRISPR/Cas DNA is now widely used to resolve the disadvantages of plant GE.

15.6.1.1 Stable Expression

Using *Agrobacterium* or biolistic systems, plants are generally transformed to stably incorporate the Cas9 and gRNAs encoding DNA into the plant genome. This device is presently being used in countless applications in plant GE. Gao and colleagues integrated a fluorescent gene into the CRISPR/Cas9 expression cassette (Gao et al. 2016). An intriguing tactic has recently been proposed wherein BARNASE and CMS have been used to destroy embryos and pollen carrying the transgene of T_0 plant offspring as suicide genes (He et al. 2018). He and his colleagues concentrated on the rice *LAZY1* gene, which is crucial in gravitropic response (Li et al. 2007). *Lazy1* mutants with a large tiller angle were functionally inactive. The apparent phenotypes of *lazy1* mutants allowed them to quantitatively analyze the editing efficiency of the constructs used for gene editing. Among the 65 T_0 seedlings, they identified 29 plants with significant tiller angle characteristics, signifying a mutation efficiency of 44%.

15.6.1.2 Transient Expression

The CRISPR transient expression module is yet another delivery technique that can be utilized to achieve transgene-free edited plants. In this method, to develop the edited offspring without integrating exogenous DNA into the plant genome, the evaluation of mutants using selectable marker genes (antibiotic/herbicide) is excluded. In 2016, for the first time in wheat, Zhang and colleagues effectively deployed this strategy to develop transgene-free plants (Zhang et al. 2016). The researcher devised a simple and efficient GE technique that included resurrecting genetically altered plants after transiently expressing CRISPR/Cas9 DNA (TECCDNA) or in vitro transcripts (IVTs) of Cas9-coding sequence and guide RNA (TECCRNA) constructs into wheat callus. The method was established for gene editing in hexaploid (*T. aestivum*) wheat (cvs: Bobwhite and Kenong199) and tetraploid (*T. durum*) wheat (cvs: Shimai11 and Yumai4). For the GE experiment, single guide RNA (sgRNA) for several genes (*TaGASR7*, *TaGW2*, *TaDEP1*, *TaNAC2*, *TaPIN1*, *TaLOX2*, *TdGASR7*) was designed. The editing efficiency/mutagenesis frequency for *TaGASR7* mutants using the TECCDNA construct was 5% (in cv Bobwhite) and 2.6% (in cv Kenong199), respectively. *TaGW2* mutants had editing efficiencies of 3.0% and 3.3%, respectively (in cv Kenong199). Analogously, the frequency of mutation in *TaDEP1* (2.0%), *TaNAC2* (2.0%), *TaPIN1* (1.0%), and *TaLOX2* (9.5%) genes in cv Kenong199 has been reported (Zhang et al. 2016). Comparably, in Shimai11 and Yumai4 cultivars of durum wheat, the editing efficiency/mutagenesis frequency for *TdGASR7* mutants utilizing the TECCDNA construct was 1.4% and 1.5%, respectively. In comparison, the editing rate of the TECCRNA construct has been reported to be 1.1% in Kenong199 cv.

15.6.2 Use of Ribonucleoproteins (RNPs)

The second important strategy to produce transgene-free plants via CRISPR/Cas9 module is foreign DNA-free GE (Veillet et al. 2019). The most efficient transgene-free editing technique has been developed by engineering the RNPs sgRNA/Cas9 tool in plants to overcome the constraints of the mRNA and plasmid-based sgRNA/Cas9 expression system (Liang et al. 2018). sgRNA/Cas9 RNPs have higher potential to generate minimal off-target level DNA-free edited plants and are more accurate than a plasmid-mediated editing method.

15.6.3 Multiplexing

Cellular functions in plants are regulated by multiple overlapping genes. Sometimes, due to the substitution action of certain genes from the same family of genes, the alteration of a particular gene does not always result in an advantageous phenotype. For multiplex gene manipulation in plants, an updated editing system with enhanced performance is therefore needed. Multiplex GE mediated by CRISPR/Cas9 enables several sgRNA cassettes to be constructed into a single-vector structure by using single or multiple promoters (Liu et al. 2017). In addition, CRISPR/Cas9-mediated gene multiplexing gives an overview of metabolic and gene regulatory mechanisms and allows complex multigenic agronomic characteristics to be engineered in plants.

15.6.4 Allelic Replacement

CRISPR-mediated HDR can accurately substitute an antecedent allele in a commercial cultivar with a superior allele, which is the "golden standard" of the vast majority of crop improvement programs (Li and Xia 2020). Despite considerable advances in recent years, CRISPR-mediated allele substitution in agricultural plants remains a significant challenge. HDR-based gene targeting (GT) is a potent technique for selective allele/gene substitution or specific genome alterations like gene tagging in higher eukaryotes (Caroll et al. 2017). The improvement in nitrogen use efficiency (NUE) in *indica* rice is attributable to a single nucleotide polymorphism (SNP) in the *NRT1.1B* gene between *japonica* and *indica* rice. Li et al. (2018) adopted a CRISPR/Cas9-mediated HDR-based methodology to correctly substitute the *japonica NRT1.1B* allele with the *indica* allele in only one generation.

15.6.5 Minimal Off-Target with Updated Versions of CRISPR/Cas9 Module

Due to the inevitable guide RNA (gRNA) mismatches, the CRISPR/Cas9 method that evolved from *Streptococcus pyogenes* has several shortcomings that obstruct its editing operation, such as several incongruent off-targets. As a consequence, several changes have been made to boost editing performance and ameliorate off-target cleavage activity of Cas9 enzymes, particularly SpCas9n (Cas9n) (Cong et al. 2013), Dead cas9 (dcas9) (Mali et al. 2013), and FokI Cas9 (fCas9) (Guilinger et al. 2014). Different bacterial strains with the novel- and stretched-PAM sequences are being used for the isolation of Cas9 proteins, which can boost non-target cleavages. SpCas9, amongst all known orthologs of Cas9, has been somewhat extensively used in GE of crop plants.

Lately, *Francisella novicida* was explored to study the CRISPR/Cpfl class II mechanism (Zetsche et al. 2017), recently called Cas13 (Zetsche et al. 2015). For cleavage and creation of cohesive ends, as opposed to Cas9, Cpfl requires a single RNA-guided structure with 4–5 nucleotides 5′-overhangs. The CRISPR/Cpfl model is being prevalently adopted in flora and fauna with few or fewer off-target effects. In 2016, the CRISPR/Cpf1 technique was successfully extended to GE in crops (Endo et al. 2016). Due to the distinctive characteristics of Cpf1, CRISPR/Cpf1 type V is often deemed to be another effective technique for plant GE (Zhang et al. 2017).

15.7 POTENTIAL CHALLENGES OF GENOME EDITING FOR DEVELOPING ABIOTIC STRESS-RESILIENT PLANTS

Even though CRISPR editing is a ground-breaking technological advance, it does have some strategic constraints. The insufficient availability of suitable candidate genes and related transformation and regeneration guidelines is the serious roadblock for the use of CRISPR/Cas GE technologies to enhance abiotic stress tolerance. The second biggest problem is off-target editing. The fusion of PAM sequence (NGG) and unique gRNA allows extremely precise editing, although there are cases of editing of many other undesirable genomic positions (Zhang et al. 2015). While it has about 20% of the binding specificity of NGG PAM, Cas9 may attach to the NAG, a tri-nucleotide sequence, where G stands for guanidine, A stands for adenine and N can be any nucleotide as PAM sequence (Hsu et al. 2013). The third major hurdle to surmount would be that CRISPR is incapable of minimizing all sequences of interest by separate nucleases due to the limited PAM requirements. CRISPR/Cas9 technology has now become an easy-to-use platform for developing non-transgenic genome-modified crop plants to contend with shifting climatic conditions and support global food security. Researchers anticipate that combating the technological and regulatory obstacles associated with rapid acceptance and the widespread implementation of GE would enable CRISPR/Cas technology to develop a new generation of high-yield, climate-smart, and abiotic stress-tolerant crops.

15.8 COMPARISON OF TRANSGENIC, TILLING, AND GENOME EDITING APPROACH

Transgenics is the process of introducing foreign DNA into a host organism's genome. The source of the integrated foreign DNA, also known as "transgene," could be from different species (transgenics) or the same species (cisgenics) or even from artificially synthesized DNA. In the case of GE, two major distinctions count in favor of GE strategies, considering the similarity between selective GE and TILLING methodology. Precision-GE has only certain drawbacks with respect to the selection of the specific location of the mutation in a gene, while TILLING is premised on spontaneous mutations around the whole genome. Furthermore, very few undesired mutations may occur in GE, whereas a larger proportion of off-target mutations is an inevitable outcome of conventional mutagenesis, which extensive backcrossing schemes in plant breeding must overcome. All three strategies are reverse genetics approaches. Transgenic technology confers new traits on the original organisms, whereas GE manipulates the genome of the organism by knocking out or replacing targeted genes, and TILLING creates a high frequency of random mutation within the genome. The major differences among these three techniques are listed in Table 15.2. In addition to mutagenesis, CRISPR/Cas9 can be used to trigger (CRISPR activation) or impede (CRISPR interference) gene expression via coupling of the deactivated Cas9 (dCas9) with a transcriptional activator or repressor. It, therefore, has the power to substitute the traditional genetic modification (GM)-based over-expression and gene silencing processes. Furthermore, transgenics can lead to undesirable phenotypic effects due to chances of multiple integration of a transgene in a single locus, or disruption of a functional non-target gene due to integration in or near the gene. This limitation is minimized in GE, as the modification is targeted. Also, GE is fast, accurate, and less expensive, and potentially can minimize off-target. Finally, for regulation, three agencies work together in the US for regulating genetically modified organisms (GMOs): the Food and Drug Administration (FDA), the Environmental Protection Agency (EPA), and the United States Department of Agriculture (USDA). Plant GMOs are regulated by USDA under the Plant Protection Act. At the EU level, the European Food Safety Authority (EFSA) conducts the risk assessment of GMOs. On the other hand, GE will not be part of the same regulation as GMO in the US, whereas the same regulation will be applied to both GE and GMO in the EU.

15.9 CONCLUSIONS

This chapter discussed the multiple aspects of GE technologies and their advances for the development of abiotic stress-resilient crops to increase yield potential. The development of technologies such as meganucleases, ZFN, TALEN, and CRISPR/Cas has made it easier for plant molecular biologists to

TABLE 15.2
Comparison of TILLING, Transgenics, and Genome Editing Approach

	TILLING	Transgenics	Genome Editing
Approach	Non-genetic engineering and reverse genetic approach; relies on chemical mutagen	Genetic engineering and process of reverse genetic engineering approach	Genetic engineering and process of reverse genetic engineering approach
Outcome	Create random mutation	Random integration in the genome	Targeted mutagenesis
Procedure	It is a conventional mutagenesis process with the use of a chemical mutagen and results in a non-chimeric population as it does not require a tissue culture method. The developed population is followed by DNA extraction, pooling, and detection of the mutation as well as validation of results	A genetic engineering approach with introduction of exogeneous gene or over-expression of endogenous gene from same or different species. It follows a tissue culture process. The molecular confirmation of the transgene from the extracted DNA samples through PCR is followed by its expression analysis and validation of the expected outcome	A genetic engineering approach that allows genetic material to be added, removed, or altered precisely at the targeted locations. It follows a tissue culture procedure. The molecular confirmation of the construct integrated into plants from the extracted DNA samples through PCR is followed by its targeted editing analysis through Sanger sequencing and validation of the expected outcome
Target plant/ Explant	The target plant does not need to be genetically transformable as it does not require a tissue culture method	The target plant needs to be genetically transformable as it requires a tissue culture method	The target plant needs to be genetically transformable as it requires a tissue culture method
Biosafety regulation	The products of TILLING create less environmental or food safety concern. The products have higher chance of acceptance with no biosafety regulation as transgenics	The product of genetic modification through transgenics in a way that would not have occurred naturally. Therefore, strict biosafety regulations are imposed on transgenics	GET refers to the plants whose modified genome is similar to one potentially generated naturally through conventional plant breeding
Off-target/ unintended effect	TILLING creates random changes in the genome. Therefore, the resulting modified crop plant might have more unknown changes than targeted editing	Undesired phenotype can be observed due to random integration of transgene. This can be overcome by generating more transgenic lines and selecting the line with less or no adverse phenotype	Genome editing can result in a few off-target mutations compared with conventional mutagenesis. Moreover, various strategies have been developed to further minimize off-target mutations by CRISPR/Cas
Limitations	If natural variation is not present, a genetic engineering approach may be the only option available	The limitations may be the unwanted altered phenotype due to random integration of transgene in the genome. Moreover, effects such as chances of horizontal gene transfer, altered phenotype, or allergic reaction	The outcome of targeted genome editing is associated with unintended additional modifications, such as deletions, partial or multiple integrations of the targeting vector, and even duplications
Efficiency	Mutation can be created at a higher rate	The efficiency of transformation depends on the different transgenic approaches used and the crop plant of interest	Higher mutation efficiency can be achieved

edit any desired gene. Among the various approaches to GE in plants, CRISPR/Cas GE is the most promising system in developing abiotic stress-resilient crops, while the other techniques are mostly expensive and involve the engineering of proteins. Further, the availability of advanced variants in CRISPR/Cas technology encourages more comprehensive application by providing better efficiency and feasibility. In plants, the application of these technologies is greatly increased due to the availability of other variants of Cas9, such as Nmcas9, Sacas9, and Stcas9, from other species. Furthermore, improvements in this area are possible with enhanced reduction of off-targets. However, due to bioethical concerns, limited and judicious application is necessary. Therefore, with the advancement in GE tools, its application for crop improvement will be a prominent area of research in the future.

REFERENCES

Ainley, W.M., L. Sastry-Dent, M.E. Welter et al. 2013. Trait stacking via targeted genome editing. *Plant Biotechnology Journal* 11(9):1126–1134.

Ali, Z., A. Abulfaraj, A. Idris et al. 2015. CRISPR/Cas9-mediated viral interference in plants. *Genome Biology* 16(1):1–11.

Baltes, N.J., A.W. Hummel, E. Konecna et al. 2015. Conferring resistance to geminiviruses with the CRISPR–Cas prokaryotic immune system. *Nature Plants* 1:15145.

Barrangou, R. and L.A. Marraffini. 2014. CRISPR-Cas systems: prokaryotes upgrade to adaptive immunity. *Molecular Cell* 54(2):234–244.

Bibikova, M., M. Golic, K.G. Golic et al. 2002. Targeted chromosomal cleavage and mutagenesis in Drosophila using zinc-finger nucleases. *Genetics* 161(3):1169–1175.

Bo, W.A., Z.H. Zhaohui, Z.H. Huanhuan et al. 2019. Targeted mutagenesis of NAC transcription factor gene, OsNAC041, leading to salt sensitivity in rice. *Rice Science* 26(2):98–108.

Boch, J. and Bonas U. 2010. Xanthomonas AvrBs3 family-type III effectors: discovery and function. *Annual Review of Phytopathology* 48:419–436.

Bouzroud, S., K. Gasparini, G. Hu et al. 2020. Down regulation and loss of auxin response factor 4 function using CRISPR/Cas9 alters plant growth, stomatal function and improves tomato tolerance to salinity and osmotic stress. *Genes* 11(3):272.

Carroll, D. 2011. Genome engineering with zinc-finger nucleases. *Genetics* 188(4):773–782.

Carroll, D. 2017. Focus: genome editing: genome editing: past, present, and future. *The Yale Journal of Biology and Medicine* 90(4):653.

Chen, G., J. Hu, L. Dong et al. 2019. The tolerance of salinity in rice requires the presence of a functional copy of FLN2. *Biomolecules* 10(1):17.

Chen, K. and C. Gao. 2014. Targeted genome modification technologies and their applications in crop improvements. *Plant Cell Reports* 33(4):575–583.

Chen, K., Y. Wang, R. Zhang et al. 2019. CRISPR/Cas genome editing and precision plant breeding in agriculture. *Annual Review of Plant Biology* 70:667–697.

Chen, S., N. Zhang, G. Zhou et al. 2021. Knockout of the entire family of AITR genes in Arabidopsis leads to enhanced drought and salinity tolerance without fitness costs. *BMC Plant Biology* 21(1):1–15.

Christian, M., T. Cermak, E.L. Doyle et al. 2010. Targeting DNA double-strand breaks with TAL effector nucleases. *Genetics* 186(2):757–761.

Christian, M., Y. Qi, Y. Zhang et al. 2013. Targeted mutagenesis of *Arabidopsis thaliana* using engineered TAL effector nucleases. *G3: Genes, Genomes, Genetics* 3(10):1697–1705.

Clasen, B.M., T.J. Stoddard, S. Luo et al. 2016. Improving cold storage and processing traits in potato through targeted gene knock-out. *Plant Biotechnology Journal* 14(1):169–176.

Cong, L., F.A. Ran, D. Cox et al. 2013. Multiplex genome engineering using CRISPR/Cas systems. *Science* 339(6121):819–823.

Curtin, S.J., Y. Xiong, J.M. Michno et al. 2018. CRISPR/cas9 and TALENs generate heritable mutations for genes involved in small RNA processing of *Glycine max* and *Medicago truncatula*. *Plant Biotechnology Journal* 16(6):1125–1137.

Donoho, G., M. Jasin and P. Berg. 1998. Analysis of gene targeting and intrachromosomal homologous recombination stimulated by genomic double-strand breaks in mouse embryonic stem cells. *Molecular and Cellular Biology* 18(7):4070–4078.

Doyon, Y., T.D. Vo, M.C. Mendel et al. 2011. Enhancing zinc-finger-nuclease activity with improved obligate heterodimeric architectures. *Nature Methods* 8(1):74.

Driedonks, N., I. Rieu and W.H. Vriezen. 2016. Breeding for plant heat tolerance at vegetative and reproductive stages. *Plant rReproduction* 29(1):67–79.

Duan, Y.B., J. Li, R.Y. Qin et al. 2016. Identification of a regulatory element responsible for salt induction of rice OsRAV2 through ex situ and in situ promoter analysis. *Plant Molecular Biology* 90(1–2):49–62.

Endo, A., M. Masafumi, H. Kaya et al. 2016. Efficient targeted mutagenesis of rice and tobacco genomes using Cpf1 from *Francisella novicida. Scientific Reports* 6(1):1–9.

Epinat, J.C., S. Arnoul, P. Chames et al. 2003. A novel engineered meganuclease induces homologous recombination in yeast and mammalian cells. *Nucleic Acids Research* 31(11):2952–2962.

Fang, Y. and L. Xiong. 2015. General mechanisms of drought response and their application in drought resistance improvement in plants. *Cellular and Molecular Life Sciences* 72(4):673–689.

FAO Land and Plant Nutrition Management Service. 2008. www.fao.org/ag/agl/agll/spush (accessed September 8, 2016).

Forner, J., A. Pfeiffer, T. Langenecker et al. 2015. Germline-transmitted genome editing in *Arabidopsis thaliana* using TAL-effector-nucleases. *PLoS ONE* 10(3):0121056.

Gaj, T., C.A. Gersbach and C.F. Barbas III. 2013. ZFN, TALEN, and CRISPR/Cas-based methods for genome engineering. *Trends in Biotechnology* 31(7):397–405.

Gaj, T., S.J. Sirk, S.L. Shui et al. 2016. Genome-editing technologies: principles and applications. *Cold Spring Harbor Perspectives in Biology* 8(12):023754.

Gao, X., J. Chen, X. Dai et al. 2016. An effective strategy for reliably isolating heritable and Cas9-free *Arabidopsis* mutants generated by CRISPR/Cas9-mediated genome editing. *Plant Physiology* 171(3):1794–1800.

Golldack, D., C. Li, H. Mohan and N. Probst. 2014. Tolerance to drought and salt stress in plants: unraveling the signaling networks. *Frontiers in Plant Science* 5:151.

Guilinger, J.P., D.B. Thompson and D.R. Liu. 2014. Fusion of catalytically inactive Cas9 to FokI nuclease improves the specificity of genome modification. *Nature Biotechnology* 32(6):577.

Han, G., F. Yuan, J. Guo, Y. Zhang, N. Sui and B. Wang. 2019. AtSIZ1 improves salt tolerance by maintaining ionic homeostasis and osmotic balance in Arabidopsis. *Plant Science* 285:55–67.

Haun, W., A. Coffman, B.M. Clasen et al. 2014. Improved soybean oil quality by targeted mutagenesis of the fatty acid desaturase 2 gene family. *Plant Biotechnology Journal* 12(7): 934–940.

He, Y., M. Zhu, L. Wang et al. 2018. Programmed self-elimination of the CRISPR/Cas9 construct greatly accelerates the isolation of edited and transgene-free rice plants. *Molecular Plant* 11(9):1210–1213.

Hsu, P.D., D.A. Scott, J.A. Weinstein et al. 2013. DNA targeting specificity of RNA-guided Cas9 nucleases. *Nature Biotechnology* 31(9):827–832.

Iba, K. 2002. Acclimative response to temperature stress in higher plants: approaches of gene engineering for temperature tolerance. *Annual Review of Plant Biology* 53(1):225–245.

Isayenkov, S.V. and F.J. Maathuis. 2019. Plant salinity stress: many unanswered questions remain. *Frontiers in Plant Science* 10:80.

Jaganathan, D., K. Ramasamy and G. Sellamuthu. 2018. CRISPR for crop improvement: an update review. *Frontiers in Plant Science* 9:985.

Jain, I., L. Minakhin, V. Mekler et al. 2019. Defining the seed sequence of the Cas12b CRISPR-Cas effector complex. *RNA Biology* 16(4):413–422.

Jankele, R. and P. Svoboda. 2014. TAL effectors: tools for DNA targeting. *Briefings in Functional Genomics* 13(5):409–419.

Jinek, M., K. Chylinski, I. Fonfara et al. 2012. A programmable dual-RNA–guided DNA endonuclease in adaptive bacterial immunity. *Science* 337(6096):816–821.

Kamburova, V.S., E.V. Nikitina, S.E. Shermatov et al. 2017. Genome editing in plants: an overview of tools and applications. *International Journal of Agronomy* 2017:1–15.

Kazama, T., M. Okuno, Y. Watari et al. 2019. Curing cytoplasmic male sterility via TALEN-mediated mitochondrial genome editing. *Nature Plants* 5(7):722–730.

Kim, D., B. Alptekin and H. Budak. 2018. CRISPR/Cas9 genome editing in wheat. *Functional & Integrative Genomics* 18(1): 31–41.

Kim, Y.G., J. Cha and S. Chandrasegaran. 1996. Hybrid restriction enzymes: zinc finger fusions to Fok I cleavage domain. *Proceedings of the National Academy of Sciences* 93(3): 1156–1160.

Klap, C., E. Yeshayahou, A.M. Bolger et al. 2017. Tomato facultative parthenocarpy results from Sl AGAMOUS-LIKE 6 loss of function. *Plant Biotechnology Journal* 15(5):634–647.

Kumar, V.S., R.K. Verma, S.K. Yadav et al. 2020. CRISPR-Cas9 mediated genome editing of drought and salt tolerance (OsDST) gene in indica mega rice cultivar MTU1010. *Physiology and Molecular Biology of Plants* 26(6):1099.

Lan, T., Y. Zheng, Z. Su et al. 2019. OsSPL10, a SBP-box gene, plays a dual role in salt tolerance and trichome formation in rice (Oryza sativa L.). *G3: Genes, Genomes, Genetics* 9(12):4107–4114.

Li, J., X. Meng, Y. Zong et al. 2016. Gene replacements and insertions in rice by intron targeting using CRISPR–Cas9. *Nature Plants* 2(10):1–6.

Li, J., X. Zhang, Y. Sun et al. 2018. Efficient allelic replacement in rice by gene editing: a case study of the NRT1.1B gene. *Journal of Integrative Plant Biology* 60(7):536–40.

Li, J.F., J.E. Norville, J. Aach et al. 2013. Multiplex and homologous recombination–mediated genome editing in *Arabidopsis* and *Nicotiana benthamiana* using guide RNA and Cas9. *Nature Biotechnology* 31(8):688–691.

Li, L., L.P. Wu and S. Chandrasegaran. 1992. Functional domains in Fok I restriction endonuclease. *Proceedings of the National Academy of Sciences* 89(10): 4275–4279.

Li, M., X. Li, Z. Zhou et al. 2016. Reassessment of the four yield-related genes Gn1a, DEP1, GS3, and IPA1 in rice using a CRISPR/Cas9 system. *Frontiers in Plant Science* 7:377.

Li, P., Y. Wang, Q. Qian et al. 2007. LAZY1 controls rice shoot gravitropism through regulating polar auxin transport. *Cell Research* 17(5):402–410.

Li, P., Y.J. Li, F.J. Zhang et al. 2017. The Arabidopsis UDP-glycosyltransferases UGT79B2 and UGT79B3, contribute to cold, salt and drought stress tolerance via modulating anthocyanin accumulation. *The Plant Journal* 89(1):85–103.

Li, Q., M. Sapkota and E. van der Knaap. 2020. Perspectives of CRISPR/Cas-mediated cis-engineering in horticulture: unlocking the neglected potential for crop improvement. *Horticulture Research* 7(1):1–11.

Li, R., C. Liu, R. Zhao et al. 2019. CRISPR/Cas9-mediated SlNPR1 mutagenesis reduces tomato plant drought tolerance. *BMC Plant Biology* 19(1):1–13.

Li, R., L. Zhang, L. Wang et al. 2018. Reduction of tomato-plant chilling tolerance by CRISPR–Cas9-mediated SlCBF1 mutagenesis. *Journal of Agricultural and Food Chemistry* 66(34):9042–9051.

Li, S. and L. Xia. 2020. Precise gene replacement in plants through CRISPR/Cas genome editing technology: current status and future perspectives. *aBIOTECH* 1(1):58–73.

Li, T., B. Liu, C.Y. Chen et al. 2016. TALEN-mediated homologous recombination produces site-directed DNA base change and herbicide-resistant rice. *Journal of Genetics and Genomics* 43(5):297–305.

Li, T., B. Liu, M.H. Spalding et al. 2012. High-efficiency TALEN-based gene editing produces disease-resistant rice. *Nature Biotechnology* 30(5):390.

Li, T., S. Huang, W.Z. Jiang et al. 2011. TAL nucleases (TALNs): hybrid proteins composed of TAL effectors and FokI DNA-cleavage domain. *Nucleic Acids Research* 39(1):359–372.

Li, T., X. Yang, Y. Yu et al. 2018. Domestication of wild tomato is accelerated by genome editing. *Nature Biotechnology* 36(12):1160–1163.

Liang, Z., K. Chen, Y. Zhang et al. 2018. Genome editing of bread wheat using biolistic delivery of CRISPR/Cas9 in vitro transcripts or ribonucleoproteins. *Nature Protocols* 13(3):413.

Liu, H., Y. Ding, Y. Zhou et al. 2017. CRISPR-P 2.0: an improved CRISPR-Cas9 tool for genome editing in plants. *Molecular Plant* 10(3):530–532.

Liu, L., H. Ji, J. An et al. 2019. Response of biomass accumulation in wheat to low-temperature stress at jointing and booting stages. *Environmental and Experimental Botany* 157:46–57.

Lou, D., H. Wang, G. Liang and D. Yu. 2017. OsSAPK2 confers abscisic acid sensitivity and tolerance to drought stress in rice. *Frontiers in Plant Science* 8:993.

Mali, P., J. Aach, P.B. Stranges et al. 2013. CAS9 transcriptional activators for target specificity screening and paired nickases for cooperative genome engineering. *Nature Biotechnology* 31(9):833–838.

Mickelbart, M.V., P.M. Hasegawa and J. Bailey-Serres. 2015. Genetic mechanisms of abiotic stress tolerance that translate to crop yield stability. *Nature Reviews Genetics* 16(4):237–251.

Miller, J.C., S. Tan, G. Qiao et al. 2011. A TALE nuclease architecture for efficient genome editing. *Nature Biotechnology* 29(2):143–8.

Mushtaq, M., Bhat, J.A., Mir, Z.A., Sakina, A., Ali, S., Singh, A.K., Tyagi, A., Salgotra, R.K., Dar, A.A. and Bhat, R., 2018. CRISPR/Cas approach: A new way of looking at plant-abiotic interactions. *Journal of Plant Physiology*, 224:156–162.

Nadolska-Orczyk, A., Rajchel, K.I., Orczyk, W. and Gasparis, S. 2017. Major genes determining yield-related traits in wheat and barley. *Theoretical and Applied Genetics* 130:1081–1098.

Nguyen, H.C., Lin, K.H., Ho, S.L., et al. 2018. Enhancing the abiotic stress tolerance of plants: from chemical treatment to biotechnological approaches. *Physiologia Plantarum* 164(4):452–466.

Nishizawa-Yokoi, A., T. Cermak, T. Hoshino et al. 2016. A defect in DNA Ligase4 enhances the frequency of TALEN-mediated targeted mutagenesis in rice. *Plant Physiology* 170(2):653–666.

Osakabe, K., Y. Osakabe and S. Toki. 2010. Site-directed mutagenesis in Arabidopsis using custom-designed zinc finger nucleases. *Proceedings of the National Academy of Sciences* 107(26):12034–12039.

Osakabe, Y., T. Watanabe, S.S. Sugano et al. 2016. Optimization of CRISPR/Cas9 genome editing to modify abiotic stress responses in plants. *Scientific Reports* 6(1):1–10.

Pabo, C.O., E. Peisach and R.A. Grant. 2001. Design and selection of novel Cys2His2 zinc finger proteins. *Annual Review of Biochemistry* 70(1):313–340.

Palpant, N.J. and Dudzinski, D., 2013. Zinc finger nucleases: looking toward translation. *Gene Therapy* 20(2):121–127.

Petolino, J.F., A. Worden, K. Curlee et al. 2010. Zinc finger nuclease-mediated transgene deletion. *Plant Molecular Biology* 73(6):617–628.

Puchta, H., B. Dujon and B. Hohn. 1993. Homologous recombination in plant cells is enhanced by in vivo induction of double strand breaks into DNA by a site-specific endonuclease. *Nucleic Acids Research* 21(22):5034–5040.

Rong, Y.S. and K.G. Golic. 2000. Gene targeting by homologous recombination in Drosophila. *Science* 288(5473):2013–2018.

Rudin, N., E. Sugarman and J.E. Haber. 1989. Genetic and physical analysis of double-strand break repair and recombination in *Saccharomyces cerevisiae*. *Genetics* 122(3):519–534.

Saikia, B., J. Debbarma, J. Maharana et al. 2020. SlHyPRP1 and DEA1, the multiple stress responsive eight-cysteine motif family genes of tomato (*Solanum lycopersicum L.*) are expressed tissue specifically, localize and interact at cytoplasm and plasma membrane in vivo. *Physiology and Molecular Biology of Plants* 26 (12):2553–2568.

Semiz, G., J.D. Blande, J. Heijari et al. 2012. Manipulation of VOC emissions with methyl jasmonate and carrageenan in the evergreen conifer *Pinus sylvestris* and evergreen broadleaf *Quercus ilex*. *Plant Biology* 14:57–65.

Shan, Q., Y. Wang, J. Li et al. 2013. Targeted genome modification of crop plants using a CRISPR-Cas system. *Nature Biotechnology* 31(8):686–688.

Shan, Q., Y. Wang, J. Li and C. Gao. 2014. Genome editing in rice and wheat using the CRISPR/Cas system. *Nature Protocols* 9(10):2395–2410.

Shen J, B. Lv, L. Luo et al. 2017. The NAC-type transcription factor OsNAC2 regulates ABA-dependent genes and abiotic stress tolerance in rice. *Science Reporter* 7:40641.

Shen, C., Z. Que, Y. Xia et al. 2017. Knock out of the annexin gene OsAnn3 via CRISPR/Cas9-mediated genome editing decreased cold tolerance in rice. *Journal of Plant Biology* 60(6):539–547.

Shi, J., H. Gao, H. Wang et al. 2017. ARGOS 8 variants generated by CRISPR-Cas9 improve maize grain yield under field drought stress conditions. *Plant Biotechnology Journal* 15(2):207–216.

Shim, J.S., N. Oh, P.J. Chung et al. 2018. Overexpression of OsNAC14 improves drought tolerance in rice. *Frontiers in Plant Science* 9:310.

Shukla, V., Gupta, M., Urnov, F. et al. 2016. U.S. Patent No. 9,523,098. U.S. Patent and Trademark Office, Washington, DC.

Silva, G., Poirot, L., Galetto, R. et al. 2011. Meganucleases and other tools for targeted genome engineering: perspectives and challenges for gene therapy. *Current Gene Therapy* 11(1): 11–27.

Surabhi, G.K., Badajena, B., and Sahoo, S.K. Genome editing and abiotic stress tolerance in crop plants. In: Wani, S.H., editor. *Recent Approaches in Omics for Plant Resilience to Climate Change*. Switzerland: Springer Nature; 2019. pp. 35–56.

Symington, L.S. and J. Gautier. 2011. Double-strand break end resection and repair pathway choice. *Annual Review of Genetics* 45:247–271.

Thierry, A. and B. Dujon. 1992. Nested chromosomal fragmentation in yeast using the meganuclease I-Sce I: a new method for physical mapping of eukaryotic genomes. *Nucleic Acids Research* 20(21):5625–5631.

Townsend, J.A., D.A. Wright, R.J. Winfrey et al. 2009. High-frequency modification of plant genes using engineered zinc-finger nucleases. *Nature* 459(7245):442–445.

Tran, M.T., D.T.H. Doan, J. Kim et al. 2021. CRISPR/Cas9-based precise excision of SlHyPRP1 domain (s) to obtain salt stress-tolerant tomato. *Plant Cell Reports* 40(6):999–1011.

Tubiello, F.N., J.F. Soussana and S.M. Howden. 2007. Crop and pasture response to climate change. *Proceedings of the National Academy of Sciences* 104(50):19686–19690.

Van Eck, J., P. Keen and M. Tjahjadi. 2019. *Agrobacterium tumefaciens*-mediated transformation of tomato. In *Transgenic Plants*, ed. S. Kumar, P. Barone and M. Smith, 225–234. Humana Press, New York, NY.

Vats, S., S. Kumawat, V. Kumar et al. 2019. Genome editing in plants: exploration of technological advancements and challenges. *Cells* 8(11):1386.

Veillet, F., L. Perrot, L. Chauvin et al. 2019. Transgene-free genome editing in tomato and potato plants using agrobacterium-mediated delivery of a CRISPR/Cas9 cytidine base editor. *International Journal of Molecular Sciences* 20(2):402.

Vu, T.V., V. Sivankalyani, E.J. Kim et al. 2020. Highly efficient homology-directed repair using CRISPR/Cpf1-geminiviral replicon in tomato. *Plant Biotechnology Journal* 18(10):2133–2143.

Wang, L., L. Chen, R. Li et al. 2017. Reduced drought tolerance by CRISPR/Cas9-mediated SlMAPK3 mutagenesis in tomato plants. *Journal of Agricultural and Food Chemistry* 65(39):8674–8682.

Wang, W.C., T.C. Lin, J. Kieber and Y.C. Tsai. 2019. Response regulators 9 and 10 negatively regulate salinity tolerance in rice. *Plant and Cell Physiology* 60(11):2549–2563.

Wang, Y., X. Cheng, Q. Shan et al. 2014. Simultaneous editing of three homoeoalleles in hexaploid bread wheat confers heritable resistance to powdery mildew. *Nature Biotechnology* 32(9):947–951.

Wright, D.A., T. Li, B. Yang et al. 2014. TALEN-mediated genome editing: prospects and perspectives. *Biochemical Journal* 462(1):15–24.

Wu, X., A.J. Kriz and P.A. Sharp. 2014. Target specificity of the CRISPR-Cas9 system. *Quantitative Biology* 2(2):59–70.

Wu, Y., D. Zhang, J.Y. Chu et al. 2012. The Arabidopsis NPR1 protein is a receptor for the plant defense hormone salicylic acid. *Cell Reports* 1(6):639–647.

Xie, K. and Y. Yang. 2013. RNA-guided genome editing in plants using a CRISPR–Cas system. *Molecular Plant* 6(6):1975–1983.

Xiong, J.S., J. Ding and Y. Li. 2015. Genome-editing technologies and their potential application in horticultural crop breeding. *Horticulture Research* 2(1):1–10.

Xu, H., T. Xiao, C.H. Chen et al. 2015. Sequence determinants of improved CRISPR sgRNA design. *Genome Research* 25(8):1147–1157.

Yin, W., Y. Xiao, M. Niu et al. 2020. ARGONAUTE2 enhances grain length and salt tolerance by activating BIG GRAIN3 to modulate cytokinin distribution in rice. *The Plant Cell* 32(7):2292–2306.

Yu, W., L. Wang, R. Zhao et al. 2019. Knock-out of SlMAPK3 enhances tolerance to heat stress involving ROS homeostasis in tomato plants. *BMC Plant Biology* 19(1):1–13.

Zeng, D.D., C.C. Yang, R. Qin et al. 2018. A guanine insert in OsBBS1 leads to early leaf senescence and salt stress sensitivity in rice (*Oryza sativa* L.). *Plant Cell Reports* 37(6):933–946.

Zetsche, B., J.S. Gootenberg, O.O. Abudayyeh et al. 2015. Cpf1 is a single RNA-guided endonuclease of a class 2 CRISPR-Cas system. *Cell* 163(3):759–771.

Zetsche, B., M. Heidenreich, P. Mohanraju et al. 2017. Multiplex gene editing by CRISPR–Cpf1 using a single crRNA array. *Nature Biotechnology* 35(1):31–34.

Zhang, A., Y. Liu, F. Wang et al. 2019. Enhanced rice salinity tolerance via CRISPR/Cas9-targeted mutagenesis of the OsRR22 gene. *Molecular Breeding* 39(3):1–10.

Zhang, H., J. Zhang, P. Wei et al. 2014. The CRISPR/Cas9 system produces specific and homozygous targeted gene editing in rice in one generation. *Plant Biotechnology Journal* 12(6):797–807.

Zhang, H., J. Zhang, Z. Lang et al. 2017. Genome editing—principles and applications for functional genomics research and crop improvement. *Critical Reviews in Plant Sciences* 36(4):291–309.

Zhang, J., S. Zhang, M. Cheng et al. 2018. Effect of drought on agronomic traits of rice and wheat: a meta-analysis. *International Journal of Environmental Research and Public Health* 15(5):839.

Zhang, M., Y. Cao, Z. Wang et al. 2018. A retrotransposon in an HKT1 family sodium transporter causes variation of leaf Na^+ exclusion and salt tolerance in maize. *New Phytologist* 217(3):1161–1176.

Zhang, S., Z. Jiao, L. Liu et al. 2018. Enhancer-promoter interaction of SELF PRUNING 5G shapes photoperiod adaptation. *Plant Physiology* 178(4):1631–1642.

Zhang, X.H., L.Y. Tee, X.G. Wang et al. 2015. Off-target effects in CRISPR/Cas9-mediated genome engineering. *Molecular Therapy-Nucleic Acids* 4:264.

Zhang, Y., F. Zhang, X. Li et al. 2013. Transcription activator-like effector nucleases enable efficient plant genome engineering. *Plant Physiology* 161(1):20–27.

Zhang, Y., Z. Liang, Y. Zong et al. 2016. Efficient and transgene-free genome editing in wheat through transient expression of CRISPR/Cas9 DNA or RNA. *Nature Communications* 7(1):1–8.

Zhao, C., Z. Zhang, S. Xie et al. 2016. Mutational evidence for the critical role of CBF transcription factors in cold acclimation in Arabidopsis. *Plant Physiology* 171(4):2744–2759.

Zhao, Y., C. Zhang, W. Liu et al. 2016. An alternative strategy for targeted gene replacement in plants using a dual-sgRNA/Cas9 design. *Scientific Reports* 6(1):1–11.

Zhou, J., K. Deng, Y. Cheng et al. 2017. CRISPR-Cas9 based genome editing reveals new insights into microRNA function and regulation in rice. *Frontiers in Plant Science* 8:1598.

16 Molecular Breeding in Rice for Abiotic Stress Resilience
The Story since 2004

Wricha Tyagi, James M., Magudeeswari P., and Mayank Rai
School of Crop Improvement, College of PG Studies in Agricultural Sciences, Central Agricultural University (Imphal), Umiam, Meghalaya, India

CONTENTS

16.1 INTRODUCTION

The year 2004 was declared as the international year of rice, coinciding with the publication of the rice genome sequence in 2005, leading to a revolution in genomics-assisted breeding in rice. Climate change poses a serious threat to global food and nutritional security. Crop yields increased by 56% worldwide during 1965–85 due to advances made in the green revolution (Thudi et al. 2020). During the first decade of this century, there was a rapid advancement in sequencing technologies and annotation tools, leading to sequencing and better sequence annotation of many crop plants. After *Arabidopsis*, rice (466 Mb), which has the smallest genome among monocots, is the model species for plant genetics and genomics research. The rice (Nipponbare) draft genome sequence was first publicly released in 2005 (International Rice Genome Sequencing Project and Sasaki 2005). Now, sequences of many cultivars, like Longdao5 (Jiang et al. 2017), 93–11 (Yu et al. 2002), Minghui 63, Zhenshan 97 (Zhang et al. 2016), and IR64 (Du et al. 2017), among many others, are available. Simultaneously, sequences of many of the wild species, like *Oryza rufipogon* (AA), *O. meridionalis* (BB) (Brozynska et al. 2017), and *O. granulate* (Shi et al. 2020), also became available. Also, a core collection of 3010 rice accessions (known as the 3K panel) collected from 89 countries was sequenced (3,000 Rice Genomes Project, 2014) (Li et al. 2014b). Identifying trait-specific markers is a pre-requisite for marker-assisted selection (MAS) and marker-assisted backcrossing (MAB).

DOI: 10.1201/9781003159636-16

Availability of the rice reference genome sequence led to rapid advancements in mapping and identifying markers in cheaper and faster ways (Thudi et al. 2020).

Various abiotic stresses occur as an unpredictable and uncontrolled factor affecting yield in agriculture. Lack of availability of fresh water and reduction in fertile cultivated land encourage the use of water-saving drought-resistant rice (WDR) and saline-alkali/acidic lands to make up for production and productivity shortages. Excess water due to flooding has adverse effects on rice growth; the optimum temperature for rice cultivation ranges from 25 to 30 °C, and any fluctuation above or below this is also a stress to the rice plant (Xiong et al. 2020). Photosynthetic efficiency and thereby yield are also affected by the intensity of light received by the rice plant during its growth and development (Simkin et al. 2019). While the resistance to many of the biotic stresses is governed by major genes, abiotic stresses are overcome by expression of many quantitative trait loci (QTLs)/genes functioning together (Najeeb et al., 2020). Therefore, the success of any breeding program targeting optimum yield under various abiotic stress conditions essentially requires the integration of precise phenotypic selection with advanced omic tools in a cost-effective manner (at both molecular and phenotyping levels). In fact, a 5G (Genome assembly; Germplasm characterization; Gene function identification; Genomic breeding (GB) methodologies; and Gene editing technologies) breeding strategy to dramatically accelerate crop genetic improvement has been recently proposed (Varshney et al. 2020). This will facilitate the identification of marker–trait associations and superior haplotypes under stress conditions.

This chapter focuses on advances made in the identification of germplasm, traits, genic regions (underlying genes), and markers for the various abiotic stresses experienced by a rice plant. Wherever available, an effort has also been made to list the markers associated with the trait and superior haplotype, if identified.

16.2 TOOLS AVAILABLE FOR MOLECULAR BREEDING IN RICE

The development of genome sequencing technologies led to SSR (simple sequence repeats) (Wu and Tanksley 1993), STS (sequence tagged sites), and SNP (single nucleotide polymorphism) (Feltus et al. 2004) based markers emerging as DNA markers used for genotyping and MAS. Use of SSR and STS for genotyping to analyze genetic diversity, QTL mapping, and molecular breeding is quite common, even though they are labor intensive and time consuming. SNPs have emerged as the marker of choice due to their abundance, stability, automation, and cost-effectiveness. There are two types of markers required for a successful breeding program: 1) for selection of trait of interest (foreground markers and flanking recombination markers) and 2) for selection of ideal genetic background (background markers). The primary idea of any breeding program is to fix the targeted foreground in a desired background as early as possible. To attain this, one has to be aware of the haplotype(s) desired, and this itself might require extensive marker survey, sequencing, or both. Once the targeted haplotype is known, markers designed for the desired alleles, which may be functional, gene based or closely linked, or a combination of these can be used for selecting the foreground. With fine mapping, cloning, and functional validation of several agronomically important genes in rice, the number of functional/allele-based markers in the molecular breeders' kit is increasing by the day. In the case of background selection, advances in sequencing technology have led to SNP assay platforms now being used for background selection (Thomson et al. 2012; Thomson et al. 2017). With the emergence of tools for high-precision phenotyping and large-scale genotyping, and ability to handle large datasets, the concept of understanding the "breeding value" through genetic gain instead of mere background selection has now emerged (Xu et al. 2017).

With the availability of a large number of sequenced individuals in a species, the "pan-genome" concept has emerged (Zhou et al. 2020). The pan-genome is obtained by overlapping genome sets of many sequenced individuals/cultivars of a species, and it provides a new dimension of genome complexity with the presence/absence variations (PAVs) of genes among these genomes (Sun et al.

2017). A pan-genome set in rice was developed for the 3K rice panel, which is available in the RPAN browser (http://cgm.sjtu.edu.cn/3kricedb/). When the sequenced 3K rice panel was aligned with the Nipponbare genome, approximately 18.9 million SNPs were identified. For large-scale genotyping, genome-wide SNP assay platforms like the BeadXpress (Chen et al. 2011; Thomson et al. 2012) and Infinium (Chen et al. 2014b; Thomson et al. 2017), developed by Illumina, and other notable platforms like Affymetrix (McCouch et al. 2016; Singh et al. 2015), KASP (Kompetitive allele specific PCR) (Cheon et al. 2018; Steele et al. 2018), and Fluidigm (Seo et al. 2020) have been developed and applied in rice genetic research and breeding. A pan-genome of *O. sativa* – *O. rufipogon* species complex (Zhao et al. 2018) is also available. Both of these are being used for characterizing allelic variations in many genes/QTLs. Now, efforts are ongoing to obtain a "platinum standard" reference genome, which will have high-quality sequenced data without gaps. In future, any sequenced or re-sequenced lines can be compared with the platinum standard reference genome to identify various allelic haplotypes.

An estimated 7,80,000 rice accessions are available worldwide (Allender 2011; Wang et al. 2018). The diverse germplasm (cultivated, wild, mutants, genetic stocks, and populations) available in the various gene banks needs to be effectively used for rice improvement programs. Use of genetically diverse natural accessions and developing mapping populations like MAGIC (multiparent advanced generation intercross), nested and mutant populations, and traditional bi-parental populations (recombinant inbred lines/ near isogenic lines), coupled with high-quality sequencing (genotyping by sequencing (GBS)), can give data on polymorphic SNPs, which can help in mining allelic variation and identification of the favorable allelic combination (haplotype) underlying a target trait. These accessions can also be a platinum reference genome in the near future.

Although many haplotypes are reported for various traits like yield, grain quality (Abbai et al. 2019), height (Angira et al. 2019), aroma (Addison et al. 2020), salinity tolerance (Babu et al. 2014), and P uptake (Tyagi et al. 2012), a suitable combination of haplotypes providing elite tailored rice (Abbai et al. 2019) or a green super rice (Wing et al. 2018) is yet to be identified. With advancements in the above-mentioned genomic tools, like next-generation sequencing (NGS), SNP chip based platforms, and concepts like the pan-genome, it will be easier to conduct QTL mapping, genomic selection (GS), and genome-wide association studies (GWAS), which in turn will lead to identification of new haplotypes and genes.

By 2021, more than 3600 genes had been cloned in rice (Yao et al. 2018), out of which around 360 (10%) genes are strictly involved in stress responses (Yao et al. 2018). Some of the major abiotic stress-tolerant genes include drought avoidance by deep rooting gene *DRO1* (Uga et al. 2013), salt tolerance by *SKC1* or *HKT8* (Ren et al. 2005), submergence or flood tolerance by *SUB1* (Xu et al. 2006), *SNORKEL1*, *SNORKEL2* (Hattori et al. 2009) for prolonged or deep water tolerance, cold-tolerant genes *qLTG3-1* (Fujino et al. 2008), *LTG1* (Lu et al. 2014) (also known as *HBD2*) and *COLD1* (Ma et al. 2015), and heat-tolerant genes *TT1* (Li et al. 2015) and *HTAS* (Liu et al. 2016).

16.3 DROUGHT

Drought is one of the major abiotic stresses, causing 9–10% yield reduction in cereal crops (Lesk et al. 2016). In rice, approximately 40% water deficit causes more than 50% yield loss (Daryanto et al. 2017). Reproductive and vegetative stages of rice are highly sensitive to drought stress. The most common traits studied under drought condition are plant height, leaf rolling, root length, root dry weight, root volume, total grain weight, grain yield, grains per panicle, panicle number, etc. (Yang et al. 2020a). In the past few years, an increase in QTL mapping and molecular-physiological studies has started to reveal the genetic basis of the traits governing this complex trait.

Mapping of QTLs helps to identify the regions controlling a complex trait like drought, and the QTLs can be further utilized with the help of linked markers in future studies by MAS, pyramiding, etc. A major limitation in the use of QTLs for breeding is the lack of repeatability of QTLs across different genetic backgrounds (Bernier et al. 2008). Meta QTL (mQTL) studies across traits for rice have been conducted to detect common QTLs across multiple and distinct genetic backgrounds. A total of 1230 QTLs reported from 159 genetic maps were compiled and used for mQTL construction by Yang et al. (2020a). Among the QTLs, a total of six QTLs for plant height (Dixit et al. 2014; Trijatmiko et al. 2014; Hemamalini et al. 2000; Babu et al. 2003), 38 QTLs for grain yield (Lanceras et al. 2004; Zou et al. 2005; Zhao et al. 2008; Vikram et al. 2012; Swamy et al. 2013; Wang et al. 2013; Verma et al. 2014; Dixit et al. 2014), 10 QTLs for grains per panicle (Trijatmiko et al. 2014; Sellamuthu et al. 2015; Wang et al. 2013), two QTLs for biomass (Bernier et al. 2007), seven QTLs for 1000 grain weight (Sangodele et al. 2014; Sellamuthu et al. 2015), six QTLs for panicle length (Liu et al. 2008; Sellamuthu et al. 2015), eight QTLs for harvest index (Dixit et al. 2014; Trijatmiko et al. 2014; Sellamuthu et al. 2015), one QTL for total grain weight (Zou et al. 2005), five QTLs for panicle fertility (Sellamuthu et al. 2015), two QTLs for peduncle length (Sellamuthu et al. 2015), five QTLs for spikelet fertility (Wang et al. 2013; Sellamuthu et al. 2015), and nine QTLs for total dry matter yield (Dixit et al. 2014; Sellamuthu et al. 2015) have been analyzed. Most of the mapped drought QTLs have a positive effect on yield. QTL mapping for higher yield under drought has resulted in the identification of several QTLs, such as qDTY 12.1 (Mishra et al. 2013), qDTY 3.1 and qDTY6.1 (Dixit et al. 2014), qDTY 1.1 (Vikram et al. 2011), qDTY 8.1 (Vikram et al. 2012), and qDTY 10.1 (Swamy et al. 2013). MAS and participatory breeding programs in target environments have led to the release of drought-tolerant rice varieties (e.g. Sahbhagidhan). Mega varieties have been introgressed with several drought-tolerant yield QTLs and released for cultivation in east and south-east Asia (Singh et al. 2016; Sandhu et al. 2019).

Even though several QTLs have been identified for drought, fine mapping and map-based cloning have not identified major genes. Around 110 genes are reported for drought tolerance in the Q-TARO database (Yonemaru et al. 2010). On the other hand, many more rice genes have been identified as key players in drought tolerance through overexpression studies (Kumar et al. 2017). These include different types of transcription factors like AT-hook motif nuclear-localized (*OsAHL1*), *OsWRKY30*, *OsNAC10*, etc.; kinases like receptors-like cytoplasmic kinases (*OsRLCK*s); and other genes like photosensitive leaf rolling 1 (*psl1*), drought responsive AP1 (*OsDRAP1*), dihydroorotate dehydrogenase (*OsDHODH1*), genes involved in production of compatible solutes, etc. (Zhang et al. 2020; Zhou et al. 2016; Shen et al. 2012; Jeong et al. 2010, Huang et al. 2018; Liu et al. 2009).

16.4 ACIDITY

Soil acidity is a major limiting factor for crop production. About 50% of the potentially arable land and 13% of the area under rice cultivation is acidic (von Uexküll and Mutert 1995). In India, around 25 million hectares of land is acidic, with a pH below 5.5, and the other quarter of land has pH in the range of 5.5 to 6.5 (Yumnam et al. 2017).

Phosphorus (P) availability is low in acid soils due to its fixation with various clay minerals at low pH. Hence, P availability is a major factor limiting crop production on acid soils (extensively reviewed in Das et al. 2017; Heuer et al. 2017; Tyagi et al. 2021). Phosphorus is one of the macronutrients crucial to the functioning of plant cells and is also a molecular constituent of DNA and RNA. Deficiency of phosphorus causes changes in plant metabolism, which ultimately leads to the reduction of plant biomass and yield.

The major QTL *Pup1* was identified from an aus rice variety Kasalath about two decades ago. The QTL was mapped on a 13.2-cM interval on the long arm of chromosome 12 with flanking markers C443 and G2140 and explaining 78% of the phenotypic variance (Wissuwa et al. 1998; Wissuwa et al. 2002). Fine mapping of the *Pup1* locus led to identification of a serine threonine kinase Phosphorus-starvation tolerance 1 (*PSTOL1*) present only in the tolerant genotype (Gamuyao et al. 2012). A set of molecular markers designed for the Pup1 region, which could differentiate between Kasalath and Nipponbare type allele (Chin et al. 2010), and dominant marker K46-2 marker along with K29 was recommended as a marker for MAS across different genetic backgrounds (Chin et al. 2011). Haplotype analysis revealed the presence of Kasalath type allele in several north-east India landraces and Indian varieties and also identified several rice genotypes that lack PSTOL1 but perform at par with Kasalath in low-P acidic soils (Tyagi et al. 2012; Yumnam et al. 2017). One such genotype, Chakhao Poreiton, showed P uptake at par with Kasalath in lowland acidic fields and maintained root length in response to 15-day low-P treatment (Tyagi and Rai 2017). It is hypothesized that suberin-mediated cell wall modification, releasing internally bound Pi, and transcriptional modifications using dehydration response-based signals are key to its low-P tolerance strategy (Tyagi et al. 2021). Recently, using Chakhao Poreiton as one of the parents, advanced rice breeding lines performing well in acidic soils have been identified (Bhutia et al. 2021). A region on chromosome 2 carrying several genes for signal transduction along with associated markers has also been identified. These lines and markers can be part of future molecular breeding strategies for low-P tolerance in acidic soils.

The Pup 1 region has been successfully introgressed using marker-assisted back cross-breeding in international varieties like IR64 and IR74; Indonesian varieties like Dodokan, Situ bagendit, and Batur (Chin et al. 2011); and improved samba mahsuri (Swamy et al. 2020).

Under low-pH conditions, upland and lowland rice experiences toxicity stress due to minerals like Al^{3+} and Fe^{2+}, respectively. In rice, tolerance to both Al and Fe is a quantitative trait (Ma et al. 2002; Das et al. 2017). Several QTLs, like AltTRG1.1, AltPRG1.1, AltPRG6.1, AltLRG9.1, Alt TRG1.1, and AltTRG12.1 (Famoso et al. 2011), as well as genes/transporters like *ART1* (Al resistance transcription factor 1) (Yamaji et al. 2009), *OsFRDL4*, *OsCDT3*, *OsMGT1*, *OsNrat1*, etc. (Ma et al. 2014; Li et al. 2014a) involved in Al detoxification, sequestration, or exclusion, are reported in rice. It is also reported that mean Al tolerance in japonica rice is nearly twice that of indica. However, a report by Tyagi et al. (2020) on Al toxicity tolerance in indica type rice, ARR09 (AR), suggests a novel mechanism, since no reported Al toxicity genes or QTLs overlapped with significant differentially expressed genes obtained in their results. Aluminum toxicity tolerance in this rice appears to be through expression of genes involved in cell signaling and energy conservation.

Three types of tolerance mechanism have been proposed in rice (reviewed in Das et al. 2017), with both root- (Wu et al. 2014) and shoot-based (Engel et al. 2012; Moore et al. 2014; Dufey et al. 2015) mechanisms known. An eQTL (expression QTL)- and mQTL-based approach has led to the identification of set of 54 genes associated with Fe toxicity tolerance (Dufey et al. 2015). Most of the genes identified appear to be involved in either avoidance mechanisms or detoxification.

16.5 SALINITY

Around 162.34 and 6.74 million ha of rice-growing area worldwide and countrywide, respectively, is salt affected. Estimates suggest that every year nearly 10% additional area becomes salinized, and by 2050, around 50% of arable land will be salt-affected (Kumar and Sharma 2020). Rice plants are highly sensitive to salt stress at seedling and reproductive stages, and plant death occurs when the soil electrical conductivity reaches 10 dSm^{-1} (Munns et al. 2006). Therefore, the rice productivity in salt-affected areas is very low – less than 1.5 tons per hectare.

Salt tolerance is a complex trait, and to date, 638 QTLs for germination, seedling, and reproductive stages spanning all the 12 chromosomes have been identified (Yang et al. 2020a). Out of the total QTLs identified, 305 QTLs for seedling stage salt stress (Rahman et al. 2019; Quan et al. 2018; Bizimana et al. 2017; Chen et al. 2020), 16 for germination stage (Tian et al. 2011), 16 for reproductive stage (Hossain et al. 2015), and six for tolerance at maturity stage (Ren et al. 2005) have been reported. A total of 21 mQTLs had a confidence interval of less than 0.5 cM (Yang et al. 2020a). A total of 15 different agronomic and physiological traits (Na^+ concentration, K^+ concentration, Na^+/K^+ ratio, SES (standard evaluation score), salt injury score (SIS), survival %, shoot length, root length, biomass shoot, biomass root, SFW (shoot fresh weight), SDW (shoot dry weight), plant height, leaf chlorophyll, and shoot root ratio) (Tian et al. 2011; Bizimana et al. 2017; Rahman et al. 2019) and four reproductive traits (panicle length, Na^+ concentration, pollen fertility, and grain yield) (Hossain et al. 2015) have been investigated in various studies.

An mQTL study for seedling stage salt tolerance led to identification of 11 mQTL regions on chromosomes 1 and 3 (Islam et al. 2019). An integrative meta-analysis approach using microarray and RNA seq led to identification of 3449 differentially expressed genes in response to salinity stress in root, shoot, seedling, and reproductive tissues (Mansuri et al. 2020). However, only 102 rice genes involved in salinity tolerance are listed in the Q-TARO database. Further utilization of these genes in breeding programs will require the identification of alleles responsible for imparting tolerance to salinity at different stages of rice growth and design of foreground markers.

A 1.5 Mb QTL region located on chromosome 1, named *saltol*, responsible for seedling stage salinity stress tolerance obtained from donor parent Pokkali and explaining 40% of the phenotypic variance for shoot Na^+/K^+ ratio (Bonilla et al. 2002) is the most well-studied QTL for salinity tolerance to date. Similarly, two large effect QTLs, *qSNC* (shoot Na^+ concentration) and *qSKC-1* (shoot K^+ concentration), explaining 48.5% and 40.1% of phenotypic variance, respectively, were obtained from Nonabokra (Lin et al. 2004). Fine mapping of *qSKC-1* revealed the gene *SKC1* located within the *saltol* region that encodes an HKT-type (high affinity K^+ transporter) Na^+ transporter (Ren et al. 2005) to be the candidate gene for seedling stage salt stress. Recent studies have reported successful incorporation of the *saltol/SKC* region into high-yielding mega varieties Pusa 44 and Sarjoo 52 using recombinant selection with flanking markers RM493 and G11a (Krishnamurthy et al. 2020).

Another major QTL, *qSE3* (Seedling establishment 3), was also cloned using a map-based strategy and is known to encode a potassium transporter, OsHAK21 (Shen et al. 2015). OsHAK21 mediates K^+ absorption by the plasma membrane and plays a crucial role in the maintenance of Na^+/K^+ homeostasis in seedlings under salinity stress (Shen et al. 2015). Na^+/H^+ antiporter (Rajgopal et al. 2007), vacuolar ATPases (Tyagi et al. 2005), and PPases are some of the key targets for manipulation for attaining salinity tolerance (Van Zelm et al. 2020). Alleles across these key genes can be mined on diverse rice genotypes and the desired haplotype(s) targeted for introgression into breeding programs using MAS. Evaluation to find the allelic constitution of *qSE3* led to identification of five haplotypes of *qSE3*, with HAP3 being most significantly associated with seedling germination under salt stress (He et al. 2019). Eight accessions, namely Sabharaj, RTS 14, Kiang-chou-chiu, Ai-Chiao-Hong, Peh-Kuh-Tsao-Tu, Taichung Native 1, Zhenshan2, and Peh-Kuh, containing HAP3 were identified. These identified donors can be used in breeding programs for salinity tolerance.

16.6 SUBMERGENCE

About 30% of people (700 million) living in abject poverty (i.e., daily income less than 1$) in Asian countries reside in flood-prone rice cultivating regions of south Asia, in Nepal, Bangladesh, and India (Oladosu et al. 2020). Nearly 30% of the rice cultivated area in India (12–14 Mha) is susceptible to flash flooding, with an average productivity of 0.5–0.8 t ha^{-1} only (Dar et al. 2017). To date, three types of submergence have been reported in rice: (i) submergence during germination or anaerobic germination (AG) (anoxia or hypoxia), which leads to poor seedling growth and death (Ismail et al. 2008), (ii) complete submergence at seedling or tillering stage for a period of 10 to 14 days, leading to complete yield loss, and (iii) prolonged deep-water flooding of 20 cm to 4 m for 1–2 months in lowland rice cultivated areas.

In response to submergence, rice plants develop more aerenchyma cells, which help in efficiently translocating oxygen to submerged roots and/or shoots (Colmer and Voesenek 2009). It is now known that auxin response factor (ARF)–mediated signaling is required for constitutive aerenchyma formation in rice roots (Yamauchi et al. 2019). Prevention of oxygen loss by deposition of suberin on cell wall forms a barrier to relative oxygen loss (Yamauchi et al. 2018). Formation of a leaf gas film helps the rice plant maintain respiration and photosynthesis (Pederson et al. 2009). Light intensity, turbidity, temperature, pH, and gas diffusion are major factors affecting aerenchyma formation and altered leaf morphology, ultimately making plants submergence susceptible. Adventitious root formation in nodes can substitute the functions of the original roots, which perform poorly under submerged conditions (Steffens and Rasmussen 2016; Kuroha et al. 2020). Reduced internode elongation under flooding (complete submergence for 1–2 weeks) is vital for survival because the elongating plants would tend to lodge as soon as the water recedes. This mechanism is suitable for flash flooding (1–2 weeks), as the limited elongation makes the plant tolerant because the plants utilize much less carbohydrate, and when the water recedes they can utilize the sugar reserves (Sarkar 1997). On the other hand, there are landraces that are tall with a thick culm and can escape deep water by elongating the internodes. Both the mechanisms are ethylene dependent: in the first case, ethylene is down-regulated and thus GA_3 (gibberellins) are not produced for internodal elongation, while it is the exact reverse for the second mechanism, where ethylene-dependent internodal elongation enables the plant to escape submergence.

Internode elongation, coleoptile elongation, and survival % were the three major traits found to be crucial for the above-mentioned three types of submergence. Overall, around 53 QTLs for the three stages of submergence across all the rice chromosomes have been reported (Hattori et al. 2007; Angaji et al. 2010; Xu et al. 1996; Septiningsih et al. 2012; Gonzaga et al. 2016), and among them five genes, *SUB1* (Xu et al. 2006), *SK1*, *SK2* (Hattori et al. 2007; Hattori et al. 2009), *SD1* (Kuroha et al. 2016) and *OsTPP7* (Kretzschmar et al. 2015), are fine mapped.

In the case of flash floods, submergence-tolerant rice varieties ("quiescent rice"/"scuba rice") have a mutation in the *SUB1* gene (Xu et al. 2006). The down-regulated *SUB1* allele can tolerate up to 17 days of submergence. Thus, when the gene is down-regulated, low ethylene accumulation occurs and hence no GA_3 are produced; therefore, the plant remains inside water as if it is in a resting stage. For deep-water submergence tolerance, three genes, *SNORKEL1* (SK1), *SNORKEL2* (SK2), and *SD1* (Kuroha et al. 2016), were identified. These function in an opposite way to the SUB1 pathway. Here, ethylene accumulates during submerged conditions, triggering the expression of the SK genes, which induce the expression of abscisic acid (ABA) degradative enzyme, resulting in reduction of ABA levels (Saika et al. 2007). Decreased ABA levels can lead to increased GA_3 response and subsequent shoot elongation in rice, resulting in the induction of internode elongation in deep-water rice. Both *SK* and *SUB1A* genes encode ERFs (Ethylene-Responsive Factors) and are connected to GA_3 production, but they have opposing functions in regulating plant growth in response to flooding (Hattori et al. 2009). In anaerobic conditions, germination of seed and coleoptile elongation are affected by the expression of *OsTPP7* (trehalose-6-phosphate phosphatase) gene. The *OsTPP7* gene modulates the T6P/sucrose ratio, activating germinating tissues, which then leads to resource allocation from the endosperm reserve to the coleoptile, thus resulting in enhanced coleoptile growth during AG.

Essentially derived varieties with SUB1 gene have shown more yield potential than their recurrent parents under field conditions (Ismail et al. 2013). In an attempt to characterize the haplotypes present in 168 landraces, eight, nine, and 61 haplotypes for *SUB1A*, *SUB1B*, and *SUB1C* have been identified, respectively (Singh et al. 2020).

16.7 COLD

The average minimum temperature over most (64%) of the total land area on Earth is less than 0 °C (Rihan et al. 2017). Cold stress can be classified as chilling (0–15 °C) and freezing (<0 °C) stress. Cold tolerance at seedling stage is important for adaptation to high-altitude locations that can experience cold temperatures at night or in the growing season. Also, cold tolerance during reproductive stages is critical for pollen survival, seed set, and grain filling to ensure maximal yield (Suh et al. 2010).

For seedling stage cold tolerance, traits for selection are high fresh weight, survival rate, new leaf emergence, seedling/leaf growth, and relative water content. Sterility percentage, total grains per panicle, rate of hollowness per panicle, extraction degree of panicle, and length of anther are the traits to be selected for booting stage cold tolerance (Zeng et al. 2006; Suh et al. 2010). Also, some physiological traits, such as increased electrolyte leakage, changes in sucrose, lipid peroxidation, proline, chlorophyll content (Kanneganti and Gupta 2008), reactive oxygen species (ROS), and malondialdehyde (MDA) level, which is an indication of cellular oxidative damage (Nakashima et al. 2007), can be studied to identify suitable cold-tolerant genotypes. A seedling growth scale, which can be used for a standard evaluation system for rice (IRRI 2002), is also available. A growth scale can also be used where a decrease in fresh weight can indicate cold stress and recovery of seedlings after cold stress can distinguish the tolerant lines. Cold stress affects membrane stability, and severe cold leads to formation of ice crystals within the cells. So, molecular targets focusing on membrane stability and anti-freezing proteins are two key areas for attaining cold stress tolerance.

More than 300 QTLs related to rice cold tolerance have been detected (www. gramene.org/db/qtl) and around 44 genes related to cold tolerance are identified, with only eight genes, *LTG3-1*, *COLD1*, *CTS-9*, *GSTZ2*, *LTG1*, *Ctb1*, *CTB4a*, and *HAN1* (Mao et al. 2019; Yun et al. 2015; Liu et al. 2018; Zhang et al. 2017), being functionally characterized. Certain genes pertain to particular growth stages, like LTG1 (Low Temperature Growth) (Fujino et al. 2008) for vegetative stage cold tolerance, CTB4 (Cold Tolerant at Booting) (Zhang et al. 2017) for booting stage cold tolerance, etc. Work in rice has revealed the role of *LTG3* and *OsSNAC* in low-temperature germination (Challam et al. 2013) and seedling stage low-temperature tolerance (Ghosh et al. 2016), respectively. Allele mining across several well-studied transcription factors like *DREB1A* and *DREB1B* has revealed a conserved pathway in rice in response to low temperature as well (Challam et al. 2015). *COLD1* (chilling tolerance divergence 1) is considered a putative cold sensor and therefore can be utilized for both seedling and vegetative stage cold stress tolerance. An mQTL analysis for cold stress has recently identified 41 mQTLs, which include 44 previously identified genes (Kong et al. 2020).

16.8 HEAT

It is estimated that climate change will affect 20 million ha of the rice-growing area in Asian countries and reduce productivity by 14% in south Asia, 10% in east Asia and the Pacific, and 15% in sub-Saharan Africa (IFPRI 2016). An increase of 1 °C night-time temperature may decrease rice productivity by 7–10% (Peng et al. 2004); similarly, a daytime increase in temperature from 28 to 34 °C will result in 7–8% yield decrease (Korres et al. 2017).

Fluctuations in traits like reduced grain filling, low amylose content resulting in chalky grains, poor-quality grains (Yamakawa et al. 2007), reduced plant height, tiller number, total dry weight (Wassmann et al. 2009), reduced anther length, size of basal pore, and early-morning flowering (Shah et al. 2011) can indicate heat stress in rice plants. Cool leaf canopy (Shah et al. 2011) can indicate that the plants do not experience heat stress. Physiological traits like decreased cell membrane thermostability, reduced relative water content, accumulation of ROS, increased levels of superoxide dismutase, catalase, peroxidase, ascorbic peroxidase, and ADP-glucose pyrophosphorylase, and presence of MDA can indicate heat stress. Also, traits like chlorophyll fluorescence, chlorophyll content, carotenoids, and gaseous exchange rates (Das et al. 2013; Zafar et al. 2017) can be used as key traits to identify tolerant lines.

More than 200 QTLs are reported across all the rice chromosomes for heat tolerance. mQTL analysis for these 200 QTLs led to identification of 35 mQTLs that confer reproductive stage heat tolerance in rice (Raza et al. 2020). Some QTLs have been fine mapped to candidate genes like BHT/HTB (Booting stage Heat Tolerance) for booting stage heat tolerance (Cao et al. 2020) and HTAS9 (Heat Tolerance at Seedling Stage) (Liu et al. 2016) for seedling stage heat tolerance. Some genes, like *TT1* (Thermo Tolerance) (Li et al. 2015), many HSP (Heat Shock Protein) (Zou et al. 2009;

Qin et al. 2014) genes, and a few genes encoding transcription factors like HSFA2 (Heat Stress Transcription Factor A2) (Liu et al. 2010) are also reported to play a vital role in heat-tolerant lines.

16.9 LOW LIGHT INTENSITY

Plants can generally utilize red light that falls in the PAR (photosynthetically active radiation) region, ranging from 400 to 700 nm. Irradiance of light more than 380 cal cm^{-2} h^{-1} provides better photosynthate production, leading to higher yield, whereas less than 350 cal cm^{-2} h^{-1} is called low light stress, affecting various yield and grain quality traits in rice (Rai et al. 2021). Low light can reduce the net photosynthetic rate; this is due to the CO_2 evolved during dark respiration (Chen et al. 2014a). Also, when planting is done in an intensified planting system, a low red to far-red light ratio (R/FR) can lead to low CO_2 assimilation in leaf (Marchiori et al. 2014).

Incidence of low light in the reproductive stages can increase grain chalkiness (Jichao et al. 2007) and reduce spikelet fertility, test weight (Liu et al. 2014), head rice recovery, amylase content, and gel consistency, which then ultimately leads to yield loss up to 55% (Liu et al. 2014). An ideal rice plant for low light tolerance is expected to have higher photosynthetic efficiency, erect leaf, delayed senescence, maximum harvest of available photons, and obviously high grain yield (Rai et al. 2021). A candidate gene–based marker–trait association study on rice germplasm from eastern India has led to the identification of eight rice genotypes, five traits, eight genes, and six markers that can be used for devising strategies to increase yield under low light intensity (Dutta et al. 2018).

For screening genotypes for low light tolerance, generally, cloth meshes of variable density were used (Murchi et al. 2002). Also, chlorophyll fluorescence and photosynthesis measurement systems can be used to measure low light–related traits (Zhang et al. 2010).

Forty-three QTLs related to low light tolerance have been identified and are available in the gramene database; 21 of these are specifically for spikelet fertility. These include 20 genes that are associated with different traits contributing to low light tolerance (Rai et al. 2021). For instance, during the vegetative stage, a gene called sgr (Stay green) maintains the photosynthesis activity of the rice plant even in low light conditions (Jiang et al. 2007) (Figure 16.1).

16.10 DONORS/GERMPLASM IDENTIFIED FOR ABIOTIC STRESS TOLERANCE

The development of tolerant varieties depends on availability of the best donor for the identified trait. Several salt-tolerant donors, including Pokkali, Nona bokra (Rahman et al. 2016), Capsule (an indica landrace grown in Bangladesh) (Rahman et al. 2016), Hasawi (aus variety) (Bimpong et al. 2014), Cheriviruppu (an indica landrace) (Hossain et al. 2015), and aromatic traditional rice landrace Kolajoha have been reported to date. Rice cultivars like Vandana, Apo, N22, Azucena, Moroberekan, tarom Molaei, INRC10192, and IRAT 177 have been identified as donors for drought. Donors like FR13A, FR43B (Vergara et al. 1976) for flash flood of 1–2 weeks, C9285 (Dowai 38/9) (Hattori et al. 2009) for internode elongation, and Khao Hlan On (Angaji et al. 2010) for AG have been identified. Other landraces, like Khao Tah Haeng 17, Leuang Pratew 123 (Singh et al. 2017), Leb Mue Nahng 111, Pin Gaew 56, Plai Ngahm, Khao Luang (floating rice), Kurkarruppan, Goda Heenati, and Thuvalu (flash), Jaladhi 2, Chinsura 21 (deep water), Rajasail, Patnai 23, and Sirembeh Puti (stagnant water), also show tolerance at different stages (Singh et al. 2017). Wild species like *Oryza rufipogon* (Hattori et al. 2009) can also tolerate deep-water submergence to threshold levels.

For cold tolerance, genotypes like LTH (Zhang et al. 2012), Oro (Maia et al. 2017), WR157 (Han et al. 2020), WR 03-35, Kongyu 131 (Guan et al. 2019), Haoannong (Najeeb et al. 2020), Andassa, Tana, and Hitomebore (Alemayehu et al. 2021) have been reported. Introduced varieties like China 1039 and China 1007 and developed lines like K65 and K39 are the most cold-tolerant rice varieties of Kashmir (Hussain and Sofi, 2018).

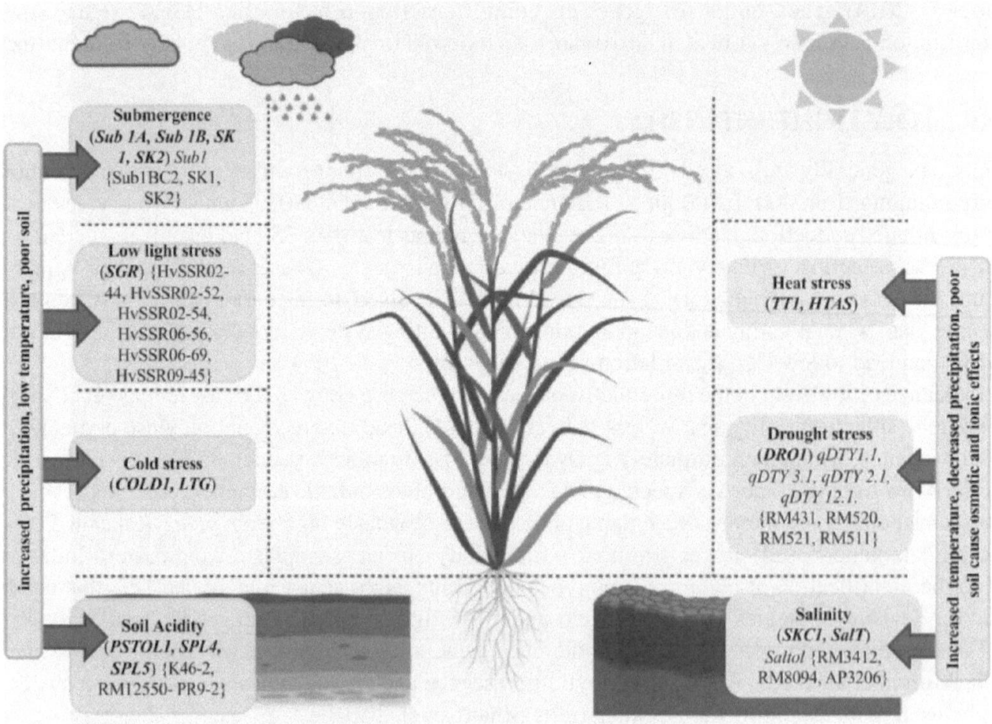

FIGURE 16.1 Representative figure showing different environmental (abiotic) stresses in rice. The stresses mentioned in the right side are the effect of increased temperature, and the left side indicates excess rainfall and poor soil problems. The genes that have been mapped and tested under field conditions for developing climate-resilient rice varieties through marker-assisted deployment are mentioned in (), with bold letters for the respective traits studied, and some major QTLs used are mentioned without bold. The peak foreground markers used for selection are given within { }.

In response to heat, several wild species of rice show differences in flowering pattern. For example, *O. eichingeri* and *O. officinalis* show early-morning flowering, *O. australiensis* shows late-afternoon flowering, and *O. alta* shows a midnight flowering phenotype (Thanh et al. 2010). *O. rufipogon* shifted flower opening time to 30 min earlier than Nipponbare. These species can be targeted for breeding for heat avoidance. Also, *O. rufipogon* and *O. nivara* have been identified for heat tolerance (Prasanth et al. 2017). Among cultivated rice, Nagina22 (Mackill et al. 1982; Prasad et al. 2006), Akitakomachi (Matsui et al. 2001, Dular (Thangapandian et al. 2010), and Teqing (Jagadish et al. 2008) are the most tolerant genotypes to heat stress reported to date. Many lines are available for low light tolerance, like IRCTN 91-84, IRCTN 91-94, Rhyllo Red, PS-4, Mahisugandh, Danteswari, Kunti, Megha Rice 1 (Dutta et al. 2018), Takanari (Ohkubo et al. 2020), Purnendu, and Swarnaprabha (Kumar et al. 2020).

The average yield potential of any of the identified landraces that can be used as donors is 2 t h^{-1} (Oladosu et al. 2020), and for wild germplasm even less. Therefore, it is imperative that introgression lines (ILs) in different genetically diverse lines be used as donors when crossing with elite cultivars so as not to compromise their yield potential. In fact, many such ILs have been developed for breeding programs; for example, lines like IR49830-7-1-2-2 and IR40931-26 (Xu and Mackill 1996) were developed from FR13A for use in submergence tolerance breeding programs. Similar ILs are now available to rice breeders for various programs targeting abiotic stress tolerance.

16.11 CROSS TALK

Rice is a staple food crop grown mostly under lowland conditions. During the different growth stages the crop has to overcome several abiotic stresses, like drought, salinity, cold, heat stress, flooding, and soil mineral stress, often simultaneously, which ultimately causes severe yield loss. Strategies to mitigate adverse effects like osmotic effects by production of compatible solutes or reduction in ROS will be effective in many stresses, like drought, salinity, acidity, and heat. However, these strategies will not provide protection against ion toxicity observed in salinity or acidity. The abiotic stress signaling and responses in plants have been extensively studied and recently well summarized (Gong et al. 2020; Tyagi et al. 2021). Plants have evolved quick and sophisticated sensory mechanisms to perceive the abiotic stress cues, convert them into cellular signals, and transmit the signals within cells and tissues. So far, several abiotic stress sensors have been identified; the majority of this work has been in *Arabidopsis*. Whether similar sensors and epigenetic changes (Liu and He 2020) involving methylation of cytosine and adenine residues work for all the abiotic stresses in rice needs to be seen. Membranes are common points of signaling and response for heat and cold stress. On the other hand, vacuoles can be actively involved in pumping out excess ions (in the case of meristematic cells) or can act as sinks for accumulation of excess sodium (salinity) or Al^{3+}/Fe^{2+} (acidity). The role of hormones like ABA in drought, salinity, and cold, and ethylene in submergence, is well documented. It is also known that certain transcription factors recognize and bind to abiotic stress-specific cis-acting elements only under stress conditions manipulating the entire signaling pathway (Rai et al. 2009; Cohen and Leach 2019). Furthermore, there are some transcription factors that play a role in tolerance to multiple abiotic stresses; for example, DST (Drought and salt tolerance) is involved in controlling stomatal aperture (Huang et al. 2009).

For example, salt stress in rice can also cause water stress and metal toxicity (Hasegawa et al. 2000). Similarly, drought is associated with other stresses like salinity, heat stress, and nutrient deficiencies. So, while breeding for high-yielding varieties, it is desirable to have multiple stress tolerance. In general, heat/salinity and drought stresses occur together; hence, breeding for one of them may often be interrelated with the other (Singh et al. 2016; Senguttuvel et al. 2020).

Varieties with the *SUB1* gene have better tolerance than normal lines when exposed to drought after the onset of flooding; this is because the gene regulates ABA responses, ROS accumulation, and other ERFs to confer dehydration tolerance (Fukao et al. 2011). SUB1A enhanced recovery from drought at the vegetative stage through reduction of leaf water loss and lipid peroxidation and increased expression of genes associated with acclimation to dehydration (Dixit et al. 2017). Recently, mega varieties like IR64, Swarna, and Sambha mahsuri have been introgressed with multiple QTLs like SUB1, SalTol, and DTY QTLs to impart multiple stress tolerance (Singh et al. 2016).

Overexpression of *OsARD1* elevates the endogenous ethylene release rate, enhances the tolerance to submergence stress, and reduces the sensitivity to drought, salt, and osmotic stresses in rice (Liang et al. 2019). Occurrence of salt and low light stress during vegetative growth increased the susceptibility of rice plants to chilling damage during panicle development (Koumoto et al. 2016).

16.12 PHENOTYPING PLATFORMS

The importance of precise, high-throughput, and automated phenotyping (phenomics) is increasingly being realized in molecular and conventional breeding programs. A High-Throughput Rice Phenotyping Facility (HRPF) has been developed in Wuhan, China, which can accommodate more than 5000 plants, where abiotic stress like drought can be applied by automated systems (Yang et al. 2014). For real-time field phenotyping, a high-throughput phenotyping (HTP) system where a tractor mounted with eight different sensory cameras comprising multispectral reflectance cameras and ultrasonic sensors was used to phenotype 1500 RIL (Recombinant Inbred Lines) population (Tanger et al. 2017). Integrating computer tomography (CT) and visible red-green-blue (RGB) imaging was used to identify a new gene, *TAC1*, governing tillering angle (Wu et al. 2019). Apart from this, other phenotyping platforms like Panomma, Gia roots 2D (3D), and Rhizoscope can be used for phenotyping various shoot, root, and grain phenes in rice (Yang et al. 2020b).

Plant phenotyping was developed to predict morphological and physiological changes by different imaging platforms. Several studies have used high-throughput non-destructive image-based phenotyping to study various plant stresses, including salinity tolerance in rice (Hairmansis et al. 2014). The response of rice to different levels of salt stress was quantified over time using RGB and fluorescence images based on total shoot area and senescent shoot area (Hairmansis et al. 2014). For drought RGB, infrared (IR) thermography and near-IR images are mainly used for stress prediction in rice (Kim et al. 2020). In addition, water use efficiency (WUE) and transpiration rate can be predicted by Drought spotter. HTP (available at International Crops Research Institute for the Semi-Arid Tropics), late sown method (dry season), polycover tents, and temperature gradient tunnel (available at International Rice Research Institute (IRRI)) are a few screening methods used for heat tolerance studies in rice. IRRI has a rainout shelter mobile sensor rack for drought physiological studies as well.

16.13 FUTURE PROSPECTS

The post-sequencing era in rice has led to a better understanding of abiotic stress tolerance. With the availability of HTP and genotyping platforms, it is hoped to develop superior varieties that will be resilient to multiple stresses. What is required are small-budget-friendly SNP/SSR-based assays using functional markers (Salgotra and Stewart 2020) for rapid foreground selection to minimize the time needed for identification of the desired foreground. This can be followed by multi-location testing and genomic/background selection using NGS-based markers to identify breeding lines with superior performance. Such identified lines can then be used as donors as well as molecular sources to understand and harness the complex genetic basis of these quantitative traits.

REFERENCES

Abbai, R., V. K. Singh, V. V. Nachimuthu, et al. 2019. Haplotype analysis of key genes governing grain yield and quality traits across 3K RG panel reveals scope for the development of tailor-made rice with enhanced genetic gains. *Plant Biotechnology Journal* 17:1612–1622.

Addison, C. K., B. Angira, M. Kongchum, et al. 2020. Characterization of haplotype diversity in the *BADH2* aroma gene and development of a KASP SNP assay for predicting aroma in US Rice. *Rice* 13:1–9.

Alemayehu, H. A., G. Dumbuya, M. Hasan, et al. 2021. Genotypic variation in cold tolerance of 18 Ethiopian rice cultivars in relation to their reproductive morphology. *Field Crops Research* 262:108042.

Allender, C. 2011. The Second Report on the State of the World's Plant Genetic Resources for Food and Agriculture. Rome: Food and Agriculture Organization of the United Nations. *Experimental Agriculture* 47:574–574.

Angaji, S. A., E. M. Septiningsih, D. J. Mackill, and A. M. Ismail. 2010. QTLs associated with tolerance of flooding during germination in rice (*Oryza sativa* L.). *Euphytica* 172:159–168.

Angira, B., C. K. Addison, T. Cerioli, et al. 2019. Haplotype characterization of the sd1 semidwarf gene in United States Rice. *The Plant Genome* 12:190010.

Babu, N. N., K. K. Vinod, S. G. Krishnan, et al. 2014. Marker based haplotype diversity of Saltol QTL in relation to seedling stage salinity tolerance in selected genotypes of rice. *Indian Journal of Genetics and Plant Breeding* 74:16–25.

Babu, R. C., B. D. Nguyen, V. Chamarerk, et al. 2003. Genetic analysis of drought resistance in rice by molecular markers: association between secondary traits and field performance. *Crop Science* 43:1457–1469.

Bernier, J., A. Kumar, V. Ramaiah, D. Spaner, and G. Atlin. 2007. A large-effect QTL for grain yield under reproductive-stage drought stress in upland rice. *Crop Science* 47:507–516.

Bernier, J., G. N. Atlin, R. Serraj, A. Kumar, and D. Spaner. 2008. Breeding upland rice for drought resistance. *Journal of the Science of Food Agriculture* 88:927–939.

Bhutia, K. L., E. L. Nongbri, T. O. Sharma, M. Rai, and W. Tyagi. 2021. A 1.84-Mb region on rice chromosome 2 carrying *SPL4, SPL5* and *MLO8* genes is associated with higher yield under phosphorus-deficient acidic soil. *Journal of Applied Genetics* 62(2):207–222.

Bimpong, I. K., B. Manneh, B. Diop, et al. 2014. New quantitative trait loci for enhancing adaptation to salinity in rice from Hasawi, a Saudi landrace into three African cultivars at the reproductive stage. *Euphytica* 200:45–60.

Bizimana, J. B., A. Luzi-Kihupi, R. W. Murori, and R. K. Singh. 2017. Identification of quantitative trait loci for salinity tolerance in rice (*Oryza sativa* L.) using IR29/Hasawi mapping population. *Journal of Genetics* 96:571–582.

Bonilla, P., J. Dvorak, D. Mackill, K. Deal, and G. Gregorio. 2002. RFLP and SSLP mapping of salinity tolerance genes in chromosome 1 of rice (*Oryza sativa* L.) using recombinant inbred lines. *University of Philippines at Los Banos* 85:68–76.

Brozynska, M., D. Copetti, A. Furtado, et al. 2017. Sequencing of Australian wild rice genomes reveals ancestral relationships with domesticated rice. *Plant Biotechnology Journal* 15:765–774.

Cao, Z., Y. Li, H. Tang, et al. 2020. Fine mapping of the qHTB1-1 QTL, which confers heat tolerance at the booting stage, using an *Oryza rufipogon* Griff. introgression line. *Theoretical and Applied Genetics* 133:1161–1175.

Challam, C., G. A. Kharshing, J. S. Yumnam, M. Rai, and W. Tyagi. 2013. Association of qLTG3-1 with germination stage cold tolerance in diverse rice germplasm from the Indian subcontinent. *Plant Genetic Resources: Characterization and Utilization* 11:206–211.

Challam, C., T. Ghosh, M. Rai, and W. Tyagi. 2015. Allele mining across DREB1A and DREB1B in diverse rice genotypes suggests a highly conserved pathway inducible by low temperature *Journal of Genetics* 94:231–238.

Chen, C. C., M. Y. Huang, K. H. Lin, W. D. Hung, and C. M. Yang. 2014a. Effects of light quality on the growth, development and metabolism of rice (*Oryza sativa* L.). *Research Journal of Biotechnology* 9:15–24.

Chen, H., H. He, Y. Zou, et al. 2011. Development and application of a set of breeder-friendly SNP markers for genetic analyses and molecular breeding of rice (*Oryza sativa* L.). *Theoretical and Applied Genetics* 123:869–879.

Chen, H., W. Xie, H. He, et al. 2014b. High density SNP genotyping array for rice biology and molecular breeding. *Molecular Plant* 7:541–553.

Chen, T., Y. Zhu, K. Chen, et al. 2020. Identification of new QTL for salt tolerance from rice variety Pokkali. *Journal of Agronomy and Crop Science* 206:202–213.

Cheon, K. S., J. Baek, Y. Cho, et al. 2018. Single nucleotide polymorphism (SNP) discovery and kompetitive allele-specific PCR (KASP) marker development with Korean japonica rice varieties. *Plant Breeding and Biotechnology* 6:391–403.

Chin, J. H., R. Gamuyao, C. Dalid, et al. 2011. Developing rice with high yield under phosphorus deficiency: Pup1 sequence to application. *Plant Physiology* 156:1202–1216.

Chin, J. H., X. Lu, S. M. Haefele, et al. 2010. Development and application of gene-based markers for the major rice QTL Phosphorus uptake 1. *Theoretical and Applied Genetics* 120:1073–1086.

Cohen, S. P., and J. E. Leach. 2019. Abiotic and biotic stresses induce a core transcriptome response in rice. *Scientific Reports* 9:6273.

Colmer, T. D., and L. A. C. J. Voesenek. 2009. Flooding tolerance: suites of plant traits in variable environments. *Functional Plant Biology* 36:665–681.

Dar, M. H., R. Chakravorty, S. A. Waza, et al. 2017. Transforming rice cultivation in flood prone coastal Odisha to ensure food and economic security. *Food Security* 9:711–722.

Daryanto, S., L. Wang, and P. A. Jacinthe. 2017. Global synthesis of drought effects on cereal, legume, tuber and root crops production: a review. *Agricultural Water Management* 179:18–33.

Das, S., W. Tyagi, M. Rai, and J. S. Yumnam. 2017. Understanding Fe2+ and P deficiency tolerance in rice for enhancing productivity under acidic soils. *Biotechnology and Genetic Engineering Reviews* 33:97–117.

Das, S., P. Krishnan, M. Nayak, and B. Ramakrishnan. 2013. Changes in antioxidant isozymes as a biomarker for characterizing high temperature stress tolerance in rice (*Oryza sativa* L.) spikelets. *Experimental Agriculture* 49:53–73.

Dixit, S., A. Singh, and A. Kumar. 2014. Rice breeding for high grain yield under drought: a strategic solution to a complex problem. *International Journal of Agronomy* 2014:1–15.

Dixit, S., A. Singh, N. Sandhu, et al. 2017. Combining drought and submergence tolerance in rice: marker-assisted breeding and QTL combination effects. *Molecular Breeding* 37:1–12.

Du, H., Y. Yu, Y. Ma, et al. 2017. Sequencing and *de novo* assembly of a near complete indica rice genome. *Nature Communications* 8:1–12.

Dufey, I., A.-S. Mathieu, X. Draye, et al. 2015. Construction of an integrated map through comparative studies allows the identification of candidate regions for resistance to ferrous iron toxicity in rice. *Euphytica* 203:59–69.

Dutta, S.S., W. Tyagi, G. Pale, et al. 2018. Marker–trait association for low-light intensity tolerance in rice genotypes from Eastern India. *Molecular Genetics and Genomics* 293:1493–1506.

Engel, K., F. Asch, and M. Becker, 2012. Classification of rice genotypes based on their mechanisms of adaptation to iron toxicity. *Journal of Plant Nutrition and Soil Science* 7:871–881.

Famoso, A. N., K. Zhao, R.T. Clark, et al. 2011. Genetic architecture of aluminum tolerance in rice (*Oryza sativa*) determined through genome-wide association analysis and QTL mapping. *PLoS Genetics* 7(8):e1002221.

Feltus, F. A., J. Wan, S. R. Schulze, J. C. Estill, N. Jiang, and A. H. Paterson. 2004. An SNP resource for rice genetics and breeding based on subspecies indica and japonica genome alignments. *Genome Research* 14:1812–1819.

Fujino, K., H. Sekiguchi, Y. Matsuda, K. Sugimoto, K. Ono, and M. Yano. 2008. Molecular identification of a major quantitative trait locus, qLTG3-1, controlling low temperature germinability in rice. *Proceedings of the Natural Academy of Sciences USA* 105:12623–12628.

Fukao, T., E. Yeung, and J. Bailey-Serres. 2011. The submergence tolerance regulator SUB1A mediates cross-talk between submergence and drought tolerance in rice. *The Plant Cell* 23:412–427.

Gamuyao, R., J. H. Chin., J. Pariasca-Tanaka, et al. 2012. The protein kinase Pstol1 from traditional rice confers tolerance of phosphorus deficiency. *Nature* 488:585–541.

Ghosh, T., M. Rai, W. Tyagi, and C. Challam. 2016. Seedling stage low temperature response in tolerant and susceptible rice genotypes suggests role of relative water content and members of *OsSNAC* gene family. *Plant Signaling and Behavior* 11(5):e1138192.

Gong, Z., L. Xiong, H. Shi, et al. (2020). Plant abiotic stress response and nutrient use efficiency. *Science China-Life Sciences* 63:635–674.

Gonzaga, Z. J. C., J. Carandang, D. L. Sanchez, D. J. Mackill, and E. M. Septiningsih. 2016. Mapping additional QTLs from FR13A to increase submergence tolerance in rice beyond SUB1. *Euphytica* 209:627–636.

Guan, S. X., Q. Xu, D. R. Ma, et al. 2019. Transcriptomics profiling in response to cold stress in cultivated rice and weedy rice. *Gene* 685:96–105.

Hairmansis, A., B. Berger, M. Tester, and S. J. Roy. 2014. Image based phenotyping for nondestructive screening of different salinity tolerance traits in rice. *Rice* 7:16.

Han, B., X. D. Ma, D. Cui, et al. 2020. Comprehensive evaluation and analysis of the mechanism of cold tolerance based on the transcriptome of weedy rice seedlings. *Rice* 13:1–14.

Hasegawa, P. M., R. A. Bressan, J. K. Zhu, and H. J. Bohnert. 2000. Plant cellular and molecular responses to high salinity. *Annual Reviews in Plant Biology* 51:463–499.

Hattori, Y., K. Miura, K. Asano, et al. 2007. A major QTL confers rapid internode elongation in response to water rise in deepwater rice. *Breeding Science* 57:305–314.

Hattori, Y., K. Nagai, S. Furukawa, et al. 2009. The ethylene response factors SNORKEL1 and SNORKEL2 allow rice to adapt to deep water. *Nature* 460:1026–1030.

He, Y. Q., B. Yang, Y. He, et al. 2019. A quantitative trait locus, *qSE3*, promotes seed germination and seedling establishment under salinity stress in rice. *Plant Journal* 97:1089–1104.

Hemamalini, G. S., H. E. Shashidhar, and S. Hittalmani. 2000. Molecular marker assisted tagging of morphological and physiological traits under two contrasting moisture regimes at peak vegetative stage in rice (*Oryza sativa* L.). *Euphytica* 112:69–78.

Heuer, S., Gaxiola, R., Schilling, R, et al. 2017. Improving phosphorus use efficiency: a complex trait with emerging opportunities. *Plant Journal* 90:868–885.

Hossain, H., M. A. Rahman, M. S. Alam, and R. K. Singh. 2015. Mapping of quantitative trait loci associated with reproductive stage salt tolerance in rice. *Journal of Agronomy and Crop Science* 201:17–31.

Huang, L., Y. Wang, W. Wang, et al. 2018. Characterization of transcription factor gene OsDRAP1 conferring drought tolerance in rice. *Frontiers in Plant Science* 9:94.

Huang, X. Y., D. Y. Chao, J. P. Gao, M. Z. Zhu, M. Shi, and H. X. Lin. 2009. A previously unknown zinc finger protein, DST, regulates drought and salt tolerance in rice via stomatal aperture control. *Genes and Development* 23:1805–1817.

Husaini, A. M., and N. R. Sofi. 2018. Rice Biodiversity in Cold Hill Zones of Kashmir Himalayas and Conservation of Its Landraces. In *Rediscovery of Landraces as a Resource for the Future* (p. 39). IntechOpen.

IFPRI. 2016. *Global Nutrition Report 2016: From Promise to Impact: Ending Malnutrition by 2030.* International Food Policy Research Institute, Washington, DC. http://dx.doi.org/10.2499/9780896295841.

International Rice Genome Sequencing Project, and T. Sasaki. 2005. The map-based sequence of the rice genome. *Nature* 436:793–800.

IRRI. 2002. *Standard Evaluation System for Rice.* International Rice Research Institute, Los Banos, Philippines. 1–45.

Islam, M., J. Ontoy, and P. K. Subudhi. 2019. Meta-analysis of quantitative trait loci associated with seedling-stage salt tolerance in rice (*Oryza sativa* L.). *Plants* 8:33.

Ismail, A. M., E. S. Ella, G. V. Vergara, and D. J. Mackill. 2008. Mechanisms associated with tolerance to flooding during germination and early seedling growth in rice (*Oryza sativa*). *Nature* 7:705–708.

Ismail, A. M., U. S. Singh, S. Singh, M. H. Dar, and D. J. Mackill. 2013. The contribution of submergence-tolerant (Sub1) rice varieties to food security in flood-prone rainfed lowland areas in Asia. *Field Crops Research* 152:83–93.

Jagadish, S. V. K., P. Q. Craufurd, and T. R. Wheeler. 2008. Phenotyping parents of mapping populations of rice for heat tolerance during anthesis. *Crop Science* 48:1140–1146.

Jeong, J. S., Y. S. Kim, K. H. Baek, et al. 2010. Root-specific expression of OsNAC10 improves drought tolerance and grain yield in rice under field drought conditions. *Plant Physiology* 153:185–197.

Jiang, H., M. Li, N. Liang, et al. 2007. Molecular cloning and function analysis of the stay green gene in rice. *The Plant Journal* 52:197–209.

Jiang, S., S. Sun, L. Bai, et al. 2017. Resequencing and variation identification of whole genome of the japonica rice variety "Longdao24" with high yield. *PLoS ONE* 12:e0181037.

Jichao, Y., Z. Y. Ding, C. Zhao, et al. 2007. Effects of sunshine-shading, leaf-cutting and spikelet-removing on yield and quality of rice in the high altitude region. *Acta Agronomica Sinica* 31:1429–1436.

Kanneganti, V., and A. K. Gupta. 2008. Overexpression of OsiSAP8, a member of stress associated protein (SAP) gene family of rice confers tolerance to salt, drought and cold stress in transgenic tobacco and rice. *Plant Molecular Biology* 66:445–462.

Kim, S. L., N. Kim, H. Lee, et al. 2020. High throughput phenotyping platform for analyzing drought tolerance in rice. *Planta* 252:38.

Kong, W., C. Zhang, Y. Qiang, H. Zhong, G. Zhao, and Y. Li. 2020. Integrated RNA-Seq Analysis and Meta-QTLs mapping provide insights into cold stress response in rice seedling roots. *International Journal of Molecular Sciences* 21:4615.

Korres, N. E., J. K. Norsworthy, N. R. Burgos, and D. M. Oosterhuis. 2017. Temperature and drought impacts on rice production: an agronomic perspective regarding short- and long-term adaptation measures. *Water Resources and Rural Development* 9:12–27.

Koumoto, T., N. Saito, N. Aoki, et al. 2016. Effects of salt and low light intensity during the vegetative stage on susceptibility of rice to male sterility induced by chilling stress during the reproductive stage. *Plant Production Science* 19:497–507.

Kretzschmar, T., M. A. F. Pelayo, K. R. Trijatmiko, et al. 2015. A trehalose-6-phosphate phosphatase enhances anaerobic germination tolerance in rice. *Nature Plants* 1:15124.

Krishnamurthy, S. L., P. Pundir, A. S. Warraich, et al. 2020. Introgressed saltol QTL lines improves the salinity tolerance in rice at seedling stage. *Frontiers in Plant Science* 11:833.

Kumar A., S., Basu, V. Ramegowda, and A. Pereira. 2017. Mechanisms of drought tolerance in rice. In *Achieving Sustainable Cultivation of Rice Volume 1: Breeding for Higher Yield and Quality*, ed. T. Sasaki. Burleigh Dodds Science Publishing Ltd, London, 131–164.

Kumar, A., D. Panda, S. Mohanty, et al. 2020. Role of sedoheptulose-1, 7 bisphosphatase in low light tolerance of rice (*Oryza sativa* L.). *Physiology and Molecular Biology of Plants* 26:1–21.

Kumar, P., and P. K. Sharma. 2020. Soil salinity and food security in India. *Frontiers in Sustainable Food Systems* 4:533781.

Kuroha, T., and M. Ashikari. 2020. Molecular mechanisms and future improvement of submergence tolerance in rice. *Molecular Breeding* 40:1–14.

Kuroha, T., K. Nagai, and R. Gamuyao. 2016. Ethylene-gibberellin signalling underlies adaptation of rice to periodic flooding. *Science* 361:181–186.

Lanceras, J. C., G. Pantuwan, B. Jongdee, and T. Toojinda. 2004. Quantitative trait loci associated with drought tolerance at reproductive stage in rice. *Plant Physiology* 135:384–399.

Lesk, C., P. Rowhani, and N. Ramankutty. 2016. Influence of extreme weather disasters on global crop production. *Nature* 529:84–87.

Li, J. Y., J. Liu, D. Dong, X. Jia, et al. 2014a. Natural variation underlies alterations in Nramp aluminum transporter (NRAT1) expression and function that play a key role in rice aluminum tolerance. *Proceedings of the National Academy of Sciences USA* 111:6503–6508.

Li, J. Y., Wang, J., and Zeigler, R. S. 2014b. The 3000 rice genome project: new opportunity and challenge for future rice research. *Giga Science* 3:2047–217X.

Li, X. M., D. Y. Chao, Y. Wu, et al. 2015. Natural alleles of a proteasome α2 subunit gene contribute to thermotolerance and adaptation of African rice. *Nature Genetics* 47:827–833.

Liang, S., W. Xiong, C. Yin, et al. 2019. Overexpression of OsARD1 improves submergence, drought, and salt tolerances of seedling through the enhancement of ethylene synthesis in rice. *Frontiers in Plant Science* 10:1088.

Lin, H. X., M. Z. Zhu, M. Yano, et al. 2004. QTLs for Na^+ and K^+ uptake of the shoots and roots controlling rice salt tolerance. *Theoretical and Applied Genetics* 108:253–260.

Liu, J., and Z. He. 2020. Small DNA methylation, big player in plant abiotic stress responses and memory. *Frontiers in Plant Science* 11:595603.

Liu, G. L., H. W. Mei, X. Q. Yu, et al. 2008. QTL analysis of panicle neck diameter, a trait highly correlated with panicle size, under well-watered and drought conditions in rice (*Oryza sativa* L.). *Plant Science* 174:71–77.

Liu, A. L., J. Zou, X. Zhang, et al. 2010. Expression profiles of Class A rice heat shock transcription factor genes under abiotic stresses. *Journal of Plant Biology* 53:142–149.

Liu, C. T., S. J. Ou, B. G. Mao, et al. 2018. Early selection of bZIP73 facilitated adaptation of japonica rice to cold climates. *Nature Communications* 9:1–12.

Liu, J., C. Zhang, C. Wei, et al. 2016. The RING finger ubiquitin E3 ligase OsHTAS enhances heat tolerance by promoting H_2O_2-induced stomatal closure in rice. *Plant Physiology* 170:429–443.

Liu, Q. H., X. Wu, B-C. Chen, et al. 2014. Effects of low light on agronomic and physiological characteristics of rice including grain yield and quality. *Rice Science* 21:243–251.

Liu, W. Y., M. M. Wang, J. Huang, et al. 2009. The OsDHODH1 gene is involved in salt and drought tolerance in rice. *Journal of Integrative Plant Biology* 51:825–833.

Lu, G., F. Q. Wu, W. Wu, et al. 2014. Rice LTG1 is involved in adaptive growth and fitness under low ambient temperature. *Plant Journal* 78:468–480.

Ma, J. F., R. Shen, Z. Zhao, et al. 2002. Response of rice to Al stress and identification of quantitative trait loci for Al tolerance. *Plant and Cell Physiology* 43:652–659.

Ma, J. F., Z. C. Chen, and R. F. Shen. 2014. Molecular mechanisms of Al tolerance in gramineous plants. *Plant and Soil* 381:1–12.

Ma, Y., X. Dai, Y. Xu, et al. 2015. COLD1 confers chilling tolerance in rice. *Cell* 160:1209–1221.

Mackill, D. J., W. R. Coffman, and J. N. Rutger. 1982. Pollen shedding and combining ability for high temperature tolerance in rice. *Crop Science* 22:730–733.

Maia, L. C. D, P. R. B. Cadore, L. C. Benitez, et al. 2017. Transcriptome profiling of rice seedlings under cold stress. *Functional Plant Biology* 44:419–429.

Mansuri, R. M., Z. S. Shobbar, N. B. Jelodar, M. Ghaffari, S. M. Mohammadi, and P. Daryani. 2020. Salt tolerance involved candidate genes in rice: an integrative meta-analysis approach. *BMC Plant Biology* 20:452.

Mao, D. H., Y. Y. Xin, Y. J. Tan, et al. 2019. Natural variation in the HAN1 gene confers chilling tolerance in rice and allowed adaptation to a temperate climate. *Proceedings of the National Academy of Sciences USA* 116:3494–3501.

Marchiori, P. E. R., E. C. Machado, and R. V. Ribeiro. 2014. Photosynthetic limitations imposed by self-shading in field grown sugarcane varieties. *Field Crops Research* 155:30–37.

Matsui, T., K. Omasa, and T. Horie. 2001. The difference in sterility due to high temperatures during the flowering period among Japonica-rice varieties. *Plant Production Science* 4:90–93.

McCouch, S., M. Wright, C. W. Tung, et al. 2016. Open access resources for genome wide association mapping in rice. *Nature Communications* 7:10532.

Mishra, K. K., P. Vikram, R. B. Yadaw, et al. 2013. qDTY12.1: a locus with a consistent effect on grain yield under drought in rice. *BMC Genetics* 14:12.

Moore, K. L., Y. Chen, A. M. L. van de Meene, et al. 2014. Combined NanoSIMS and synchrotron X-ray fluorescence reveal distinct cellular and subcellular distribution patterns of trace elements in rice tissues. *New Phytologist* 201:104–115.

Munns, R., R. A. James, and A. Läuchli. 2006. Approaches to increasing the salt tolerance of wheat and other cereals. *Journal of Experimental Botany* 57:1025–1043.

Murchi, E. H., S. Hubbart, Y. Chen, et al. 2002. Acclimation of rice photosynthesis to irradiance under field conditions. *Plant Physiology* 130:1999–2010.

Najeeb, S., J. Ali, A. Mahender, et al. 2020. Identification of main-effect quantitative trait loci (QTLs) for low-temperature stress tolerance germination-and early seedling vigor-related traits in rice (*Oryza sativa* L.). *Molecular Breeding* 40:1–25.

Nakashima, K., L. S. Tran, D. V. Nguyen, et al. 2007. Functional analysis of a NAC-type transcription factor OsNAC6 involved in abiotic and biotic stress responsive gene expression in rice. *Plant Journal* 51:617–630.

Ohkubo, S., Y. Tanaka, W. Yamori, and S. Adachi. 2020. Rice cultivar Takanari has higher photosynthetic performance under fluctuating light than Koshihikari, especially under limited nitrogen supply and elevated CO_2. *Frontiers in Plant Science* 11:1308.

Oladosu, Y., M. Y. Rafii, F. Arolu, et al. 2020. Submergence tolerance in rice: review of mechanism, breeding and future prospects. *Sustainability* 12:1632.

Pedersen, O., S. M. Rich, and T. D. Colmer. 2009. Surviving floods: leaf gas films improve O_2 and CO_2 exchange, root aeration, and growth of completely submerged rice. *Plant Journal* 58:147–156.

Peng, S., J. Huang, J. E. Sheehy, et al. 2004. Rice yields decline with higher night temperature from global warming. *Proceedings of National Academy of Sciences USA* 101:9971–9975.

Prasad, P. V. V., K. J. Boote, L. H. Allen, J. E. Sheehy, and J. M. G Thomas. 2006. Species, ecotype and cultivar differences in spikelet fertility and harvest index of rice in response to high temperature stress. *Field Crops Research* 95:398–411.

Prasanth, V. V., M. S. Babu, R. K. Basava, et al. 2017. Trait and marker associations in *Oryza nivara* and *O. rufipogon* derived rice lines under two different heat stress conditions. *Frontiers in Plant Science* 8:1819.

Qin, D., F. Wang, X. Geng, et al. 2014. Overexpression of heat stress-responsive TaMBF1c, a wheat (*Triticum aestivum* L.) multiprotein bridging factor, confers heat tolerance in both yeast and rice. *Plant Molecular Biology* 87:31–45.

Quan, R. D., J. Wang, J. Hui, et al. 2018. Improvement of salt tolerance using wild rice genes. *Frontiers in Plant Science* 8:2269.

Rahman, M. A., M. J. Thomson, M. De Ocampo, et al. 2019. Assessing trait contribution and mapping novel QTL for salinity tolerance using the Bangladeshi rice landrace Capsule. *Rice* 12:63.

Rahman, M. A., M. J. Thomson, M. S. Alam, M. De Ocampo, J. Egdane, and A. M. Ismail. 2016. Exploring novel genetic sources of salinity tolerance in rice through molecular and physiological characterization. *Annals of Botany* 117:1083–1097.

Rai, M., C. K. He, and R. Wu. 2009. Comparative functional analysis of three abiotic stress inducible promoters in transgenic rice. *Transgenic Research* 18:787–799.

Rai, M., S. S. Dutta, and W. Tyagi. 2021. Molecular breeding strategies for enhancing rice yields under low light intensity. In *Molecular Breeding for Rice Abiotic Stress Tolerance and Nutritional Quality*, ed. M. A. Hossain, L. Hassan, K. Md. Ifterkharudduala, A. Kumar and R. Henry, 201–214. Oxford: John Wiley.

Rajagopal, D., P. Agarwal, W. Tyagi, S. L. Singla-Pareek, M. K. Reddy, and S. K. Sopory. 2007. *Pennisetum glaucum* Na^+/H^+ antiporter confers high level of salinity tolerance in transgenic *Brassica juncea. Molecular Breeding* 19:137–151.

Raza, Q., A. Riaz, K. Bashir, and M. Sabar. 2020. Reproductive tissues-specific meta-QTLs and candidate genes for development of heat-tolerant rice cultivars. *Plant Molecular Biology* 104:97–112.

Ren, Z. H., J. P. Gao, L. G. Li, et al. 2005. A rice quantitative trait locus for salt tolerance encodes a sodium transporter. *Nature Genetics* 37:1141–1146.
Rihan, H. Z., M. Al-Issawi, and M. P. Fuller. 2017. Advances in physiological and molecular aspects of plant cold tolerance. *Journal of Plant Interactions* 12:143–157.
Saika, H., M. Okamoto, K. Miyoshi, et al. 2007. Ethylene promotes submergence induced expression of *OsABA8ox1*, a gene that encodes ABA 8′-hydroxylase in rice. *Plant Cell Physiology* 48:287–298.
Salgotra, R. K., and C. N. Stewart. 2020. Functional markers for precision plant breeding. *International Journal of Molecular Sciences* 21:4792.
Sandhu, N., S. Dixit, B. P. M. Swamy, et al. 2019. Marker assisted breeding to develop multiple stress tolerant varieties for flood and drought prone areas. *Rice* 12:1–16.
Sangodele, E. A., R. R. Hanchinal, N. G. Hanamaratti, V. Shenoy, and V. M. Kumar. 2014. Analysis of drought tolerant QTL linked to physiological and productivity component traits under waterstress and non-stress in rice (*Oryza sativa* L.). *International Journal of Current Research and Academic Review* 2:108–113.
Sarkar, R. K. 1997. Saccharide content and growth parameters in relation with flooding tolerance in rice. *Biologia Plantarum* 40:597–603.
Sellamuthu, R., C. Ranganathan, and R. Serraj. 2015. Mapping QTLs for reproductive-stage drought resistance traits using an advanced backcross population in upland rice. *Crop Science* 55:1524–1536.
Senguttuvel, P., V. Jaldhani, N. S. Raju, et al. 2020. Breeding rice for heat tolerance and climate change scenario; possibilities and way forward. A review. *Archives of Agronomy and Soil Science* 68:1–18.
Seo, J., G. Lee, Z. Jin, et al. 2020. Development and application of indica–japonica SNP assays using the Fluidigm platform for rice genetic analysis and molecular breeding. *Molecular Breeding* 40:1–16.
Septiningsih, E. M., D. L. Sanchez, N. Singh, et al. 2012. Identifying novel QTLs for submergence tolerance in rice cultivars IR72 and Madabaru. *Theoretical and Applied Genetics* 124:867–874.
Shah, F., J. Huang, K. Cui, et al. 2011. Impact of high-temperature stress on rice plant and its traits related to tolerance. *Journal of Agricultural Sciences* 149:545–556.
Shen, H., C. Liu, Y. Zhang, et al. 2012. *OsWRKY*30 is activated by MAP kinases to confer drought tolerance in rice. *Plant Molecular Biology* 80:241–253.
Shen, Y., L. Shen, Z. Shen, et al. 2015. The potassium transporter *OsHAK*21 functions in the maintenance of ion homeostasis and tolerance to salt stress in rice. *Plant Cell Environment* 38:2766–2779.
Shi, C., W. Li, Q. J. Zhang, et al. 2020. The draft genome sequence of an upland wild rice species, *Oryza granulata*. *Scientific Data* 7:1–12.
Simkin, A. J., P. E. López-Calcagno, and C. A. Raines. 2019. Feeding the world: improving photosynthetic efficiency for sustainable crop production. *Journal of Experimental Botany* 70:1119–1140.
Singh, A., M. E. Septiningsih, H. S. Balyan, N. K. Singh, and V. Rai. 2017. Genetics, physiological mechanisms and breeding of flood-tolerant rice (*Oryza sativa* L.). *Plant and Cell Physiology* 58:185–197.
Singh, R., Y. Singh, S. Xalaxo, et al. 2016. From QTL to variety-harnessing the benefits of QTLs for drought, flood and salt tolerance in mega rice varieties of India through a multi-institutional network. *Plant Science* 242:278–287.
Singh, A., Y. Singh, A. K. Mahato, et al. 2020. Allelic sequence variation in the Sub1A, Sub1B and Sub1C genes among diverse rice cultivars and its association with submergence tolerance. *Scientific Reports* 10:1–18.
Singh, N., P. K. Jayaswal, K. Panda, et al. 2015. Single-copy gene based 50K SNP chip for genetic studies and molecular breeding in rice. *Scientific Reports* 5:11600.
Steele, K. A., M. J. Quinton-Tulloch, R. B. Amgai, et al. 2018. Accelerating public sector rice breeding with high-density KASP markers derived from whole genome sequencing of indica rice. *Molecular Breeding* 38:1–13.
Steffens, B., and A. Rasmussen. 2016 The physiology of advantageous adventitious roots. *Plant Physiology* 170:603–617.
Suh, J. P., J. U. Jeung, J. I. Lee, et al. 2010. Identification and analysis of QTLs controlling cold tolerance at the reproductive stage and validation of effective QTLs in cold-tolerant genotypes of rice (*Oryza sativa* L.). *Theoretical and Applied Genetics* 120:985–995.
Sun, C., Z. Hu, T. Zheng, et al. 2017. RPAN: rice pan-genome browser for ~ 3000 rice genomes. *Nucleic Acids Research* 45:597–605.
Swamy, B. P. M., H. U. Ahmed, A. Henry, et al. 2013. Genetic, physiological, and gene expression analyses reveal that multiple QTL enhance yield of rice mega-variety IR64 under drought. *PLoS ONE* 8:e62795.

Swamy, M. H. K., M. Anila, R. R. Kale, et al. 2020. Marker assisted improvement of low soil phosphorus tolerance in the bacterial blight resistant, fine grain type rice variety, improved Samba Mahsuri. *Scientific Reports* 10:21143.

Tanger, P., S. Klassen, J. P. Mojica, et al. 2017. Field-based high throughput phenotyping rapidly identifies genomic regions controlling yield components in rice. *Scientific Reports* 7:1–8.

Thangapandian, R., T. Jayaraj, E. D. Redona, and S. Sheryl. 2010. Development of heat tolerant breeding lines and varieties in rice. *Indian Journal of Plant Genetic Resource* 23:1–3.

Thanh, P. T., P. D. T. Phan, R. Ishikawa, and T. Ishii. 2010. QTL analysis for flowering time using backcross population between *Oryza sativa* Nipponbare and *O. rufipogon*. *Genes and Genetic Systems* 85:273–279.

Thomson, M. J., K. Zhao, M. Wright, et al. 2012. High-throughput single nucleotide polymorphism genotyping for breeding applications in rice using the BeadXpress platform. *Molecular Breeding* 29:875–886.

Thomson, M. J., N. Singh, M. S. Dwiyanti, et al. 2017. Large-scale deployment of a rice 6K SNP array for genetics and breeding applications. *Rice* 10:1–13.

Thudi, M., R. Palakurthi, J. C. Schnable, et al. 2020. Genomic resources in plant breeding for sustainable agriculture. *Journal of Plant Physiology* 257:153351.

Tian, L., L. B. Tan, F. X. Liu, H. W. Cai, and C. Q. Sun. 2011. Identification of quantitative trait loci associated with salt tolerance at seedling stage from *Oryza rufipogon*. *Journal of Genetics and Genomics* 38:593–601.

Trijatmiko, K. R., P. J. Supriyanta, M. J. Thomson, C. M. Vera Cruz, S. Moeljopawiro, and A. Pereira. 2014. Meta-analysis of quantitative trait loci for grain yield and component traits under reproductive-stage drought stress in an upland rice population. *Molecular Breeding* 34:283–295.

Tyagi, W., and M. Rai. 2017. Root transcriptomes of two acidic soil adapted Indica rice genotypes suggest diverse and complex mechanism of low phosphorus tolerance. *Protoplasma* 254:725–736.

Tyagi, W., D. Rajagopal, S. L. Singla-Pareek, M. K. Reddy, and S. K. Sopory. 2005. Cloning and regulation of stress-inducible c subunit of vacuolar H^+ ATPase gene from *Pennisetum glaucum* and characterization of its promoter that expresses in shoot hairs and floral organs. *Plant Cell Physiology* 46:1411–1422.

Tyagi, W., M. Rai, and A. Dohling. 2012. Haplotype analysis for Pup1 locus in rice genotypes of North-Eastern and Eastern India to identify suitable donors tolerant to low phosphorus. *SABRAO Journal of Breeding and Genetics* 44:398–405.

Tyagi, W., E. N. Nongbri, and M. Rai. 2021. Harnessing tolerance to low phosphorus in rice: recent progress and future perspectives. In *Molecular Breeding for Rice Abiotic Stress Tolerance and Nutritional Quality*, ed. M. A. Hossain, L. Hassan, K. Md. Ifterkharudduala, A. Kumar and R. Henry, 215–233. Oxford: John Wiley.

Tyagi, W., J. S. Yumnam, D. Sen, et al. 2020. Root transcriptome reveals efficient cell signaling and energy conservation key to aluminum toxicity tolerance in acidic soil adapted rice genotype. *Scientific Reports* 10:1–14.

Uga, Y., K. Sugimoto, S. Ogawa, et al. 2013. Control of root system architecture by DEEPER ROOTING 1 increases rice yield under drought conditions. *Nature Genetics* 45:1097–1102.

Van Zelm, E., Y. Zhang, and C. Testerink. 2020. Salt tolerance mechanisms of plants. *Annual Review of Plant Biology* 71:403–433.

Varshney, R. K., P. Sinha, V. K. Singh, A. Kumar, Q. Zhang, and J. L. Bennetzen. 2020. 5Gs for crop genetic improvement. *Current Opinion in Plant Biology* 56:190–196.

Vergara, B. S., B. Jackson, and S. K. De Datta. 1976. Deep water rice and its response to deep water stress. In Climate and Rice. International Rice Research Institute, Los Banos, Philippines, 301–319.

Verma, S. K., R. R. Saxena, M. S. Xalxo, and S. B. Verulkar. 2014. QTL for grain yield under water stress and non-stress conditions over years in rice (*Oryza sativa* L.). *Australian Journal of Crop Science* 8:916–926.

Vikram, P., B. P. M. Swamy, S. Dixit, et al. 2012. Bulk segregant analysis: an effective approach for mapping consistent-effect drought grain yield QTLs in rice. *Field Crops Research* 134:185–192.

Vikram, P., B. P. M. Swamy, S. Dixit, H. U. Ahmed, M. Teresa Sta Cruz, and A. Kumar. 2011. qDTY1.1, a major QTL for rice grain yield under reproductive-stage drought stress with a consistent effect in multiple elite genetic backgrounds. *BMC Genetics* 12:89.

Von Uexküll, H. R., and E. Mutert. 1995. Global extent, development and economic impact of acid soils. *Plant and Soil* 171:1–15.

Wang, W., R. Mauleon, Z. Hu, et al. 2018. Genomic variation in 3,010 diverse accessions of Asian cultivated rice. *Nature* 557:43–49.

Wang, Y., J. Zang, Y. Sun, J. Ali, J. L. Xu, and Z. Li. 2013. Back ground-independent quantitative trait loci for drought tolerance identified using advanced backcross introgression lines in rice. *Crop Science* 53:430–441.

Wassmann, R., S. V. K. Jagadish, K. Sumfleth, et al. 2009. Regional vulnerability of climate change impacts on Asian rice production and scope for adaptation. *Advances in Agronomy* 102:91–133.

Wing, R. A., M. D. Purugganan, and Q. Zhang. 2018. The rice genome revolution: from an ancient grain to Green Super Rice. *Nature Reviews Genetics* 19:505–517.

Wissuwa, M., J. Wegner, N. Ae, and M. Yano. 2002. Substitution mapping of Pup1: a major QTL increasing phosphorus uptake of rice from a phosphorus deficient soil. *Theoretical and Applied Genetics* 105:890–897.

Wissuwa, M., M. Yano, and N. Ae. 1998. Mapping of QTLs for phosphorus deficiency tolerance in rice (*Oryza sativa* L.). *Theoretical and Applied Genetics* 97:777–783.

Wu, D., Z. Guo, J. Ye, et al. 2019. Combining high-throughput micro-CT-RGB phenotyping and genome-wide association study to dissect the genetic architecture of tiller growth in rice. *Journal of Experimental Botany* 70:545–561.

Wu, L.-B., M. Y. Shhadi, G. Gregorio, et al. 2014. Genetic and physiological analysis of tolerance to acute iron toxicity in rice. *Rice* 7:1.

Wu, K. S. and S. D. Tanksley. 1993. Abundance, polymorphism and genetic mapping of microsatellites in rice. *Molecular and General Genetics* 241:225.

Xiong, L., Y. Uga, and Y. Li. 2020. Rice functional genomics: theories and practical applications. *Molecular Breeding* 40:72.

Xu, K., and D. J. Mackill. 1996. A major locus for submergence tolerance mapped on rice chromosome 9. *Molecular Breeding* 2:219–224.

Xu, Y., P., Li, C. Zou, et al. 2017. Enhancing genetic gain in the era of molecular breeding. *Journal of Experimental Botany* 68:2641–2666.

Xu, K., X. Xu, T. Fukao, et al. 2006. Sub1A is an ethylene-response-factor-like gene that confers submergence tolerance to rice. *Nature* 442: 705–708.

Yamaji, N., C. F. Huang, S. Nagao, et al. 2009. A zinc finger transcription factor ART1 regulates multiple genes implicated in aluminum tolerance in rice. *Plant Cell* 21:3339–3349.

Yamakawa, H., T. Hirose, M. Kuroda, and T. Yamaguchi. 2007. Comprehensive expression profiling of rice grain filling-related genes under high temperature using DNA microarray. *Plant Physiology* 44:258–277.

Yamauchi, T., T. D. Colmer, O. Pedersen and M. Nakazono. 2018. Regulation of root traits for internal aeration and tolerance to soil waterlogging-flooding stress. *Plant Physiology* 176:1118–1130.

Yamauchi, T., A. Tanaka, H. Inahashi, et al. 2019. Fine control of aerenchyma and lateral root development through AUX/IAA- and ARF dependent auxin signaling. *Proceedings of the National Academy of Sciences* USA 116:20770–20775.

Yang, W., Z. Guo, C. Huang, et al. 2014. Combining high-throughput phenotyping and genome-wide association studies to reveal natural genetic variation in rice. *Nature Communications* 5:1–9.

Yang, L., L. Lei, H. Liu, J. Wang, H. Zheng, and D. Zou. 2020a. Whole-genome mining of abiotic stress gene loci in rice. *Planta* 252:1–20.

Yang, W., H. Feng, X. Zhang, et al. 2020b. Crop phenomics and high-throughput phenotyping: past decades, current challenges, and future perspectives. *Molecular Plant* 13:187–214.

Yao, W., G. Li, Y. Yu, and Y. Ouyang. 2018. Fun Rice Genes dataset for comprehensive understanding and application of rice functional genes. *Giga Science* **7**:1–9.

Yonemaru, J., T. Yamamoto, S. Fukuoka, Y. Uga, K. Hori, and M. Yano. 2010. Q-TARO:QTL Annotation Rice Online Database. *Rice* 3:194 (Q-TARO database: http://qtaro.abr.affrc.go.jp/ accessed on January 20, 2021).

Yu, J., S. Hu, J. Wang, et al. 2002. A draft sequence of the rice genome (*Oryza sativa* L. ssp. indica). *Science* 296:79–92.

Yumnam, J. S., M. Rai, and W. Tyagi. 2017. Allele mining across two low-P tolerant genes PSTOL1 and PupK20-2 reveals novel haplotypes in rice genotypes adapted to acidic soils. *Plant Genetic Resources: Characterization and Utilization* 15:221–229.

Yun, M., X. Dai, Y. Xu, et al. 2015. *COLD1* confers chilling tolerance in rice. *Cell* 160:1209–1221.

Zafar, S. A., A. Hameed, A. S. Khan, and M. Ashraf. 2017. Heat shock induced morpho-physiological response in indica rice (*Oryza sativa* L.) at early seedling stage. *Pakistan Journal of Botany* 49:453–463.

Zeng, Y. W., S. C. Li, X. Y. Pu, et al. 2006. Ecological difference and correlation among cold tolerance traits at the booting stage for core collection of rice landrace in Yunnan, China. *Chinese Journal of Rice Science* 20:265–271.

Zhang, D., X. Pan, G. Mu, and J. Wang. 2010. Toxic effects of antimony on photosystem II of Synechocystis sp. as probed by in vivo chlorophyll fluorescence. *Journal of Applied Phycology* 22:479–488.

Zhang, G., X. Hou, L. Wang, et al. 2020. P*HOTO-SENSITIVE LEAF ROLLING 1* encodes a polygalacturonase that modifies cell wall structure and drought tolerance in rice. *New Phytologist* 229:890–901.

Zhang, J., L. L. Chen, F. Xing, et al. 2016. Extensive sequence divergence between the reference genomes of two elite indica rice varieties Zhenshan 97 and Minghui 63. *Proceedings of the National Academy of Sciences USA* 113:E5163–E5171.

Zhang, T., X. Q. Zhao, W. S. Wang, et al. 2012. Comparative transcriptome profiling of chilling stress responsiveness in two contrasting rice genotypes. *PLoS ONE* 7:e43274.

Zhang, Z., J. Li, Y. Pan, et al. 2017. Natural variation in CTB4a enhances rice adaptation to cold habitats. *Nature Communications* 8:14788.

Zhao, Q., Q. Feng, H. Lu, et al. 2018. Pan-genome analysis highlights the extent of genomic variation in cultivated and wild rice. *Nature Genetics* 50:278–284.

Zhao, X. Q., J. L. Xu, M. Zhao, et al. 2008. QTLs affecting morph-physiological traits related to drought tolerance detected in overlapping introgression lines of rice (*Oryza sativa* L.). *Plant Science* 174:618–625.

Zhou, L., Z. Liu, Y. Liu, et al. 2016. A novel gene *OsAHL1* improves both drought avoidance and drought tolerance in rice. *Scientific Reports* 6:30264.

Zhou, Y., D. Chebotarov, D. Kudrna, et al. 2020. A platinum standard pan-genome resource that represents the population structure of Asian rice. *Scientific Data* 7:1–11.

Zou, G., H. Mei, H. Liu, et al. 2005. Grain yield responses to moisture regimes in a rice population: association among traits and genetic markers. *Theoretical and Applied Genetics* 112:106–113.

Zou, J., A. Liu, X. Chen, et al. 2009. Expression analysis of nine rice heat shock protein genes under abiotic stresses and ABA treatment. *Journal of Plant Physiology* 166:851–861.

17 Nanotechnology in Developing Abiotic Stress Resilience in Crops
A Physiological Implication

Satyen Mondal[1,2], Md. Ruhul Quddus[1], Tuhin Halder[1], M. Ashik Iqbal Khan[1], Guanglong Zhu[3], M. Rafiqul Islam[4], and Tofazzal Islam[2]*
[1]Bangladesh Rice Research Institute, Gazipur, Bangladesh
[2]Bangabandhu Sheikh Mujibur Rahman Agricultural University. Gazipur, Bangladesh
[3]Yangzhou University, Yangzhou, Jiangsu Province, China
[4]International Rice Research Institute, DAPO Box 7777 Metro Manila, Philippines
*Correspondence: satyen1981@gmail.com; satyen@bsmrau.edu.bd

CONTENTS

17.1 INTRODUCTION

Nanotechnology is the application of the science of matter on an atomic, molecular, and supermolecular scale for various purposes. Nanoparticles (NPs; 1 to 100 nm in size) are frequently used in agriculture for ensuring global food security (Nakache et al., 1999; Hossain et al., 2020). The potential applications of nanotechnology include improvement of genetic properties of plants, soil texture, pest and pathogen management, nutrient delivery, and abiotic stress tolerance, which are linked to plant growth and development (Ghormade et al., 2011). Nanochelating technology is a novel approach applied in synthesizing efficient nanostructures in various disciplines. The efficiency of nanochelated fertilizers is well established. For example, nanochelated nitrogen fertilizer is more

DOI: 10.1201/9781003159636-17

resistant to leaching and has higher nitrogen use efficiency to increase crop yield compared with urea (Zareabyaneh and Bayatvarkeshi, 2014).

Agro-ecosystems frequently suffer from several environmentally induced unfavorable crop farming situations, usually termed abiotic stresses. These stresses are the main hindrance to agricultural production. The world population is rapidly increasing and is estimated to reach 9.7 billion by the year 2050. However, agricultural production is not rising at a parallel pace. Doubling productivity is a challenge to address the sustainable development goals (SDGs), as the area under cultivation is likely to remain constant or even decrease due to increasing pressure on land for nonagricultural uses. While increased investment and technological breakthroughs can improve availability, these may not necessarily translate into increased accessibility and absorption of food. With climate change closing in, abiotic stresses are considered to be a major constraint to sustaining crop productivity. According to one estimate, approximately 70% of yield reduction of crops is directly or indirectly influenced by abiotic stresses (Acquaah, 2007). Abiotic stress leads to a series of morphological, physiological, biochemical, and molecular changes that adversely affect plant growth and productivity (Mohi-Ud-Din et al., 2021). Drought, salinity, and extreme temperature conditions are the most prevalent abiotic stresses threatening global food security. The development of stress-tolerant plants can be a worthwhile strategy to overcome the problem of decreasing global food production.

Nanotechnology is considered one of the most promising technologies to promote crop production and agricultural sustainability (Scrinis et al., 2007). The application of nanoparticles or nanodevices has both positive and negative impacts on plant growth and development. Nanotechnology comprises novel properties of nanomaterials that facilitate agricultural research in crop improvement programs as well as alleviation of abiotic stresses (Carmen et al., 2003). Although plenty of literature is available on improvement of stress tolerance in plants by nanotechnology, a comprehensive chapter on this important topic is lacking. The multifunctional roles of nanomaterials in plant tolerance to various abiotic stresses are summarized and illustrated in Figure 17.1. This chapter provides an

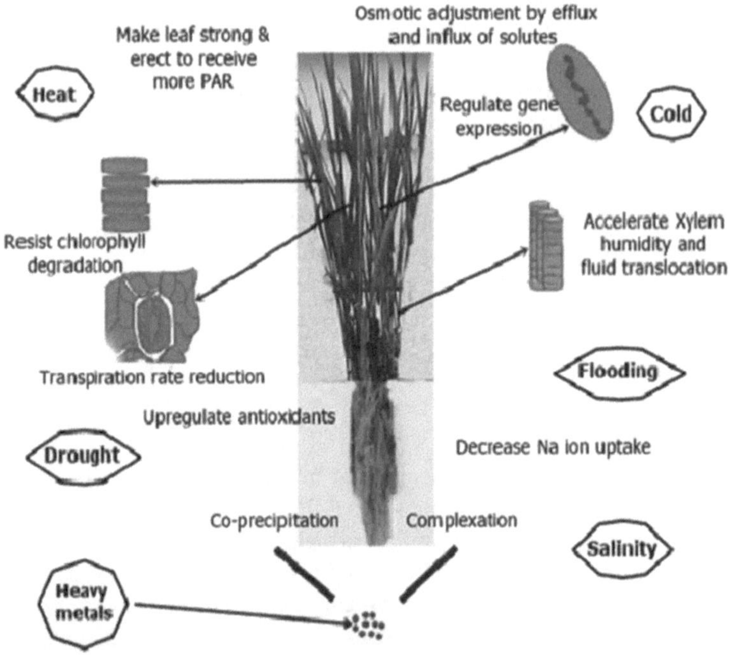

FIGURE 17.1 Roles of nanomaterials in alleviation of various kinds of abiotic stresses on plants, including drought, salinity, cold, heat, flooding, and heavy metals.

(Redrawn and adapted from Elsakhawy et al., 2018.)

update on progress and potential applications of nanotechnology in promoting climate-smart sustainable crop production. Challenges and promises of the application of nanotechnology to abiotic stress resilience in crops are also described.

17.2 NANOTECHNOLOGY-MEDIATED ABIOTIC STRESS TOLERANCE IN PLANTS

17.2.1 Opportunities of Nanotechnology in Developing Flooding Stress Resilience in Crops

An effective flood-escaping strategy to secure the grain harvest is the cultivation of a short-duration crop variety. Abdel-Aziz et al. (2016) reported that wheat plants treated with nanofertilizer show 23.5% shorter growth duration as compared with conventional fertilizer treatment (130 days in comparison to 170 days). Treatment of soybean with 50 ppm Al_2O_3 nanoparticles promoted plant growth by controlling cell death and energy metabolism under flooded conditions (Mustafa et al., 2015). Protein abundance analysis among flood-stressed Al_2O_3 nanoparticle-treated soybean revealed that 172 energy metabolism-related proteins significantly changed in abundance (Figure 17.2). Al_2O_3 nanoparticle-responsive proteins related mRNA expression analysis showed the downregulation of flavodoxin-like quinone (Mustafa et al., 2015). It is also reported that silver (Ag) nanoparticles boost soybean growth during flooding-stressed conditions by facilitating the detoxification of toxic byproducts of glycolysis (Mustafa et al., 2015).

In mung bean, glycolysis mechanism is important for waterlogging tolerance (Sairam et al., 2009). Low-oxygen situations induce the production of anaerobic proteins to contribute to sugar metabolism, glycolysis, and fermentation in rice (Mondal et al., 2020). Reduction in anaerobic protein levels induced by Al_2O_3 nanoparticles may result in a metabolic change towards aerobic pathways, which will aid in submergence tolerance (Mustafa et al., 2015).

At the time of post-flooding recovery in plants, a rise in the major latex proteins (MLP)-like proteins has been reported (Khan et al., 2014). Downregulation of MLP-like protein 43 at the transcriptional level was observed in flooding-stressed soybean treated with Al_2O_3 nanoparticles. These outcomes suggested that Al_2O_3 nanoparticles might affect the action of protease inhibitors by regulating MLP-like protein 43 in soybean under flooding-stressed conditions (Mustafa et al., 2015).

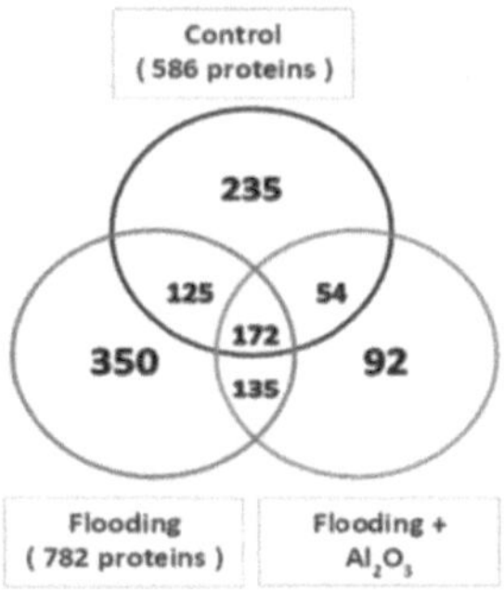

FIGURE 17.2 Venn diagram showing protein abundance analysis among flooding-stressed Al_2O_3 nanoparticle-treated soybean, flooding-stressed soybean, and control reveals that 172 proteins, mainly related to energy metabolism, significantly changed in abundance. Al_2O_3 nanoparticle-responsive proteins related mRNA expression analysis showed the downregulation of flavodoxin-like quinone reductase and upregulation of NmrA-like negative transcriptional regulator.

(Mustafa et al., 2015.)

Flooding stress causes proteasome/ubiquitin-mediated proteolysis, which results in loss of the root tip (Yanagawa and Komatsu, 2012). Notably, Al_2O_3 nanoparticle-treated soybean experiences less cell death during 2 days of flooding stress, advising that these nanoparticles induce plant acclimation to flooding stress.

17.2.2 Use of Nanoparticles under Drought Stress

Drought is a key abiotic stress that affects plant growth and crop yield. Among various approaches used to ameliorate drought stress in plants, the application of nanoparticles is effective and promising. Zinc oxide nanoparticle (ZnO NP) application improved the percentage of soybean seed germination during water stress (Sedghi et al., 2013). Likewise, application of ZnO NPs to the culture medium stimulated somatic embryo development and plant regeneration and improved stress tolerance in banana plantlets grown in vitro (Helaly et al., 2014). Cu and Zn NPs successfully reduced drought effects on *Triticum aestivum* (wheat) crop by stimulating antioxidant enzyme activities as well as relative water content, decreasing thiobarbituric acid reactive substance (TBARS) accumulation, and stabilizing photosynthetic pigment content in leaves (Taran et al., 2017). Several transcellular membrane sensors, mainly Ca^{2+}-binding proteins and Ca^{2+} channels, act as the key to plants response towards environment stress factors (Thapa et al., 2011). *ERD1* (*Early Responsive to Dehydration 1*) and *RD20A* (*Response to Desiccation 20A*) are among the stress marker genes that are most responsive in crops (Huang et al., 2012; Neves-Borges et al., 2012). Transcription factor (TF) gene families such as DREB (dehydration-responsive element-binding factor), bZIP (basic leucine zipper), NAC (NAM (no apical meristem), CUC (cup-shaped cotyledon)), WRKY (composed of a conserved WRKYGQK motif), ATAF (Arabidopsis transcription activation factor), MYB (myeloblastosis), ABF/AREB (abscisic acid responsive element-binding factor), and ERF (ethylene response factor) are linked with environmental stress-related plant adaptation (Golldack et al.,2011; Jin et at al., 2013). From the bZIP family, *GmFDL19* overexpression led to water shortage tolerance in transgenic soybean (Li et al., 2017). In the same way, *GmDREB2* overexpression improved the drought and salinity stress tolerance of transgenic Arabidopsis plants (Chen et al., 2007). Lately, *GmWRKY27*, *GmMYB174*, and *GmMYB118* were observed to be highly upregulated in soybean under abiotic stress, especially drought (Wang et al., 2015; Du et al., 2018). Upregulation of marker genes of drought tolerance (*GmDREB2*, *GmRD20A*, *GmERD1*, *GmNAC11*, *GmFDL19*, *GmMYB118*, *GmWRKY27*, and *GmMYB174*) during induced drought conditions was observed depending on the particular plant tissue, the gene, and the nanoparticles used. Moreover, under induced drought conditions, Cu NP application increased the transcript product level for GmERD1, GmDREB2, GmMYB174, and GmNAC11 in roots and for GmERD1, GmDREB2, GmNAC11, GmMYB174, GmWRKY27, and GmMYB118 in leaves, whereas Co NP application amplified the expression level of GmERD1, GmRD20A, GmWRKY27, and GmNAC11 in roots and GmWRKY27, GmERD1, GmMYB174, and GmMYB118 in leaves. A few drought-tolerant marker genes were observed to be upregulated in both leaves (*GmMYB174*, *GmERD1*, and *GmMYB118*) and roots (*GmNAC11*, *GmERD1*, and *GmDREB2*) when ZnO NP treatment was applied to the plants. Copper (Cu), cobalt (Co), iron (Fe), and zinc oxide nanoparticles increased the expression of *GmWRKY27* (8–10-fold), *GmMYB118* (6–20-fold), and *GmMYB174* (2–8-fold) in leaves (Linh et al., 2020).

When Cu, Co, Fe, and zinc oxide nanoparticles were applied to the soybean cultivar DT26 under drought stress, an increase in the drought tolerance level was reflected in the measurements of relative water content (RWC), plant growth, biomass reduction rate, and debt-to-income (DTI) (Linh et al., 2020). Root length and root length per unit dry mass are among the root phenotypic traits linked with sustained productivity during drought stress (Fitter et al., 2002), and these traits were observed to be in good condition when nanoparticles were applied during drought stress (Linh et al., 2020).

Metal-based NP-derived positive effects have been reported for drought tolerance in a number of plant species. Addition of Fe NPs to an in vitro culture of *Fragaria* × *ananassa* Duch increased drought tolerance (Mozafari et al., 2018). TiO_2 NPs with gibberellin (GA3) application in *Ocimum basilicum* improved the endurance against drought stress (Kiapour et al., 2015) . Fe and Cu NP treatment increased wheat production under water stress conditions through the rise in sugar content and superoxide dismutase (SOD) action (Yasmeen et al., 2017). Cu and Zn nanoparticles encouraged a rise in RWC by 8%–10% in seedling leaves of two separate wheat varieties in drought stress conditions (Taran et al., 2017). Moreover, such physiological traits were repeatedly used for screening of water stress-tolerant genotypes in soybean (Chowdhury et al., 2018), finger millet (Mukami et al., 2019), and wheat (Yadav et al., 2019). Fe NP was reported to be the most efficient NP in increasing tolerance against drought in soybean (DTI increased seven-fold compared with control), followed by Co NP (DTI increased four-fold) (Linh et al., 2020). DTI is considered a more accurate index for drought tolerance evaluation because it pools all tolerance parameters, i.e., recovery index, tolerance index, and recovery factor (Linh et al., 2020). *GmMYB174*, *GmMYB118*, and *GmWRKY27* showed increased expression in leaves, which proves that NP application does activate plant response against drought (Linh et al., 2020).

The plant TF gene superfamily WRKY includes the gene *GmWRKY27* and is involved in several developmental and physiological processes in plants, such as hormone catabolism, signaling, phosphate acquisition, secondary biosynthesis, seed germination, lignin biosynthesis, and stress responses (Rushton et al., 2010) The WRKY TF family plays a vital role in the stress response and plant defense regulation. A wide range of abiotic stress conditions (drought, cold, flooding, salt, heat, low humidity, heavy metal toxicity; and osmotic, oxidative, and UV stress) were reported to be regulated through WRKY proteins (Rushton et al., 2010). Increased expression of *GmWRKY27* gene in the leaves was reported in NP-treated plants during drought conditions, but not in roots, implying that NPs may be associated with the control of stomatal function and abscisic acid (ABA) biosynthesis under drought conditions (Linh et al., 2020). ABA biosynthesis and stomatal opening and closure are important to regulate water loss from transpiration during water deficit conditions (Lim et al., 2015). *GmWRKY27* overexpression was also reported to make soybean plants tolerant against drought and salinity (Wang et al., 2015; Du et al., 2018). It is claimed that GmMYB118 TF acts on improving tolerance to salt stress and drought by increasing the gene expression associated with stress (Du et al., 2018). The application of several NPs promoted the expression of *GmMYB118* gene in leaves, whereas Fe NP application increased its expression in the tissues of both roots and leaves at the time of drought conditions (Linh et al., 2020). *ERD1* plays a vital role at the early stage in the pathway of drought response (Neves-Borges et al., 2012). Upregulation of *GmERD1* was considered important to plants lacking the characters required for plant adaptation to water stress (Wang et al., 2015) *GmERD1* upregulation in both leaves and roots under induced drought conditions was observed in NP-treated plants.

This gene is associated with a cascade of reactions and directly acts on the abiotic stress response (Stolf-Moreira et al., 2011; Shinozaki and Yamaguchi-Shinozaki, 2006). *GmERD1* was strongly expressed in a cultivar of drought-sensitive soybean during water deficit conditions, but not in the tolerant counterpart (Neves-Borges et al., 2012; Shinozaki and Yamaguchi-Shinozaki, 2006). Soybean plants were grown in a pot soil system, and drought stress was induced, whereupon *RD20A* gene expression was increased. But abrupt drought stress induction did not increase the *RD20A* gene expression (Neves-Borges et al., 2012). In another experiment, *RD20A* expression was curiously increased only in roots when Fe or Co NP treatment was applied in plants grown in pot soil. The results reflected that Fe and Co NPs had positive effects on the expression of drought marker genes in roots; consequently, morphological alterations of roots and DTI values were observed in Fe and Co NP-treated plants (Linh et al., 2020). Besides *RD20A*, all NP treatments showed enhanced expression of *GmNAC11* in roots, while *DREB2* increment was observed in roots of the ZnO, Fe, and Cu NP-treated plants (Linh et al., 2020). Moreover, NPs also play a role in reactive

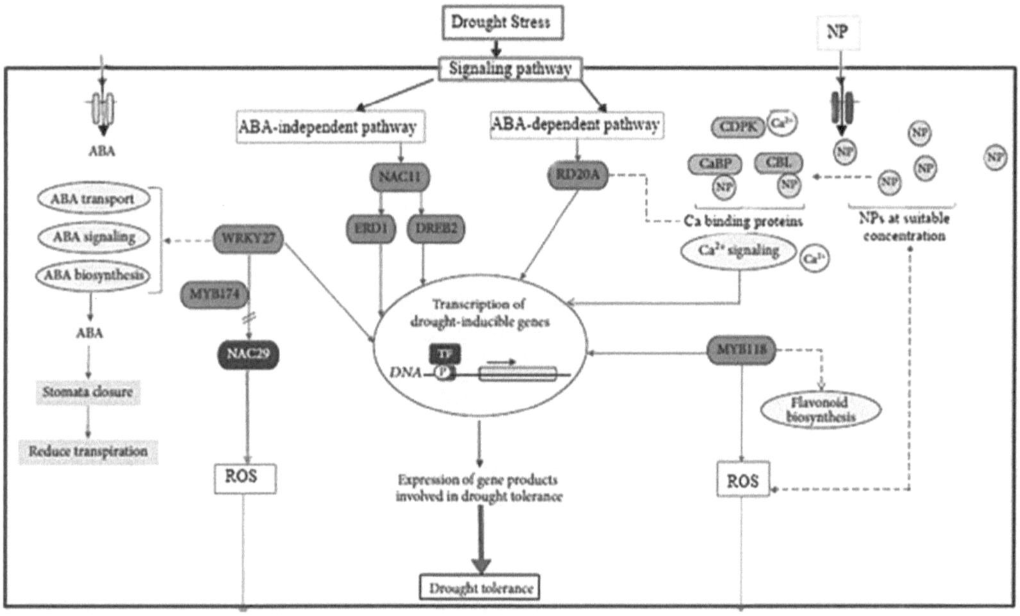

FIGURE 17.3 Simplified model for drought-responsive mechanism in soybean cells triggered by nanoparticles.

CaBP, CDPK, CBL: calcium binding proteins; NP: nanoparticle; ROS: reactive oxygen species; TF: transcription factor; P: promoter. Solid arrows indicate characterized pathways. Dotted arrows and lines indicate possible functions or speculated pathways.

(Linh et al., 2020.)

oxygen species (ROS) scavenging through encouraging antioxidative enzyme activities. Sun et al. (2020) delivered strong proof for the above statement in which ascorbate peroxidase (APX), catalase (CAT), Fe/Mn SOD, and Cu/Zn SOD expression were significantly increased under drought in ZnO NP-treated plants. The drought tolerance mechanism induced by NP application in plants is illustrated in Figure 17.3.

From this discussion we can conclude that the application of NPs successfully helped different plant species to adapt to drought stress. Moreover, morphology improvement of shoot and root, improvement in drought tolerance indices, and significantly enhanced expressions of marker genes of drought tolerance under drought situations in NP-treated plants were empirically discussed (Table 17.1). Thus, NPs might increase plant endurance to drought stress through inducing drought-associated gene expression. The utility of NP application to cope with drought stress in different plant species of commercial interest warrants further research.

17.2.3 Nanomaterials against Salinity Stress

Presence of salts in the soil and water sources, including groundwater and surface water, is a well-recognized threat to crop productivity. Agricultural productivity can be reduced as a result of toxic effects produced by high salt content, which restricts plant uptake of essential nutrients and water. Salinity intrusion produces many soil-related problems, such as poor porosity, high Na content, waterlogging, hydraulic constraints, and loss of nutrients. Salinity increases sodium and chloride ion content, which induces quite a few disorders in plant physiological processes (Tavakkoli et al., 2010). Salinity leads to ionic and osmotic stress and causes enzymatic inhibition, nutrient imbalance, and

TABLE 17.1
Responses of Plants after Application of Nanotechnology under Drought Stress

Nanomaterial	Plant Species	Stress Responses	References
Nano TiO_2	Wheat (*Triticum aestivum* L.)	Expediting growth, yield, gluten and starch content of wheat.	Jaberzadeh et al. (2013)
Nano TiO_2	Flax (*Linum usitatissimum* L.)	Enhancing chlorophyll and carotenoids content, improving crop growth and yield attributes, diminishing H_2O_2 and malondialdehyde (MDA) content.	Aghdam et al. (2016)
Nano TiO_2	Basil (*Ocimum basilicum* L.)	Improving the detrimental effects of drought stress on basil plants.	Kiapour et al. (2015)
Nano Zero valent Fe	*Arabidopsis thaliana* L.	Activation of plasma membrane H^+-ATPase, stomatal opening, increasing Chl content and dry matter, maintaining normal growth under drought, increasing CO_2 assimilation in plants.	Kim et al. (2015)
Nano SiO_2	*Crataegus* sp.	A positive significant effect on photosynthetic rate, stomatal conductance, and dry matter production; nonsignificant effect on chlorophyll and carotenoid content.	Ashkavand et al. (2015)
Nano ZnO	Soybean (*Glycine max* L.)	Promoting germination process and increasing germination rate, decreasing in seed residual moisture content of soybean.	Mahmoodzadeh et al. (2013)
SiO_2	Hawthorns (*Crataegus* sp.)	Increased plant biomass and xylem water potential and decreased MDA content, especially under drought conditions.	Ashkavand et al. (2015)
Silicon	Sorghum (*S. bicolor*)	Increase in leaf area index (LAI), specific leaf weight (SLW), chlorophyll content (SPAD), leaf dry weight (LDW), shoot dry weight (SDW), root dry weight (RDW), total dry weight (TDW).	Ahmed et al. (2011)
TiO_2 and SiO_2	Cotton	Increased total phenolic compounds, total soluble proteins, total free amino acids, proline content, total reducing power, total antioxidant capacity, and catalase, peroxidase, and SOD activity under stress.	Magdy et al. (2016)

membrane damage (Hasanuzzaman et al., 2012). By the middle of the 21st century, salinity intrusion will cause the loss of half of global arable lands (Mahajan et al., 2005).

Nanosilicon application can decrease the antagonistic effects of salinity on fava bean (*Vicia faba* L.) plants by increasing the action of antioxidant enzymes (Qados, 2015). The same article demonstrated that NaCl treatments produced a rise in proline content and in some enzyme actions and a reduction in Chl a and b and carotenoids. Application of nano-Si triggered a significant escalation in the action of CAT, peroxidase (POD), and APX in plant leaves, whereas it caused a reduction in the action of SOD compared with unstressed plants. Salinity stress-induced oxidative damage seemed to reduce in accordance with an increase in the activity of antioxidant enzymes during nano-Si treatments,; thus, salinity tolerance was achieved (Qados, 2015). Salinity treatment inhibited root formation in *Gardenia jasminoides* (Ellis plant) in Murashige and Skoog (MS) medium supplemented with 6-benzyladenine (BA) and naphthaleneacetic acid (NAA) (in vitro propagation medium), while the addition of 5 mg/l of Fe or Zn nanoparticles to the salinity-treated

MS media induced rooting capability, shoot proliferation, and elongation (Taha et al., 2020). Nano-SiO_2 increases plant dry weight and fresh weight, improves seed germination, and increases chlorophyll content with proline accumulation in squash plants and tomato under NaCl stress (Haghighi et al., 2012, Siddiqui et al., 2014). Sunflower cultivars showed a positive reaction to salinity stress tolerance when nano-iron sulfate ($FeSO_4$) was applied by foliar means (Torabian et al., 2017). It is also reported that nano-$FeSO_4$ application increased net carbon dioxide (CO_2) assimilation rate, leaf area, chlorophyll content, sub-stomatal CO_2 concentration (Ci), shoot dry weight, maximum photochemical efficiency of photosystem II (Fv/Fm), and iron (Fe) content; while sodium (Na) content in leaves was significantly decreased (Torabian et al., 2017). Antioxidant enzymes like glutathione peroxidase (GPX), APX, SOD, and CAT are scavengers of ROS and suppress the detrimental effects of ROS on plant cells (Gharsallah et al., 2016). It is well recognized that NPs can alter the action of antioxidant enzymes in plants, usually positively, because of their ability to trigger the antioxidant defense system (Ruotolo et al., 2018). The application of citric acid-coated CeO_2 NPs increases CAT enzyme activity in tomato leaves (Barrios et al., 2015). Salinity pressure on plants generates an overproduction of ROS, which causes harm due to the oxidative stress in the cells (Reddy et al., 2016). Use of Cu NPs can enhance salinity stress tolerance in tomato plants by regulating oxidative and ionic stress, perhaps due to the expression of JA (jasmonic acid) and SOD genes (Hernández-Hernández et al., 2018). Djanaguiraman et al. (2018) suggested that the achievement of tolerance to abiotic stress is closely associated with the exclusion of ROS, so the observed escalation in antioxidant enzyme activity is a pointer to reduction in salinity stress in plants.

Nanosilica acts on water translocation and xylem humidity and increases turgor pressure; thus, RWC in leaf and water use efficiency are increased (Elsakhawy et al., 2018). Silicon deposition in plant tissues aids in mitigating water stress by reducing the rate of transpiration and boosts the photosynthesis process by improving light acceptance efficiency and retaining the leaf rigid and erect, inhibiting chlorophyll degradation (Ali et al., 2013; Siddiqui et al., 2014). Nanosilicon is accountable for a number of improvements, which lead ultimately to enhancement of plant growth and productivity under the stress of both salinity (Table 17.2) and drought. This includes the modification of gas exchange attributes, homeostasis of nutrient elements, stimulation of antioxidant enzymes, osmotic adjustment by regulating the synthesis of compatible solutes, and gene expression in plants (Qados, 2015). The application of nano-silver has been found effective in increasing resistance against salinity at the time of germination of cumin (Ekhtiyari and Moraghebi, 2011) and fennel (Ekhtiyari et al., 2011). It is stated that exposure to Ag NPs alleviated the harmful effects of salinity stress and improved the root length, germination, seedlings, and fresh and dry weight of tomato seeds under NaCl stress (Almutairi, 2016b). Furthermore, similar results were also reported in a study conducted by Wahid et al. (2020), where Ag NPs aided in alleviating salt stress-induced adversities in wheat by positively regulating the physiological machinery under salt stress. Moreover, salt-affected soil management also can be achieved using these nanomaterials (Table 17.3). In line with this, many researchers have found nanomaterials to be effective in soil amelioration (Patra et al., 2016). Harmful effects of salinity stress on cotton plants were successfully mitigated by 200 ppm nano-Zn foliar application. This result opened the gate to cultivating cotton in saline regions where diluted seawater can be used as an irrigation source (Hussein et al., 2018).

17.2.4 Nanomaterials to Combat Cold Stress

Cold stress affects plant cells through oxidative processes. ROS initiate these oxidative processes. ROS interact with cellular components and trigger peroxidative reactions that result in damage to nucleic acids, proteins, lipids and membranes (Mittler et.al., 2004). Environmental signals for transportation, perception, and transduction are governed by the crucial role of the plasma membrane. Environmental changes cause a cell either to adapt or to kill itself. To adapt to environmental changes, organisms use different factors to modulate the physical state of the cell membrane (Heidarvand and

TABLE 17.2
Effect of Various Nanoparticles on Plant Growth and Productivity under Salinity (NaCl) Stress

Nanoparticles	Plant	Responses under Stress	References
Silica nanoparticles	Basil (*Ocimum basilicum*)	Upregulate Chlorophyll-a and b and proline, and increase shoot dry matter.	Kalteh et al. (2014)
Silicon nanoparticles	Fava bean (*Vicia faba*)	Membrane characteristics positive response, photosynthetic pigments, soluble sugars, proline, enzymatic antioxidants increased.	Qados (2015)
Silica ions and nanosilica	Rice (*Oryza sativa* L.)	Ca and K content were increased under silica ion. Proline content, free amino acid content, and total carbohydrate content also increased by silica ion under salt stress. MDA content, PMP, and H_2O content were controlled by silica ion, slight increase in antioxidative enzymatic activities, shikimic acid content decreased, phenolic compounds content increased, hormone content amplified.	Abdel-Haliem et al. (2017)
Silica nanoparticles	Common bean (*Phaseolus vulgaris*)	Accelerate seed germination process, shoot and root length, increase plant biomass.	Alsaeedi et al. (2017)
Cerium oxide nanoparticles (CeO_2-NPs)	*Brassica napus* L. (canola) cv. "Dwarf Essex"	Improve growth attributes, leaf pigments, chlorophyll fluorescence, proline, photosynthetic activity, mineral nutrients, cerium uptake.	Rossi et al. (2016)
Cerium oxide nanoparticles	*Brassica napus* L.	Plant dry mass showed no significant effect, formation of apoplastic barrier, significantly higher values of ratio of variable fluorescence to maximum fluorescence (Fv/Fm), increase in sodium concentration of leaves and roots.	Rossi et al. (2017)
ZnO nanoparticles	Tomato (*Lycopersicon esculentum* Mill)	Upregulate the mRNA levels of SOD and glutathione peroxidase (GPX) genes.	Alharby et al. (2016)
ZnO nanoparticles	Lupine (*Lupinus termis*)	Growth attributes, photosynthetic pigments, organic solutes (soluble sugar, soluble protein, total free amino acids, and proline), total phenols, malondialdehyde (MDA), antioxidant enzyme activities, ascorbic acid, Na and Zn content.	Latef et al. (2017)
ZnO and Fe_3O_4 magnetic nanoparticles	*Moringa peregrina*	Under synthetic seawater salt stress, improve growth parameters, chlorophyll content, total carbohydrates, macro- and micronutrients, and crude proteins. Increase enzymatic and nonenzymatic antioxidants.	Soliman et al. (2015)
Iron nanoparticles	Grape	Increase the total protein content and reduce proline, enzymatic antioxidant activity and hydrogen peroxide, increase membrane stability index and reduce of MDA content, lower the sodium content, and increase the potassium content.	Mozafari et al. (2018)
Nitric oxide chitosan nanoparticles	*Zea mays*	If plants exposed to NaCl with mercaptosuccinic acid stress, higher leaf S-nitrosothiols content, increase in photosystem-II activity, chlorophyll content, and growth of maize plants.	Oliveira et al. (2016)

Source: Ahmad and Akhtar (2019).

TABLE 17.3
Different Beneficial Effects Resulting from Nanotechnology Intervention in Salt-Affected Soils

Processes	Details
Reducing salt concentration in soil solution	Enhancement of Na removal, soil stability, and hydraulic characteristics using polymeric carriers in nano-reclaimants via clay binding processes
Improving soil drainage	Removing Na^+ from soil solution leads to improved crop growth by improving soil structure in subsurface and sub-soils using nano-Ca^{2+} and nanoferrites as well as biofriendly nanopolymers
Replacing Na^+ by Ca^{2+}	Nano-Ca^{2+}, nano-Mg^{2+}, and nano-K^+ can be used in removing Na^+ from soil because of the high selectivity and spontaneous reactions (negative ΔG0) for Ca^{2+} over Na^+ for all clay minerals
Changing carbonate chemistry	Capping/encapsulating Na_2CO_3 with nanopolymers and nanocomposites selectively may yield products that are insoluble and may be leached through preferential flows; formation of nano-organic carbonates could be another possibility
Prevention of Na_2CO_3 formation	Nanomaterials including nanocalcium carbonates, nano-Ca^{2+}, nano-Mg^{2+}, nano-K^+, and nano-iron oxides may be used in preventing Na_2CO_3 formation in soils
Addition of K^+	Some nanomaterials, such as nano-K^+ in clay minerals like illite, could be used in accelerating ion exchange reactions (Ca^{2+}, Mg^{2+} and Na^+) to reduce exchangeable sodium saturation
Solubilizing of $CaCO_3$	Calcium carbonate could possibly be solubilized by the application of nano-Ca^{2+}, nano-Mg^{2+}, and nano-Fe oxides
Precipitation	Nanopolymers and nano-organic substances may be used in forming complexes or insoluble salts with harmful ions and precipitated
Common ion effect	Nano-iron oxides, nano-Ca^{2+}, and nano-Mg^{2+} may be applied to counter adverse effect of Na^+ as a well-known phenomenon

Source: Mukhopadhyay and Kaur (2016); Patra et al. (2016).

Maali, 2010). The lipid peroxidation of unsaturated fatty acids produces malondialdehyde (MDA), which is a marker of membrane damage evaluation. TiO_2 NPs protect the cell membrane from damage during cold stress by inducing plant antioxidant systems to alleviate the increase in MDA content and electrolyte leakage (Lei et al., 2008). The application of TiO_2 NPs promotes the actions of antioxidant enzymes like APX, CAT, and SOD (Lei et al., 2008; Nazari et al., 2012). TiO_2 NPs protected chickpea genotypes from oxidative damage and ameliorated cold stress-induced damage when cold stress treatment was applied (Mohammadi et al., 2013).

17.2.5 Nanoparticles for Heavy Metal Remediation

Pollutants like heavy metals are capable of hampering the normal growth and yield of crops. Several researchers have found nanomaterials to be effective in remediation or detoxification of heavy metals. Cd accumulation in rice plants was regulated by the foliar application of a 2.5 mM concentration of nano-Si, which significantly increased Cd stress tolerance (Wang et al., 2015). They also showed that nano-Si is effective for Cu, Pb, and Zn with Cd. It seems that nano-Si fertilizers can putatively have an benefit over traditional fertilizers in decreasing heavy metal buildup (Wang et al., 2016). In another study, Cd translocation was inhibited and the antioxidant potential was stimulated, thereby mitigating Cd toxicity and improving the growth of rice plants, using foliar application of TiO_2 and SiO_2 nanoparticles. Nano-Si improves plant growth during heavy metal stress. The suggested mechanisms include reducing active ions of heavy metals in growth media, chelating and stimulating antioxidant systems in plants, producing complex, co-precipitating toxic metals with Si, compartmenting, and regulating the expression of metal transport genes. These mechanisms might be related to genotypes, plant species, growth conditions, metal elements, duration of the stress enforced, and so on. Consequently, the generalization of Si-facilitated mitigation of metal toxicity should be made carefully (Adrees et al., 2015). Several studies mentioned the ameliorative effect of Si treatment on plants under the stress of heavy metals or hazardous toxic elements, like Cr, Cu, Al, Cd, or Pb (Shen et al., 2014; Keller et al., 2015). Zinc (Zn) is one of the important micronutrients vital for the optimal growth and development of plants, which contributes to different metabolic reactions in plant cells to encourage growth and development. Nano-Zn also acts to reduce toxic heavy metal uptake by crops, thereby protecting crops from the toxicity of heavy metals like Cd (Baybordi, 2005). The application of nano-TiO_2 can decrease cadmium (Cd) stress; the suggested mechanism includes the creation of new bonds with the Cd/nano-TiO_2 particles in the plant tissue (Singh and Lee, 2016).

17.3 NANOFERTILIZERS IN CROP PHYSIOLOGY AND PRODUCTION

Nutritional deficiency is the one of the most important limiting factors for plant growth. Plant roots and microorganisms can directly take up nutrient ions from the solid phase of minerals. Nanotechnology has played a vital role in enhancing plant nutrient uptake. In fact, nanotechnology has opened up new opportunities to improve nutrient use efficiency and minimize costs to protect the environment in the form of nanofertilizers, which could be an alternative way to enhance nutrient use efficiency and to mitigate the serious problems of eutrophication. Nanofertilizers are fertilizers that are made with the help of nanotechnology and used to improve soil fertility, productivity, and quality of agricultural products (Brunnert et al., 2006).

Nanofertilizer has a greater surface area, mainly due to the minute structure of the particles, which provide more sites to assist with different metabolic processes in the plant system, resulting in the production of more photosynthates. Due to their minute nature, they have high reactivity with other compounds. They have high solubility in different solvents, especially water. The particle size of nanofertilizers is less than 100 nm, which enables greater penetration of nanoparticles into the plant from application on surfaces such as soil or leaves. Smaller particle size results in increased specific surface area and number of particles per unit area of a fertilizer, providing more opportunity for contact with nanofertilizers, which leads to greater penetration and uptake of the nutrient (Liscano et al., 2000). It is also encapsulated in nanoparticles, which will increase the availability and uptake of nutrients by the crop plants (Rai et al., 2012).

Encapsulation of nutrients within the nanoparticles is performed in different ways, particularly:

i) The nutrient can be encapsulated inside nonporous materials,
ii) Coated with thin polymer film, and
iii) Delivered as particles or emulsion of nanoscale dimensions (Rai et al., 2012).

In these ways, the losses of nutrients in soil, water, and air are decreased or reduced due to direct insertion into crop plants, and the interactions of nutrients with soil, micro-organisms, water, and air are prevented (De Rosa et al., 2010). Nanofertilizers coated and bound with nano and sub-nano compounds are able to regulate the release of nutrients from the fertilizer capsule (Liu et al., 2006). This improves soil by reducing the toxic effects associated with over-application of fertilizer.

Nanofertilizers are beneficial for many reasons, mostly their penetration ability, size, and very high surface area, which normally differ from the same material found in bulk form. In practice, nanoparticles show a very high surface: volume ratio. Hence, the reactive surface area is proportionally over-represented in nanoparticles compared with larger particles. Particle surface area increases with decreasing particle size, and the surface free energy of the particle is a function of its size (Liscano et al., 2000).

Among the fertilizers, nitrogen fertilizer is lost in the environment due to volatilization, denitrification, leaching, and run-off. Application and harvesting methods enhance these losses. Consequently, it is necessary to substitute these incorrect methods by a readily available fertilizer to make the nutrients available to the plants. Sources of nitrogen include ammonia, ammonium nitrate, diammonium phosphate, ammonium sulfate, calcium nitrate, calcium cyanamide, sodium nitrate, and urea (Daniels and Schmidt, 1997). Nitrogen is a widely used fertilizer because of its availability and quick response. The role of this element in the environment is complex. Due to its high solubility, nitrogen fertilizers can damage the plants and the surroundings. In the case of nitrogen fertilizers, gentle release will be beneficial for plants, because growers can frequently apply the minimum amount, providing the nutrients slowly and steadily.

Nanoporous zeolite was used with urea, and there was a notable increase in the uptake of nitrogen-efficient urea with controlled release (Manikandan and Subramanian, 2015). Zeolites, naturally occurring minerals with a honeycomb-like covered crystal structure, are another strategy for increasing fertilizer use efficiency (Chinnamuthu and Boopathi, 2009). The zeolite generally helps in slow delivery of the fertilizer to the plant; in this way, rather than minimal uptake of nutrition, the plant grabs the entire amount of nutrients from the supplied fertilizer. Since zeolite has a high surface area, abundant molecules can fit into it and be released whenever the plant needs.

Zeolite can be loaded with nitrogen and potassium, combined with other slowly dissolving ingredients containing phosphorus, calcium, and a complete suite of minor and trace nutrients. Zeolite acts as a reservoir for nutrients that are slowly released "on demand." Fertilizer particles can be coated with nanomembranes that facilitate slow and steady release of nutrients.

Significant increases in yield have been observed due to foliar application of phosphorus nanoparticles as fertilizer (Tarafdar et al., 2012a; Tarafdar et al., 2012b). It was shown that 640 mg ha^{-1} foliar application (40 ppm concentration) of nanophosphorus gave 80 kg ha^{-1} P equivalent yield of clusterbean and pearl millet in an arid environment. Currently, research is underway to develop nanocomposites to supply all the required essential nutrients in suitable proportions.

Micronutrients are those elements that are required by the plant in trace/low quantities, but are essential to maintain vital metabolic processes in plants. They are very important for crop growth, and deficiency shows negative effects resulting in the formation of fruits and vegetables with poor nutritional value (López-Valdez et al., 2018).

The effect of foliar application of two micronutrient nanofertilizers of Zn and boron (B) at three different concentrations was evaluated, and it was found that lower amounts of B (34 mg $tree^{-1}$) or Zn nanofertilizers (636 mg $tree^{-1}$) increased fruit yield by 30% in pomegranate trees (*Punica granatum* cv. Ardestani) (Khot, 2012). Also, a notable yield increase was found when using Zn nanoparticles as a nutrient source in rice, wheat, maize, potato, sunflower, and sugarcane (Monreal, 2016).

Iron (Fe) and manganese (Mn) are other important nutrients essential for plants in minute quantities for maintaining proper growth and development. Although they are needed in trace amounts, deficiency or excess causes impairment in key plant metabolic and physiological processes, thereby leading to reduced yield. The strategy of fertilizer application influences their efficiency and impact on plant systems. There are several methods used for nanofertilizer delivery to plants.

Aeroponics was the technique first reported by Weathers and Zobel (1992). In this technique, the roots of the plant are suspended in air and the nutrient solution is sprayed continuously. By this method, the gaseous environment around the roots can be restricted. However, it requires a high level of nutrients to sustain quick plant growth, so the use of aeroponics is not widespread.

Hydroponics is another method first introduced by Gericke (1937) for dissolved inorganic salts. The method is also commonly known as "solution culture" as the plants are grown with their roots immersed in a liquid nutrient solution (without soil), and nutrient solution, maintenance of oxygen demands, and pH are factors that need attention while using this method of nutrient delivery.

The most widely used is the soil application method of nutrient supplements using chemical and organic fertilizers. The application methods of fertilizers depend on different factors such as how long the fertilizer will last in the soil, soil texture, soil salinity, and plant sensitivities to salts, salt content, and pH of the amendment. Among ions, nitrate remains mobile in the soil solution and is vulnerable to leaching by water moving through the soil. Due to the positively charged Fe^{2+}, Fe^{3+} , and Al^{3+} having OH groups, they can bind to soil particles containing phosphate ions; as a result, phosphate can be tightly bound, and its lack of mobility and availability in soil can limit plant growth (Taiz and Zeiger, 2010).

In the foliar application method, liquid fertilizers are directly sprayed onto leaves. It is generally used for the supply of micronutrients. Foliar application can limit the time lag between application and uptake by plants during the rapid growth phase of the crop. It can also avoid the problem of restricted uptake of a nutrient from soil; it is more efficient to apply manganese and copper by this method as compared with soil application, where they become adsorbed on soil particles and hence are less available to the root system (Taiz and Zeiger, 2010).

In conclusion, nanotechnology in crop nutrition fertilizers has played a vital role in enhancing plant nutrient uptake as well as increasing yield performance. In fact, nanofertilizers have opened up new opportunities to improve nutrient use efficiency and minimize costs to protect the environment.

17.4 NANOPARTICLES IN MOLECULAR BREEDING AND GENE EXPRESSION

Nanotechnology-based carriers can deliver a variety of bioactive molecules into the cells. Several lines of evidence suggest that nanomaterials have been extensively utilized in animal cells and organs as delivery vehicles for targeted drugs, cancer therapy, and the treatment of genetic diseases (Gu et al., 2011). Nonetheless, despite increasing interest among plant researchers, their use in plant systems lags behind (Wang et al., 2016). Considerable study indicates the use of nanomaterials in delivering plasmid DNA (pDNA), double-stranded RNA (dsRNA), short interfering RNA or silencing RNA (siRNA), proteins, and phytohormones into either cell wall-free protoplasts or intact cells. A vitally important breakthrough is the development of honeycomb-like mesoporous silica nanoparticles (MSNs) as carriers to deliver pDNA and protein into the intact cells associated with a biolistic method (Martin-Ortigosa et al., 2014; Torney et al., 2007). In that approach, the pDNA and its chemical inducer are loaded simultaneously into MSNs, and the ends of the MSNs are then capped using gold nanoparticles as a gatekeeper (Torney et al., 2007). Once the MSNs are provided into the cells, the gold nanoparticles are uncapped by using a gate-opening trigger (i.e., opening in response to changes in chemical or physiological conditions) to allow controlled release of the biogenic molecules. Further development has functionalized MSNs to deliver foreign pDNA into the epidermis, cortex, and endodermis of intact *Arabidopsis* roots without the application of any mechanical force (Chang et al., 2013). In addition to MSNs, other nanomaterial-based delivery systems have been developed, including carbon nanotubes (CNTs), layered double hydroxide (LDH) clay nanosheets, DNA nanostructures, and magnetic nanoparticles (Bao et al., 2017; Zhao et al., 2017; Mitter et al., 2017; Kwak et al., 2019; Demirer et al., 2019a; Zhang et al., 2019). Although remarkable progress has been made in this respect, little is known about how these nanomaterials interact with plant cells having a series of barriers at cellular and subcellular levels (Gu et al., 2011). For example, functionalized CNTs can be passively internalized into intact and walled cells of a range of plant species. This approach then enables exogenous functional genes in the form of pDNA or siRNA to be delivered into cells, resulting in strong expression of the exogenous gene (Demirer et al., 2019a) or highly efficient gene silencing (Demirer et al., 2019b). Meanwhile, CNTs can also protect polynucleotides from nuclease degradation for a desired period, providing a time-window to allow gene expression or silencing in the absence of transgene integration into the plant genome. Such approaches are capable of both subcellular and tissue targeting. For example, in a remarkable study, CNTs were designed to selectively deliver pDNA into chloroplasts through a lipid exchange envelope-penetration mechanism without external biolistic or chemical aid (Kwak et al., 2019). This elegant methodology enables pH-dependent release of loaded pDNA (e.g., only releasing DNA strands at pH(7.5), thereby allowing release in the chloroplast stroma (ca. pH 8) but not in the plant cell cytosol (ca pH 5.5). It has also been demonstrated that LDH nanosheets can facilitate delivery of dsRNA into walled plant cells of the model plant *Nicotiana tabacum*, thereby enabling gene silencing of homologous RNA and providing sustained protection against plant viruses (Mitter et al., 2017). Finally, magnetic nanoparticles can be used to drive exogenous DNA into pollen through pollen apertures in the presence of a magnetic field. This facilitates the creation of transgenic plants, which is particularly useful for plant species that are difficult to genetically transform using conventional methods (Zhao et al., 2017). These pioneering studies have made important progress in developing nanomaterial-based carrier systems to deliver exogenous biomolecules into intact and walled plant cells, paving the way toward future applications in plant biotechnology. The Clustered Regularly Interspaced Short Palindromic Repeats (CRISPR)-Cas (CRISPR-associated protein) genome editing technology has shown great promise for quickly addressing emerging challenges in agriculture (Haque et al., 2018). The revolutionary CRISPR-Cas technology has emerged as a powerful toolkit that relies on Cas/sgRNA (single guide RNA) ribonucleoprotein complexes (RNPs) to target and edit DNA in plants. Recently, a systemic nanoparticle delivery of CRISPR-Cas9 ribonucleoproteins for effective tissue specific genome editing has been demonstrated (Wei et al., 2020).

17.5 CONCLUSION AND FUTURE PROSPECTS

Global climate change is plunging world crop production towards various abiotic stresses. Flooding, drought, salinity, cold, and heavy metals cause interconnected problems in crop production. Flooding fields often produces toxic compounds and gases, which may kill crop plants. If crop plants experience flooding during the vegetative stage, they may also suffer from terminal drought during the reproductive stage, depending on the ecosystem, even if those crop plants are tolerant to flooding. Likewise, periodic drought conditions may upregulate the existing salinity stress through intensification of a salt layer on the upper surface of the soil. In contrast, loss of fluidity of membranes in plant cells and leakage of solutes are the distinct effects of cold stress. Salinity stress downregulates the ability of plants to take up water and nutrients, aside from causing nutritional imbalance, which hinders growth and yield. Almost all abiotic stresses enhance the generation of ROS, which damage the cellular membranes, proteins, and nucleic acids that are vital to plant survival, growth, and yield. In this situation, it imperative to develop an improved and sustainable farming technology and also cultivars resistant to all these hazards, taking into account the need to address food-security issues effectively. So, there is a dire need to adopt a potential nanomaterial to address the problems of abiotic stresses for future crop breeding and production as well.

It is speculated that nanotechnology has four distinct generations of advancement. We are currently experiencing the first or maybe the second generation of nanomaterials. The first generation is all about material science, with enhancement of properties achieved by the incorporation of "passive nanostructures". This can be in the form of coatings and/or the use of carbon nanotubes to strengthen plastics. The second generation makes use of an active nanoframe, for instance, providing a bioactive drug at a specific target cell or tissue. This could be done by coating the nanoparticle with specific proteins. There is further complexity in the third and fourth generations, Starting with an advanced nanosystem, for example, nanorobotics, and moving on to a molecular nanosystem to control the growth of artificial organs in the fourth generation of nanotechnology. The development of the "Safe-by-design" concept for nanomaterials is currently under investigation by scientists. The basic premise is: rather than testing the safety of nanomaterials after they are put on the market, the safety assessment should be incorporated into the design and innovation stage of a nanomaterial's development to expedite crop production, ensuring food security through exploitation of natural resources.

REFERENCES

Abdel-Aziz, H.M.M., M.N.A. Hasaneen and A.M. Omer. 2016. Nano chitosan-NPK fertilizer enhances the growth and productivity of wheat plants grown in sandy soil. *Spanish Journal of Agricultural Research* 14:17.

Abdel-Haliem, M.E.F., H.S. Hegazy, N.S. Hassan and D.M. Naguib. 2017. Effect of silica ions and nano silica on rice plants under salinity stress. *Ecological Engineering* 99:282–289.

Acquaah, G. 2007. *Principles of Plant Genetics and Breeding*. Oxford: Blackwell.

Adrees, M., S. Ali, M. Rizwan, et al. 2015. The effect of excess copper on growth and physiology of important food crops: a review. Environmental Science and Pollution Research 22:8148–8162. https://doi.org/10.1007/s11356-015-4496-5

Aghdam, M.T.B., H. Mohammadi and M. Ghorbanpour. 2016 Effects of nanoparticulate anatase titanium dioxide on physiological and biochemical performance of *Linum usitatissimum* (Linaceae) under well watered and drought stress conditions. *Brazilian Journal of Botany* 39:139–146. DOI: 10.1007/s40415-015-0227-x.

Ali, S., M.A. Farooq, T. Yasmeen, S. Hussain, M.S. Arif, F. Abbas and G. Zhang. 2013. The influence of silicon on barley growth, photosynthesis and ultra-structure under chromium stress. *Ecotoxicology and Environmental Safety* 89:66–72.

Almutairi, Z.M. 2016b. Influence of silver nano-particles on the salt resistance of tomato (*Solanum lycopersicum* L.) during germination. *International Journal of Agriculture and Biology* 18(2). DOI: 10.17957/IJAB/15.0114.

Alsaeedi, A.H., H. El-Ramady, T. Alshaal, M. El-Garawani, N. Elhawat and M. Almohsen. 2017. Engineered silica nanoparticles alleviate the detrimental effects of Na^+ stress on germination and growth of common bean (*Phaseolus vulgaris*). *Environmental Science and Pollution Research* 24:21917–21928.

Ashkavand, P, M. Tabari, M. Zarafshar, I. Tomášková and D. Struve. 2015. Effect of SiO_2 nanoparticles on drought resistance in hawthorn seedlings. *Leśne Prace Badawcze/Forest Research Papers Grudzień* 76(4):350–359.

Bao, W., Y. Wan and F. Baluška. 2017. Nanosheets for delivery of biomolecules into plant cells. *Trends in Plant Science* 22:445–447.

Barrios, A.C., C.M. Rico, J. Trujillo-Reyes, I.A. Medina-Velo, J.R. Peralta-Videa and J.L. Gardea-Torresdey. 2016. Effects of uncoated and citric acid coated cerium oxide nanoparticles, bulk cerium oxide, cerium acetate, and citric acid on tomato plants. *Science of the Total Environment* (563–564):956–964.

Baybordi, A. 2005. Effect of zinc, iron, manganese and copper on wheat quality under salt stress conditions. Iranian Journal of Soil and Water Sciences 140:150–170.

Booze-Daniels, N.B. and R.E. Schmidt. 1997. Use of slow release nitrogen fertilizers on roadside – A literature review, Dept of Crop and Soils Environmental Science Virginia Polytechnic Institute and State University, Va 24061-0404 (540) 231–7175.

Brunner, T.J., P. Wick, P. Manser, P. Spohn, R.N. Grass, L.K. Limbach, A. Bruinink and W.J. Stark. 2006. In vitro cytotoxicity of oxide nano-particles: Comparison to asbestos, silica and effects of particle solubility. *Environ*mental *Sci*ence & *Technology* 40:4374–4381.

Carmen, I.U., P. Chithra, Q. Huang, P. Takhistov, S. Liu and J.K. Kokini. 2003. Nanotechnology: A new frontier in food science. *Food Technology* 57:24–29.

Chang, F.P., L-Y. Kuang, C-A. Huang, W-N. Jane, Y. Hung, Y-I.C. Hsing and C-Y. Mou. 2013. A simple plant gene delivery system using mesoporous silica nanoparticles as carriers. *Journal of Materials Chemistry B* 1:5279–5287.

Chen, M., Q.Y. Wang and X.G. Cheng. 2007. GmDREB2, a soybean DRE-binding transcription factor, conferred drought and high-salt tolerance in transgenic plants. *Biochemical and Biophysical Research Communications* 353(2):299–305.

Chinnamuthu, C.R. and P.M. Boopathi. 2009. Nanotechnology and agroecosystem. *Madras Agricultural Journal* 96:17–31.

Chowdhury, J., M. Karim, Q. Khaliq, A. Ahmed and A.M. Mondol. 2018. Effect of zinc, iron, manganese and copper on wheat quality under salt stress conditions. *SAARC Journal of Agriculture* 15(2):163–175.

Demirer, G.S. et al. 2019a. High aspect ratio nanomaterials enable delivery of functional genetic material without DNA integration in mature plants. *Nature Nanotechnology* 14:456–464.

Demirer, G.S. et al. 2019b. Nanotubes effectively deliver siRNA to intact plant cells and protect siRNA against nuclease degradation. bioRxiv Published online March 1 https://doi.org/10.1101/564427

De Rosa, M.R., C. Monreal, M. Schnitzer, R. Walsh and Y Sultan. 2010. Nanotechnology in fertilizers. *Nature Nanotechnology* 5(2):91.

Djanaguiraman, M., N. Belliraj, S.H. Bossmann and P.V.V Prasad. 2018. High-temperature stress alleviation by selenium nanoparticle treatment in grain sorghum. *ACS Omega* 3(3):2479–2491.

Du, Y.T., M.J. Zhao and C.T. Wang. 2018. Identification and characterization of GmMYB118 responses to drought and salt stress. *BMC Plant Biology* 18(1):320.

Ekhtiyari, R. and F. Moraghebi. 2011. The study of the effects of nano silver technology on salinity tolerance of cumin seed (*Cuminum cyminum* L.). *Plant and Ecosystem* 7(25):99–107.

Ekhtiyari, R., H. Mohebbi and M. Mansouri. 2011. The study of the effects of nano silver technology on salinity tolerance of (*Foeniculum vulgare* Mill.). *Plant and Ecosystem* 7(27):55–62.

Elsakhawy, T., A. Omara, T. Alshaal and H. El-Ramady. 2018. Nanomaterials and plant abiotic stress in agroecosystems. *Environment, Biodiversity and Soil Security* 2(2018):73–94. doi: 10.21608/jenvbs.2018.3897.1030

Fitter, D.W., D.J. Martin, M.J. Copley, R.W. Scotland and J.A. Langdale. 2002. GLK gene pairs regulate chloroplast development in diverse plant species. *The Plant Journal* 31(6):713–727.

Gericke, W.F. 1937. Hydroponics – crop production in liquid culture media. *Science* 85(2198):177–178. DOI: 10.1126/science.85.2198.177. PMID: 17732930.

Gharsallah, C., H. Fakhfakh, D. Grubb and F. Gorsane. 2016. Effect of salt stress on ion concentration, proline content, antioxidant enzyme activities and gene expression in tomato cultivars. *AoB Plants* 8:plw055.

Ghormade, V., M.V. Deshpande and K.M. Paknikar. 2011. Perspectives for nano-biotechnology enabled protection and nutrition of plants. *Biotechnology Advances* 29: 792–803, https://doi.org/10.1016/j.biotechadv.2011.06.007.

Golldack, D., I. Lüking and O. Yang. 2011. Plant tolerance to drought and salinity: stress regulating transcription factors and their functional significance in the cellular transcriptional network. *Plant Cell Reports* 30(8):1383–1391.

Gu, Z., A. Biswas, M. Zhao and Y. Tang. 2011. Tailoring nanocarriers for intracellular protein delivery. *Chemical Society Reviews* 40(7):3638. DOI:10.1039/c0cs00227e

Haghighi, M., Z. Afifipou and M. Mozafariyan. 2012. The effect of N-Si on tomato seed germination under salinity levels. *Journal of Biological and Environmental Sciences* 6:87–90.

Haque, E., H .Taniguchi, M.M. Hassan, P. Bhowmik, M.R. Karim, M. Śmiech, K. Zhao, M. Rahman and T. Islam. 2018. Application of CRISPR/Cas9 genome editing technology for the improvement of crops cultivated in tropical climates: recent progress, prospects, and challenges. *Frontiers in Plant Science* 9:617. DOI: 10.3389/fpls.2018.00617

Hasanuzzaman, M., M.A. Hossain and M. Fujita. 2012. Exogenous selenium pretreatment protects rapeseed seedlings from cadmium-induced oxidative stress by up regulating the antioxidant defense and methylglyoxal detoxification systems. *Biological Trace Element Research* 149:248–261. DOI:10.1007/s12011-012-9419-4

Heidarvand, L. and A.R. Maali. 2010. What happens in plant molecular responses to cold stress. *Acta Physiologiae Plantarum* 32:419–431.

Helaly, M.N., M.A. El-Metwally, H. El-Hoseiny, S.A. Omar and N.I. El-Sheery. 2014. Effect of nanoparticles on biological contamination of "in vitro" cultures and organogenic regeneration of banana. *Australian Journal of Crop Science* 8(4):612–624.

Hernández-Hernández, H., A. Juárez-Maldonado, A. Benavides-Mendoza, H. Ortega-Ortiz, G. Cadenas-Pliego, D. Sánchez-Aspeytia and S. González-Morales. 2018. Chitosan-PVA and copper nanoparticles improve growth and overexpress the SOD and JA genes in tomato plants under salt stress. *Agronomy* 8:175.

Hossain, A., R.G. Kerry, M. Farooq, N. Abdullah and M.T. Islam. 2020. Application of Nanotechnology for Sustainable Crop Production Systems. In: Thangadurai et al. (eds.) *Nanotechnology for Food, Agriculture, and Environment*. Springer. pp. 135–159. doi.org/10.1007/978-3-030-31938-0_7

Huang, G.T., S.L. Ma and L.P. Bai. 2012. Signal transduction during cold, salt, and drought stresses in plants. *Molecular Biology Reports* 39(2):969–987.

Hussein, M.M. and N.H. Abou-Baker. 2018. The contribution of nano-zinc to alleviate salinity stress on cotton plants. *Royal Society Open Science* 5:171809. http://dx.doi.org/10.1098/rsos.171809

Jaberzadeh, A., P. Moaveni, H.R.T. Moghadam and H. Zahedi. 2013. Influence of bulk and nanoparticles titanium foliar application on some agronomic traits, seed gluten and starch contents of wheat subjected to water deficit stress. *Notulae Botanicae Horti Agrobotanici Cluj-Napoca* 41:201–207.

Jin, J., H. Zhang, L. Kong, G. Gao, and J. Luo. 2013. PlantTFDB 3.0: a portal for the functional and evolutionary study of plant transcription factors. *Nucleic Acids Research* 42(D1): D1182–D1187.

Kalteh, M., Z.T. Alipour, S. Ashraf, M.M. Aliabadi and A.F. Nosratabadi. 2014. Effect of silica nanoparticles on basil (*Ocimum basilicum*) under salinity stress. *Journal of Chemical Health Risks* 4:49–55.

Keller, C., M. Rizwan, J.C. Davidian, O.S. Pokrovsky, N. Bovet, P. Chaurand and J.D. Meunier. 2015. Effect of silicon on wheat seedlings (Triticum turgidum L.) grown in hydroponics and exposed to 0 to 30 μM Cu. Planta 241(4):847–860.

Khan, M.N., K. Sakata, S. Hiraga and S. Komatsu. 2014. Quantitative proteomics reveals that peroxidases play key roles in post-flooding recovery in soybean roots. *Journal of Proteome Research* 13:5812–5828.

Khot, L.R., S. Sankaran, J.M. Maja, R. Ehsani and E.W. Schuster. 2012. Applications of nanomaterials in agricultural production and crop protection: a review. *Crop Protection* 35:64–70.

Kiapour, H., P. Moaveni, D. Habibi and B. Sani. 2015. Evaluation of the application of gibbrellic acid and titanium dioxide nanoparticles under drought stress on some traits of basil (*Ocimum basilicum* L.). *International Journal of Agronomy and Agricultural Research* 6(4):138–150.

Kim, J.H., Y. Oh, H. Yoon, I. Hwang and Y.S. Chang. 2015. Iron nanoparticle-induced activation of plasma membrane H^+-ATPase promotes stomatal opening in *Arabidopsis thaliana. Environmental Science & Technology* 4:1113–1119. DOI: 10.1021/es504375t.

Kwak, S.Y. et al. 2019. Chloroplast-selective gene delivery and expression in planta using chitosan-complexed single-walled carbon nanotube carriers. *Nature Nanotechnology* 14:447–455.

Latef, A.A.H.A., M.F.A. Alhmad and K.E. Abdelfattah. 2017. The possible roles of priming with ZnO nanoparticles in mitigation of salinity stress in lupine (*Lupinus termis*) plants. *Journal of Plant Growth Regulation* 36:60–70.

Lei, Z, M.Y. Su, X. Wu, C. Liu, C.X. Qu, L. Chen, H. Huang, X.Q. Liu and F.S. Hong. 2008. Antioxidant stress is promoted by nano-anatase in spinach chloroplasts under UV-Beta radiation. *Biological Trace Element Research* 121:69–79.

Li, Y., Q. Chen and H. Nan. 2017. 2017. Over-expression of GmFDL19 enhances tolerance to drought and salt stresses in soybean. *PLoS ONE* 12(6):0179554.

Lim, C., W. Baek, J. Jung, J.H. Kim and S. Lee. 2015. Function of ABA in stomatal defense against biotic and drought stresses. *International Journal of Molecular Sciences* 16(12):15251–15270.

Linh, T.M., C.M. Nguyen, T.H. Pham, Q.L. Lien, K.B. Ninh, T.T.H. Le, H.C. Nguyen and T.V. Nguyen. 2020. Metal-based nanoparticles enhance drought tolerance in soybean. *Journal of Nanomaterials*. Article ID 4056563, https://doi.org/10.1155/2020/4056563.

Liscano, J.F., C.E. Wilson, R.J. Norman and N.A. Slaton. 2000. Zinc availability to rice from seven granular fertilizers. *Arkansas Agricultural Experiment Station Bulletin* 963:1–31. Fayetteville, AR: University of Arkansas.

Liu, X., Feng, Z., Zhang, F., Zhang, S. and He, X. 2006. Preparation and testing of cementing and coating nano-subnanocomposites of slow/controlled-release fertilizer. *Agricultural Sciences in China* 5:700–706.

López-Valdez, F., M. Miranda-Arámbula, A.M. Ríos-Cortés, F. Fernández-Luqueño and V.D.L. Luz. 2018. Nanofertilizers and their controlled delivery of nutrients. In: López-Valdez, F. and Fernández-Luqueño, F. (eds) *Agricultural Nanobiotechnology*. Cham: Springer. https://doi.org/10.1007/978-3-319-96719-6_3

Magdy, A.S, M.M.H. Hazem, A.M.N Alia and A.I. Alshaimaa. 2016. Effects of TiO_2 and SiO_2 nanoparticles on cotton plant under drought stress. *Research Journal of Pharmaceutical, Biological and Chemical Sciences Biochemical and Physiological* 7(4):1540.

Mahajan, S. and N. Tuteja. 2005. Cold, salinity and drought stresses: an overview. *Archives of Biochemistry and Biophysics* 444:139–158. doi:10.1016/j.abb.2005.10.018

Mahmoodzadeh, H., M. Nabavi and H. Kashefi. 2013. Effect of nanoscale titanium dioxide particles on the germination and growth of canola (*Brassica napus*). *Journal of Ornamental and Horticultural Plants* 3:25–32.

Manikandan, A. and K. Subramanian. 2015. Evaluation of zeolite based nitrogen nano-fertilizers on maize growth, yield and quality on inceptisols and alfisols. *International Journal of Plant & Soil Science* 9:1–9. 10.9734/IJPSS/2016/22103

Martin-Ortigosa, S. et al. 2014. Mesoporous silica nanoparticle-mediated intracellular Cre protein delivery for maize genome editing via loxP site excision. *Plant Physiology* 164:537–547.

Mitter, N. et al. 2017. Clay nanosheets for topical delivery of RNAi for sustained protection against plant viruses. *Nature Plants* 3:16207.

Mittler, R., S. Vanderauwera, M. Gollery and F. van Breusegem. 2004. Reactive oxygen gene network of plants. *Trends in Plant Sciences* 9:490–498.

Mohammadi, R., R. Maali-Amiri and A. Abbasi. 2013. Effect of TiO_2 nanoparticles on chickpea response to cold stress. *Biological Trace Element Research* 152:403–410. DOI: 10.1007/s12011-013-9631-x

Mohi-Ud-Din, M., M.N. Siddiqui, M.M. Rohman, S.V.K. Jagadish, J.U Ahmed, M.M. Hassan, A. Hossain and T. Islam. 2021. Physiological and biochemical dissection reveals a trade-off between antioxidant capacity and heat tolerance in bread wheat (*Triticum aestivum* L.) *Antioxidants* 10:351. https://doi.org/ 10.3390/ antiox10030351

Mondal, S., M.I.R. Khan, F. Entila, S. Dixit, P.C.S. Cruz, M.P. Ali, B. Pittendrigh, E.M. Septiningsih and A.M. Ismail. 2020. Responses of *AG1* and *AG2* QTL introgression lines and seed pre-treatment on growth and physiological processes during anaerobic germination of rice under flooding. *Scientific Reports* 10:1021. https://doi.org/10.1038/s41598-020-67240-x.

Monreal, C.M., M. Derosa, S.C. Mallubhotla, P.S. Bindraban and C. Dimkpa. 2016. Nanotechnologies for increasing the crop use efficiency of fertilizer-micronutrients. *Biology and. Fertility of Soils* 52:423–437.

Mozafari, A., A. Ghadakchiasl and N. Ghaderi. 2018. Grape response to salinity stress and role of iron nanoparticle and potassium silicate to mitigate salt induced damage under *in vitro* conditions. *Physiology and Molecular Biology of Plants* 24:25–35.

Mukami, A., A. Ngetich, C. Mweu, R.O. Oduor, M. Muthangya and W.M. Mbinda. 2019. Differential characterization of physiological and biochemical responses during drought stress in finger millet varieties. *Physiology and Molecular Biology of Plants* 25(4):837–846.

Mukhopadhyay, S.S. and N. Kaur. 2016. Nanotechnology in Soil-Plant System. In C. Kole et al. (eds.), *Plant Nanotechnology*. Springer International Publishing. Switzerland. pp. 329–348. DOI: 10.1007/978-3-319-42154- 4_13

Mustafa, G., K. Sakata and S. Komatsu. 2015. Proteomic analysis of flooded soybean root exposed to aluminum oxide nanoparticles. *Journal of Proteomics* 128:280–297.

Nakache, E., N. Poulain, F. Candau, A.M. Orecchioni and J.M. Irache. 1999. Biopolymer and Polymer Nanoparticles and Their Biomedical Applications. In Nalwa, H.S. (ed.), *Handbook of Nanostructured Materials and Nanotechnology*. New York: Academic Press. p. 3461.

Nazari, M., R. MaaliAmiri, F.H. Mehraban and H.Z. Khaneghah. 2012. Change in antioxidant responses against oxidative damage in black chickpea following cold acclimation. *Russian Journal of Plant Physiology* 59:183–189. DOI:10.1134/S102144371201013X

Neves-Borges, A.C., F. Guimarães-Dias and F. Cruz. 2012. Expression pattern of drought stress marker genes in soybean roots under two water deficit systems. *Genetics and Molecular Biology* 35(1 suppl 1):212–221.

Oliveira, H.C., B.C.R. Gomes, M.T. Pelegrino and A.B. Seabra. 2016. Nitric oxide-releasing chitosan nanoparticles alleviate the effects of salt stress in maize plants. *Nitric Oxide* 61:10–19.

Patra, A.K., T. Adhikari and A.K. Bhardwaj. 2016. Enhancing Crop Productivity in Salt-Affected Environments by Stimulating Soil Biological Processes and Remediation Using Nanotechnology. In J.C. Dagar et al. (eds.), *Innovative Saline Agriculture*. Springer. India. pp 83–103. DOI: 10.1007/978-81-322-2770-0_4,

Qados, A.M.A. 2015. Mechanism of nano silicon-mediated alleviation of salinity stress in faba bean (*Vicia faba* L.) plants. *American Journal of Experimental Agriculture* 7(2):78.

Rai, V., S. Acharya and N.J. Dey. 2012. Implications of nanobiosensors in agriculture. *Journal of Biomaterials and Nanobiotechnology* 3(2A):315–324.

Reddy, P.V.L., J.A. Hernandez-Viezcas, J.R. Peralta-Videa and J.L. Gardea-Torresdey. 2016. Lessons learned: are engineered nanomaterials toxic to terrestrial plants? *Science of the Total Environment* 568:470–479.

Rossi, L., W. Zhang, L. Lombardini and X. Ma. 2016. The impact of cerium oxide nanoparticles on the salt stress responses of *Brassica napus* L. *Environment and Pollution* 219:28–36.

Rossi, L., W. Zhang and X. Ma. 2017. Cerium oxide nanoparticles alter the salt stress tolerance of *Brassica napus* L. by modifying the formation of root apoplastic barriers. *Environment and Pollution* 229:132–138.

Ruotolo, R., E. Maestri, L. Pagano, M. Marmiroli, J.C. White and N. Marmiroli. 2018. Plant response to metal-containing engineered nanomaterials: an omics-based perspective. *Environmental Science and Technology* 52:2451–2467.

Rushton, P.J., I.E. Somssich, P. Ringler and Q.J. Shen. 2010. WRKY transcription factors. *Trends in Plant Science* 15(5):247–258.

Sairam, R.K., K. Dharmar, V. Chinnusamy and R.C. Meena. 2009. Waterlogging-induced increase in sugar mobilization, fermentation, and related gene expression in the roots of mung bean (*Vigna radiata*). *Journal of Plant Physiology* 166:602–616.

Scrinis, G. and K. Lyons. 2007. The emerging nano-corporate paradigm: nanotechnology and the transformation of nature, food and agri-food systems. *International Journal of Sociology of Agriculture and Food* 15:22–44.

Sedghi, M., H. Mitra and T. Sahar. 2013. Effect of nano zinc oxide on the germination of soybean seeds under drought stress. *Annals of West University of Timisoara: Series of Biology* 16(2):73–78.

Shen, X., X. Xiao, Z. Dong and Y. Chen. 2014. Silicon effects on antioxidative enzymes and lipid peroxidation in leaves and roots of peanut under aluminum stress. Acta Physiologiae Plantarum 36(11):3063–3069.

Shinozaki, K. and K.Y. Shinozaki. 2006. Gene networks involved in drought stress response and tolerance, *Journal of Experimental Botany* 58(2): 221–227.

Siddiqui, M.H. and M.H. Al-Whaibi. 2014. Role of nano-SiO_2 in germination of tomato (*Lycopersicum esculentum* seeds Mill.). *Saudi Journal of Biological Sciences* 21:13–17.

Siddiqui, M.H., M.H. Al-Whaibi, M. Faisal and A.A. Al Sahli. 2014. Nano-silicon dioxide mitigates the adverse effects of salt stress on *Cucurbita pepo* L. *Environmental Toxicology and Chemistry* 33(11):2429–2437.

Singh, J., and B.K. Lee. 2016. Influence of nano-TiO_2 particles on the bioaccumulation of Cd in soybean plants (*Glycine max*): A possible mechanism for the removal of Cd from the contaminated soil. *Journal of Environmental Management* 170:88–96.

Soliman, A.S., S.A. El-feky and E. Darwish. 2015. Alleviation of salt stress on *Moringa peregrina* using foliar application of nanofertilizers. *Journal of Horticulture and Forestry* 7:36–47.

Stolf-Moreira, R., E.G.M. Lemos and L. Carareto-Alves. 2011. Transcriptional profiles of roots of different soybean genotypes subjected to drought stress. *Plant Molecular Biology Reporter* 29(1):19–34.

Sun, L., F. Song and J. Guo. 2020. Nano-ZnO-induced drought tolerance is associated with melatonin synthesis and metabolism in maize. *International Journal of Molecular Sciences* 21(3):782.

Taha, L.S., N. Youssef, I. El-Sayed and W.H. Salama. 2020. Nanotechnology application for salinity tolerance enhancement of micropropagated *Gardenia jasminoides* ellis plant. *Plant Archives* 20(2):8450–8456.

Taiz, L. and E. Zeiger. 2010. *Plant Physiology*. Massachusetts: Sinauer Associates Inc., 5th edition. 781 pp.

Tarafdar, J.C., A. Agrawal, R. Raliya, P. Kumar, U. Burman and R.K. Kaul. 2012a. ZnO nanoparticles induced synthesis of polysaccharides and phosphatases by Aspergillus fungi. *Advanced Science, Engineering and Medicine* 4:1–5.

Tarafdar, J.C., R. Raliya and I Rathore. 2012b. Microbial synthesis of phosphorus nanoparticles from tri-calcium phosphate using *Aspergillus tubingensis* TFR-5. *Journal of Bionanoscience* 6:84–89.

Taran, N., V. Storozhenko, N. Svietlova, L. Batsmanova, V. Shvartau and M. Kovalenko. 2017. Effect of zinc and copper nanoparticles on drought resistance of wheat seedlings. *Nanoscale Research Letters* 12(1):60.

Tavakkoli, E., P. Rengasamy and G.K. McDonald. 2010. High concentrations of Na^+ and Cl^- ions in soil solution have simultaneous detrimental effects on growth of faba bean under salinity stress. *Journal of Experimental Botany* 61:4449–4459. doi:10.1093/jxb/erq251

Thapa, G., M. Dey, L. Sahoo and S.K. Panda. 2011. An insight into the drought stress induced alterations in plants. *Biologia Plantarum* 55(4):603–613.

Torabian, S., M. Zahedi and A.H. Khoshgoftar. 2017. Effects of foliar spray of nano-particles of $FeSO_4$ on the growth and ion content of sunflower under saline condition. *Journal of Plant Nutrition* 40:615–623.

Torney, F. et al. 2007. Mesoporous silica nanoparticles deliver DNA and chemicals into plants. *Nature Nanotechnology* 2:295–300.

Wahid, I., S. Kumari, R. Ahmad, S.J. Hussain, S. Alamri, M.H. Siddiqui and M.I.R. Khan. 2020. Silver nanoparticle regulates salt tolerance in wheat through changes in ABA concentration, ion homeostasis, and defense systems. *Biomolecules* 10:1506.

Wang, F., H.W. Chen and Q.T. Li. 2015. GmWRKY 27 interacts with GmMYB 174 to reduce expression of GmNAC 29 for stress tolerance in soybean plants. *The Plant Journal* 83(2):224–236.

Wang, P. et al. 2016. Nanotechnology: a new opportunity in plant sciences. *Trends in Plant Sciences* 21:699–712.

Weathers, P.J. and R.W. Zobel. 1992. Aeroponics for the culture of organisms, tissues and cells. *Biotechnology Advances* 10:93–115.

Wei, T., Q. Cheng, Y.L. Min, E.N. Olson and D.J. Siegwart. 2020. Systemic nanoparticle delivery of CRISPR-Cas9 ribonucleoproteins for effective tissue specific genome editing. *Nature Communications* 11:3232. https://doi.org/10.1038/s41467-020-17029-3

Yadav, A.K., A.J. Carroll, G.M. Estavillo, G.J. Rebetzke and B.J. Pogson. 2019. Wheat drought tolerance in the field is predicted by amino acid responses to glasshouse-imposed drought. *Journal of Experimental Botany* 70(18):4931–4948.

Yanagawa, Y. and S. Komatsu. 2012. Ubiquitin/proteasome mediated proteolysis is involved in the response to flooding stress in soybean roots, independent of oxygen limitation, *Plant Science* 185–186:250–258.

Yasmeen, F., N.I. Raja, A. Razzaq and S. Komatsu. 2017. Proteomic and physiological analyses of wheat seeds exposed to copper and iron nanoparticles. *Biochimica et Biophysica Acta (BBA) – Proteins and Proteomics* 1865(1):28–42.

Zareabyaneh, H. and M. Bayatvarkeshi. 2014. The effect of nanofertilizers on nitrate leaching and its distribution in soil profile with an emphasis on potato yield. *Nanoscience and Nanotechnology* 8(6):1–11.

Zhang, H. et al. 2019. DNA nanostructures coordinate gene silencing in mature plants. *Proceedings of the National Academy of Sciences of the U. S. A.* 116: 7543–7548.

Zhao, X. et al. 2017. Pollen magnetofection for genetic modification with magnetic nanoparticles as gene carriers. *Nature Plants* 3:956–964.

Index

Note: Page locators in **bold** represent tables and page number in *italics* represents figures on the corresponding pages.

B

C

D

E

F

G

H

I

J

K

L

M

N

O

S

X

Y

Z